Analytical Biotechnology

Analytical Biotechnology

Editor: Kane Lloyd

R CALLISTO
REFERENCE
www.callistoreference.com

Callisto Reference,
118-35 Queens Blvd., Suite 400,
Forest Hills, NY 11375, USA

Visit us on the World Wide Web at:
www.callistoreference.com

ISBN: 978-1-64116-192-3 (Hardback)

Trademark Notice: Registered trademark of products or corporate names are used only for explanation and identification without intent to infringe.

Cataloging-in-Publication Data

Analytical biotechnology / edited by Kane Lloyd.
    p. cm.
Includes bibliographical references and index.
ISBN 978-1-64116-192-3
1. Analytical biotechnology. 2. Biotechnology. I. Lloyd, Kane.
TP248.24 .A53 2019
660.6--dc23

Table of Contents

Preface

Biotechnology is the science of developing or modifying products or processes using organisms and living systems for a specific purpose. This field encompasses the use of artificial selection and hybridization to improve practices like domestication of animals and cultivation of plants. There are various branches of biotechnology such as, red biotechnology, bioinformatics, white biotechnology and blue biotechnology, among many others. Bioanalysis is a sub-discipline of analytical chemistry. It is involved in the quantitative measurement of xenobiotics and biotics in biological systems. Such studies are guided by a number of bioanalytical techniques which include liquid chromatography-mass spectrometry, capillary electrophoresis–mass spectrometry, gas chromatography, supercritical fluid chromatography, etc. This book traces the progress in the field of analytical biotechnology and highlights some of its key concepts and applications. Most of the topics introduced herein cover new techniques of this discipline. This book is a resource guide for experts as well as students.

Various studies have approached the subject by analyzing it with a single perspective, but the present book provides diverse methodologies and techniques to address this field. This book contains theories and applications needed for understanding the subject from different perspectives. The aim is to keep the readers informed about the progresses in the field; therefore, the contributions were carefully examined to compile novel researches by specialists from across the globe.

Indeed, the job of the editor is the most crucial and challenging in compiling all chapters into a single book. In the end, I would extend my sincere thanks to the chapter authors for their profound work. I am also thankful for the support provided by my family and colleagues during the compilation of this book.

Editor

Adjustment of Matrix-Assisted Laser Desorption/Ionization for Glycolipids

Yusuke Suzuki[1*], Akira Okamoto[1], Anila Mathew[1,2] and Yasunori Kushi[1]

[1]College of Science and Technology, Nihon University, Tokyo, Japan

[2]Anila Mathew's current address is Bio-Nano Electronics Research Centre, Toyo University, 2100, Kujirai, Kawagoe, Saitama 350-8585, Japan

Abstract

Gangliosides isolated from biological sources are usually detected as sodium/potassium adduct ions and these ions are easily fragmented by dissociation of labile glycosidic bonds in the positive ion mode Mass Spectrometry (MS). A large number of conditions for fragment suppression of these acidic glycolipids by using non-acidic matrices or changing metal additions have been demonstrated. Compared to sodium/potassium adduct ions, the cesium adduct ions suppress the fragmentation of these acidic glycolipids. On the contrary, lithium adduct ions induce the fragmentation, but generate more informative fragment ions of glycolipids than other alkali metal adduct ions in post-source decay, MS/MS as well as MS spectra. To suppress the fragmentation of labile glycosidic bond and generate more informative fragmentation, we have examined and established the optimal condition for the detection of $[M+Li]^+$ of gangliosides (GM3 and GM2) using the matrix-assisted laser desorption/ionization time-of-flight MS by adjusting matrix and alkali metal salt combinations and concentrations.

Keywords: Ganglioside; MALDI-TOF MS; Alkali metal adduct ion; MS/MS

Abbreviations: MALDI-TOF MS: Matrix-Assisted Laser Desorption/Ionization Time-Of-Flight Mass Spectrometry; GSL: Glycosphingolipid; DHB: 2,5-Dihydroxybenzoic Acid

Introduction

Gangliosides are sialic acid-containing glycosphingolipids, which are components of plasma membrane. It is well known that gangliosides/glycophospholipids play important roles in a wide variety of cellular functions, including cell-cell interactions, cell growth and differentiation, and signaling [1-3]. A detailed structural characterization is required to resolve their functions. A number of analytical methods are available for the characterization of the gangliosides/glycophospholipids. Matrix-Assisted Laser Desorption/Ionization Time-of-Flight Mass Spectrometry (MALDI-TOF MS) is one of the indispensable tools for the rapid structural characterization [4,5]. In the positive ion mode MS spectra of biological samples, sodium/potassium adduct (Na^+/K^+) ions are usually detected and the labile glycosidic/phosphate bonds are easily fragmented [5]. Since the detection of molecular ions is required for the determination of whole molecular weight and simple interpretation of MS pattern, the suppression of the acidic glycolipid fragments are important and necessary. Furthermore, informative Na^+/K^+ fragments are often hardly obtained by further mass analysis, such as Post-Source Decay (PSD) or MS/MS. Therefore, the detailed structural information of glycolipids requires fragment acceleration. As described above, an understanding of the reciprocal events between fragment suppression and acceleration is necessary for characterization of glycolipids.

In the MS spectra, the stabilities and properties of ions are drastically changed by alkali metal species, and the cationization is easily changed by the addition of alkali metal salts to the sample. Lithium salts accelerate fragmentation of the labile compound, and hence the relative intensities of molecular ions are decreased in the MS spectra. However, lithium adduct (Li^+) ions of these gangliosides or phospholipids generate more informative fragment ions than other alkali metal ions in the MS/MS [6-10]. On the other hand, cesium salts suppress fragmentation of the labile compound, and the relative intensities of molecular ions are increased in the MS spectra. However, cesium adduct (Cs^+) ions do not generate sufficient fragment ions for characterization of the acidic glycolipids in MS/MS compared to other

alkali metal ions [11]. As described above, molecular ions could be obtained by changing the molecular ions from Na^+/K^+ to Cs^+ adducts for MS analysis, and detailed fragment ions could be obtained by changing molecular ions to Li^+ adducts for MS/MS analysis. Li^+ and Cs^+ have compatible properties, and combining the advantageous properties of lithium and cesium salts helps in controlling the ionizing condition of these glycolipids.

To suppress the fragmentation of labile compounds, the use of various kinds of matrices are also reported including non-acidic and liquid matrices for MALDI-TOF MS analysis [12-17]. So far, 2,5-Dihydroxybenzoic Acid (DHB) is the most widely used matrix for glycoconjugate analysis. Compared to the other common matrices such as α-cyano-4-hydroxycinnamic acid (CHCA) and sinapinic acid, DHB allows softer desorption. Under 337 nm radiation, non-acidic matrices such as 2-amino-3-hydroxypyridine, norharman, 6-aza-2-thiothymine, and 2'4'6'-trihydroxyacetophenone, were used for the fragment suppression of acid-labile compounds like synthetic polymer, acidic oligosaccharide, glycopeptides, and other glycoconjugates [18]. Furthermore, liquid matrix of DHB with butylamine (DHBB) suppressed neuraminic acid fragmentation of monosialyl-monofucosyl-lacto-N-neohexaose, ganglioside, and 3'-sialyllactose [16]. Liquid matrix of 2-(4-hydroxyphenylazo)benzoic acid (HABA) with 1,1,3,3-tetramethylguanidine and spermine were used for heparin sulfate [17]. Recently, Liang et al., showed that frozen solution of aqueous acetonitrile containing oligosaccharides and DHB generates more oligosaccharide ions and less fragments from PSD spectra [19].

As described above, a number of trials for regulating the fragment pattern of labile compounds including gangliosides using alkali-metal addition and non-acidic/liquid matrix have been reported. The

*Corresponding authors: Yusuke Suzuki, Department of Materials and Applied Chemistry, College of Science and Technology, Nihon University, Chiyoda-ku, Japan
E-mail: suzuki.yuusuke@nihon-u.ac.jp

fragment patterns were regulated by the species and concentrations of matrices and alkali-metal salts. In this study, we established an optimum condition for the analysis of GM3 and GM2 gangliosides using alkali-metal salts and non-acidic matrices.

Materials and Methods

GM3 (Neu5Ac 2-3Gal 1-4Glc 1-1'Cer) prepared from bovine brain was purchased from HyTest, Ltd. (Turku, Finland). Lithium chloride and cesium chloride were purchased from Nacalai tesque, Inc. (Kyoto, Japan). 2,5-Dihydroxy benzoic acid was purchased from Bruker Daltonics (Leipzig, Germany).

MALDI-TOF MS analysis of ganglioside GM3

MALDI-TOF MS was performed on a Voyager RP-PRO MALDI-TOF mass spectrometer (Life Tech., California, USA) equipped with a 337 nm nitrogen laser. MS spectra was calibrated externally using a peptide calibration standard mixture containing bradykinin ([M+H]+, 757.40), angiotensin II ([M+H]+, 1045.64), and human adrenocorticotrophic hormone (ACTH, fragments 18-39) ([M+H]+, 2465.20) as 10 pmol/μL solutions. The ganglioside GM3 were dissolved in chloroform/methanol (C/M, 1:1, v/v) as 10 mg/mL solutions. The matrices used in this study were dissolved in water at concentrations of 10 mg/mL. The alkali metal salts were also dissolved in water at various concentrations. The matrix, alkali metal salts, and GM3 or GM2 were mixed in a glass tube and placed on the target plate for crystallization. Crystallization was accelerated by a gentle stream of cold air.

Results and Discussion

Confirmation of MALDI-TOF MS spectrum of ganglioside GM3

Figure 1 shows the structures of matrices used in this study. To determine the optimal condition for ganglioside analysis, various concentrations of GM3 (1 nmol, 100 pmol, 10 pmol, and 1 pmol) in DHB matrix were analyzed by MALDI-TOF MS in positive ion mode (Figure 2A-2D). In the MS spectra, the molecular ions were detected at m/z 1205 and 1233, and the ions derived from loss of sialic acid from GM3 were detected as sodium adducts at m/z 913 and 941. The ions derived by the loss of carbon dioxide from GM3 were also detected as sodium adducts at m/z 1161 and 1189. On the other hand, the background peaks at m/z 768 and 1017 were detected, and poor signal to noise ratios were observed in the lower concentrations (1 and 10 pmol) of GM3 were observed (Figure 2C and 2D). Therefore, further experiments were carried out using 100 pmol of GM3.

Effect of alkali metal salt on MALDI-TOF MS pattern of GM3

To confirm the effect of alkali metal salts on MALDI-TOF MS pattern of GM3, we analyzed MALDI-TOF MS spectra of GM3 after preparation with DHB and various concentrations of lithium salts (400 nmol, 40 nmol, 4 nmol, and 400 pmol) (Figure 3A-3D). In the MS spectra, the molecular ions were detected at m/z 1194 and 1221 (Figure 3B and 3C), and the ions derived from loss of sialic acid from GM3 were detected as lithium adducts at m/z 897 and 925 (Figure 3A-3D). The ions derived by the loss of carbon dioxide from GM3 were also detected at m/z 1144 and 1172 as lithium adducts. Furthermore, the sodium adduct ions were detected at m/z 912, 940, 1160, and 1188 at lower concentration of LiCl (400 pmol) (Figure 3D).

Next, we analyzed MALDI-TOF MS spectra of GM3 after preparation with DHB and various concentrations of cesium salts (400 nmol, 40 nmol, 4 nmol, and 400 pmol) (Figure 3E-3H). In the

MS spectra, the molecular ions were detected as cesium adducts at m/z 1315 and 1343 and as double cesium adducts at m/z 1446 and 1474 (Figure 3F and 3G). The sodium adduct ions were detected at m/z 912, 940, 1160, 1188, 1205, and 1233 at lower concentration of CsCl (400 pmol) (Figure 3H). The ratios of cationization changed from sodium to single and/or double cesium adducts in a CsCl concentration-dependent manner (Figure 3E-3H). Although the detection of GM3 derived ions were suppressed at high concentration of CsCl (400 nmol), the ions derived from loss of sialic acid or carbon dioxide from GM3 were almost suppressed at high concentration of CsCl (Figure 3E-3G).

These results indicate that lithium ions induce and cesium ions suppress the fragmentation of gangliosides, compared to sodium/potassium adduct ions which are usually detected in MALDI-TOF MS, as reported previously. Therefore, we concluded that the optimal concentration of these salts were few nmols for changing cationization.

Effect of matrices on MALDI-TOF MS pattern of GM3

Non-acidic/liquid matrices are well known to suppress

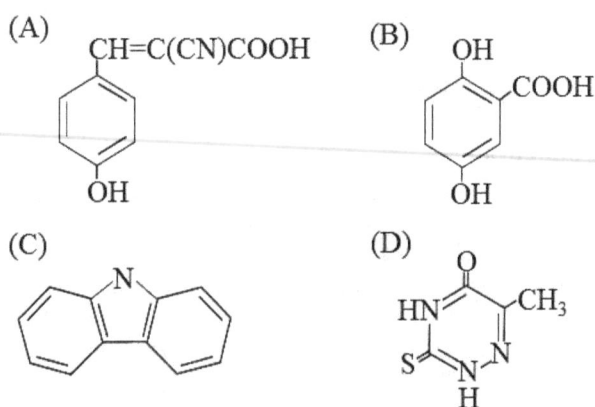

Figure 1: The structures of matrices used in this study. (A) a-cyano-4-hydroxycinnamic acid (CHCA), (B) 2,5-dihydroxybenzoic acid (DHB), (C) 9H-pyrido[3,4-b] indole (Norharman), (D) 6-aza-2-thiothymin (6-AZA).

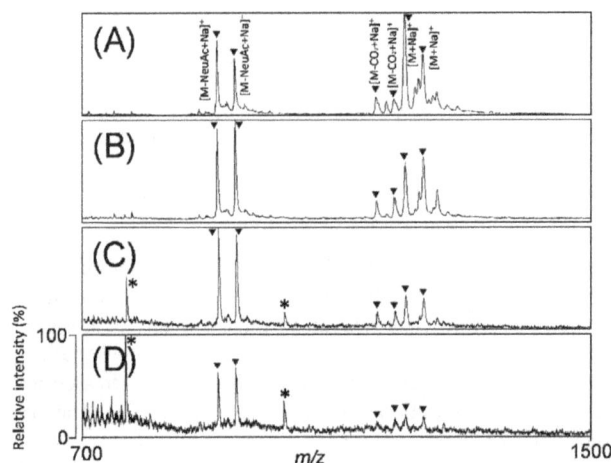

Figure 2: MALDI-TOF MS spectra of various concentrations of ganglioside GM3 after preparation with DHB. GM3 concentrations - (A) GM3 1 nmol, (B) 100 pmol, (C) 10 pmol, and (D) 1 pmol. ▼ and * indicate sodium adduct ions and background ions, respectively.

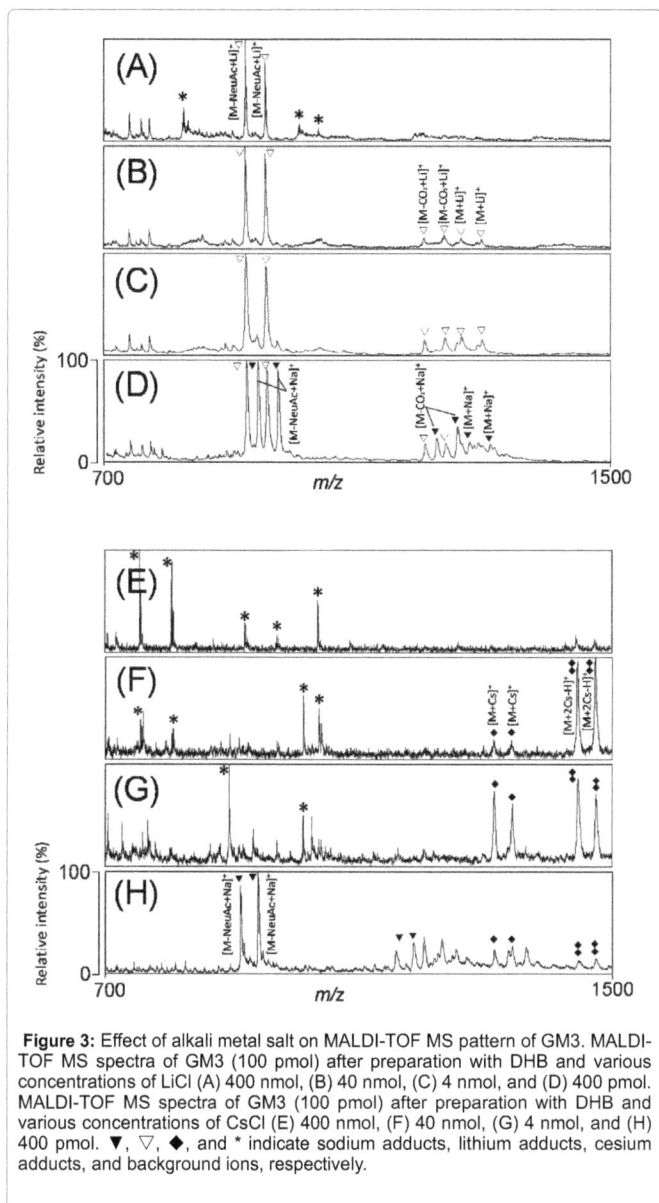

Figure 3: Effect of alkali metal salt on MALDI-TOF MS pattern of GM3. MALDI-TOF MS spectra of GM3 (100 pmol) after preparation with DHB and various concentrations of LiCl (A) 400 nmol, (B) 40 nmol, (C) 4 nmol, and (D) 400 pmol. MALDI-TOF MS spectra of GM3 (100 pmol) after preparation with DHB and various concentrations of CsCl (E) 400 nmol, (F) 40 nmol, (G) 4 nmol, and (H) 400 pmol. ▼, ▽, ◆, and * indicate sodium adducts, lithium adducts, cesium adducts, and background ions, respectively.

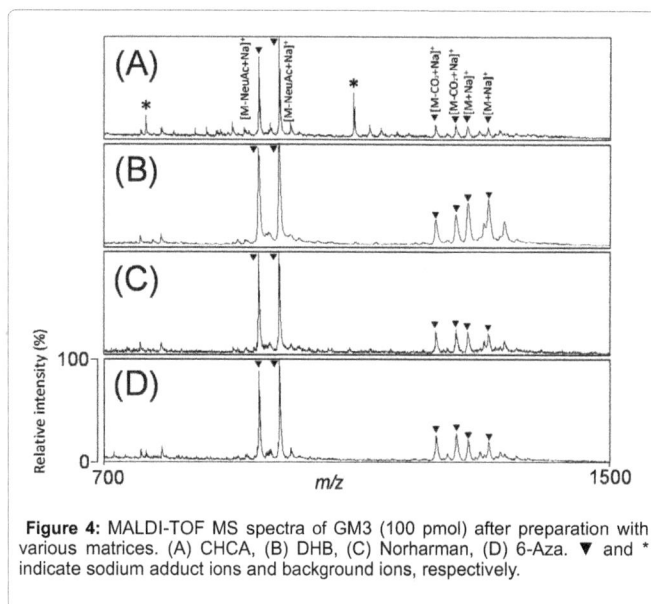

Figure 4: MALDI-TOF MS spectra of GM3 (100 pmol) after preparation with various matrices. (A) CHCA, (B) DHB, (C) Norharman, (D) 6-Aza. ▼ and * indicate sodium adduct ions and background ions, respectively.

fragmentation of gangliosides [11-17, 20]. We analyzed MALDI-TOF MS spectra of GM3 after preparation with various matrices having both acidic and non-acidic properties (Figure 4). Although GM3 derived ions were detected in all MALDI-TOF MS spectra, the ratio of molecular ions to N-acetylneuraminic acid (NeuAc)-dissociated ions was higher after preparation with Norharman, than with other matrices. Therefore, we compared the MS spectra of GM3 after preparation with DHB and Norharman for further analyses.

Combinational effect of matrices and alkali metal salts on MALDI-TOF MS pattern of GM3

We analyzed MALDI-TOF MS spectra of GM3 after preparation with DHB and 4 nmol of LiCl, and various concentrations of CsCl. The GM3 derived ions were detected as single or double cesium adducts as well as sodium adducts under high concentration of CsCl (400 nmol) (Figure 5A). On the other hand, the GM3 derived ions were detected as sodium and lithium adducts under low concentrations of CsCl (40, 4 nmol, or 400 pmol) (Figure 5B-5D). After preparation with DHB, LiCl,

and CsCl, fragmentation of GM3 was not suppressed, and the lithiated molecular ions of GM3 ([M+Li]⁺) could not be detected (Figure 5A-5D).

Norharman do not contain any acidic functional group, and is a more non-acidic matrix than DHB. Therefore, Norharman can be used for acid-labile samples and fragment suppression of sialic acid. We analyzed MALDI-TOF MS spectra of GM3 after preparation with Norharman and 4 nmol of LiCl, and various concentrations of CsCl. Although single and double cesiated molecular ions of GM3 were detected under high concentration of CsCl (400 nmol) (Figure 5E), lithiated molecular ions of GM3 were detected (Figure 5F-5H). Furthermore, NeuAc-dissociated ions were remarkably suppressed (Figure 5E-5H). These results indicated that the optimal analytical conditions for gangliosides using Norharman, LiCl, and CsCl inhibit dissociation of sialic acid, while allowing the detection of lithiated molecular ions of GM3.

MALDI-TOF MS analysis of GM2 using DHB combinational conditions of Norharman, LiCl, and CsCl

To confirm whether the established method can be applied for other ganglioside analyses, we analyzed MALDI-TOF MS spectra of GM2, which has a more complicated structure than GM3. In the MS spectrum with DHB, molecular ions of GM2 were detected at m/z 1408 and 1436, and the ions derived from loss of sialic acid from GM2 were detected at m/z 1115 and 1143 as sodium adducts (Figure 6A). Next, we analyzed MALDI-TOF MS spectra of GM2 after preparation with DHB and 4 nmol of LiCl-CsCl. The molecular ions of GM2 were detected as single or double cesium adducts at m/z 1517 and 1545, or at m/z 1649 and 1677 (Figure 6B). However, the low relative ratio of [M+Li]⁺ derived from GM2 was observed (Figure 6B). We analyzed MALDI-TOF MS spectra of GM2 after preparation with Norharman and 4 nmol of LiCl-CsCl. Although [M+Cs]⁺ or loss of sialic acid of GM2 as lithium adducts were detected, [M+Li]⁺ were detected as the major peaks (Figure 6C).

These results indicated that, our approach of adjusting the matrix, and alkali metal salt combinations and concentrations are powerful tools for changing the alkali metal cation species in gangliosides during analysis by MALDI-TOF MS.

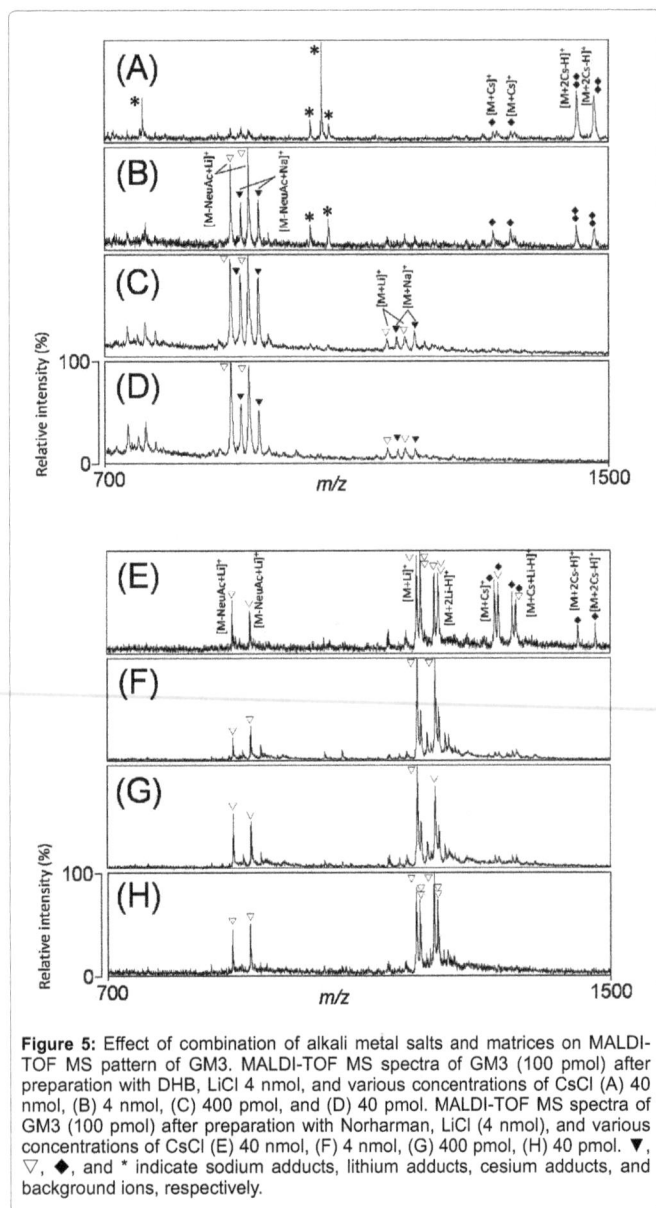

Figure 5: Effect of combination of alkali metal salts and matrices on MALDI-TOF MS pattern of GM3. MALDI-TOF MS spectra of GM3 (100 pmol) after preparation with DHB, LiCl 4 nmol, and various concentrations of CsCl (A) 40 nmol, (B) 4 nmol, (C) 400 pmol, and (D) 40 pmol. MALDI-TOF MS spectra of GM3 (100 pmol) after preparation with Norharman, LiCl (4 nmol), and various concentrations of CsCl (E) 40 nmol, (F) 4 nmol, (G) 400 pmol, (H) 40 pmol. ▼, ▽, ◆, and * indicate sodium adducts, lithium adducts, cesium adducts, and background ions, respectively.

Alkali metal cations were investigated for their effectiveness in glycolipid analysis by MALDI-TOF MS [5-8]. The fragmentation yields for oligosaccharides or glycolipids were found to be related to alkali metal ion size and electrostatic charge density. Li adducts increase fragmentations of labile glycosidic bonds as well as ceramide moieties of glycolipids [5,8,10,21]. Cs adducts decrease fragmentations of glycolipids, therefore molecular ions can be detected in the MS spectra [11,18]. Although the fragmentation of sialic acid is not efficiently suppressed, the molecular ions of GM3 were changed from Na$^+$ to Cs$^+$ after preparation with Li/Cs alkali metal salts and DHB. Finally, the conditions for fragment suppression of sialic acid and detection of [M+Li]$^+$ ions were established by the combinations of these alkali salts and Norharman.

Recently, new types of mass spectrometries, such as MALDI ion-mobility MS, have been developed and they have become powerful tools for the ganglioside analyses [22]. Our established method is quite simple and easily adjustable; therefore it is also thought to be useful for the analysis of other glycolipids as well as gangliosides by regardless

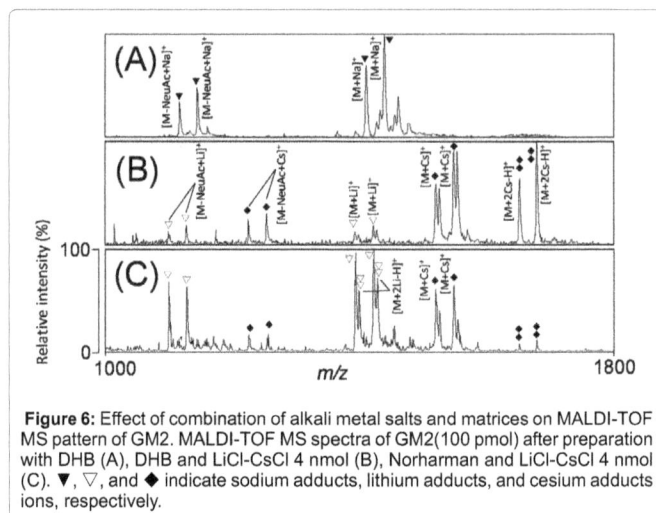

Figure 6: Effect of combination of alkali metal salts and matrices on MALDI-TOF MS pattern of GM2. MALDI-TOF MS spectra of GM2(100 pmol) after preparation with DHB (A), DHB and LiCl-CsCl 4 nmol (B), Norharman and LiCl-CsCl 4 nmol (C). ▼, ▽, and ◆ indicate sodium adducts, lithium adducts, and cesium adducts ions, respectively.

of mass types. Furthermore, 2,6-dihydroxyacetophenone (DHAP) has been reported as a new matrix for acidic glycolipid analysis [23]. In our preliminary data, we confirmed the fragmentation suppression of ganglioside GM3 using DHAP more effectively than Norharman, but the cesium adducted molecular ions were detected as the major peaks under the established condition. In the future, we will examine the applicability of this established method in the analysis of other acidic glycolipids, and will confirm the optimal condition of DHAP preparation and effectiveness of liquid matrices on the glycolipid analysis using this method.

In this study, we established the optimal condition for the detection of [M+Li]$^+$ of gangliosides (GM3 and GM2) using the MALDI-TOF MS by adjusting matrix and alkali metal salt combinations and concentrations. This is the first study reporting a quite simple method for the detection of lithiated molecular ion of ganglioside in the positive ion MS. In the near future, we would expect that more detailed information of ganglioside is determined by MS/MS with selection of [M+Li]$^+$. Furthermore, the optimization of ionization and Cs/Li salt combinations is quite simple and easily adjustable method, as shown in this manuscript. Therefore, we also expect that the method would be useful to ganglioside as well as other glycolipid analyses even using other mass spectrometries.

Acknowledgement

This work was supported by Nihon University College of Science and Technology Grant-in Aid for Fundamental Science Research (YS), and partly supported by Grant-in-Aid for Young Scientists (B) No. 24750164 from MEXT Japan (YS). The authors declare no conflicts of interest.

References

1. Murate M, Hayakawa T, Ishii K, Inadome H, Greimel P, et al. (2010) Phosphatidylglucoside forms specific lipid domains on the outer leaflet of the plasma membrane. Biochemistry 49: 4732-4739.

2. Yu RK, Suzuki Y, Yanagisawa M (2010) Membrane glycolipids in stem cells. FEBS Lett 584: 1694-1699.

3. Nozaki H, Itonori S, Sugita M, Nakamura K, Ohba K, et al. (2010) Invariant Valpha 14 natural killer T cell activation by edible mushroom acidic glycosphingolipids. Biol Pharm Bull 33: 580-584.

4. Jackson SN, Colsch B, Egan T, Lewis EK, Schultz JA, et al. (2011) Gangliosides' analysis by MALDI-ion mobility MS. Analyst 136: 463-466.

5. Suzuki Y, Suzuki M, Ito E, Goto-Inoue N, Miseki K, et al. (2006) Convenient structural analysis of glycosphingolipids using MALDI-QIT-TOF mass spectrometry with increased laser power and cooling gas flow. J Biochem 139: 771-777.

6. Arigi E, Singh S, Kahlili AH, Winter HC, Goldstein IJ, et al. (2007) Characterization of neutral and acidic glycosphingolipids from the lectin-producing mushroom, Polyporus squamosus. Glycobiology 17: 754-766.

7. Calvano CD, De Ceglie C, Aresta A, Facchini LA, Zambonin CG (2013) MALDI-TOF mass spectrometric determination of intact phospholipids as markers of illegal bovine milk adulteration of high-quality milk. Anal Bioanal Chem 405: 1641-1649.

8. Hsu FF, Turk J (2001) Structural determination of glycosphingolipids as lithiated adducts by electrospray ionization mass spectrometry using low-energy collisional-activated dissociation on a triple stage quadrupole instrument. J Am Soc Mass Spectrom 12: 61-79.

9. Hsu FF, Turk J (2010) Electrospray ionization multiple-stage linear ion-trap mass spectrometry for structural elucidation of triacylglycerols: assignment of fatty acyl groups on the glycerol backbone and location of double bonds. J Am Soc Mass Spectrom 21: 657-669.

10. Hsu FF, Turk J, Stewart ME, Downing DT (2002) Structural studies on ceramides as lithiated adducts by low energy collisional-activated dissociation tandem mass spectrometry with electrospray ionization. J Am Soc Mass Spectrom 13: 680-695.

11. Griffiths RL, Bunch J (2012) A survey of useful salt additives in matrix-assisted laser desorption/ionization mass spectrometry and tandem mass spectrometry of lipids: introducing nitrates for improved analysis. Rapid Commun Mass Spectrom 26: 1557-1566.

12. Giménez E, Benavente F, Barbosa J, Sanz-Nebot V (2010) Ionic liquid matrices for MALDI-TOF-MS analysis of intact glycoproteins. Anal Bioanal Chem 398: 357-365.

13. Laremore TN, Murugesan S, Park TJ, Avci FY, Zagorevski DV, et al. (2006) Matrix-assisted laser desorption/ionization mass spectrometric analysis of uncomplexed highly sulfated oligosaccharides using ionic liquid matrices. Anal Chem 78: 1774-1779.

14. Laremore TN, Zhang F, Linhardt RJ (2007) Ionic liquid matrix for direct UV-MALDI-TOF-MS analysis of dermatan sulfate and chondroitin sulfate oligosaccharides. Anal Chem 79: 1604-1610.

15. Li YL, Gross ML (2004) Ionic-liquid matrices for quantitative analysis by MALDI-TOF mass spectrometry. J Am Soc Mass Spectrom 15: 1833-1837.

16. Mank M, Stahl B, Boehm G (2004) 2,5-Dihydroxybenzoic acid butylamine and other ionic liquid matrixes for enhanced MALDI-MS analysis of biomolecules. Anal Chem 76: 2938-2950.

17. Przybylski C, Gonnet F, Bonnaffé D, Hersant Y, Lortat-Jacob H, et al. (2010) HABA-based ionic liquid matrices for UV-MALDI-MS analysis of heparin and heparan sulfate oligosaccharides. Glycobiology 20: 224-234.

18. Rashidzadeh H, Wang Y, Guo B (2000) Matrix effects on selectivities of poly(ethylene glycol)s for alkali metal ion complexation in matrix-assisted laser desorption/ionization. Rapid Commun Mass Spectrom 14: 439-443.

19. Liang CW, Chang PJ, Lin YJ, Lee YT, Ni CK (2012) High ion yields of carbohydrates from frozen solution by UV-MALDI. Anal Chem 84: 3493-3499.

20. Bonnel D, Franck J, Mériaux C, Salzet M, Fournier I (2013) Ionic matrices pre-spotted matrix-assisted laser desorption/ionization plates for patient maker following in course of treatment, drug titration, and MALDI mass spectrometry imaging. Anal Biochem 434: 187-198.

21. Levery SB, Toledo MS, Doong RL, Straus AH, Takahashi HK (2000) Comparative analysis of ceramide structural modification found in fungal cerebrosides by electrospray tandem mass spectrometry with low energy collision-induced dissociation of Li$^+$ adduct ions. Rapid Commun Mass Spectrom 14: 551-563.

22. Laremore TN, Linhardt RJ (2007) Improved matrix-assisted laser desorption/ionization mass spectrometric detection of glycosaminoglycan disaccharides as cesium salts. Rapid Commun Mass Spectrom 21: 1315-1320.

23. Colsch B, Woods AS (2010) Localization and imaging of sialylated glycosphingolipids in brain tissue sections by MALDI mass spectrometry. Glycobiology 20: 661-667.

Comparative Adsorption Studies of Cd(II) on EDTA and Acid Treated Activated Carbons from Aqueous Solutions

Magda A Akl[1]*, Ali M Abou-Elanwar[2], Magda D Badri[2] and Abdel-fattah M Youssef[1]

[1]*Chemistry Department, Faculty of Science, Mansorua University, Egypt*
[2]*National Research Centre, Dokki, Giza, Egypt*

Abstract

This study plans to functionalize activated carbon with oxygen and nitrogen containing groups to improve removal of Cd(II) ions from aqueous solutions. Activated carbon (AC) was treated with nitric acid giving MC-HNO$_3$ that was pursued by modifying with EDTA to form MC-EDTA. AC, MC-HNO$_3$ and MC-EDTA were analyzed using FTIR, SEM, TEM, Boehm titration, point of zero charge and N$_2$ adsorption-desorption analysis. The batch technique sorption studies of Cd(II) onto sorbents were conducted and the factors controlling the adsorption process of Cd(II) were tested. Langmuir and Freundlich models were used to analyze the isotherm data. The equilibrium data fitted well Langmuir isotherm for all adsorbents. Pseudo-first order, pseudo-second order, intraparticle diffusion and the Boyd equations were used to analyze the kinetic data. The rate constants, equilibrium capacities and related correlation coefficients (R^2) for each kinetic model were evaluated and discussed. The second order model is the best fit model for the three sorbents. Though intraparticle diffusion has essential role in rate-controlling step in the adsorption process of Cd(II) onto the investigated sorbents, film diffusion is also governing this process. The thermodynamic parameters ΔG°, ΔH° and ΔS° for the sorption processes of Cd(II) onto the adsorbents were estimated, and spontaneity of adsorption was deduced from the negative sign of ΔG°. Desorption study highlighted the cost saving due to the easiness of regeneration for both MC-HNO$_3$ and MC-EDTA.

Keywords: Water; Activated carbon; Adsorption; Cd(II); Nitric acid; EDTA

Introduction

Heavy metal pollution is a critical environmental issue due to its harmful impacts and accumulation throughout the food chain and therefore in the human body. Cadmium is considered as one of the highly harmful heavy metals pollutants whereas Cd(II) is listed as the 7th most hazardous substance by the Agency for Toxic Substances and Disease Registry [1]. A number of chronic and acute diseases caused by Cd(II) exposure, such as itai-itai disease, renal damage, emphysema, hypertension and testicular atrophy [2]. Cd(II) toxicity is likely related to its high tendency to form bonds with thiol functional groups in some enzymes which lead to the displacement of biologically vital metals [3]. Cd(II) enter the water bodies through diverse ways comprising erosion of natural deposits, metal refinery discharges, and electronic waste runoff [1,4,5]. To keep water healthy, the Cd (II) should be preserved below certain limits. The permissible limit for Cd(II), provided by Environmental Protection Agency (EPA), is 0.005 mg/l and the current guideline value of potable water defined by the World Health Organization (WHO) is 0.003 mg/l [4,6] therefore Cd(II) concentration should not pass these limits to keep water safe. Purification of Cd(II) ions from waste water can be accomplished by applying different conventional techniques that are generally recognized as inefficient and/or expensive [7]. Adsorption by activated carbon is known as one of the extremely efficient methods for the removal of heavy metals [8]. Although the adsorption efficacy of a carbon is essentially related to the pore structure and surface area of the carbon, also the nature and density of surface functional groups have of important role in the adsorption of ionic or polar species [9]. The present work targets to functionalize the surface of activated carbon by oxidation with nitric acid and treatment with EDTA as a chelating agent for the improvement of heavy metals removal from water. The influence of parameters e.g., solution pH, sorbent dose and metal concentration, adsorption isotherms, kinetics modes and thermodynamic parameters of Cd(II) adsorption have been studied and discussed. The experimental equilibrium adsorption data

are analyzed by both Freundlich and Langmuir isotherm models. The desorption studies of sorbents were also examined.

Materials and Methods

Materials and apparatus

All activated carbon (Ubichem Ltd., UK) used in the study was in the granular form. The analytical grade reagents and EDTA disodium salt were purchased from Sigma-Aldrich. 1.7909 g of CdCl$_2$.2H$_2$O was dissolved in 1 liter of acidified bi-distilled to prepare 1000 ppm Cd(II) solution that used to prepare the other diluted concentrations. Buffer solution pH=5.5 was adjusted using CH$_3$COOH and NaOH. Glacial acetic acid, nitric acid (65%), hydrochloric acid (36%) were brought from Merck and used without any purification. The concentration of the Cd(II) was analyzed using flame atomic absorption (Analyst 300 Perkin Elmer FAAS). An Analyst 300 Perkin Elmer (FAAS) was used for the quantitative determination. A pH meter (Hi 931401, HANNA Portugal) was used for pH measurements. The adsorbents were weighed using analytical balance. Shaking Water Bath (NE5, Nickel-Electro Ltd., UK) used for adsorption and desorption experiments.

Functionalization of activated carbon

The functionalization of activated carbon was processed based in a

---

*Corresponding author: Magda A Akl, Chemistry Department, Faculty of Science, Mansorua University, Egypt, E-mail: magdaakl@yahoo.com

previous work [10]. 50 g of activated carbon (0.8-1.25 mm) was stirred in 1000 ml round bottom flask containing 500 ml concentrated nitric acid for 4 h then the AC was filtered off and washed several times with water until the pH of filtrate became 5.5. The AC was air dried overnight at room temperature and then in oven for 24 h at 110°C. This oxidized AC was labeled as MC-HNO₃. 20 g of MC-HNO₃ was refluxed in 1000 ml round bottom flask containing 500 ml 0.15 M EDTA disodium salt at 80°C for 4 h. Then filtered and washed with hot distilled water for many times to get rid of excess nonreacted EDTA and the carbon was dried for 24 h at 110°C and the product was labeled as MC- EDTA.

Characterization of the adsorbents

Fourier transform infrared spectrophotometer (FTIR), Jasco, Model 6100- Japan, was used to investigate the functional groups existed on the sorbents. The oxygen-containing groups were estimated using Boehm titration [11-13]. Also point of zero charge pH_{PZC} and surface pH were measured [10,14]. Surface Area and Pore Size Analyzer (Quantachrome-Nova 2000 Series) was applied at 77K to investigate N_2 adsorption onto the adsorbents and to measure the BET surface areas (S_{BET}) [15]. The α_s method was applied to analyze the N_2 adsorption isotherms to calculate: S^α (total surface area), S_n^α (non-microporous surface area) and the volume of micropores (V_m^α) [16]. Also the average pore radius r (nm) and the total pore volume (V_T) were estimated [10]. Scanning electron microscope JSE-T20 (JEOL, Japan) at 40 kW and high resolution transmission electron microscope (JOEL JEM2010 HR-TEM) at 200 kV were used to analyze the morphological features of the adsorbents.

Adsorption experiments

Batch adsorption studies of Cd(II) ions onto AC, MC-HNO₃ and MC-EDTA were performed using 100 ml to 1 L conical flasks and shaker water bath at 150 rpm. The influence of pH solution on Cd(II) adsorption was investigated by shaking 0.1 g sorbent with 25 ml of 150 ppm Cd(II) solutions for 24 h at 298.15K from 2 to 8. The pH of Cd(II) ion solution was fitted using 0.1M HCl or 0.1M NaOH. The isotherm studies were executed by shaking 0.1 g of adsorbents with 25 ml solution of various Cd(II) concentrations from 10 to 300 ppm for 24 h at 298.15K. Kinetics studies were performed by agitating 0.5 g of the adsorbents in 500 ml containing 100 ppm Cd(II) solutions at certain time periods from 5 min to 48 h then samples were withdrawn from the solution by fast filtration. The influence of sorbent doses was examined by equilibrating 25 ml of 350 ppm metal solutions and different adsorbents doses ranging from 0.05 g to 0.3 g. The effect of the temperature was investigated by shaking 0.1 g of sorbents with 25 ml solution at certain concentration of Cd(II) ranging from 10 to 300 ppm at 298.15, 308.15 and 318.15K.

The adsorption capacities were evaluated as follow:

$$q_t = \frac{(C_o - C_t).V}{m} \qquad (1)$$

Where, q_t (mg adsorbate/g adsorbent) is the adsorption capacity at certain time t; C_o (mg/L) is the initial concentration of Cd(II) ion ; C_t is the remaining concentration of Cd(II) after adsorption for certain time t (mg/L); V(L) is volume of Cd(II) solution and m(g) is mass of the sorbents. The removal % of Cd(II) ions from solution is estimated as follows:

$$Removal\,(\%) = \frac{(C_o - C_t).100}{C_o} \qquad (2)$$

The desorption percentages of Cd(II) from adsorbents were examined by equilibrating 0.1 g of the carbons into 50 ml solution whose concentration is)80 ppm Cd(II) with pH=5.6). After equilibrium, the total adsorbed Cd (II) concentration was calculated, and then the solution was filtrated using a membrane to recover the sorbents. These sorbents were dried for 6 h at 80°C and then placed into 50 ml distilled water. The pH of the dispersed solutions was tuned from 1.5 to 5.5 using suitable amounts of HCl. After equilibrium, Cd (II) concentration was analyzed. These adsorption/desorption procedures were performed for three times, to further ascertain the desorption capability of adsorbents. The Cd(II) desorption % was evaluated by the following equation:

$$Desorption\,(\%) = \frac{Amount\,released\,to\,solution\,(ppm)}{Total\,Adsorbed\,(ppm)} \times 100\% \qquad (3)$$

Results and Discussion

Characterization of the adsorbents

FT-IR: Figure 1 illustrated alike FT-IR spectra for the three sorbents. The wide bands at (3251-3430 cm⁻¹) appeared for all sorbents were due to stretching model of O-H of hexagonal group. All sorbents possessed shoulders at (2845-2955 cm⁻¹) that may be assigned to the C-H Aliphatic [17]. The signals appeared around 1700 cm⁻¹ were attributed to stretching vibrations (C=O) of carboxyl groups, ketones lactones or aldehydes [18]. The clear peak appeared at about 1645 cm⁻¹ for the samples may be due to C=O moieties in quinine. Peaks ranging from 1000-1200 cm⁻¹ is problematic due to numerous overlapping broad bands and hence they were very hard to be described as simple motion of precise chemical bonds or functional [19]. Many researchers attributed this interfering band to phenolic and epi-oxide structures and ether stretching vibrations presented in several structural backgrounds [19-21].

Surface acidity and Boehm titration: Boehm titration results are shown in Table 1. The total number of acidic functional groups was increased significantly after treatment with nitric acid and EDTA. The total number of basic functional groups were found to be smaller than that for acidic ones and these numbers were agreed the surface pH values and the pH_{pzc}. The MC-EDTA indicted minor increment in the total acidic groups and big increment of basic groups in compared with MC-HNO₃.

Nitrogen adsorption-desorption data: The nitrogen adsorption-desorption isotherms for the AC, MC-HNO₃ and MC-EDTA are shown in Figure 2. The porous features of the AC, MC-HNO₃ and MC-EDTA estimated from the exploration of N_2 adsorption isotherms are indicated in Table 2 [22]. The S_{BET}, S^α, S_n^α, S_m^α surface areas and pore volumes (V_T, V_n^α, V_m^α) for the activated carbon were declined after treatment with nitric acid and EDTA with widening of the Average pore radius (r). This could interpreted by the difficult accessibility of nitrogen molecules into the internal adsorption sites because of the increased functional groups existed on MC-HNO₃ and MC-EDTA at the pore entrance leading to electrostatic repelling of surface probe molecules (N_2). Also perhaps by the destruction of some walls with narrow pores that occurred during the oxidation of the AC [23,24]. Additional treatment of MC-HNO₃ with EDTA blocked the pore entrance with further functional groups resulting in extra electrostatic repelling and thinner pores thus causing to additional decrement in surface areas as well as the average pore radius with enlarging in micro surface area (S_m^α).

SEM and TEM: Obvious changes in the morphology of the adsorbents were noticed from SEM as shown in Figure 3. MC-HNO₃ illustrated a broader pores than that of AC which ensuring the eroding action of nitric acid and collapsing of large grains into small ones and might form a very large micropores as a result of the destruction in the

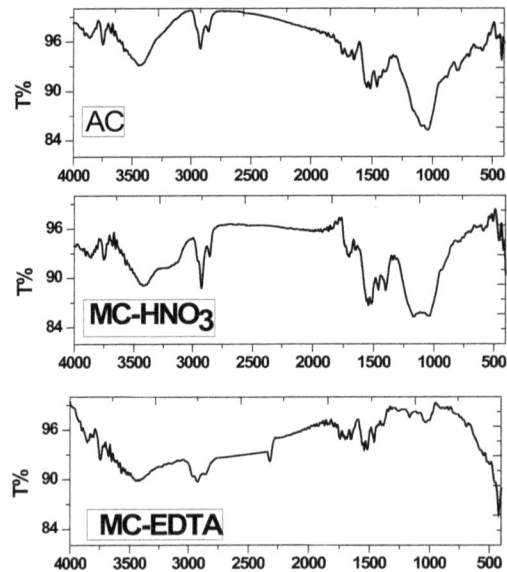

Figure 1: Fourier transform infrared spectra for AC, MC-HNO₃ and MC-EDTA.

Adsorbent	Carboxyli (mmol/g)	Lactonic (mmol/g)	Phenolic (mmol/g)	Total Acidic group (mmol/g)	Basic Groups (mmol/g)	Moisture content %	Ash content %	Surface pH	Point of zero charge
AC	0.078	0.0225	0.0456	0.1461	0.125	10	9	6.5	7
MC-HNO₃	0.585	0.0476	0.2523	0.8849	0.084	–	6.38	3.5	3.5
MC-EDTA	0.4038	0.489	0.0966	0.9894	0.26	–	3.7	7.2	7.45

Table 1: Surface properties of the adsorbents.

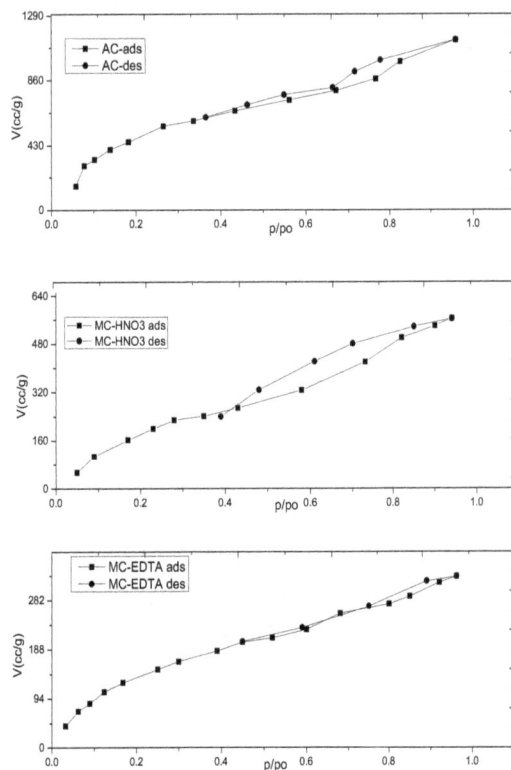

Figure 2: N₂ adsorption- desorption isotherms at 77K of AC, MC-HNO₃, MC-EDTA.

adjacent micropores due oxidation by nitric acid. MC-EDTA showed alike micrographs with that of MC-HNO₃ but with smaller number of wider pores. Some slight alterations in the surface are noticed from TEM images as shown in Figure 4 and porous structure of MC-HNO₃ and MC-EDTA are demonstrated in comparing with AC.

Adsorption and desorption studies

Effect of the solution pH: The pH of the solution governs the charge of the surface charge, ionization extent of the functional groups for the sorbents as well as the form of the metal ions in the aqueous solution. Therefore pH of the solution seems to be the most important parameter affecting the sorption process. The pH of the Cd(II) solutions was altered from 2 to 7.4 which is below the pH necessary for precipitation. It was apparent from Figure 5 that the uptake rate is rapid below pH 5.6 but above this pH the increasing in the uptake became slow and the adsorption capacity is near the maximum one. At lower pH, the adsorption quantity of Cd(II) was quite low that is assigned to the repelling between positive Cd(II) ions and positive function groups on the adsorbents surface. However, with raising the pH, the adsorbents surfaces accepted further negative charges from the ionization of the function groups, that could improve the electrostatic attractions of Cd(II) ions with negatively charged adsorbents surfaces so enhance sorption process. It is noticed that the adsorption capacities for MC-EDTA and MC-HNO₃ were larger than that of AC over the studied pH range which indicate the adsorption capacity is increased by the introduction of function groups on the surface of the activated carbon after treatment with HNO₃ and EDTA. A pH=5.6 was selected as the peak for further experiments.

Effect of initial concentrations: The equilibrium isotherms for AC, MC-HNO₃ and MC-EDTA are shown in Figure 6 indicated that

	SBET (m²/g)	S⁴ (m²/g)	S_n^α (m²/g)	S_m^α (m²/g)	V_T (ml/g)	V_m^α (ml/g)	V_n^α (ml/g)	$\bar{r}$ (nm)
AC	2114.6	1765.68	1181.25	584.43	1.7515	0.35	1.40	1.657
MC-HNO₃	934.4	720.944	576.332	144.611	0.87575	0.20	0.67	1.874
MC-EDTA	587.8	516.874	287.196	229.678	0.5084	0.16	0.35	1.730

Table 2: Nitrogen adsorption-desorption data.

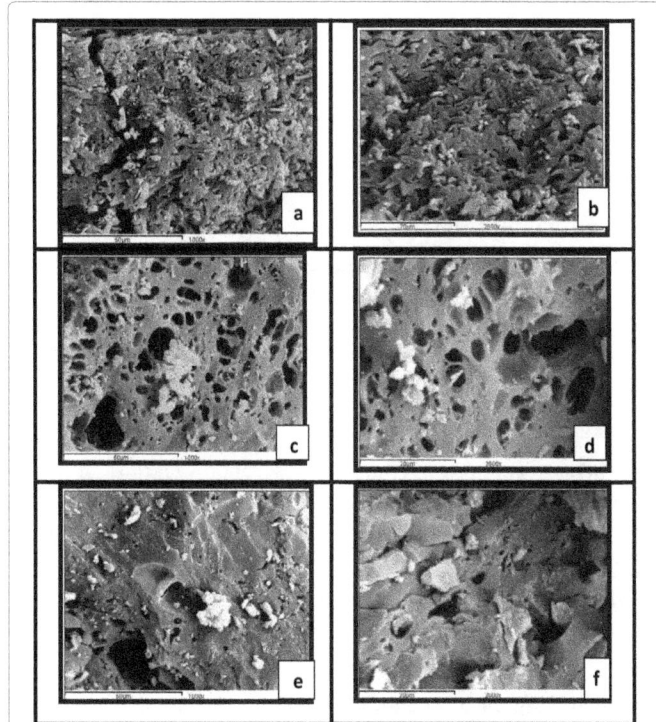

Figure 3: SEM images (a) and (b) for AC, (c) and (d) for MC-HNO₃ and (e) and (f) for MC-EDTA.

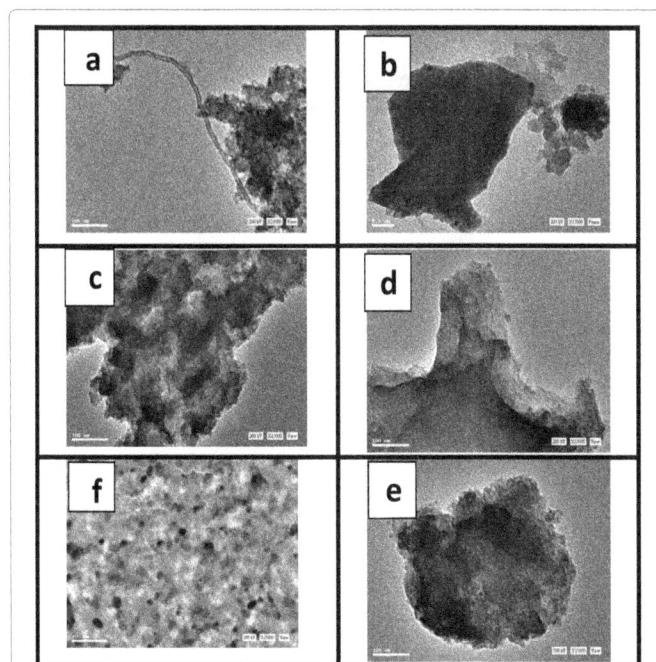

Figure 4: TEM images (a) and (b) for AC, (c) and (d) for MC-HNO₃ and (e) and (f) for MC-EDTA.

the adsorption capacities improved with raising the initial Cd(II) concentrations. This is due to the increment in the driving force produced from the concentration gradient [25,26]. This figure showed that the treated AC seemed to be more efficient under higher and lower Cd(II) concentration whereas the maximum adsorption capacity increased from 42 mg/g in case of AC to 82 mg/g for MC- HNO₃ (1.95 times as AC)and 89 mg/g for MC-EDTA(2.12 times as AC). It was observed, the surface treatments with nitric and EDTA had created new sites for binding with Cd(II) ions, which was responsible for enhancing the adsorption.

Effect of contact time: The influence of contact time was investigated for the adsorbents are shown in Figure 7. It was observed that the adsorption of Cd(II) ions increased rapidly until the first 6 hr. The rapid uptake at the initial period may attribute to the sufficient presence of active sites on the surface of the sorbents. After 10 hr, the adsorption was steady therefore it could be selected as the equilibrium time for Cd(II) adsorption.

Effect of adsorbent dose: Figure 8 showed the effect of sorbent dose on the adsorption process. With raising the sorbent dosage from 0.05 g/25 ml to 0.25 g/25 ml, the % of Cd(II) ions uptake improved from 30.2 to 34 for AC, from 51.7% to around 95.1% for MC-HNO₃ while increased from 54% to 100% in case of MC-EDTA. It was noticed that the adsorption per unit mass declined by raising the sorbent concentration. The reduction in adsorption density with raising in the adsorbent dosage is mostly because of the unsaturated of adsorption sites during the adsorption process [27,28]. This behavior could also resulted from the aggregation due the particle interaction and these aggregation might resulted in decrement in total surface area of the carbons and an increment in diffusional path length [28].

Equilibrium adsorption studies: The equilibrium adsorption isotherms are very useful for studying and designing of sorption modules. Langmuir, Sips, Freundlich and Redlich-Peterson models are commonly applied to describe the adsorption of solutes onto solid adsorbents [29]. Langmuir and Freundlich models describe the relation between the concentration of the adsorbate solution at equilibrium and the adsorption capacity at a certain temperature. Langmuir and Freundlich models are used to interpret the adsorption isotherms using regression analysis. The Langmuir equation (linear formula) is indicated as follows:

$$\frac{1}{q_e} = \frac{1}{bq_m} + \frac{1}{c_e} \qquad (4)$$

Where, q_e (mg/g) is the amount adsorbed of Cd(II) per unit mass of adsorbent, C_e is the equilibrium liquid phase concentration of Cd(II), b (l/mg) is Langmuir equilibrium constant, q_m(mg/g) is the monolayer adsorption capacity. Both b and q_m are estimated by plotting C_e/q_e against C_e as shown in Figure 9a. Langmuir isotherm is frequently evaluated by a separation factor, R_L, which expressed as follows:

$$R_L = \frac{1}{1 + bC_o} \qquad (5)$$

Where, C_o in that situation is the maximum initial concentration of Cd(II). The type of the isotherm and the nature of the adsorption

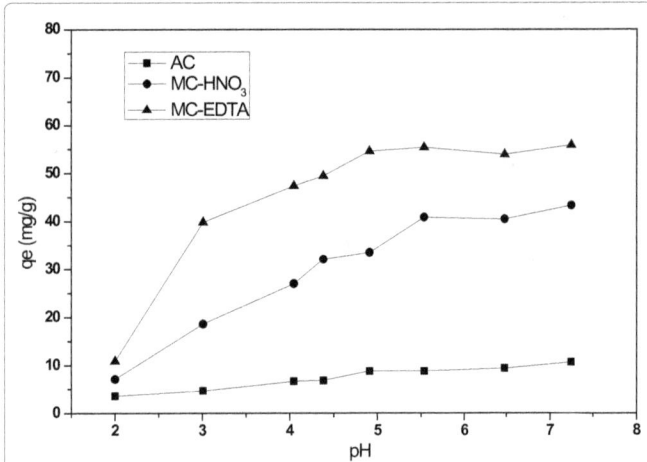

Figure 5: Effect of pH on adsorption of Cd(II) on AC , MC-HNO$_3$, MC-EDTA.

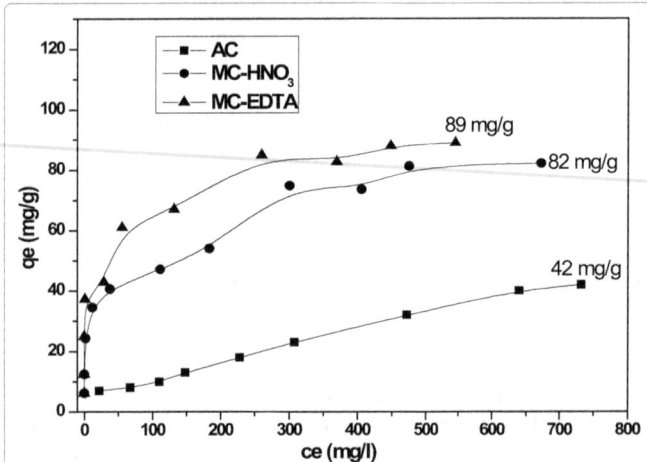

Figure 6: Adsorption isotherm of Cd(II) on AC, MC-HNO$_3$, MC-EDTA.

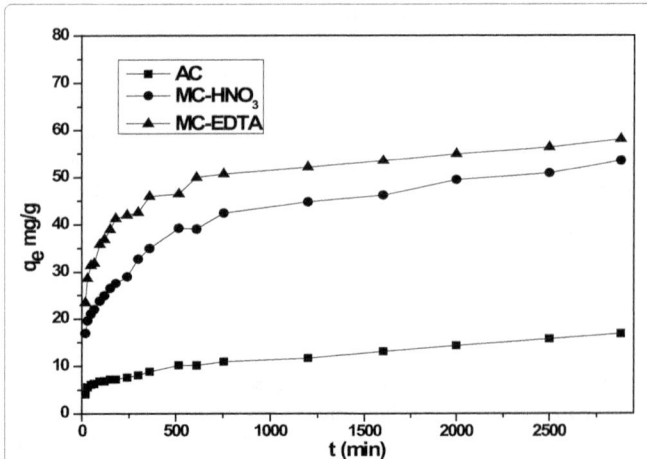

Figure 7: Effect of contact time on adsorption of Cd(II) on AC, MC-HNO$_3$, MC-EDTA.

process could be described according to the value of the separation factor. The adsorption can be irreversible R_L=0, unfavorable if R_L>1, linear when R_L=1 or favorable when 0<R_L<1 [30]. In all the studied case, the R_L values were existed between 0<R_L<1 which ensured favorability of the adsorption for Cd (II) ions. From the R_L values the favorability

of the adsorption ordered as follow: MC-EDTA>MC-HNO$_3$>AC. Freundlich model is completely empirical and it well interprets the adsorption process on heterogeneous surfaces [31]. Freundlich model (linear formula) is expressed as follow:

$$\log q_e = \log K_F + \frac{1}{n}\log C_e \qquad (6)$$

Where K_F (l/g) is Freundlich constant whereas n is Freundlich exponent. These constants are evaluated by plotting log q_e against log C_e as shown in Figure 9b. The isotherm constants for the adsorption of Cd(II) ions onto the adsorbents are illustrated in Table 3. It was observed that the Langmuir adsorption isotherm well fitted the experimental data for AC (r^2>0.9857), MC-HNO$_3$ (r^2>0.9764) and MC-EDTA (r^2>0.9941).

Adsorption kinetics

Pseudo-first-order and pseudo-second-order: Pseudo-first-order and pseudo-second-order were used to explore the adsorption kinetics of Cd (II) on the surface of the carbons. The pseudo-first-order model is commonly used to analyze kinetic data [32]. Its linear form is given as follows:

$$\log(q_e - q_t) = \log q_e - \frac{k_1}{2}.303t \qquad (7)$$

Where, q_e and q_t (mg/g) are the adsorption capacities of Cd(II) at equilibrium and at certain time t and k_1 (min^{-1}) is the equilibrium rate constant which was estimated from the slopes by plotting $\ln(q_e$-$q_t)$ against t (as shown in Figure 10a).

The pseudo-second-order kinetic model is defined as follows:

$$\frac{t}{q_t} = \frac{1}{k_2 q_e^{\,2}} + \frac{1}{q_e}t \qquad (8)$$

Where, k_2 (g/mg min) is the equilibrium rate constant whereas q_e is evaluated by plotting of t/q_t against t (Figure 10b) [33]. The correlation coefficients for both kinetic models are shown in Table 4. For AC, MC-HNO$_3$ and MC-EDTA: The second order model provides the best fit for practical kinetic data because the value of the calculated q_e agreed very well with the practical data and r^2 is higher than 0.992 for all carbons.

Intraparticle diffusion equation and Boyd equation: The intraparticle diffusion and Boyd models were used to identify diffusion mechanisms [34-36]. Intraparticle diffusion equation is expressed as follows:

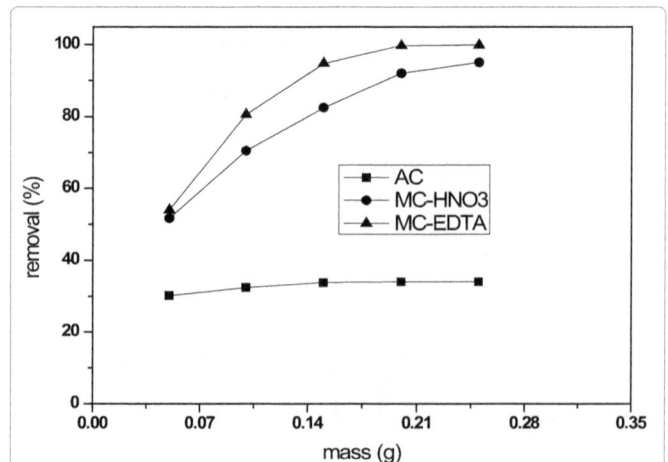

Figure 8: The effect of mass on adsorption of Cd(II) on AC, MC-HNO$_3$, MC-EDTA.

Adsorbents	Langmuir constants				Freundlich constants		
	q_m (mg/g)	b (L/mg)	R^2	R_L	K_F	1/n	R^2
AC	102.77	0.0010	0.9857	0.526316	0.8520	0.5755	0.9208
MC-HNO$_3$	85.84	0.0207	0.9764	0.050942	19.3335	0.2193	0.9550
MC-EDTA	91.91	0.0376	0.9941	0.028703	32.8746	0.1540	0.8720

Table 3: Langmuir and Freundlich isotherm constants for Cd(II) adsorption on AC ,MC-HNO$_3$ and MC-EDTA.

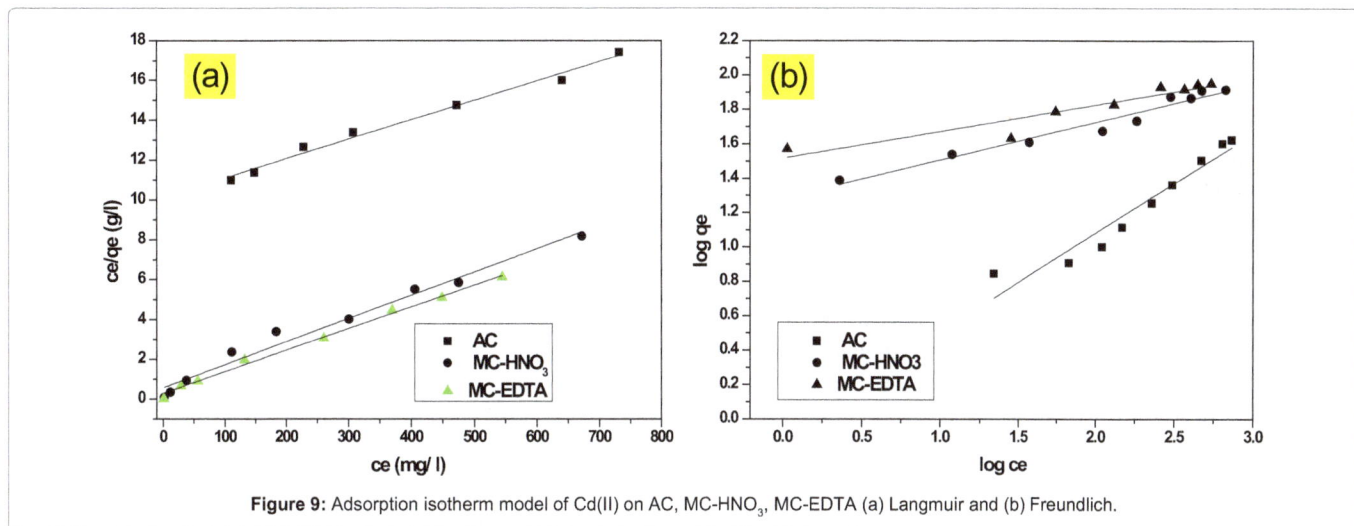

Figure 9: Adsorption isotherm model of Cd(II) on AC, MC-HNO$_3$, MC-EDTA (a) Langmuir and (b) Freundlich.

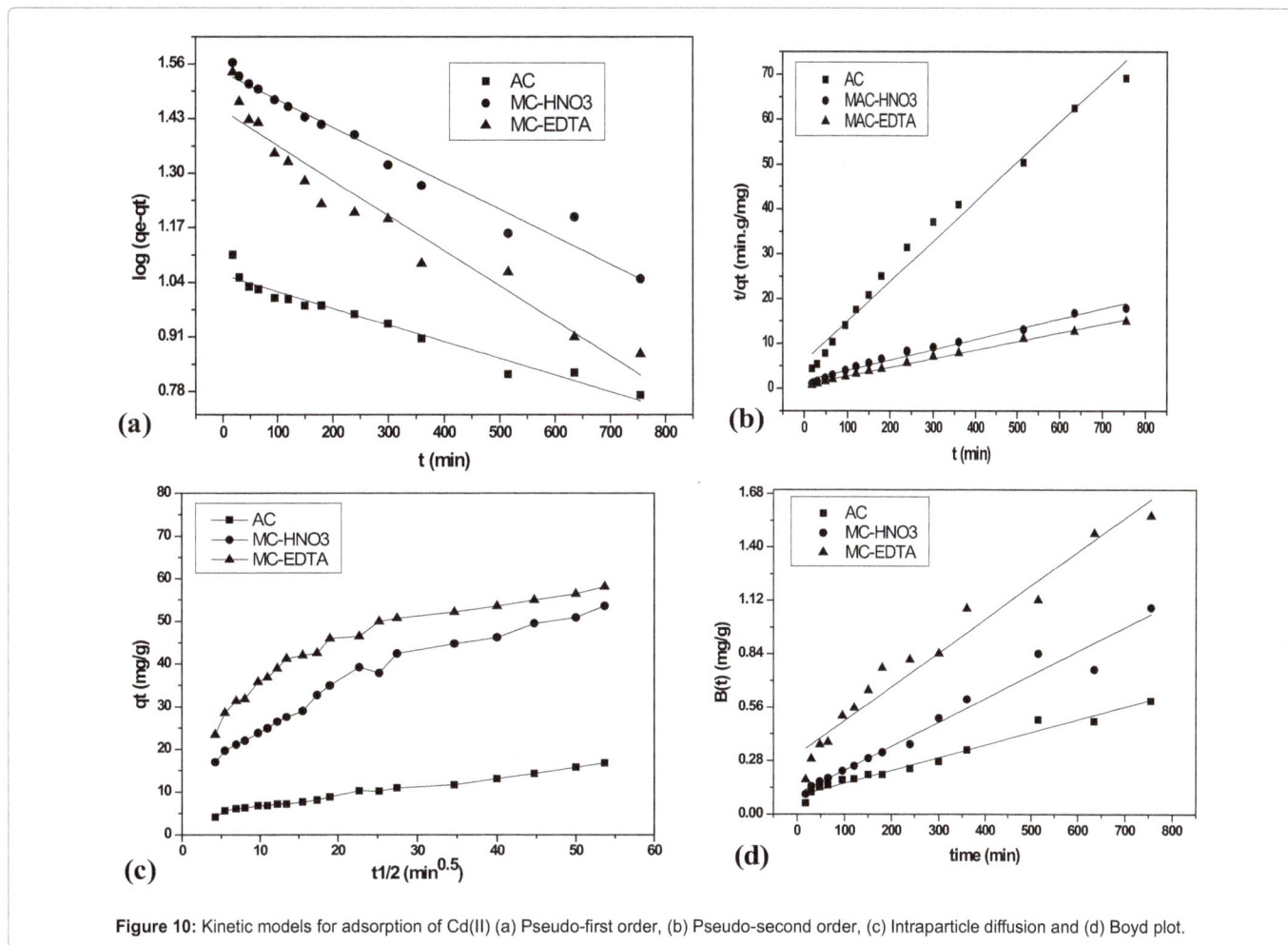

Figure 10: Kinetic models for adsorption of Cd(II) (a) Pseudo-first order, (b) Pseudo-second order, (c) Intraparticle diffusion and (d) Boyd plot.

$$q_t = k_{int} t^{0.5} + C \qquad (9)$$

where k_{int} (g/mg min$^{1/2}$) is the intraparticle rate constant for Cd(II) adsorption, C is the intercept and can be estimated by plotting q_t against $t^{1/2}$ (Figure 10c) [37]. Multi-linearity existed in such graphs [38,39], illustrating that two or more stages occurred. The first stage is sharper part and caused by the external surface or instantaneous adsorption. The second part is the gradual sorption stage, while intraparticle diffusion is rate-determined and from it K_{int} is estimated. The third part is the last equilibrium step while intraparticle diffusion starts to become steady due to extremely small solute concentrations of aqueous solutions.

Also Boyd model was used to investigate the kinetic data [40] to find whether adsorption undergoes to an external diffusion or intraparticle diffusion mechanism, which is defined as follows:

$$F(t) = 1 - \frac{6}{\pi^2} \sum_{n=1}^{n} \frac{1}{n^2} \exp(-n^2 Bt) \qquad (10)$$

Where F is the fractional of equilibrium at different periods (t), and B (t) is mathematical function of F. while n is integer that describes the infinite series solution and F is the fractional attainment of equilibrium at time t and is given as follows:

$$F = \frac{q_t}{q_e} \qquad (11)$$

where q_t and q_e is the adsorbed amount of Cd(II) ion at certain time (t) and equilibrium. Reichenberg [41] gave the next approximations:

When F > 0.85; $B\ (t) = -0.4977 - ln\ (1 - F) \qquad (12)$

When F < 0.85; $B(t) = \left(\sqrt{\pi} - \sqrt{\pi - (\frac{\pi^2 F(t)}{3})} \right)^2 \qquad (13)$

Figure 10c showed that the intraparticle diffusion plot did not passed through the origin point which indicated that both film diffusion (boundary layer diffusion) and intraparticle diffusion affect the rate-determining step of Cd(II) adsorption process for all the studied adsorbents. Also the same result could be deduced from the Boyd plots (Figure 10d) because the plots did not pass with the origin point.

Thermodynamic studies: The different values of the Langmuir constants b (l/mol) at different temperature used to calculate the change in the free energy ÄG° for Cd(II) adsorption using the following expression:

$$ÄG° = -RT\ lnb \qquad (15)$$

Where R is general gas constant (8.314 J/mol/K), T is the temperature in K^0. The Standard enthalpy (ÄH°) and entropy (ÄS°) for Cd(II) adsorption were calculated from Van't Hoff equation:

$$\ln b = \frac{-\Delta H^o}{RT} + \frac{\Delta S^o}{R} \qquad (16)$$

Figure 11 illustrated the relations between ln b against 1/T were linear and from the slopes the enthalpies ΔH^o were evaluated while the entropies ΔS^o were estimated from the intercept. Spontaneity of Cd(II) adsorption was deduced from the negative sign of $\Delta G°$ for all the adsorbents as shown in Table 5. The amount of the adsorbed Cd(II) at equilibrium was raised with raising the temperature for all cases. The adsorption processes were found to be endothermic due to the positive values of enthalpies ΔH^o for all cases. The affinity of Cd(II) adsorption by the carbons were predicted from positive signs of ΔS^o also proposed some structural variations in the activated carbons and Cd(II).

Desorption study: The cost saving of the adsorption systems that applied for water treatment is related to both the adsorption capacity and reversibility of the adsorption process. Therefore identical adsorbents should possess high adsorption capacity and easiness for desorption. Figure 12 illustrated the Cd (II) desorption % with regard to solutions at different pH values. It was found that the Cd (II) desorption % improved with decreasing the solution pH. The Cd(II) desorption % raised rapidly from pH=4.5 to pH=1.6 and peaked at pH=1.6 with values above 99.5% for all cases. These results indicated that the adsorbed Cd(II) by the carbons could be simply desorbed and can be applied currently in water decontamination from Cd(II) ions. Moreover, the recycling easiness illustrated that the adsorption mechanism is mainly controlled by ion exchange.

Conclusion

Treatment of AC with HNO$_3$ and EDTA highly functionalized its surface with oxygen and nitrogen containing groups which accompanied with significant alteration in its textural and morphological features of the activated carbon. Cd (II) adsorption capacities were controlled significantly by pH, metal concentration, equilibration time, sorbent dose, temperature. Functionalization of AC improved the adsorption capacity of AC from 42 mg/g to 82 mg/g in case of MC-HNO$_3$ and 89 mg/g for MC-EDTA. The adsorption isotherms of the adsorbents were well described by Langmuir model. Pseudo second-order well fitted the kinetic data AC, MC-HNO$_3$ and MC-EDTA. Both external and intraparticle diffusions controlled the rate-determining step for Cd(II) adsorption process. The three sorbents adsorbed the

		AC	MC-HNO$_3$	MC-EDTA
	qe, exp (mg/g)	16.85	53.6	58.1
First-order kinetic equation	q_1 (mg/g)	11.4256	34.659	28.17
	k_1 (min^{-1}) × 10^{-3}	0.91	1.5	1.92
	R_1^2	0.95011	0.96403	0.93294
Second-order kinetic equation	q_2 (mg/g)	17.31	55.13	58.96
	k_2 [g/(mg min)] × 10^{-3}	3.01	1.02	2.03
	R_2^2	0.962	0.989	0.9974
Intraparticle diffusion equation	k_{int} [mg/(g min1/2)]	0.22248	0.48327	0.32747
	C	3	20	34
	R_{int}^2	0.99059	0.93791	0.89998
Boyd equation	Intercept	0.0952	0.1059	0.3121
	R^2	0.9720	0.9678	0.9528

Table 4: Kinetic parameters for the adsorption of for Cd(II) adsorption on AC ,MC-HNO$_3$ and MC-EDTA.

	AC			MC-HNO$_3$			MC-EDTA		
T (k)	298.15	308.15	318.15	298.15	308.15	318.15	298.15	308.15	318.15
ΔG⁰(KJ/mol)	-11.604	-13.258	-14.498	-17.702	-18.928	-19.977	-18.931	-20.174	-22.083
ΔH⁰(KJ/mol)		31.6024			16.2539			27.9456	
ΔS⁰(J/mol/k)		145.133			113.98			156.877	

Table 5: Thermodynamic parameters for the adsorption of Cd(II) on on AC, MC-HNO$_3$ and MC-EDTA.

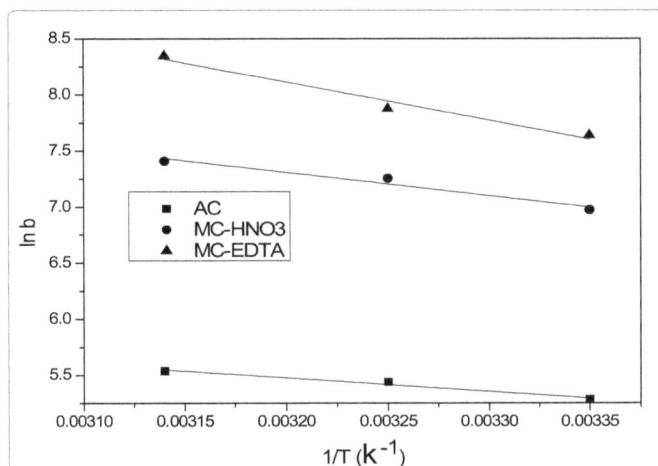

Figure 11: Van't Hoff isotherm of adsorption for Cd(II) AC, MC-HNO$_3$, MC-EDTA.

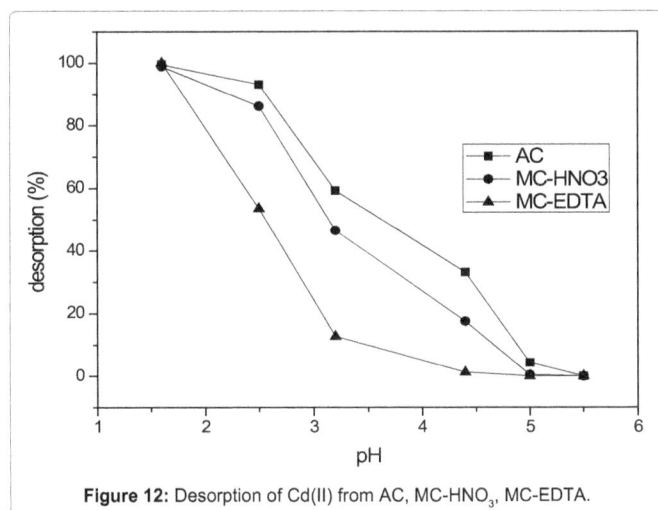

Figure 12: Desorption of Cd(II) from AC, MC-HNO$_3$, MC-EDTA.

Cd(II) spontaneously that was deduced from negative values of ΔG°. Desorption study illustrated the cost saving due to the easiness of regeneration for both MC-HNO$_3$ and MC-EDTA.

Acknowledgements

The author(s) express a grateful acknowledgement to The *Academy of Scientific Research and Technology (ASRT)* of Egypt for the financial support. Grant name: Scientist for Next Generation grant, cycle 3, group B Grant code: ASRT/TRG/B/2010-16.

References

1. ATSDR (2011) Priority List of Hazardous Substances.

2. Kadirvelu K, Thamaraiselvi K, Namasivayam C (2001) Removal of heavy metals from industrial wastewaters by adsorption onto activated carbon prepared from an agricultural solid waste. Bioresource Technology 76: 63-65.

3. Baes CF, Mesmer RE (1976) The hydrolysis of cations. Wiley, New York, USA.

4. USEPA (2009) National Primary Drinking Water Regulations.

5. Cadmium C, Nickel N (2007) Monitored Natural Attenuation of Inorganic Contaminants in Ground Water.

6. WHO (2011) Guidelines for drinking-water quality.

7. Mahmoud ME, Hafez OF, Alrefaay A, Osman MM (2010) Performance evaluation of hybrid inorganic/organic adsorbents in removal and preconcentration of heavy metals from drinking and industrial waste water. Desalination 253: 9-15.

8. Tian Y, Yin P, Qu R, Wang C, Zheng H, et al. (2010) Removal of transition metal ions from aqueous solutions by adsorption using a novel hybrid material silica gel chemically modified by triethylenetetraminomethylenephosphonic acid. Chemical Engineering Journal 162: 573-579.

9. Vasu AE (2008) Surface Modification of Activated Carbon for Enhancement of Nickel(II) Adsorption. E-Journal of Chemistry 5: 814-819.

10. Youssef AM, Dawy MB, Akland MA, Abou-Elanwar AM (2013) EDTA Versus Nitric Acid Modified Activated Carbon For Adsorption Studies of Lead (II) From Aqueous Solutions. Journal of Applied Sciences Research 9: 16.

11. Goertzen SL, Thériault KD, Oickle AM, Tarasuk AC, Andreas HA (2010) Standardization of the Boehm titration. Part I. CO$_2$ expulsion and endpoint determination. Carbon 48: 1252-1261.

12. Boehm HP (1994) Some aspects of the surface chemistry of carbon blacks and other carbons. Carbon 32: 759-769.

13. Boehm HP (1966) Chemical Identification of Surface Groups, in Advances in Catalysis. Eley HPDD, Paul BW (Eds), Academic Press. pp: 179-274.

14. Rivera-Utrilla J, Bautista-Toledo I, Ferro-García MA, Moreno-Castilla C (2001) Activated carbon surface modifications by adsorption of bacteria and their effect on aqueous lead adsorption. Journal of Chemical Technology & Biotechnology 76: 1209-1215.

15. Brunauer S, Emmett PH, Teller E (1938) Adsorption of Gases in Multimolecular Layers. Journal of the American Chemical Society 60: 309-319.

16. Selles-Perez MJ, Martin-Martinez JM (1991) Application of α and n plots to N2 adsorption isotherms of activated carbons. Journal of the Chemical Society, Faraday Transactions 87: 1237-1243.

17. Kennedy LJ, Vijaya JJ, Sekaran G (2005) Electrical conductivity study of porous carbon composite derived from rice husk. Materials Chemistry and Physics 91: 471-476.

18. Kennedy LJ, Vijaya JJ, Sekaran G (2004) Effect of Two-Stage Process on the Preparation and Characterization of Porous Carbon Composite from Rice Husk by Phosphoric Acid Activation. Industrial & Engineering Chemistry Research 43: 1832-1838.

19. Painter P, Starsinic M, Coleman M (1985) Fourier Transform Infrared Spectroscopy. Academic Press, New York. pp: 169-189.

20. Gómez-Serrano V, Acedo-Ramos M, López-Peinado AJ, Valenzuela-Calahorro C (1994) Oxidation of activated carbon by hydrogen peroxide. Study of surface functional groups by FT-IR. Fuel 73: 387-395.

21. Fanning PE, Vannice MA (1993) A DRIFTS study of the formation of surface groups on carbon by oxidation. Carbon 31: 721-730.

22. Brunauer S, Deming LS, Deming WE, Teller E (1940) On a Theory of the van der Waals Adsorption of Gases. Journal of the American Chemical Society 62: 1723-1732.

23. Salame II, Bagreev A, Bandosz TJ (1999) Revisiting the Effect of Surface Chemistry on Adsorption of Water on Activated Carbons. The Journal of Physical Chemistry B 103: 3877-3884.

24. Choma J, Burakiewicz-Mortka W, Jaroniec M, Li Z, Klinik J (1999) Monitoring Changes in Surface and Structural Properties of Porous Carbons Modified by Different Oxidizing Agents. Journal of Colloid and Interface Science 214: 438-446.

25. Ozacar M, Sengil IA (2005) Adsorption of metal complex dyes from aqueous solutions by pine sawdust. Bioresour Technol 96: 791-795.

26. Sun G, Xu X (1997) Sunflower Stalks as Adsorbents for Color Removal from Textile Wastewater. Industrial & Engineering Chemistry Research 36: 808-812.

27. Yu LJ, Shukla SS, Dorris KL, Shukla A, Margrave JL (2003) Adsorption of chromium from aqueous solutions by maple sawdust. J Hazard Mater 100: 53-63.

28. Shukla A, Zhang YH, Dubey P, Margrave JL, Shukla SS (2002) The role of sawdust in the removal of unwanted materials from water. J Hazard Mater 95: 137-152.

29. Ania CO, Parra JB, Pis JJ (2002) Influence of oxygen-containing functional groups on active carbon adsorption of selected organic compounds. Fuel Processing Technology 79: 265-271.

30. Karagöz S, Tay T, Ucar S, Erdem M (2008) Activated carbons from waste biomass by sulfuric acid activation and their use on methylene blue adsorption. Bioresour Technol 99: 6214-6222.

31. Ng C, Losso JN, Marshall WE, Rao RM (2002) Freundlich adsorption isotherms of agricultural by-product-based powdered activated carbons in a geosmin-water system. Bioresour Technol 85: 131-135.

32. Hameed BH, Ahmad AL, Latiff KNA (2007) Adsorption of basic dye (methylene blue) onto activated carbon prepared from rattan sawdust. Dyes and Pigments 75: 143-149.

33. Franca AS, Oliveira LS, Ferreira ME (2009) Kinetics and equilibrium studies of methylene blue adsorption by spent coffee grounds. Desalination 249: 267-272.

34. Özacar M (2003) Equilibrium and kinetic modelling of adsorption of phosphorus on calcined alunite. Adsorption 9: 125-132.

35. Wu FC, Tseng RL, Juang RS (2001) Adsorption of dyes and phenols from water on the activated carbons prepared from corncob wastes. Environ Technol 22: 205-213.

36. Annadurai G, Juang RS, Lee DJ (2002) Use of cellulose-based wastes for adsorption of dyes from aqueous solutions. J Hazard Mater 92: 263-274.

37. Altenor S, Carene B, Emmanuel E, Lambert J, Ehrhardt JJ, et al. (2009) Adsorption studies of methylene blue and phenol onto vetiver roots activated carbon prepared by chemical activation. J Hazard Mater 165: 1029-1039.

38. Qi L, Xu Z (2004) Lead sorption from aqueous solutions on chitosan nanoparticles. Colloids and Surfaces A: Physicochemical and Engineering Aspects 251: 183-190.

39. Yan G, Viraraghavan T (2003) Heavy-metal removal from aqueous solution by fungus Mucor rouxii. Water Res 37: 4486-4496.

40. Boyd GE, Adamson AW, Myers LS (1947) The Exchange Adsorption of Ions from Aqueous Solutions by Organic Zeolites. II. Kinetics. Journal of the American Chemical Society 69: 2836-2848.

41. Reichenberg D (1953) Properties of Ion-Exchange Resins in Relation to their Structure. III. Kinetics of Exchange. Journal of the American Chemical Society 75: 589-597.

3

Definition of System Suitability Test Limits on the Basis of Robustness Test Results

Prafulla Kumar Sahu*

Department of Pharmaceutical Analysis and Quality Assurance, Raghu College of Pharmacy, Dakamarri, Bheemunipatnam, Visakhapatnam-531 162, Andhra Pradesh, India

Abstract

A chromatographic method newly optimized to identify and assay four antihypertensive drugs in tablet dosage forms was complemented by a robustness test. The best system suitability criteria for numerous responses were evaluated on the basis of the robustness test results. Generally speaking, it is difficult to achieve a total satisfactory solution. Situations may also become ambiguous if the system suitability limits for few responses of a robust method are violated. In this context, it becomes crucial to redefine these limits based on the robustness test results. In the present study, the extreme experimental (worst-case) conditions that offer worst result but still acceptable and likely to occur were predicted from the robustness test effects. Eventually, replicated experiments were executed in such worst conditions and the system suitability test (SST) limits were determined.

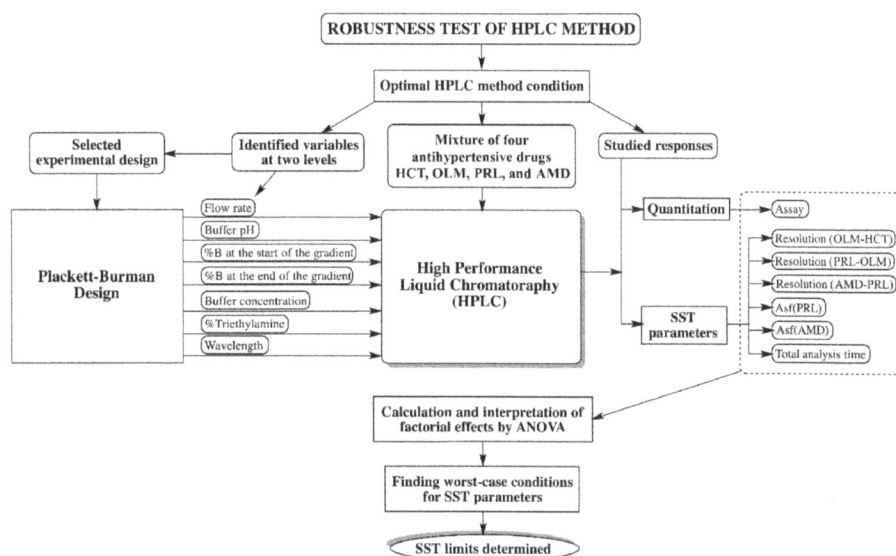

Graphical Abstract

Keywords: HPLC method validation; Plackett-Burman design; Robustness test; System suitability test; Worst-case condition

Abbreviations: SST: System suitability test; ICH: International Conference on Harmonisation; PBD: Plackett–Burman design; AMD: Amlodipine; OLM: Olmesartan; HCT: Hydrochlorothiazide; PRL: Propanolol; CAN: Acetonitrile; ANOVA: Analysis of variance.

Introduction

When a new method is optimized it is important to establish how robust it is. As defined by the International Conference on Harmonisation (ICH) [1], robustness is the ability of an analytical method to stay unbiased by small, but deliberately introduced variations in the method variables. ICH guidelines prescribe that the robustness of a method should be assessed during the development phase (or at the beginning of the validation), and not at the end of method validation [2].

Robustness can be evaluated by statistical experimental design to examine simultaneously the influence of the variation in several method variables, e.g., mobile phase flow rate, temperature, type of column, slope of the gradient, buffer pH, ionic strength, detector wavelength, additives type and concentration, etc., on the outcome (response) of a method [3-5]. Based on the objective, two strategies can be adapted for robustness studies. If the investigation only meant to verify that the already validated method is robust, screening designs such as

***Corresponding author:** Prafulla Kumar Sahu, Department of Pharmaceutical Analysis and Quality Assurance, Raghu College of Pharmacy, Dakamarri, Bheemunipatnam, Visakhapatnam-531 162, Andhra Pradesh, India
E-mail: kunasahu1@rediffmail.com

Plackett–Burman design (PBD) [5-7], fractional factorial design [5,8,9] or supersaturated design is employed. In case of an experimental model need to determine the robust domain (tolerable variations) via response surfaces, preferably optimization designs such as central composite design or Box-Behnken design can be considered [10]. The variables are examined usually at two levels [low (-1) and high (+1)] situated around the nominal one. The nominal levels are the optimal conditions as stated in the assay procedure. In general, by performing the robustness test of a method we can identify the critical variables that might significantly influence the outcome of the studied responses when the method is repeated at different conditions or laboratory. The knowledge of these critical variables is necessary if a "precautionary statement" [1] is included in the method description when transferred to another laboratory.

The chromatographic variables studied in a system suitability test (SST) such as resolution, efficiency, capacity factor, peak asymmetry factors, etc., can also be viewed as responses in a robustness test. It is worth noticing that, it is possible to define system suitability limits based on the evaluation of the robustness because it explores the most extreme variations in the variables that may occur. Since it is rare to get a globally satisfactory solution, this procedure would avoid ambiguous situations. Vander Heyden and coworkers [11] used a PBD in robustness testing and defined the experimental conditions giving the worst result that still is acceptable and probable to occur; this way the system suitability limits are redefined from replicated experiments in such conditions. A stepwise guidance in setting-up and interpreting a robustness test was reported combined with derivation of system suitability limits from robustness test results based on worst-case condition [12].

The use of antihypertensive agents in combination is common. To decrease the pill burden and improve patient compliance, combination antihypertensive therapies compile two or more active drugs. Thiazide diuretics are frequently recommended as one of the first-line therapy for the treatment of hypertension in combination with other class of antihypertensive drugs, i.e., angiotensin-converting enzyme inhibitor, angiotensin receptor blocker, calcium channel blocker and β-blocker. Furthermore, generic preparations containing combination of amlodipine (AMD) [13] with olmesartan (OLM) or atenolol or telmisartan, OLM with hydrochlorothiazide (HCT), propanolol (PRL) with HCT are the more commonly prescribed heart and cardiovascular medications to lower the prescription costs [14]. Hence, a suitable analytical method is highly desirable for simultaneous determination of these drugs in bulk and pharmaceutical formulations.

The numerous analytical methods dealt with assay of HCT, OLM, PRL, and AMD available [15-20] are impaired by inefficient or time-consuming procedures and lack of statistical evaluation of significant variables. No HPLC method has been developed for simultaneous analysis of the four drugs so far. The present study particularly focuses on the robustness testing of the newly developed HPLC procedure for the assay of the analytes in tablet dosage forms. In this work, the SST limits for several chromatographic parameters were established on the base of the robustness test results. It is emphasized that chemometry, usually used for experimental design, is crucial for method validation.

Experimental

Apparatus used

A binary gradient HPLC system equipped with two LC-20AD pumps, a SPD-M20A diode array detector with a manual injector (all from Shimadzu, Kyoto, Japan) were used. A reverse-phase Grace Alltima HP Amide (150 × 4.6 mm; 3 μm) was used for chromatographic

separation of the four drugs. The chromatographic analysis and data integration were recorded on a computer system using LC-Solution data acquiring software (Shimadzu, Kyoto, Japan).

Chemicals

Standard hydrochlorothiazide, olmesartan, propranolol and amlodipine besylate were a kind gift from Sun Pharmaceuticals, Ahmedabad, India. HPLC grade methanol was purchased from Merck Private Limited, Mumbai; and acetonitrile (ACN) from Finer chemicals limited, Ahmedabad.

Preparation of solutions

Reference solution: Accurately 100 mg of OLM, 25 mg of AMD, 62.5 mg of HCT, 100 mg of PRL and 20 ml of methanol were transferred into a 100 ml volumetric flask. The mixture was shaken and sonicated for 15 mins.

Sample solution (Formulation): The sample solution was prepared by taking marketed formulation, Triolmezest tablets (Sun Pharma Pvt Ltd) containing OLM, AMD and HCT and Inderal tablets (Abbott Health Care Pvt. Ltd) containing PRL. 20 tablets each were weighed and finely powdered. The powder equivalent to 100 mg OLM, 25 mg AMD, 62.5 mg HCT and 100 mg PRL was taken in 100 ml volumetric flask and then dissolved in 20 ml methanol and makeup to the volume with water and the mixture was mechanically shaken for 30 min, and filtered through a durapore HVLP 0.45 μm filter paper.

Blank solution: A mixture of methanol and water (50:50 v/v) was used as a blank solution.

Preparation of buffer: The aqueous phase of the HPLC solvent system consists of equimolar mixture of sodium dihydrogen o-phosphate dihydrate and disodium hydrogen o-phosphate dihydrate buffer (22 mM and 18 mM). Different amount of triethylamine (0.3% and 0.5%) as organic modifier was added according to the study designs. The final volume was made up with HPLC grade water to get the desired buffer following pH adjustment to 6.7 and 7.3. The buffer was filtered through 0.25 μm membrane filter and degassed for 30 min in an ultrasonic bath.

Chromatographic conditions

The method prescribes a 150 mm length, 4.6 mm I.D. column, packed with Grace Alltima HP Amide, 3 μ particle size. The substances are eluted at a flow-rate of 0.8 ml/min. The solvent gradient used is shown in Table 1. The injection volume was 20 μl. UV detection was observed at 272 nm. The optimization was a one variable at a time procedure.

Calculations and software

Chromatographic responses were acquired using Shimadzu data acquiring software, "LC-Solution". The choice of the experimental design and runs was done by the software package Design-Expert 9.0.3 trial version for Windows (Stat-Ease Inc.). The calculation of effects and their statistical interpretation for the current optimization study was also performed with the same software. Calculation of standard deviation, % coefficient of variance for various validation parameters of the chromatographic method was made using MS-Excel.

Time (min)	0.01	3.49	3.50	4.00
%ACN	42	42	60	40

aComposition of the mobile phase during the solvent gradient as % ACN.
Table 1: LC gradient time programminga.

Robustness test

PBD, a supersaturated design was employed to evaluate the robustness of the developed HPLC method. The influence of the deliberately introduced small variations in the method variables, i.e., mobile phase flow rate, buffer pH, Percent acetonitrile at the start and end of the gradient, buffer concentration, percent triethylamine and detection wavelength were simultaneously investigated for resolutions [Rs (OLM-HCT), Rs (PRL-OLM) and Rs (AMD-PRL)], tailing factors [Asf(PRL) and Asf(AMD)] and total analysis time [t_R(AMD)]. The variables are examined usually at two levels [low (-1) and high (+1)] situated around the nominal one. The nominal levels are the optimal conditions as stated in the assay procedure.

Results and Discussion

The robustness evaluation of the high-performance liquid chromatography (HPLC) method for identification and assay of OLM, AMD, HCT and PRL in tablet simulations investigated the variables summarized in Table 2, while (i) the studied responses (ii) the expected values under nominal conditions and (iii) the SST limits that were established before the robustness test was applied are detailed in Table 3.

The low and high levels of the quantitative variables in Table 2 were selected based on the uncertainty with which a variable level can be set. Some were chosen as a constant percentage above (+) and below (-) the nominal level. The seven variables were examined in a PBD for 11 variables requiring 12 experiments (Table 4). In the 4 spare columns (randomly selected) dummy variables are entered. These are imaginary variables whose change from one level to the other does not cause a physical change in the responses. The effects estimated from these dummies represent the experimental error and are crucial for the statistical analysis.

For each of the 12 experiments a blank injection, two injections of the reference solution and an injection of the sample solution were performed. The second injection of the reference solution was used to determine the system suitability test parameters. The two reference injections were used to estimate the % recovery of the four drugs in the sample solution.

Figure 1 shows a typical chromatogram for a reference solution obtained at nominal conditions. Table 5 illustrates results for the experimental responses that are studied by PBD. As can be observed the recoveries of HCT range from 100.09 to 103.90%, OLM from 97.79 to 102.83%, PRL from 98.49 to 103.98% and AMD from 98.01 to 102.72%, the resolution (OLM-HCT) from 1.857 to 5.964, resolution (PRL-OLM) from 1.131 to 8.972, resolution (AMD-PRL) from 1.576 to 6.870, the tailing factor (PRL) from 1.193 to 2.640, tailing factor (AMD) from 1.055 to 2.785 and the total analysis time [t_R(AMD)] from 4.460 to 8.576 min. The chromatograms for selected runs are depicted in Figure 2. The effect of a variable on a response is calculated according to the following equation [11]:

$$E_x = \frac{\sum Y(+1)}{n} - \frac{\sum Y(-1)}{n} \qquad (1)$$

Where, E_x is the effect of variable X; $\sum Y(+1)$ and $\sum Y(-1)$ are the sums of the responses where variable X was at level (+1) and at level (-1), respectively and n is the number of runs in which X was at level (+1) or at level (-1), usually equal to $N/2$ with N, the number of design experiments.

Analysis of Variance (ANOVA)

The significant effects of each variable on the studied responses were interpreted with the aid of analysis of variance (ANOVA) [5,8]. The sum of squares $(SS)_X$ for a variable can be calculated as $[E_X N/2]^2/N$ while total error is estimated from the sum of the sums of squares from the dummies. The mean square for a variable, $(MS)_X$ is obtained from the ratio of $(SS)_X$ and the degrees of freedom (df). The F ratio is calculated by dividing $(MS)_X$ by $(MS)_{total\ error}$ and the P value gives an indication of the significance of an effect because it represents the probability of being wrong when accepting that an effect is significant. For example, if the P value is below the considered level of confidence α, an effect may be considered to be statistically significant. For example, when $P<0.01$ then an effect is significant at $\alpha =0.01$.

Table 6 features the effects of the different variables on the considered responses. The variables having a statistically significant effect on a response, at significance levels of 5% (P, 0.05) and of 10%

Variables	Limits	Nomial	-1	+1
Flow of the mobile phase	± 0.1	0.8	0.7	0.9
pH of the buffer	± 0.3	7.0	6.7	7.3
Percentage organic solvent (% B) ACN in the mobile phase at the start of the gradient	± 1	42	41	43
% B in the mobile phase at the end of the gradient	± 2	40	38	42
Concentration of the buffer (%)	± 10	20	18	22
Percentage of Triethylamine	± 0.1	0.4	0.3	0.5
Detection wavelength	± 5 nm	272	267	277

Table 2: Variables investigated in the design.

Response	Substances considered	Expected value at nominal levels	SST limits
Resolution (Rs)	OLM-HCT	3.797	3.671
Resolution (Rs)	PRL-OLM	2.913	2.821
Resolution (Rs)	AMD-PRL	2.639	2.486
Tailing factor (Asf)	PRL	2.257	2.624
Tailing factor (Asf)	AMD	2.074	2.350
Total analysis time (Rt); min.	AMD	5.502	5.494

Table 3: Responses studied.

Runs	Flow	Dum 1	pH	Dum 2	%B start	%B end	Dum 3	Buffer conc.	% TEA	Wavelength	Dum 4
1	0.7	1	6.7	1	43	38	1	22	0.5	267	-1
2	0.7	1	7.3	-1	43	42	1	18	0.3	267	1
3	0.7	-1	6.7	-1	41	38	-1	18	0.3	267	-1
4	0.9	-1	7.3	1	43	38	-1	18	0.5	267	1
5	0.7	1	7.3	1	41	38	-1	22	0.3	277	1
6	0.9	-1	7.3	1	41	42	1	22	0.3	267	-1
7	0.9	1	6.7	-1	41	42	-1	22	0.5	267	1
8	0.7	-1	6.7	1	41	42	1	18	0.5	277	1
9	0.9	1	7.3	-1	41	38	1	18	0.5	277	-1
10	0.9	1	6.7	1	43	42	-1	18	0.3	277	-1
11	0.7	-1	7.3	-1	43	42	-1	22	0.5	277	-1
12	0.9	-1	6.7	-1	43	38	1	22	0.3	277	1

Abbreviations: Flow, flow of the mobile phase (ml/min); Dum1 to Dum 4, dummy variables; pH, pH of the buffer; % B start, percentage ACN in the mobile phase at the start of the gradient; % B end, percentage ACN in the mobile phase at the end of the gradient; Buffer conc., concentration of the buffer in mM; % TEA, percentage triethylamine; Wavelength, wavelength of the detector in nm.

Table 4: Plackett-Burman design.

Run	Rs(OLM-HCT)	Rs(PRL-OLM)	Rs(AMD-PRL)	Asf(PRL)	Asf(AMD)	t_R(AMD) min	% HCT	% OLM	% PRL	% AMD
1	4.498	5.138	1.576	2.367	2.785	8.224	100.09	97.79	102.3	99.45
2	2.362	6.021	2.504	2.640	1.672	8.296	102.78	97.87	102.47	98.01
3	5.964	1.617	3.731	2.140	2.076	6.708	101.88	99.95	100.09	99.58
4	3.210	4.098	4.449	1.470	1.055	8.145	100.74	100.06	99.17	99.66
5	5.547	4.779	2.282	2.483	2.100	6.463	102.95	99.13	98.49	96.71
6	1.857	5.654	2.000	2.352	2.157	4.995	103.41	101.09	103.98	102.72
7	2.535	2.938	6.870	1.795	1.073	7.229	101.65	98.33	101.25	98.34
8	3.293	1.857	2.697	2.080	1.943	5.293	100.90	102.83	102.43	99.17
9	5.151	3.852	2.322	1.193	2.605	5.800	103.90	99.60	98.65	100.62
10	3.036	2.612	2.834	2.270	1.897	4.460	100.70	99.95	99.95	99.17
11	3.490	8.972	5.863	1.578	1.785	8.576	101.65	101.08	100.3	101.4
12	5.229	1.131	3.423	2.066	1.889	4.923	102.45	98.53	99.48	99.52
MEAN							101.95	99.68	100.71	99.53
RSD							1.1	1.4	1.7	1.5

Abbreviations: Rs(OLM-HCT), resolution between OLM and HCT; Rs(PRL-OLM), resolution between PRL and OLM; Rs(AMD-PRL), resolution between AMD and PRL; Asf(PRL), tailing factor of PRL; Asf(AMD), tailing factor of AMD; t_R(AMD), retention time of AMD; % HCT, percentage of HCT; % OLM, percentage of OLM; %PRL, percentage of PRL; % AMD, percentage of AMD.

Table 5: Results of the experiments.

Responses	Variables										
	Flow rate	Dum 1	pH	Dum 2	%B start	%B end	Dum 3	Buffer conc	%TEA	Wave length	Dum 4
%HTC	+0.22	-	+0.65	-0.46	-0.52	-	+0.33	-	-0.44	-	-
%OLM	-	-0.91	-	+0.46	-0.47	+0.51	-	-0.36	+0.26	+0.50	-0.23
%PRL	-	-	-	-	-	+1.02	+0.84	-	-	-0.83	-
%AMD	+0.48	-0.81	+0.32	-0.049	-	+0.27	+0.39	+0.16	+0.24	-0.097	-0.96
Rs(OLM-HCT)	-0.35	-	-0.25	-0.27	-0.20	-1.08	-0.12	+0.020	-0.16	+0.43	-0.14
Rs(PRL-OLM)	-0.69	+0.15	+1.49	-	+0.63	+0.63	-0.13	+0.73	+0.41	-0.20	-0.57
Rs(AMD-PRL)	-	-	-	-	-	-	-1.10	-	-	-	-
Asf(PRL)	-0.29	-	-0.14	+0.22	+0.14	+0.23	+0.028	+0.16	-0.44	-0.24	+0.16
Asf(AMD)	-	-	-0.39	+0.42	+0.36	-	-	+0.39	-	+0.39	-
t_R(AMD)	-0.67	-	+0.45	-0.33	+0.51	-	-0.34	-	+0.62	-0.67	-

Insignificant terms were excluded.

Table 6: Effects of the variables on the different responses.

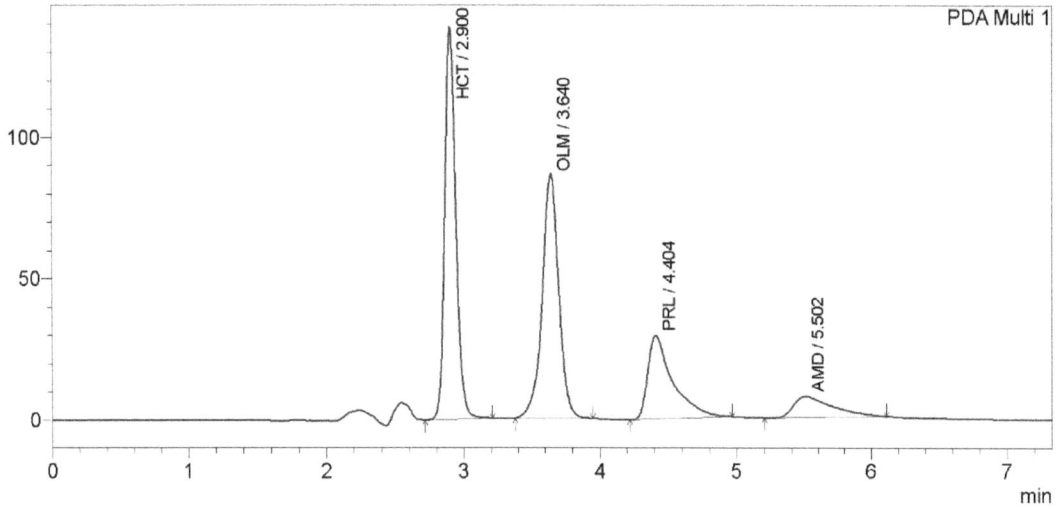

Figure 1: A typical chromatogram for a reference solution obtained at nominal conditions.

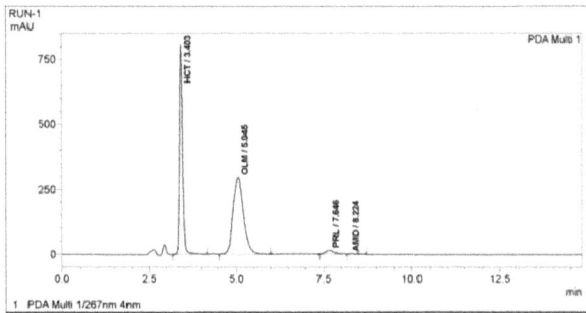

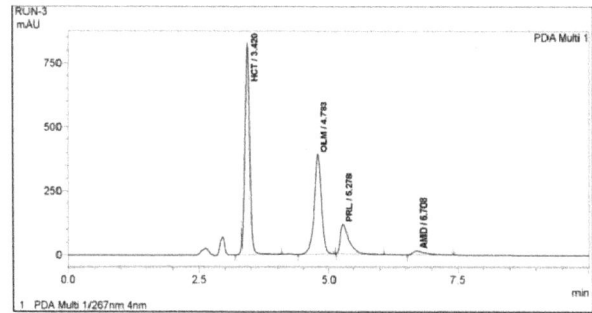

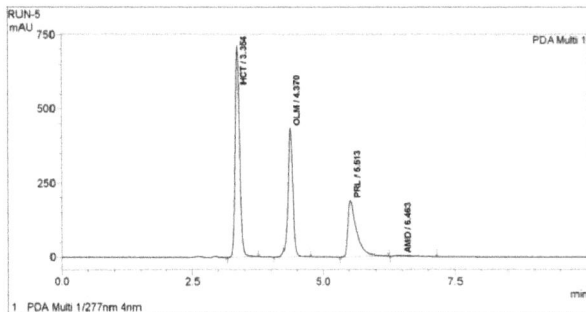

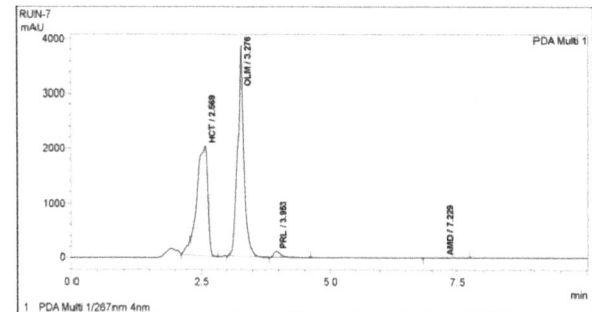

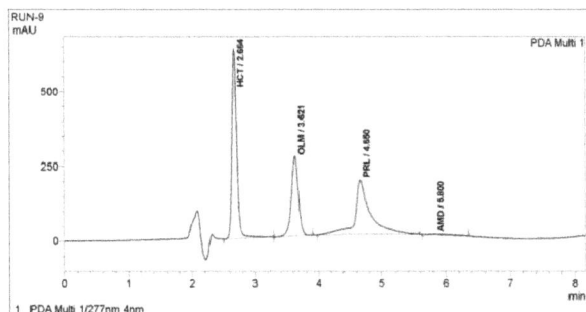

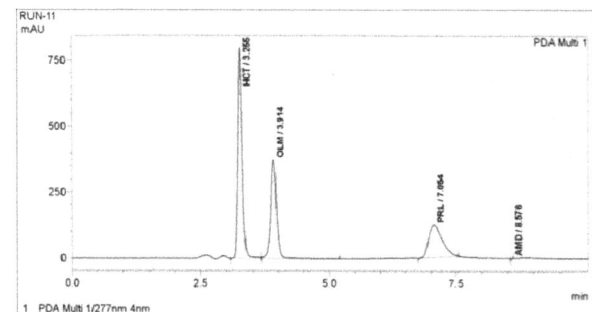

Figure 2: Represented chromatograms for selected runs of PBD.

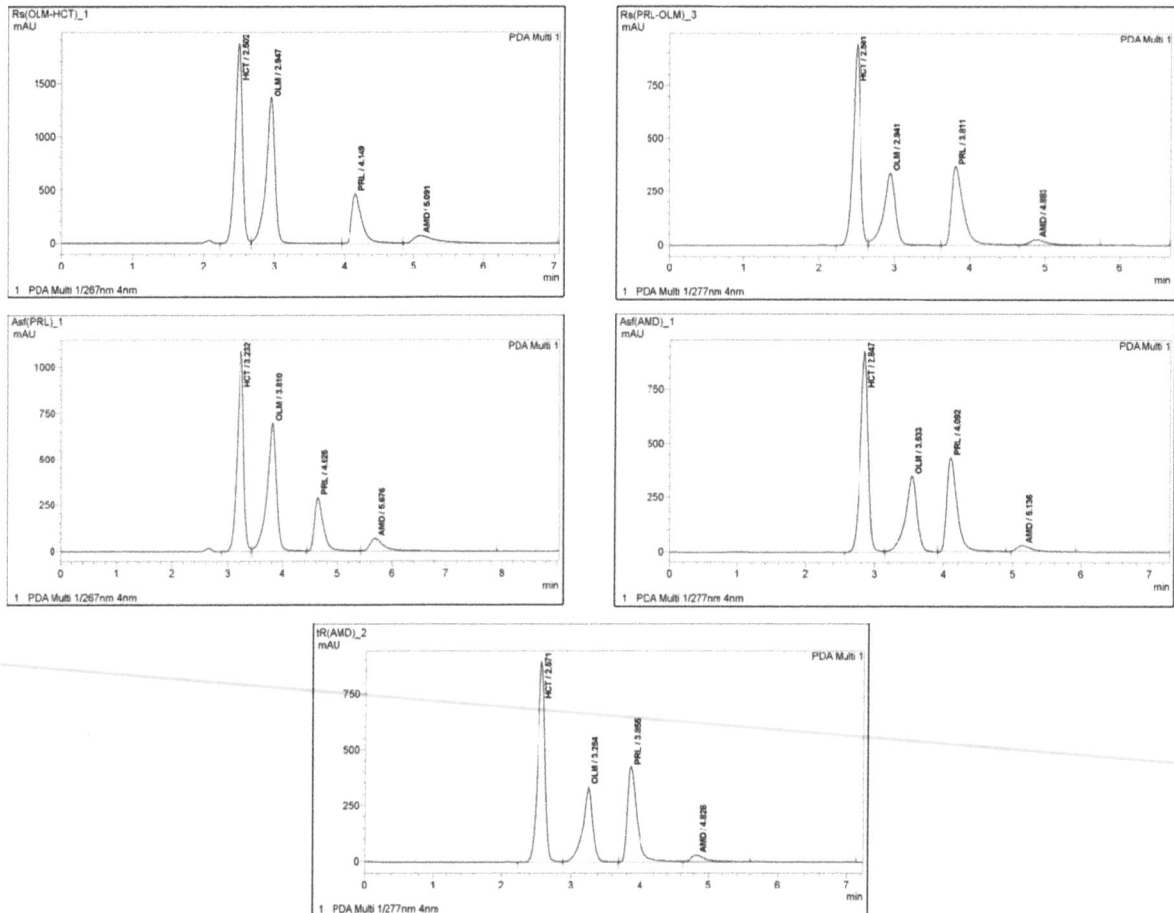

Figure 3: Chromatograms from the worst-case experiments.

(P, 0.1), were indicated in Table 7. It is clear that none of the variables has a significant effect on the determination of the recovery of the four compounds, whose ranges from Table 5 were narrow and with small percent relative standard deviations (1.1%, 1.4% 1.7% and 1.5% for HCT, OLM, PRL and AMD respectively). Based on these facts, the method for assay of Triolmezest and Inderal tablets can be considered robust as regards recoveries. However, the factorial effects on the other responses that demonstrate the method performance under the different design conditions evidenced that several effects are significant (Table 6). The responses, Rs(OLM-HCT), Rs(PRL-OLM), and Asf(PRL) are affected by most of the tested variables. From Table 6, it was also evidenced that none of the variables could affect Rs(AMD-PRL) and was excluded for further study.

Finding the worst-case variable level combinations

It is worth noticing that a statistical significant effect on a response is not always chromatographically relevant; to assess this relevance, the most extreme results from the design experiments have to be considered and compared with the existing SST limits. The most extreme resolution of 1.857 (OLM-HCT) and 1.131 (PRL-OLM); tailing factor of 2.640 (PRL) and 2.785 (AMD) from the design results are not within the SST specifications, namely below 3.671, 2.821; and above 2.624, 2.350 respectively. However, the most extreme design results are not necessarily the worst results since these could be given by a combination of variables not necessarily executed in the design.

To decide on the conditions of this worst-case experiment only the statistically significant effect on a response, at significance levels of 5% (P, 0.05) and of 10% (P, 0.1), were considered. These variables were included because they are able to cause a systematic change in a response when changed from one level to the other. The variables with a P>0.1 were considered as negligible and related only to experimental error. As the PBD is a saturated two-level design, it can only account for linear effects in the prediction of the worst-case situation. This is acceptable because in robustness testing the domain is restricted and only linear effects are important. The variable level combination leading to the worst result for a response Y is predicted by the equation [11]:

$$Y = E_{F1}F_1 + E_{F2}F_2 \ldots\ldots\ldots\ldots E_{FK}F_K \tag{2}$$

E_{F1} represents the effect of the variable considered for the worst-case experiment and F_i the level of this variable. Non-important variables (P>0.1) are kept at nominal value. Table 8 details the worst-case variable-level combinations for the different responses. The worst-case experiment was run in triplicate and the mean result was then compared with the system suitability limit by a one-sided t-testto find out if the system suitability limit is statistically violated. Table 9 illustrates the results of the worst-case experiments for the different responses and of the t-tests. Figure 3 depicts the resultant chromatograms for the worst-case experiments. For the resolution (PRL-OLM), the worst case results are not significantly smaller than the SST limit at nominal condition (H$_1$:R_s(PRL-OLM)=3.670>H$_0$:R_s(PRL-OLM)=2.821). The

Responses	Variables						
	Flow rate	pH	%B start	%B end	Buffer conc.	%TEA	Wave length
Rs(OLM-HCT)	**0.0024*	**0.0034*	**0.0042*	**0.0008*	**0.0420*	**0.0053*	**0.0020*
Rs(PRL-OLM)	**0.0166*	**0.0078*	**0.0185*	**0.0182*	**0.0159*	**0.0285*	*0.0568*
Asf(PRL)	**0.0054*	**0.0117*	**0.0116*	**0.0069*	**0.0102*	**0.0036*	**0.0065*
Asf(AMD)	-	*0.0505*	*0.0650*	-	**0.0473*	-	**0.0486*
t_R(AMD)	**0.0082*	**0.0296*	**0.0202*	-	-	**0.0107*	**0.0079*

**=Significance at α=0.10 level, *=significance at α=0.05 level.

Table 7: P values obtained for these effects.

Responses	Variables						
	Flow rate	pH	%B start	%B end	Buffer conc	%TEA	Wave length
Rs(OLM-HCT)	+1 (0.9)	+1 (7.3)	+1 (43)	+1 (42)	-1 (18)	+1 (0.5)	-1 (267)
Rs(PRL-OLM)	+1 (0.9)	-1 (6.7)	-1 (41)	-1 (38)	-1 (18)	-1 (0.3)	+1 (277)
Asf(PRL)	-1 (0.7)	-1 (6.7)	+1 (43)	+1 (42)	+1 (22)	-1 (0.3)	-1 (267)
Asf(AMD)	0 (0.8)	-1 (6.7)	+1 (43)	0 (40)	+1 (22)	0 (0.4)	+1 (277)
t_R(AMD)	+1 (0.9)	-1 (6.7)	-1 (41)	0 (40)	0 (20)	-1 (0.3)	+1 (277)

Table 8: Predicted worst-case variable-level combinations for the different responses.

Run	Rs (OLM-HCT)	Rs (PRL-OLM)	Asf (PRL)	Asf (AMD)	t_R(AMD), min
1	1.993	3.804	1.818	1.736	4.927
2	2.073	3.767	1.799	1.728	4.767
3	1.990	3.702	1.804	1.748	4.828
Mean	2.018	3.757	1.807	1.7373	4.8406
SD	0.047078	0.051	0.009849	0.01	0.080
Normal SST limits	3.671	2.821	2.624	2.350	5.494
SST limits from worst case results					
	$2.018 - 2.92 \times \dfrac{0.047}{\sqrt{3}} = 1.939$	$3.757 - 2.92 \times \dfrac{0.051}{\sqrt{3}} = 3.67$	$1.807 + 2.92 \times \dfrac{0.009}{\sqrt{3}} = 1.823$	$1.737 + 2.92 \times \dfrac{0.01}{\sqrt{3}} = 1.754$	$4.840 - 2.92 \times \dfrac{0.080}{\sqrt{3}} = 4.704$

Table 9: Results of the worst-case experiments for the different responses.

tailing factors are not found to be significantly larger than the limit at nominal condition (H_1: Asf (PRL)=1.823<H_0:Asf(PRL)=2.624; H_1:Asf (AMD)=1.754<H_0:Asf(PRL)=2.350).

However, it is possible that when the method is transferred to another laboratory, some SST criteria may be violated. In our example, it could be the case for the resolution (OLM-HCT) [H_1:R_s(OLM-HCT)=1.939<H_0:R_s(OLM-HCT)=3.671]. However, results of the robustness test indicate that the method is robust. It follows that a more or less arbitrary selection of system suitability test parameter limits can lead to problems not related to quality and therefore highly undesirable.

Hence it is better to derive the system suitability limits from the results of the experimental design, as proposed by Mulholland et al. [3,21] who make use of the extreme results to define the SST limits. Since, these extreme results may not be the worst we propose to use the worst-case situations to define the SST limits. This way ambiguous situations may be avoided and SST limits are established, as recommended by the ICH guidelines, from the robustness test. The SST limit could be the upper (for tailing factor) or lower (for resolutions and total analysis time) limit from the one-sided 95% confidence interval [22] around the worst-case mean.

The lower limit is $\overline{X}_{worst-case} - t_{\alpha,n} \times \left(\dfrac{s}{\sqrt{n}}\right)$ while the upper one is $\overline{X}_{worst-case} + t_{\alpha,n} \times \left(\dfrac{s}{\sqrt{n}}\right)$. This would lead to system suitability limits of 1.939 for the Rs(OLM-HCT), 3.670 for the Rs(PRL-OLM), 1.823 for the Asf(PRL), 1.754 for the Asf(AMD), and 4.704 for the total analysis time (Table 9). Noteworthy this approach give some SST limits stricter than the previously used ones [resolution Rs(OLM-HCT)].

The rationale for this approach is beside the recommendation of the ICH guidelines. Defining SST limits on the basis of a robustness test results can be endorsed also for practical reasons. For example, when a separation method was properly optimized, the quantitative results did not change significantly, although some of the SST limits (selected rather arbitrarily and independently from the results of a robustness test) were frequently violated. This may happen if they were set too strictly during method optimization.

Conclusion

The conclusion for the robustness test of the chromatographic method for the analysis of Triolmezest and Inderal film-coated tablets is that the method is robust concerning the analysis results of the four compounds.

Defining system suitability limits based on the worst-case results for which the conditions were predicted from the robustness test, allows to avoid an undesirable situation where a method is found to be robust for its quantitative aspect while some externally defined system suitability criteria are violated.

References

1. ICH Harmonized Tripartite Guideline Q2(R1) (2005) Validation of Analytical Procedures: Text and Methodology, International Conference on Harmonisation of Technical Requirements for registration of Pharmaceuticals for Human Use.

2. Youden WJ, Steiner EH (1975) Statistical manual of the Association of Official Analytical Chemists; Statistical techniques for collaborative tests, planning and analysis of results of collaborative tests. p: 96.

3. Mulholland M (1988) Ruggedness testing in analytical chemistry. Trends Anal Chem 7: 383-389.

4. Van Leeuwen JA, Buydens LMC, Vandeginste BGM, Kateman G, Schoenmakers PJ, et al. (1991) RES, an expert system for the set-up and interpretation of a ruggedness test in HPLC method validation. Part 1: The ruggedness test in HPLC method validation. Chemomet Intell Lab Systems 10: 337-347.

5. Heyden YV, Massart DL (1996) Robustness of Analytical Methods and Pharmaceutical Technological Products. Elsevier, Amsterdam, pp: 79-147.

6. Jimidar M, Niemeijer N, Peeters R, Hoogmartens J (1998) Robustness testing of a liquid chromatography method for the determination of vorozole and its related compounds in oral tablets. J Pharm Biomed Anal 18: 479-485.

7. Plackett RL, Burman JP (1946) The design of optimum multifactorial experiments. Biometrika 33: 305-325.

8. Morgan E (1991) Chemometrics-Experimental Design. Analytical Chemistry by Open Learning. Wiley, Chichester.

9. Jimidar M, Khots MS, Hamoir TP, Massart DL (1993) Application of a fractional factorial experimental design for the optimization of fluoride and phosphate separation in capillary zone electrophoresis with indirect photometric detection. Quim Anal 12: 63-68.

10. Dejaegher B, Dumarey M, Capron X, Bloomfield MS, Heyden YV (2007) Comparison of Plackett–Burman and supersaturated designs in robustness testing. Anal Chim Acta 595: 59-71.

11. Heyden YV, Jimidar M, Hund E, Niemeijer N, Peeters R, et al. (1999) Determination of system suitability limits with a robustness test. J Chromatogr A 845: 145-154.

12. Heyden YV, Nijhuis A, Smeyers-Verbeke J, Vandeginste BGM, Massart DL (2001) Guidance for robustness/ruggedness tests in method validation. J Pharm Biomed Anal 24: 723-753.

13. Levine CB, Fahrbach KR, Frame D, Connelly JE, Estok RP, et al. (2003) Effect of amlodipine on systolic blood pressure. Clin Ther 25: 35-57.

14. Liew D, Liu L, Jeffers BW, Foody J (2014) PW177 Literature review of the cost and cost effectiveness of amlodipine in the treatment of hypertension.

Glob Heart 9: e293-e294.

15. Dubey N, Jain A, Raghuwanshi AK, Jain DK (2012) Simultaneous Determination and Validation of OlmesartanMedoxomil, Amlodipine Besilate and Hydrochlorothiazide in Combined Tablet Dosage Form Using RP-HPLC Method. Asian J Chem 24: 4535-4537.

16. da Silva Sangoi M, Wrasse-Sangoi M, de Oliveira PR, Todeschini V, Rolim CMB (2011) Rapid simultaneous determination of aliskiren and hydrochlorothiazide from their pharmaceutical formulations by monolithic silica HPLC column employing experimental designs. J Liq Chrom Rel Technol 34: 1976-1996.

17. Li H, Wang Y, Jiang Y, Tang Y, Wang J, et al. (2007) A liquid chromatography/tandem mass spectrometry method for the simultaneous quantification of valsartan and hydrochlorothiazide in human plasma. J Chromatogr B 852: 436-442.

18. Vignaduzzo SE, Castellano PM, Kaufman TS (2011) Development and validation of an HPLC method for the simultaneous determination of amlodipine, hydrochlorothiazide, and valsartan in tablets of their novel triple combination and binary pharmaceutical associations. J Liq Chrom Rel Technol 34: 2383-2395.

19. Sharma M, Kothari C, Sherikar O, Mehta P (2014) Concurrent Estimation of Amlodipine Besylate, Hydrochlorothiazide and Valsartan by RP-HPLC, HPTLC and UV–Spectrophotometry. J Chromatogr Sci 52: 27-35.

20. Hemke AT, Bhure MV, Chouhan KS, Gupta KR, Wadodkar SG (2010) UV Spectrophotometric Determination of Hydrochlorothiazide and Olmesartan Medoxomil in Pharmaceutical Formulation. J Chem 7: 1156-1161.

21. Mulholland M, Waterhouse J (1987) Development and evaluation of an automated procedure for the ruggedness testing of chromatographic conditions in high-performance liquid chromatography, J Chromatogr A 395: 539-551.

22. Massart DL, Vandeginste BGM, Buydens LMC, De Jong S, Lewi PJ, et al. (1997) Handbook of Chemometrics and Qualimetrics - Part A. Elsevier, Amsterdam.

4

Electrochemical Determination of Folic Acid at Sodium Alpha Olefin Sulphonate Modified Carbon Paste Electrode: A Voltammetric Study

TS Sunil Kumar Naik, BE Kumara Swamy*, CC Vishwanath and Mohan Kumar

Department of PG Studies and Research in Industrial Chemistry, Kuvempu University, JnanaSahyadri, Shankaraghatta, Shivamoga, Karnataka, India

Abstract

Sodium alpha olefin sulphonate (SAOS) was used for the modification of carbon paste electrode (CPE) to determine the electrochemical behavior of folic acid (FA) in 0.2M phosphate buffer solution (PBS) at pH 7.4 with the scan rate of 50 mVs^{-1}. The effects of scan rate, concentration and simultaneous determination of FA at modified carbon paste electrode (MCPE) were studied. The effect of interference of dopamine was carried out and real sample analysis of FA was studied at MCPE. From the scan rate and concentration shows that, the overall electrode process was found to be diffusion-controlled at SAOSMCPE and detection limit was found to be 28.8 µM. The modified electrode (SAOSMCPE) exhibits good electrocatalytic activity towards the determination of folic acid when compared to BCPE. The same method can also be applied for other drug analysis.

Keywords: Folic acid; Sodium alpha olefin sulphonate; Modified carbon paste electrode; Cyclic voltammetry

Introduction

Vitamins are a group of organic compounds, essential in small amount for the normal functioning of the body and regulate the metabolic activity [1]. Deficiency of vitamins may result in often painful and potentially harmful diseases. As a result vitamins play an important role in our body. Folic acid (FA) (vitamin BC, vitamin M or vitamin B$_9$) is a water soluble vitamin and was first discovered in Spinach [2]. During metabolism FA involved in single carbon transfer reactions and it is the precursor of the active tetrahydrofolic acid coenzyme [3]. FA is a nutrient of great importance, especially for women planning for pregnancy. To reduce significantly the incidence and reoccurrence of neural tube defects periconceptual supplementation of folic acid has been demonstrated [4]. Moreover FA is usually employed in the treatment or prevention of megaloblastic anemia during pregnancy [5]. The US Food and Drug Administration introduced mandatory reinforcement of cereal-grain products with folic acid at a concentration of 140 mg/100 g in January 1998 [6]. The Department of Health in the UK proposed fortification of flour with folic acid at 240 mg/100 g [7]. Deficiency of FA gives rise to the macrocytic anemia [8]. Hence FA determination is often required in pharmaceutical, clinical and food samples. There are several methods were reported for the determination of FA including liquid chromatography/tandem mass spectrometry (LC/MS/MS) [9], High Performance Liquid Chromatography (HPLC) [10], Capillary Electrophoresis [11], Microemulsion Electrokinetic Chromatography (MEEKC) [12] and Enzyme linked Immunosorbent assay (ELISAs) [13] and electrochemical methods [14]. Among these techniques electrochemical method is significant one because of its convenience and low cost [15]. FA (Scheme 1) is one of the electroactive species and accordingly we employed electrochemical method for the determination and some of the electrochemical methods for other electro-active species have been reported [16-20].

Dopamine (Scheme 2) is one of the most significant catecholamine, functioning as a neurotransmitter in the central nervous system and deficiency of DA leads to the Parkinson's disease [21-22]. Changes in DA concentration in biological sample indicate the possibility of abnormalities or diseases in human being. Therefore, the determination of DA plays an important role. DA possesses high electrochemical activity and has been widely studied by electroanalytical techniques to significantly advantage towards biosciences [23-25]. Dopamine has been determined using various electrochemical methods [26]. One of the most significant methods is the determination of analytes by using modified carbon paste electrode through the voltammetic technique, which has the ability to eliminate the interfering substances from DA. The modification can be done by adding different types of modifiers [27-32].

In this study, the surfactant (sodium alpha olefin sulphonate (Scheme 3)) is used as a modifier for the electrochemical determination of FA and DA. The term surfactant is derived from surface active agent and is a compound that contains a hydrophilic (attracted to water) and a hydrophobic (repelled by water) segments. Because of their unique molecular structures, surfactant has been extensively used in the fields of electrochemistry and electroanalytical chemistry [33-36] for various purposes. To improve the detection limits of some biomolecules Hu's group [37-39] has introduced surfactants to electroanalytical chemistry.

In present work, a simple and sensitive voltammetric method was

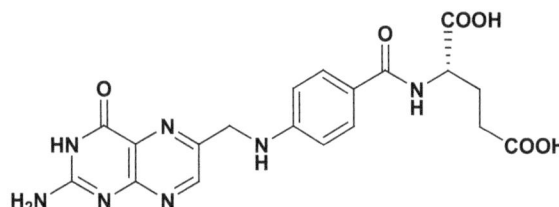

Scheme 1: Folic Acid.

---

*Corresponding author: Kumara Swamy BE, Department of PG Studies and Research in Industrial Chemistry, Kuvempu University, JnanaSahyadri, Shankaraghatta-577451, Shivamoga, Karnataka, India
E-mail: kumaraswamy21@yahoo.com

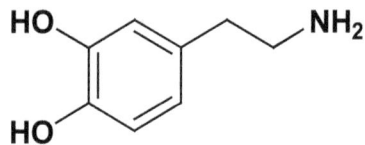

Scheme 2: Dopamine.

Scheme 3: Sodium alpha olefin sulphonate.

presented for the electrochemical determination of folic acid at SAOS modified carbon paste electrode. The modified electrode showed an excellent electrocatalytic activity for the oxidation of FA in presence of dopamine.

Experimental

Reagents and chemicals

Chemicals used in present work were Folic acid (FA) from Merck chemicals. Stock solution of FA was prepared in 0.1M NaOH. Disodium hydrogen phosphate (Na_2HPO_4), Sodium dihydrogen orthophosphate (NaH_2PO_4) and Sodium alpha olefin sulphonate were purchased from Himedia chemicals. Spectrally pure graphite powder (particle size <50 mm) from Merck and high viscous paraffin oil (density=0.88 gcm^{-3}) from Fluka were used for the preparation of the carbon paste electrode. Phosphate buffer (0.2M pH 7.4) was used as optimum measurements. All chemicals used in this experiments were analytical grade and used without any further purification.

Apparatus

Cyclic voltammetric (CV) measurements were performed with a model CHI-660c (CH Instrument-660 electrochemical workstation). All electrochemical experiments were performed in a standard three electrode cell. The bare and sodium alpha olefin sulphonate modified carbon paste electrode (SAOSMCPE) was used as working electrode, Platinum electrode as counter electrode and saturated Calomel electrode (SCE) as the reference electrode. All potentials reported were versus the SCE.

Preparation of bare and modified carbon paste electrode (SAOSMCPE)

The bare CPE was prepared by hand mixing graphite powder and silicone oil in the ratio 70:30 (w/w) for about 30 minutes in an agate mortar to produce a homogeneous mixture. The paste was then packed into the homemade cavity. Sodium alpha olefin sulphonate modified carbon paste electrode (SAOSMCPE) was prepared by immobilizing 15 μL of sodium alpha olefin sulphonate (SAOS) on to the surface of bare carbon paste electrode for 5 minutes.

Results and Discussion

Effect of sodium alpha olefin sulphonate (SAOS) on the surface of CPE

In optimization of modified CPE towards the $K_4[Fe(CN)_6]$, the bare CPE was immobilized on to the surface of the carbon paste electrode in

the different concentrations (5, 10, 15, 20, 25 and 30 μL) of SAOS and their electrochemical response towards 1mM $K_4[Fe(CN)_6]$ in presence of 1M KCl as supporting electrolyte was studied. The obtained result illustrates the increase in the quantity of SAOS the corresponding anodic peak current increases upto 15 μL and afterwards the current goes on decreasing (Figure 1), as a result the 15 μL of SAOS immobilized modified CPE was preferred for further electrochemical investigation.

Electrocatalytic behavior of SAOSMCPE at potassium ferrocyanide

In order to standardize the fabricated SAOSMCPE, potassium ferrocyanide was used for the electrochemical parameters. Figure 2 shows the cyclic voltammograms of 1mM $K_4[Fe(CN)_6]$ at bare (dotted line) and SAOSMCPE (solid line) in presence of 1M KCl as supporting electrolyte with a scan rate of 50 mVs^{-1}. Comparatively SAOSMCPE shows increase in redox peak currents of 1mM $K_4[Fe(CN)_6]$ than BCPE. The probable mechanism is the SAOS surfactant molecules diffuses into

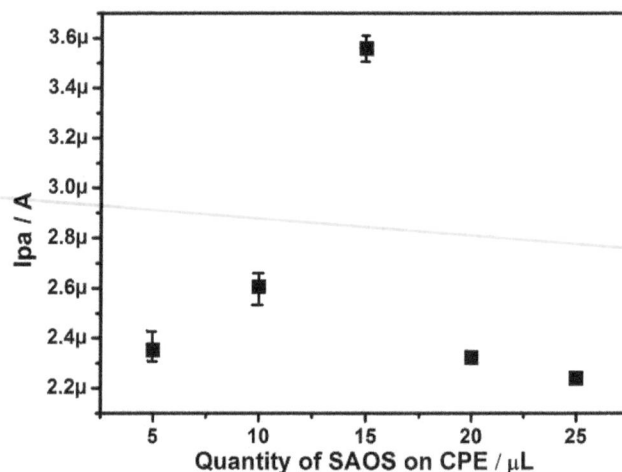

Figure 1: The effect of quantity of SAOS (μL) on the peak current of 1mM $K_4[Fe(CN)_6]$ at the scan rate of 50 mVs^{-1} in presence of 1M KCl as supporting electrolyte.

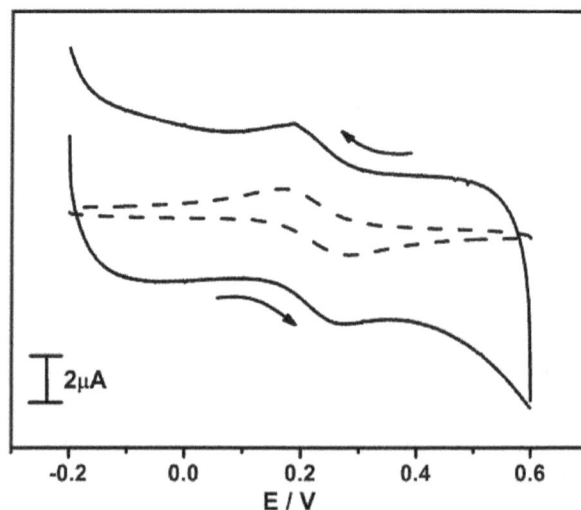

Figure 2: Cyclic voltammograms of 1mM $K_4[Fe(CN)_6]$ at bare (dotted line) and SAOSMCPE (solid line) in 1M KCl at the scan rate of 50 mVs^{-1}.

the carbon paste electrode along with the potassium ferrocyanide and causes increase in the redox peak signals [40]. The difference in redox peak potential (ΔE_p) for BCPE was found to be 0.114 V and 0.084 V for SAOS modified carbon paste electrode. The ΔE_p is a function of the rate of electron transfer, hence lower the ΔE_p value higher will be the electron transfer rate [41]. The proposed modified electrode shows lower ΔE_p value than the BCPE and exhibits great enhancement in the redox peak current. Figure 3a shows the cyclic voltammograms of potassium ferrocyanide at different scan rates. The redox peak current of potassium ferrocyanide goes on increases with increase in scan rate. The plot of Ipa versus scan rate shows good linearity with correlation coefficient value $R^2 = 0.99958$ as shown in Figure 3b. The plot of Ipa v/s square root of scan rate also shows good linearity with the correlation coefficient value $R^2 = 0.9965$ as shown in Figure 3c. From the scan rate study the electrode process was found to be both adsorption and diffusion controlled. Based on the above observation SAOSMCPE shows favorable electrocatalytic behavior and it might be used as a chemically modified electrode to explore electroanalytical application.

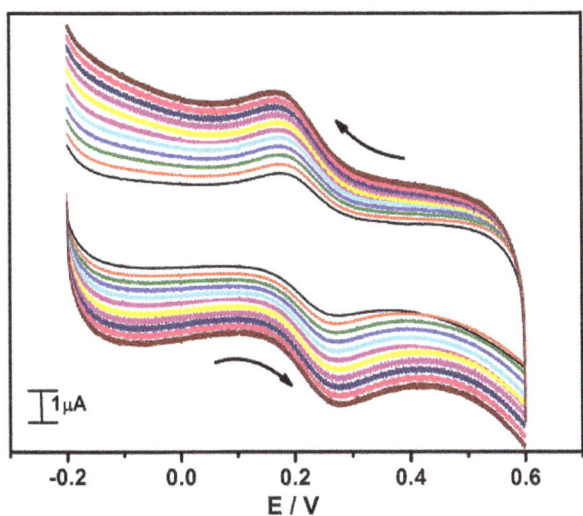

Figure 3c: Graph of anodic peak current (Ipa) v/s square root of scan rate ($v^{1/2}$) of 1mM $K_4[Fe(CN)_6]$ in 1M KCl.

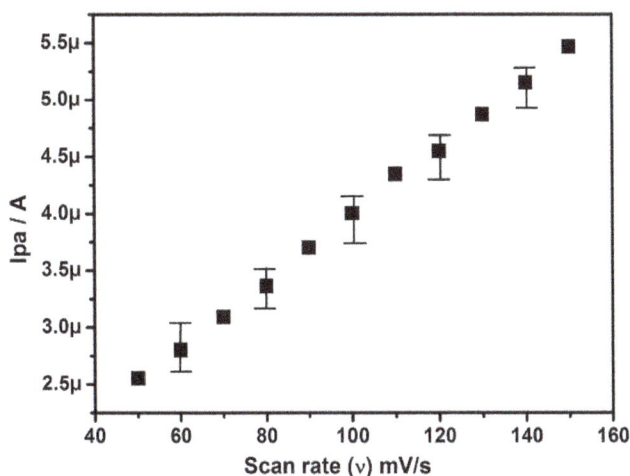

Figure 3a: Cyclic voltammograms obtained for 1mM $K_4[Fe(CN)_6]$ at different scan rates (50 to 150 mVs⁻¹) containing 1M KCl.

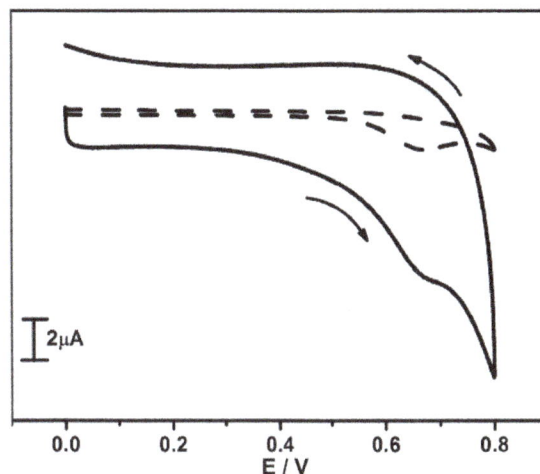

Figure 4: Cyclic voltammograms of 0.5 × 10⁻⁴M folic acid at BCPE (dashed line) and SAOSMCPE (solid line) at scan rate 50 mVs⁻¹ using 0.2M PBS (pH 7.4).

Electrochemical response of folic acid at SAOSMCPE

The electrochemical response of folic acid (FA) was studied at SAOS modified carbon paste electrode. The Figure 4 shows the cyclic voltammograms of 0.5 × 10⁻⁴M folic acid at BCPE (dotted line) and SAOSMCPE (solid line) in 0.2M PBS at pH 7.4 with the scan rate of 50 mVs⁻¹. Comparatively SAOSMCPE shows great enhancement in the anodic peak current (Ipa) than BCPE. Thus remarkable improvement in electrochemical sensitivity towards folic acid at SAOSMCPE gives an evidence for the catalytic effect of proposed electrode.

Effect of scan rate on SAOSMCPE

In order to investigate the kinetics of electrode reaction, effect of scan rate (v) was studied at SAOSMCPE. Figure 5a shows the cyclic voltammograms of 0.5 × 10⁻⁴M folic acid at different scan rate in 0.2M phosphate buffer solution (pH 7.4). The result shows that, with the increase in the scan rate (10-100 mVs⁻¹) the anodic peak current (Ipa) of folic acid goes on increases. The plot of anodic peak current (Ipa)

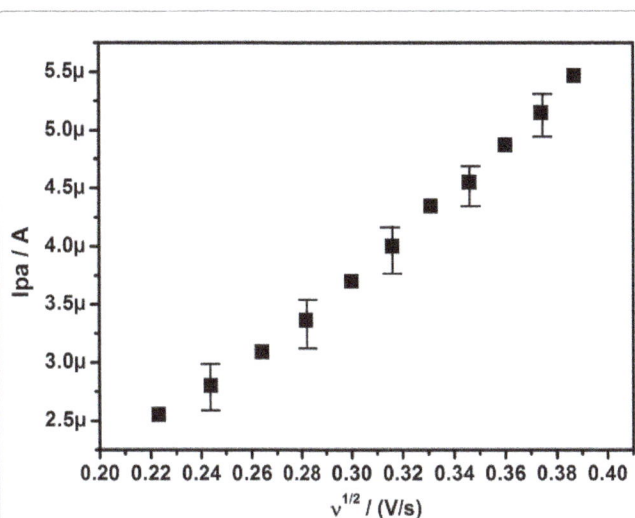

Figure 3b: Graph of anodic peak current (Ipa) v/s scan rate (v) of 1mM $K_4[Fe(CN)_6]$ in 1M KCl.

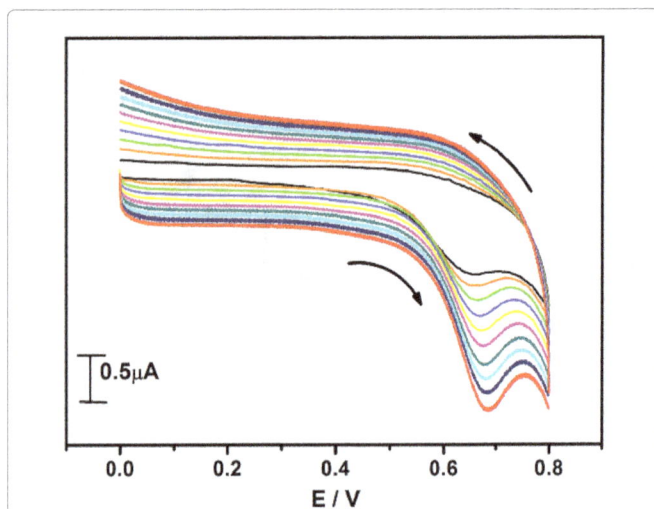

Figure 5a: Cyclic voltammograms of 0.5 × 10⁻⁴M folic acid at SAOSMCPE with different scan rates (10-100 mVs⁻¹) using 0.2M PBS (pH 7.4).

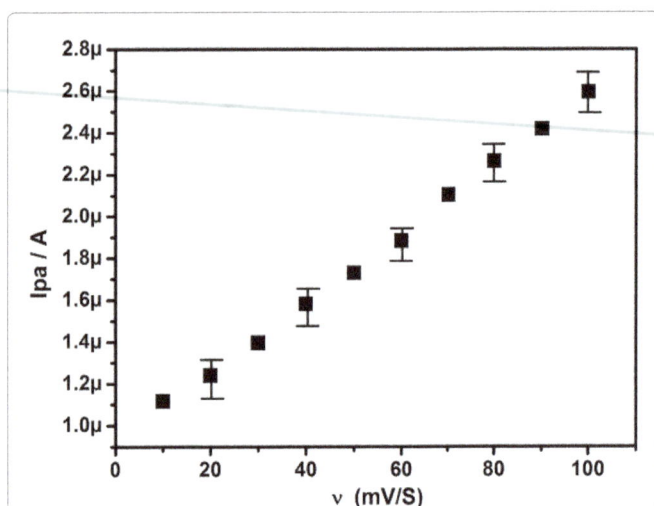

Figure 5b: Graph of anodic peak current (Ipa) v/s scan rate (v) of 0.5 × 10⁻⁴M folic acid in 0.2M PBS (pH 7.4).

v/s scan rate (v) shows good linearity with correlation coefficient value R^2=0.99916 as shown in Figure 5b. Along with the plot of anodic peak current (Ipa) versus square root of scan rate ($v^{1/2}$) was also studied and correlation coefficient value was found to be R^2=0.9846 indicating the modified electrode process was both adsorption and diffusion controlled electrode reactions (Figure 5c).

Effect of folic acid concentration at SAOSMCPE

The electrochemical oxidation of folic acid at different concentration using SAOSMCPE was studied. Figure 6a shows the cyclic voltammograms obtained for the folic acid at different concentration in presence of 0.2M PBS (pH 7.4) with the scan rate of 50 mVs⁻¹. With the increase in the concentration (1.0-3.5 × 10⁻⁴M) the anodic peak current (Ipa) of folic acid goes on increasing. The plot of anodic peak current (Ipa) versus concentration (Figure 6b) shows good linearity with the correlation coefficient value R^2=0.9958. The limit of detection (LOD) was calculated by using the formula (1) [42-45] and it was found to be 2.88 × 10⁻⁵M and the limit of quantification (LOQ) was calculated by using the formula (2) and it was found to be 9.6 × 10⁻⁵M.

$$LOD=3S/M \quad\quad (1)$$

$$LOQ=10S/M \quad\quad (2)$$

Where, S is the standard deviation and M is the slope

Simultaneous electrochemical determination of FA and DA at SAOSMCPE

The cyclic voltammetry was employed for the simultaneous electrochemical determination of FA and DA. The Figure 7 shows the cyclic voltammograms obtained for the mixture of 0.5 × 10⁻⁴M FA and 0.5 × 10⁻⁴M DA in the presence of 0.2M PBS (pH 7.4) at the scan rate of 50 mVs⁻¹. The Dotted line shows cyclic voltammograms obtained for the mixture of FA and DA at BCPE. The solid line shows cyclic voltammogram obtained for the mixture of FA and DA at SAOSMCPE. The modified electrode shows great enhancement in peak current (Ip) than BCPE and showing good voltammetric separation for both DA and FA and proved its electrocatalytic behaviour in simultaneous determination. Hence the proposed modified electrode (SAOSMCPE) has the ability for the simultaneous determination of FA and DA.

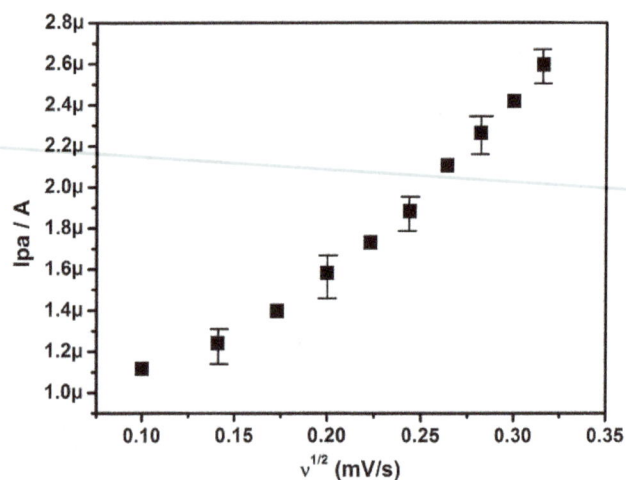

Figure 5c: Graph of anodic peak current (Ipa) v/s square root of scan rate ($v^{1/2}$) of 0.5 × 10⁻⁴M folic acid in 0.2M PBS (pH 7.4).

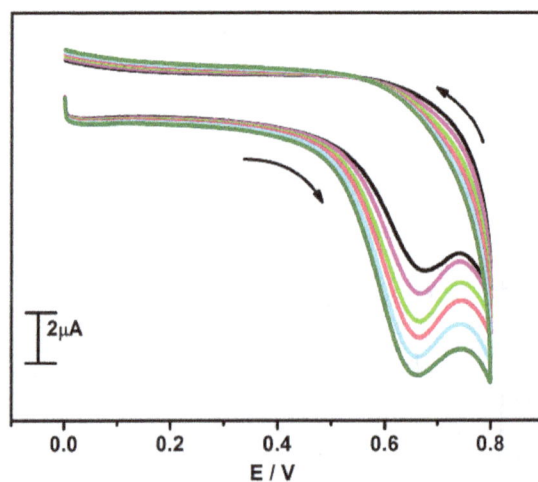

Figure 6a: Cyclic voltammograms of folic acid at SAOSMCPE with different concentration (1.0, 1.5, 2.0, 2.5, 3.0 and 3.5 × 10⁻⁴M) using 0.2M PBS (pH 7.4) at scan rate 50 mVs⁻¹.

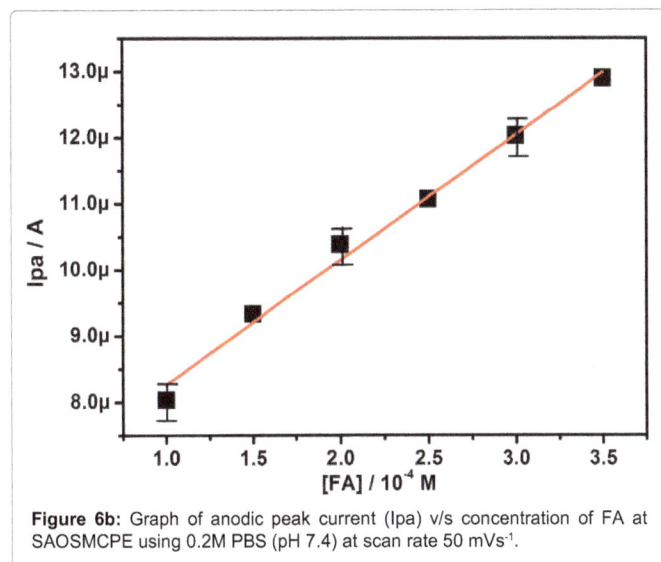

Figure 6b: Graph of anodic peak current (Ipa) v/s concentration of FA at SAOSMCPE using 0.2M PBS (pH 7.4) at scan rate 50 mVs⁻¹.

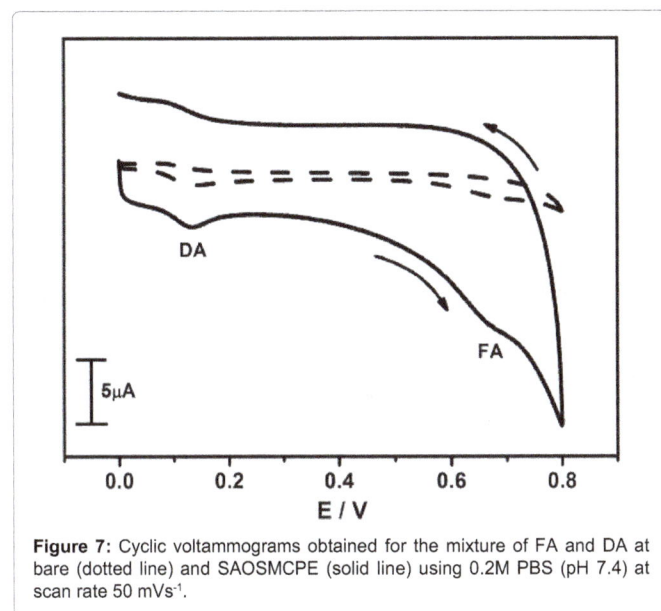

Figure 7: Cyclic voltammograms obtained for the mixture of FA and DA at bare (dotted line) and SAOSMCPE (solid line) using 0.2M PBS (pH 7.4) at scan rate 50 mVs⁻¹.

Sample	Content	Added in mL	Found	Recovery
Tablet	5 mg	3.5	8.038 ± 0.414	92.2%
Tablet	5 mg	4.0	1.039 ± 0.031	99.2%

Table 1: Voltammetric signals and their results.

Determination of folic acid in tablet

To investigate the capability of the modified electrode for the determination of the folic acid in the tablet. By using standard addition method, the recovery test was obtained for the voltammetric signals and the observed results were shown in Table 1.

Conclusion

For the electrochemical determination of folic acid, an anionic surfactant sodium alpha olefin sulphonate was used as the modifier. The modified electrode exhibits good electrocatalytic response towards the determination of folic acid. The electrode process of modified electrode was investigated and found to be both adsorption and diffusion controlled and it shows good detection limit. To know the capability of the modified electrode folic acid was determined in the tablet. Finally, it concludes that the proposed electrode (SAOSMCPE) shows very good electrocatalytic properties and can be applied to the electrochemical determination of folic acid.

References

1. Ball GFM (2008) Vitamins-Their Role in the Human Body. John Wiley & Sons, UK.

2. Mitchell HK, Snell EE, Williams RJ (1988) Journal of the American Chemical Society, Vol. 63, 1941: The concentration of "folic acid" by Herschel K. Mitchell, Esmond E. Snell, and Roger J. Williams. Nutr Rev 46: 324-325.

3. Blakley RL (1969) The Biochemistry of Folic Acid and Related Pteridines. North-Holland Publishing Company, New York, USA.

4. Gujska E, Kuncewicz A (2005) Determination of folate in some cereals and commercial cereal-grain products consumed in Poland using trienzyme extraction and high-performance liquid chromatography methods. European Food Research and Technology 221: 208-213.

5. Al-Shammary FJ, Al-Rashood KA, Mian NA, Mian MS (1990) Analytical Profile of Folic Acid. Anal Profiles Drug Sub 19: 221-259.

6. Rockville MD (1996) Food standards: amendment of standards of identity for enriched grain products to require addition of folic acid. Food and Drug Administration 61: 8781-8797.

7. Wright AJA, Finglas PM, Southon S (2001) Proposed mandatory fortification of the UK diet with folic acid: have potential risks been underestimated. Trends Food Sci Technol 12: 313-321.

8. Chong WM (1990) Cardiology: A Socratic Approach. Orient Longman Pvt Ltd, India.

9. Nelson BC, Sharpless KE, Sander LC (2006) Quantitative determination of folic acid in multivitamin/multielement tablets using liquid chromatography/tandem mass spectrometry. J Chromatogr A 1135: 203-211.

10. Rodríguez-Bernaldo de Quirós A, Castro de Ron C, López-Hernández J, Lage-Yusty MA (2004) Determination of folates in seaweeds by high-performance liquid chromatography. J Chromatogr A 1032: 135-139.

11. Zhao S, Yuan H, Xie C, Xiao D (2006) Determination of folic acid by capillary electrophoresis with chemiluminescence detection. J Chromatogr A 1107: 290-293.

12. Aurora-Prado MS, Silva CA, Tavares MF, Altria KD (2004) Determination of folic acid in tablets by microemulsion electrokinetic chromatography. J Chromatogr A 1051: 291-296.

13. Hoegger D, Morier P, Vollet C, Heini D, Reymond F, et al. (2007) Disposable microfluidic ELISA for the rapid determination of folic acid content in food products. Anal Bioanal Chem 387: 267-275.

14. Kumar M, Swamy KBE (2014) ZnO Modified Carbon Paste Electrode for the Detection of Folic Acid and Paracetamol: A Cyclic Voltammetric Study. Journal of Chemical Engineering and Research 2: 121-126.

15. Alvarez JMF, Garcia AC, Oridieres AJM, Blanco PT (1987) Adsorptive stripping voltammetric behaviour of folic acid. J Electroanal Chem 225: 241-253.

16. Prasad BB, Madhuri R, Tiwari MP, Sharma PS (2010) Electrochemical sensor for folic acid based on a hyperbranched molecularly imprinted polymer-immobilized sol-gel-modified pencil graphite electrode. Sens Actuators B Chem 146: 321-330.

17. Wei SH, Zhao FQ, Xu ZY, Zeng BZ (2006) Voltammetric Determination of Folic Acid with a Multi-Walled Carbon Nanotube-Modified Gold Electrode. Microchim Acta 152: 285-290.

18. Xiao F, Ruan C, Liu L, Yan R, Zhao F, et al. (2008) Single-walled carbon nanotube-ionic liquid paste electrode for the sensitive voltammetric determination of folic acid. Sens Actuators B Chem 134: 895-901.

19. Ensafi AA, Karimi-Maleh H (2010) Modified multiwall carbon nanotubes paste electrode as a sensor for simultaneous determination of 6-thioguanine and folic acid using ferrocenedicarboxylic acid as a mediator. J Electroanal Chem 640: 75-83.

20. Beitollahi H, Sheikhshoaie I (2011) Electrocatalytic oxidation and determination of epinephrine in the presence of uric acid and folic acid at multi walled carbon

nanotubes/molybdenum(VI) complex modified carbon paste electrode. Analytical Methods 3: 1810-1814.

21. Smith TE (1992) Text Book of Biochemistry with Clinical Correlations. Devlin TM (Ed), Wiley/Liss, New York, USA 929.

22. O'Neill RD (1994) Microvoltammetric techniques and sensors for monitoring neurochemical dynamics in vivo. A review. Analyst 119: 767-779.

23. Junter GA (1988) Electrochemical Detection Techniques in the Applied Biosciences. Halsted, New York.

24. Wang Q, Dong D, Li N (2001) Electrochemical response of dopamine at a penicillamine self-assembled gold electrode. Bioelectrochemistry 54: 169-175.

25. Chicharroa M, Sancheza A, Zapardiel A, Rubianesc MD, Rivasc G (2004) Capillary electrophoresis of neurotransmitters with amperometric detection at melanin-type polymer-modified carbon electrodes. Anal Chim Acta 523: 185-191.

26. Xue KH, Tao FF, Xu W (2005) Selective determination of dopamine in the presence of ascorbic acid at the carbon atom wire modified electrode. J Electroanal Chem 578: 323-239.

27. Gilbert O, Chandra U, Swamy BEK, Pandurangachar M, Nagaraj C, et al. (2008) Poly (Alanine) Modified Carbon Paste Electrode for Simultaneous Detection of Dopamine and Ascorbic Acid. International Journal of Electrochemical Sciences 3: 1186-1195.

28. Raj CR, Tokuda K, Ohsaka T (2001) Electroanalytical applications of cationic self-assembled monolayers: square-wave voltammetric determination of dopamine and ascorbate. Bioelectrochemistry 53: 183-191.

29. Ganesh PS, Swamy BEK (2015) Voltammetric Resolution of Dopamine in Presence of Ascorbic Acid and Uric Acid at Poly (Brilliant Blue) Modified Carbon Paste Electrode. J Anal Bioanal Tech 5: 229.

30. Downard AJ, Roddick AD, Bond AM (1995) Covalent modification of carbon electrodes for voltammetric differentiation of dopamine and ascorbic acid. Anal Chim Acta 317: 303-310.

31. Zhang P, Wu FH, Zhao GC, Wei XW (2005) Selective response of dopamine in the presence of ascorbic acid at multi-walled carbon nanotube modified gold electrode. Bioelectrochemistry 67: 109-114.

32. Lokesh SV, Sherigara BS, Jayadev, Mahesh HM, Mascarenhas JR (2008) Electrochemical Reactivity of C60 Modified Carbon Paste Electrode by Physical Vapor Deposition Method. International Journal of Electrochemical Sciences 3: 578-587.

33. Hu C, Hu S (2004) Electrochemical characterization of cetyltrimethyl ammonium bromide modified carbon paste electrode and the application in the immobilization of DNA. Electrochim Acta 49: 405-412.

34. Rusling JF (1991) Controlling electrochemical catalysis with surfactant microstructures. Acc Chem Res 24: 75-81.

35. Kumar M, Swamy BEK, Reddy S, Sathisha TV, Manjanna J (2013) Synthesis of ZnO and its surfactant based electrode for the simultaneous detection of dopamine and ascorbic acid. Analytical Methods 5: 735- 740.

36. Hu SS, Wu KB, Yi HC, Cui DF (2002) Voltammetric behavior and determination of estrogens at Nafion-modified glassy carbon electrode in the presence of cetyltrimethylammonium bromide. Anal Chim Acta 464: 209-216.

37. Hu S, Yan Y, Zhao Z (1991) Determination of progesterone based on the enhancement effect of surfactants in linear sweep polarography. Anal Chim Acta 248: 103-108.

38. Yi H, Wu K, Hu S, Cui D (2001) Adsorption stripping voltammetry of phenol at Nafion-modified glassy carbon electrode in the presence of surfactants. Talanta 55: 1205-1210.

39. Zhang S, Wu K, Hu S (2002) Voltammetric determination of diethylstilbestrol at carbon paste electrode using cetylpyridine bromide as medium. Talanta 58: 747-754.

40. Niranjana E, Naik RR, Swamy BEK, Sherigara BS, Jayadevappa H (2007) Studies on Adsorption of Triton X-100 at Carbon Paste and Ceresin Wax Carbon Paste Electrodes and the Enhancement Effect in Dopamine Oxidation by Cyclic Voltammetry. International Journal of Electrochemical Sciences 2: 923-934.

41. Gilbert O, Swamy BEK, Chandra U, Sherigara BS (2009) Simultaneous detection of dopamine and ascorbic acid using polyglycine modified carbon paste electrode: A cyclic voltammetric study. J Electroanal Chem 636: 80-85.

42. Shankar SS, Swamy BEK, Pandurangachar M, Chandra U, Chandrashekar BN, et al. (2010) Electrocatalytic Oxidation of Dopamine on Acrylamide Modified Carbon Paste Electrode : A Voltammetric Study. International Journal of Electrochemical Sciences 5: 944-954.

43. Chandra U, Swamy BEK, Mahanthesha KR, Vishwanath CC, Sherigara BS (2013) Poly(malachite green) film based carbon paste electrode sensor for the voltammetric investigation of dopamine. Chemical Sensors 3.

44. Chitravathi S, Swamy BEK, Mamatha GP, Sherigara BS (2013) Determination of salbutamol sulfate by Alcian blue modified carbon paste electrode: A cyclic voltammetric study. Chemical Sensors 3.

45. Mahanthesha KR, Swamy BEK, Chandra U, Sathish Reddy, Pai KV (2014) Sodium dodecyl sulphate/polyglycine/phthalamide/carbon paste electrode based voltammetric sensors for detection of dopamine in the presence of ascorbic acid and uric acid. Chemical Sensors 4.

5

Method for Discovery of Peptide Reagents Using a Commercial Magnetic Separation Platform and Bacterial Cell Surface Display Technology

Deborah A. Sarkes[1], Brandi L. Dorsey[2], Amethist S. Finch[1] and Dimitra N. Stratis-Cullum[1]*

[1]U.S. Army Research Laboratory, Sensors and Electron Devices Directorate, Adelphi MD, USA
[2]Federal Staffing Resources, Annapolis MD, USA

Abstract

Biopanning by bacterial display has many advantages over yeast and phage display, including the speed to discovery of affinity reagents and direct amplification of bound cells without the need to elute and reinfect. However, widespread use is limited, in part due to poor performance achieved using manual Magnetic-Activated Cell Sorting (MACS) methods, and an absence of widely-available, low cost, high-performance sorting alternatives. Here, we have developed a methodology for bacterial cell sorting using the semi-automated autoMACS® Pro Separator for the first time, and have produced a complete method for sorting of bacteria displaying 15-mer peptides on their cell surface using this device, including downstream bioinformatic analysis of candidates for binding to a target of interest. Two autoMACS® programs designed for isolation of target cells with low frequency were evaluated and adapted to bacterial biopanning, using protective antigen (PA) of Bacillus anthracis as the model system. In contrast to manual MACS, the bacterial display library was preferentially enriched by autoMACS® sorting, yielding several promising candidates after only three rounds of biopanning and bioinformatic analysis. Individual candidates were evaluated for relative binding to fluorescently-labeled PA target or streptavidin negative control using Fluorescence-Activated Cell Sorting (FACS). The top thirteen peptide candidates from the autoMACS® sort demonstrate binding to PA with low cross-reactivity to streptavidin, while only two of eighteen candidates from the manual sort showed binding to PA, and both demonstrated greater cross-reactivity to streptavidin. Overall, the autoMACS® platform quickly harvested higher affinity peptide candidates with demonstrated specificity to the PA target. Peptide candidates produced with this method contained the previously reported PA consensus WXCFTC, further validating this method and the commercially available autoMACS® platform as the first low cost, semi-automated biopanning approach for bacterial display that is widely accessible and more reliable than the MACS/FACS standard protocol.

Keywords: PA; Peptide; Bacterial display; autoMACS˙; Bio-threat; Affinity reagent; Biocombinatorial; eCPX

Introduction

Bacterial display sorting is a powerful emerging technology that offers a rapid, high-throughput approach to discovery of robust affinity reagents [1-3]. With increased use of peptides as therapeutics, the application space for this discovery platform has expanded in recent years [4-6]. As compared with yeast and phage display, bacterial display is ideal because of the fast doubling time of bacteria, about 20 minutes for Escherichia coli (E. coli) [7] versus about 2 hours for Saccharomyces cerevisiae (S. cerevisiae) [8,9], and direct amplification of the bound bacterial cells containing plasmid DNA encoding the displayed peptide responsible for binding, without elution and reinfection. With phage display, abrasive chemicals are often necessary to elute the tightest binders for reinfection, and low pH conditions are required at a minimum when other known ligands for the target are unavailable. Therefore, the resulting pool of candidates may be biased to lower affinity binders, or stronger binders that are able to survive the harsh conditions or that can be isolated by physical methods [10-14]. The mode of binding can also be a factor in successful elution with low pH; binders interacting by non-electrostatic interactions may not be easily eluted at low pH, and a bias for positively charged sequences can result [15]. In addition to bacterial display's advantages of speed and direct replication, E. coli is relatively easy to manipulate with high efficiency, allowing for generation of customized libraries for development of biocombinatorial discovery methods for both biological and inorganic-binding peptide reagents [16-23].

Ideally, a semi-automated bacterial display sorting method should be used for both speed and reproducibility. It is absolutely imperative that the biopanning method allow for selective enrichment of a rare target population amongst a large population of very similar bacteria, differing only in variations across short peptide sequences displayed on the surface scaffold. Microfluidic screening of bacterial display libraries was previously demonstrated for epitope mapping [24], and we have shown that the engineered bacterial display library eCPX, displaying random 15-amino acid peptides on the bacterial cell surface [25], can be combined with Micro-Magnetic Separation (MMS) for use as a simple, semi-automated discovery platform [18,19,26]. In addition to the rapid discovery of peptide affinity reagents using the MMS method (less than one-week), the method produced a family of peptide reagents to protective antigen (PA) of Bacillus anthracis, an emerging biological threat [19]. Most recently, the eCPX E. coli display library and MMS platform were utilized in the discovery of peptide affinity reagents against staphylococcal enterotoxin B (SEB) as well, further demonstrating that this semi-automated methodology is a valuable tool for the detection of biological threats [18]. The peptides isolated by the MMS using PA as a target showed sequence consensus (WXCFTC) and exhibited similar or better peptide interaction with the PA protein target than with a streptavidin negative control, measured through

*Corresponding authors: Dimitra N. Stratis-Cullum, U.S. Army Research Laboratory, Sensors and Electron Devices Directorate, 2800 Powder Mill Road, Adelphi, MD 20783, USA, E-mail: dimitra.n.stratis-cullum.civ@mail.mil

Fluorescence-Activated Cell Sorting (FACS) assays and compared to peptides isolated by conventional Magnetic-Activated Cell Sorting (MACS)/FACS sorting, where a library is pre-enriched by several rounds of MACS to reduce the number of cells to be sorted using FACS [19]. FACS alone is not sufficient for earlier rounds of sorting when the diversity is above 10^8 members because these devices are limited to sorting 10^7 to 10^8 cells [27]. Although a single candidate with the WXCFTC consensus was isolated by MACS/FACS sorting in Kogot *et al.*, the enrichment of peptides with this binding consensus was poor using this method as compared to the MMS. Additionally, the FACS portion of the MACS/FACS method may not be an option for some due to the higher cost of obtaining and maintaining FACS devices capable of sorting cells for downstream use, typically $350,000-$500,000 to purchase [3], not to mention that they require highly skilled personnel to operate.

Although eCPX bacterial display technology has been shown to be a powerful approach to biopanning and the study of genetically engineered peptides, a widely available, low cost, semi-automated method is lacking because the previously characterized MMS platform is not commercially available for routine discovery. The need for a fast, inexpensive, reliable, reproducible method for discovery of affinity reagents to emerging bio-threats necessitates investigating other platforms, such as the autoMACS® Pro Separator, commercially available from Miltenyi Biotec. The autoMACS® Pro Separator is under $50,000 and is very simple to use. Methods have been published on this platform, as well as with an earlier model that was lower-throughput, for cell sorting using yeast surface display [28-31], as well as for sorting tumor epithelial cells for downstream screening of interacting scFVs by phage display [32]. However, to the best of our knowledge, the work herein represents the first utilizing a bacterial system, let alone a bacterial display library for peptide discovery. *E. coli* bacteria is approximately 1 μm in diameter, 2 μm in length, and 1 μm³ in volume while *Saccharomyces cerevesiae (S. cerevisiae)* yeast is approximately 3-6 μm in diameter and 30-40 μm³ in volume [33-37]. The smaller size of bacteria creates unique challenges over separation of yeast in a machine primarily used for separation of eukaryotic cells, where bacteria is a contaminant to be avoided, as highlighted by the use of sodium azide in the autoMACS® Running Buffer - MACS® Separation Buffer.

In this paper, we discovered several peptide capture candidates using the autoMACS® Pro Separator and a bacterial display library for the first time. We investigated the applicability of the system for discovery of peptide reagents for PA as a model system, allowing for a benchmark of comparison to previously published work using the MMS platform [19], and included important considerations to the number of rounds of biopanning, sequence analysis, and cross-reactivity.

Materials and Methods

Biopanning bacterial display library

Four rounds of biopanning were performed using a Dynabeads® MPC®-S magnetic particle concentrator (Life Technologies, Grand Island, NY, USA) for manual MACS, or the autoMACS® Pro Separator (Miltenyi Biotec, San Diego, CA, USA) for autoMACS®, similarly to the previously described protocols for manual MACS and the MMS [18,19,26]. See supplementary protocol for detailed adaptation and optimization for our method (Biopanning bacterial display library for Protective Antigen binders using manual MACS or autoMACS®). The bacterial display peptide library used was eCPX 3.0 (Cytomx Therapeutics, San Francisco, CA, USA) and was pre-depleted of streptavidin binders to avoid direct binding of non-specific peptides

to the beads [18,19,26]. The target was recombinant protective antigen (PA; List Biological Laboratories, Campbell, CA, USA) biotinylated using No-Weigh Sulfo-NHS-LC-Biotin (Thermo Scientific, Rockford, IL, USA; PA-Biotin) and tested for biotinylation using the Pierce Biotin Quantitation Kit (Thermo Scientific, Rockford, IL, USA), and the paramagnetic beads used were Dynabeads® MyOne Streptavidin T1 Beads (Invitrogen, Carlsbad, CA, USA).

Manual MACS: Each sample containing PA-Biotin-bound cells and Streptavidin T1 beads was placed in a magnetic particle concentrator and allowed to separate for 5 minutes. The supernatant was removed and the beads washed 3 times with 1 mL PBS with 0.5% w/v Bovine Serum Albumen (PBS-B) by inverting the tube 3-4 times in the absence of the magnet, then returning the sample to the magnet between washes to retain the beads and remove the supernatant. After three washes, the positive fraction containing cells and beads were processed for the next sorting round and for spot plating and FACS analysis, as described below and in the supplemental protocol. The positive fraction was the washed beads resuspended in 1 mL PBS-B.

autoMACS®: For each round, the sample containing PA-Biotin-bound cells with Streptavidin T1 beads was moved to a 15 mL conical tube for sorting on the autoMACS® Pro Separator, and an additional 500 μL of PBS-B was added to the sample after using it to recover remaining beads from the microcentrifuge tube. The samples, and empty 15 mL conical tubes for positive and negative fraction collection, were placed in a cold Chill 15 rack (Miltenyi Biotec, San Diego, CA, USA), as described in the manufacturer's instructions, and run through one of two pre-loaded separation programs, "Posseld" or "Posselds," designed for positive selection of target cells with low frequency [38] and named "Program D" and "Program DS" herein for simplicity. PBS-B was used in place of the Miltenyi Biotec autoMACS® wash and running buffers for both methods to avoid exposing the bacteria to detergents and sodium azide, and because it worked well with the MMS [18,19,26]. A rinse step was added in between and at the end of all samples, then the system was returned to its recommended run and wash buffers and a sleep step was run with 70% ethanol before turning off the machine, to prevent bacterial growth in the tubing and pump. See supplemental protocol (Biopanning bacterial display library for Protective Antigen binders using manual MACS or autoMACS®) for more detail.

Spot plating

The positive fraction was collected after separation with either manual MACS or autoMACS® and a small amount serially diluted 1:10 for spot plating [39] on Luria Broth (LB) Agar plates containing 25 μg/mL chloramphenicol (LB Cm$_{25}$ Agar plates). In triplicate, 10 μl spots were plated for 10^{-2}-10^{-7} dilutions of Colony Forming Units (CFU)/mL, as compared to the undiluted positive fraction, such that a 10 μl spot of undiluted sample constitutes a 10^{-2} "dilution" of CFU/mL. The spots were allowed to dry before inverting the plates and incubating overnight at 37°C, and the colonies in each spot were counted the next day, and replicate spots averaged. A spot with about 10-20 colonies was ideal for estimating the total number of cells/mL in the positive fraction.

FACS analysis of sorting rounds

BD FACS Canto™ II and BD FACSDiva™ Software (BD, Franklin Lakes, New Jersey, USA) were used to assess the level of PA binding after each round of biopanning, using induced samples saved on ice after each round of sorting. For each sample, 5 μL of induced cells was added to 25 μL of cold PBS alone or containing 150 nM YPet Mona positive control (ARL, Adelphi, MD, USA) [21], 900 nM PA labeled with Dylight 488 NHS Ester (Thermo Scientific, Rockford, IL, USA; PA488), or 900 nm Streptavidin, R-Phycoerythrin conjugate (SAPE, Life Technologies, Grand Island, NY, USA), and incubated for 45 minutes on ice. The cells were centrifuged at 5,000 xg for 5 minutes and the supernatant removed. The pellet was resuspended in 500 μL ice cold, filtered BD FACSFlow™ (BD, Franklin Lakes, New Jersey, USA), mixed thoroughly, and read immediately. The samples were then run on the FACS and analyzed by gating and comparing to the PBS alone sample for measuring the percentage of bound cells falling outside of the gate [19,40], and by the Normalized Median Fluorescence Intensity (nMFI) of the total population for each sample, normalized to the MFI of a negative control sample expressing eCPX with no 15-mer peptide, incubated with the same fluorophore [41].

Sequence analysis of potential PA binders

For rounds 2-4 of autoMACS° and for round 4 of manual MACS, 144 bacterial colonies were sent to a local Genewiz facility (Frederick, MD, USA) for DNA sequencing using their pBad Forward primer. Resulting sequences were proof-read when necessary, and multiple sequence alignment performed using ClustalW2 (**The EMBL-European Bioinformatics Institute,** Hinxton, Cambridge, UK) and Jalview with Clustal_X windows interface [42,43], available online [44]. Gap penalties were set to 100 for both pairwise and multiple sequence alignment, but the default settings were otherwise used. This was done for all sequences in each sorting round tested, as well as for individual sets of sequences described in the results and discussion.

FACS analysis of individual isolates

For each individual isolate of interest, including all repeats, a 5 ml culture of LB containing 25 μg/mL chloramphenicol (LB Cm$_{25}$) and supplemented with 0.2% w/v D-glucose was inoculated and grown overnight at 37°C in an orbital shaker at 225 RPM. The next day, the cultures were diluted 1:50 in 3 ml LB Cm$_{25}$ without glucose and grown to OD$_{600}$ of 0.5-0.55, then bacterial display was induced by adding 0.04% w/v L-arabinose plus 2 mM ethylenediaminetetraacetic acid (EDTA, for facilitation of peptide display [45]) and incubating for 45 minutes at 37°C with shaking. Induced cultures were placed on ice, and 5 μL of induced cells was added to 25 μL of ice cold PBS alone or containing 150 nM YPet Mona, 250 nM PA488, or 250 nM SAPE, and incubated for 45 minutes on ice. Labeled cells were centrifuged at 5,000xg for 5 minutes and the supernatant removed. The cell pellet was resuspended in 500 μL of ice cold, filtered BD FACSFlow™ and mixed thoroughly. Each sample was then loaded on the FACS immediately after mixing and analyzed as described for sorting rounds. Four independent experiments were performed for manual MACS repeats and random colonies, and three independent experiments were performed for the autoMACS° best binders, as determined by an initial FACS screen including all repeats. The average and standard deviation were calculated and graphed using Prism Software (GraphPad, La Jolla, CA, USA).

Results and Discussion

Biopanning is a powerful method utilizing a biocombinatorial library of candidate binders, with several rounds of exposure of this library to a target of interest, isolating and amplifying the pool of interacting library members with each round. Enrichment in the percentage of the population that binds to the target is generally observed over several rounds. The end result is discovery of a smaller pool of candidates that contain the property of interest, in this case the relative binding performance of isolated peptide candidates to the model target of interest, protective antigen (PA) [3,46]. In order to investigate the feasibility of using the autoMACS° Pro Separator for

biopanning a bacterial display library, it was necessary to develop a supporting method framework that considers the cell surface density of the displayed peptide library, reproducibility of isolation, enrichment throughout the discovery process, and the number of sorting rounds necessary to generate peptide candidates. However, equally important is the development of an analysis approach to guide the understanding of peptide consensus based on physio-chemical properties including cross-platform consensus, as well as a methodology framework for down-selection and analysis of the candidate pool.

Manual magnetic sorting versus autoMACS° sorting and FACS analysis of sorting rounds

To compare the enrichment of peptide binders for PA isolated from a bacterial display library using manual and semi-automated methods, biopanning via manual MACS was run in parallel and compared to the results obtained from adapting two programs for isolation of rare cells on the autoMACS° Pro Separator: "Posseld" and "Posselds," named "Program D" and "Program DS" herein for simplicity. Both autoMACS° programs allow positive selection of labeled target cells and both use two magnetic columns to capture target cells with less than 5% frequency in the initial population. However, Program DS should be more sensitive for weak antigen expression, with a slower flow rate through column 1 to increase exposure time [38]. In order to prescreen running buffer and wash buffer compatibility and performance, the recommended manufacturer buffers were evaluated and compared to PBS-B, which was used in previous bacterial cell sorting studies [18,19,26]. Through a comparison of spot plates, both Program D and Program DS yielded similar numbers of viable cells in the positive fraction throughout the rounds of sorting using PBS-B or the autoMACS° buffers purchased from the manufacturer, about 10^6 cells for round 1 in both cases. However, the overnight growth in LB Cm$_{25}$ solution and downstream analysis were problematic using the autoMACS° Wash and Running buffers, likely due to the presence of sodium azide and detergents in the recommended buffers. PBS-B was therefore chosen for further study, and allowed for a more direct comparison with manual MACS and the previous MMS study.

Figure 1 shows FACS analysis of each sorting round for the discovery of peptide capture candidates for protective antigen (PA) using manual MACS (Figure 1A) and autoMACS° separation programs (Figure 1B and 1C). For each round of sorting, the nMFI for binding to 900 nM PA488 using autoMACS° was found to be substantially higher than for manual MACS, with very little evidence of enrichment found using the manual methodology. Specifically, the nMFI for manual MACS reached only 0.70 by round 4, and was therefore similar to the nMFI of 0.85 reached for manual MACS in round 1. Similarly, when looking at the FACS data by percent bound, where the gated population represents unbound cells in buffer only [19,40], there was a slight increase in binding from 0.5% of the population in round 1 to 5.7% in round 4. This was substantially lower than the level of binding observed using the autoMACS°, determined by both nMFI and percent binding, but the percent binding using manual MACS is similar to the enrichment obtained using a MiniMACS™ separation column and magnet for cell sorting with bacterial display [47], although both methods may be user dependent. PA488 binding for Program D increased from nMFI of 1.23 (0.3% bound) in round 1 to nMFI of 11.07 (65.2% bound) in round 4, and for Program DS increased from nMFI of 1.25 (1.6% bound) in round 1 to nMFI of 9.66 (61.6% bound) in round 4. Differences in the rate of increase between the earlier rounds were clearly evident upon comparison of the two automated methods (Figure 1B and 1C). Program D had a substantial increase in binding between rounds 1 and 2, and quickly leveled off (nMFI of 11.17 at round 2 and 13.75 at round 3), while Program DS exhibited a slower rate of enrichment across the four rounds of biopanning (nMFI of 1.45 at round 2 and 7.31 at round 3). Program D therefore had the highest nMFI at round 3, while Program DS increased with every subsequent round and never reached the higher nMFI of 13.75 seen for Program D. It is possible that the population of cells sorted by Program DS could have reached a higher nMFI with a fifth round of sorting, further excluding low- and non-binders. Downstream analysis shows that this is not necessary to obtain candidates with affinity and specificity for PA (Figure 2). At rounds 2 and 3, the enrichment in the percentage of cells capable of binding PA using Program D was more comparable to the MMS and MACS/FACS sorting methods previously described (Figure 1B) [19].

Bioinformatic analysis of isolated PA binders and further characterization by FACS

In order to investigate individual candidate binders, single bacterial colonies were plated for sequencing. A bioinformatics analysis on the resulting peptide sequences was performed, and relative binding of isolated binders was accomplished via FACS analysis. Since the early rounds of manual MACS showed little promise by FACS analysis (Figure 1A), only candidates from round 4 were sequenced. For each of the autoMACS* programs, however, rounds 2 through 4 were sequenced. Any sequences that repeated at least once in round 4 were screened by FACS using 250 nM PA488 and 250 nM SAPE. Screening against

SAPE was essential to assess specificity for the target since streptavidin is used in the process of isolating peptides from the bacterial display library through the use of streptavidin-conjugated paramagnetic beads. From this initial screen, the top PA binders from each autoMACS* method, thirteen in total, were selected for their high target (PA488) to background (SAPE ratio). The sequence, frequency, and PA488 nMFI to SAPE nMFI ratio of these top candidate PA binders are shown in Table 1. The prefixes "D" or "DS" in the peptide names signify that they were discovered using Program D or Program DS, respectively. Three of these sequences, D/DS-A28, D/DS-E32, and D/DS-B43, were present in the sequenced colonies from both autoMACS* methods. None of these thirteen sequences were present when sequencing the round 4 pool of candidates isolated using manual MACS.

In general, all of the top candidates listed in Table 1 were represented in round 3 of autoMACS* sorting, although their presence continued to increase in round 4. Therefore, three rounds of sorting are sufficient to obtain PA binders with potential for future applications, using either Program D or Program DS. This is consistent with the result obtained using the MMS [19], and means that peptide candidates are available for study after only three days. The PA488 signal quickly saturated in Program D but not Program DS (Figure 1), so it is not surprising that some of these sequences were already evident in round 2 for Program D, but not Program DS (Table 1). The sequences of the top candidates from Table 1 were aligned using the online proteomics tools ClustalW2

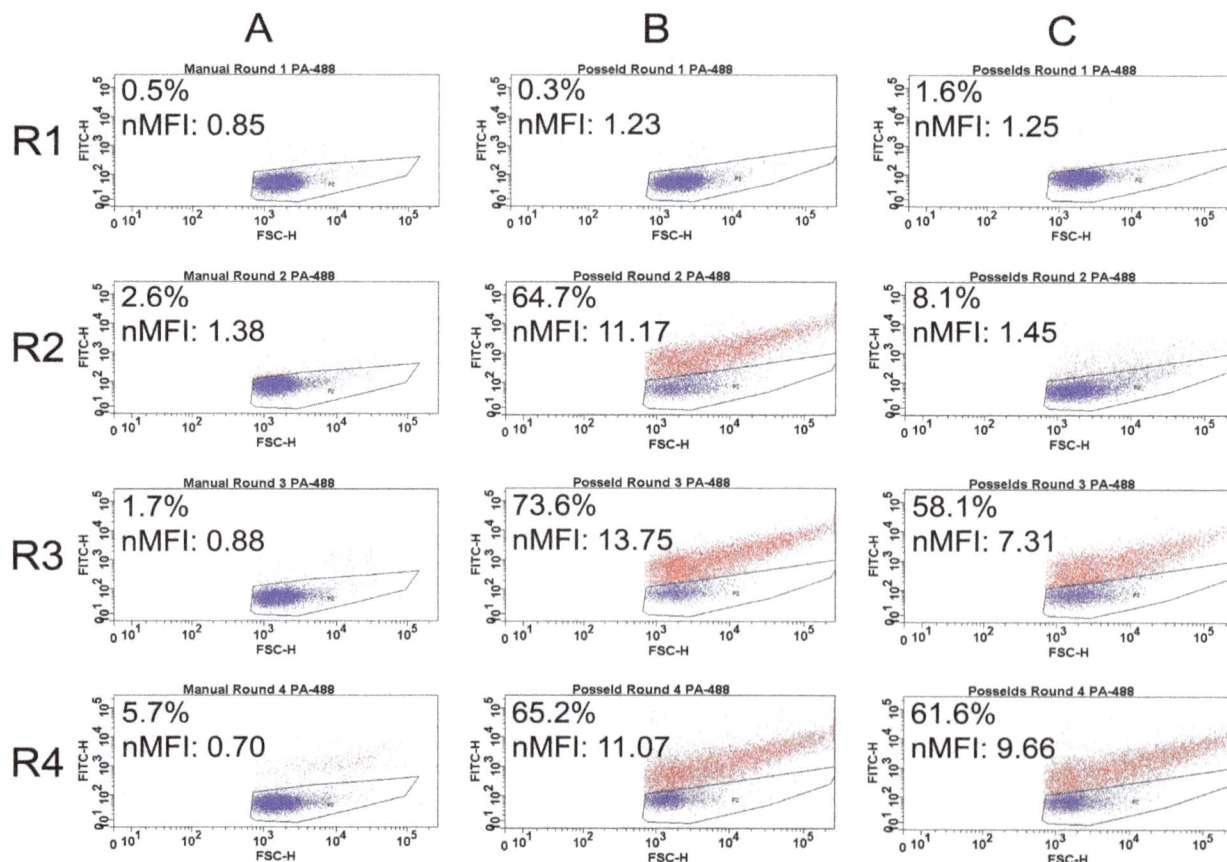

Figure 1: FACS analysis of sorting rounds for discovery of peptide candidates to protective antigen (PA). FITC-H *vs.* FSC-H of cells bound to 250 nM PA488 for rounds 1-4 (R1-R4) of (A) manual MACS sorting, (B) autoMACS® Pro Separator Program D sorting, or (C) autoMACS® Pro Separator Program DS sorting of a bacterial peptide display library. The percentage of cells falling outside of the P2 gate is shown in red and in the top left corner of each box. The Normalized Median Fluorescence Intensity (nMFI) of the entire population is also shown.

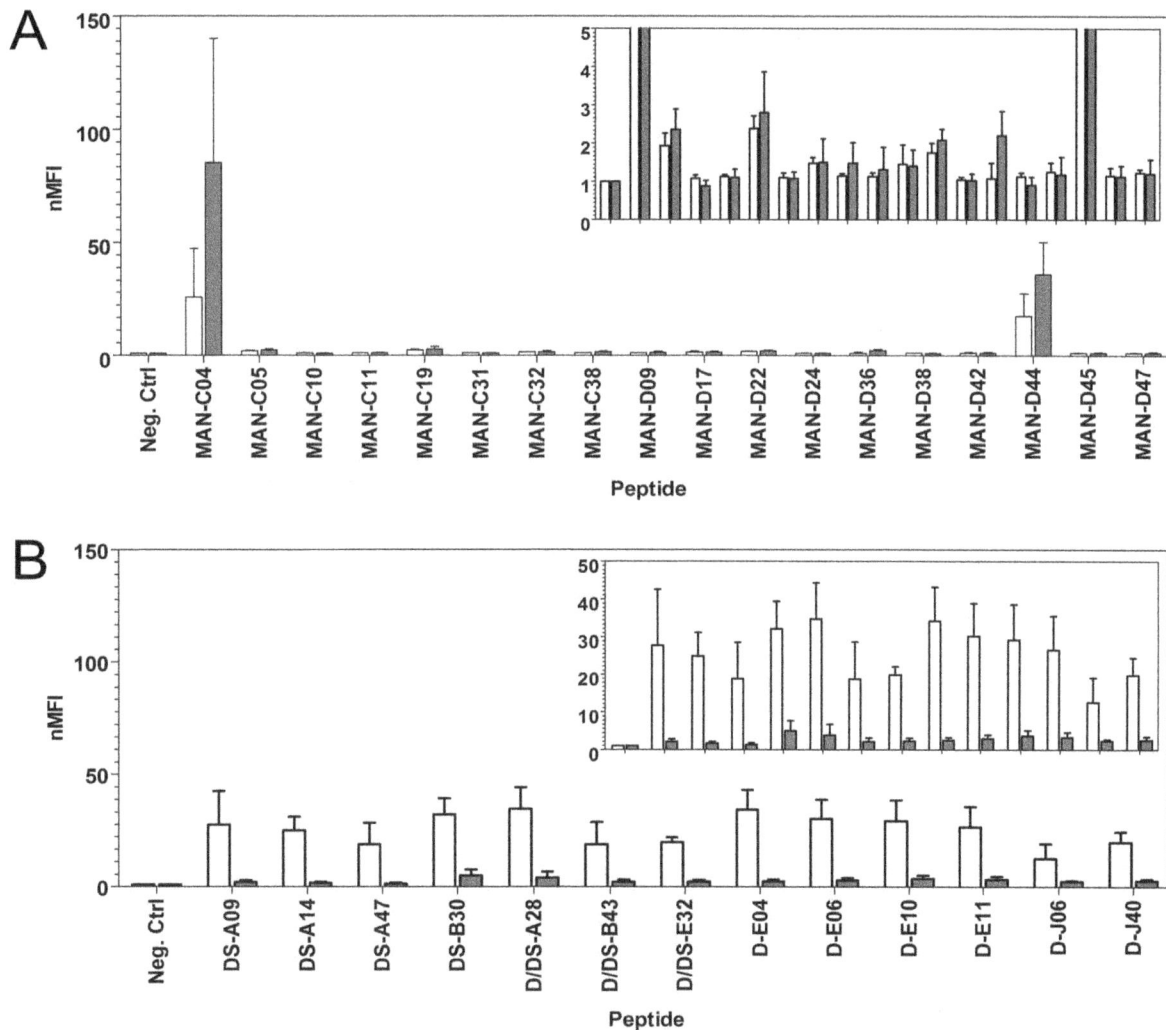

Figure 2: Normalized Median Fluorescence Intensity (nMFI) of potential PA binders. (A) Manual round 4 on-cell peptides analyzed by FACS. Inlet shows the graph with maximum nMFI of 5 to highlight the consistency between the PA488 and SAPE signal throughout. (B) Miltenyi autoMACS® round 4 top candidates among repeating sequences, chosen for high PA488:SAPE by FACS. Inlet shows the graph with maximum nMFI of 50 to highlight that the magnitude of the PA488 signal is 10-fold higher than most manual candidates, and the SAPE binding is low.

and Jalview, and these results are shown in Figure 3. Note the high level of conservation leading to the consensus WXCFTC, the same consensus previously shown for binding to PA488 using the MMS platform [19], further demonstrating the utility of this approach. Phenylalanine and tryptophan are found interchangeably in both the first and fourth positions of this consensus, although tryptophan is usually in the first and phenylalanine in the fourth position. All sequences obtained from each manual or autoMACS° round tested were also aligned using ClustalW2 and Jalview, and the same WXCFTC consensus was seen in the total sequenced population (144 sequences for each) after 3 or 4 rounds of sorting using either autoMACS° program, but not for either autoMACS° program at round 2 or for the manual sort at round 4 (data not shown), further indicating that 3 rounds of sorting was sufficient with either autoMACS° program. Two of the top PA binders, D-E06 and D-J40, did not contain the entire WXCFTC consensus, but did contain tryptophan and phenylalanine residues present in the consensus. The most notably different of these was D-J40, which had a high nMFI for PA488 at 19.8 and low nMFI for SAPE at 2.6, with very little similarity to the consensus aside from the tryptophan and

phenylalanine residues (Table 1, Figures 2 and 3B). For comparison, the DS-A14 peptide showed the highest signal (24.9 nMFI) to SAPE background (1.7 nMFI) with a ratio of 14.8. The peptide with the highest PA488 overall, D/DS-A28, had nMFI for PA488 of 34.7 and nMFI for SAPE of 3.9 with a similar signal to SAPE background of 8.9. Peptide D-J40 demonstrated similar PA and streptavidin binding, even without the entire consensus. The same was true of D-E06, with PA488 nMFI of 30.2 and SAPE nMFI of 2.9. D-E06 was lacking the cysteine residues of the WXCFTC consensus but otherwise followed the same pattern, although it contained a second tryptophan instead of phenylalanine. These aromatic residues, spaced two residues apart, appear to be the most important part of the WXCFTC consensus.

To further compare manual MACS to autoMACS°, the three repeat sequences from the manual round 4 sort, MAN-C04, MAN-D36, and MAN-D09, and fifteen additional, non-repeating sequences, were tested by FACS for binding to 250 nM PA488 and 250 nM SAPE. Most of the fifteen additional sequences were chosen at random, other than avoiding stop codons, due to a lack of repeating candidates and the WXCFTC consensus. However, several of these sequences were also

Peptide Sequence	Peptide Name	Program D			Program DS			PA: SAPE
		Round 2	Round 3	Round 4	Round 2	Round 3	Round 4	
WFCFTCPSSSDVIKG	DS-A09	0	0	0	0	1	2	13.2
YTD**FVCFTC**TMPQLQ	DS-A14	0	0	0	0	1	2	14.8
WSCFTCDHGAETLVS	DS-A47	0	0	0	0	1	2	13.9
T**WFCFTC**YKAPVKHD	DS-B30	0	0	0	0	9	10	6.5
SY**WSCFTC**TTLSGFS	D/DS-A28	0	4	13	0	3	4	8.9
FTN**WSCFTC**SSSTNA	D/DS-B43	0	3	1	0	1	3	8.9
SN**WICFTC**AFPRETA	D/DS-E32	0	9	5	0	2	0	9.0
PGISEVQ**WSCFTC**IV	D-E04	0	12	12	0	0	0	13.8
VV**WI**P**WT**VWTVAPET	D-E06	0	2	2	0	0	0	10.5
STL**FYCFTC**LSSVGS	D-E10	2	13	23	0	0	0	7.9
SS**WLCFTC**LQAPAIS	D-E11	3	6	5	0	0	0	8.1
Y**WHCWTC**NSVNTDSR	D-J06	0	5	2	0	0	0	5.4
PFSYLGTLYIP**WESF**	D-J40	0	1	2	0	0	0	7.6

E. coli cells displaying peptides on an eCPX scaffold were sorted using the autoMACS® separation programs shown and 144 colonies were sequenced for sorting rounds 2 through 4. The number of sequenced colonies expressing each top PA binding peptide candidate is shown for each round. Consensus sequence WXCFTC is shown in bold. PA:SAPE is the ratio of the average nMFI for each, as determined by FACS from 3 independent experiments.

Table 1: Frequency of top PA binding candidates sorted by autoMACS®.

selected for a noted trend of arginine richness, and MAN-D44 was specifically selected for its double tryptophan residues, a trend also present in two αPA peptides previously reported [48,49]. All manual sequences tested by FACS are listed in the peptide alignment in Figure 3A. Note that there is poor consensus when using the manual method, unlike the autoMACS' sequences that exhibited the strong consensus WXCFTC. Many of the individual sequences, and the manual "consensus" (or lack thereof) determined by Jalview, have a high frequency of leucine and serine residues, most likely due to the higher number of codons available for translating these amino acids rather than interactions with PA. The most abundant sequence in the manual sort, MAN-C04, repeated 12 times and was tyrosine-rich, with 6 tyrosine residues out of 15 total amino acids. Tyrosine residues are aromatic like tryptophan and phenylalanine, so it seemed a promising candidate. This sequence was the best PA-binding candidate from the manual sort but was non-specific, with greater than 3 fold higher binding to SAPE (nMFI 85.2) than to PA488 (nMFI 25.7) (Figure 2A). MAN-D36, which repeated 3 times in the 144 sequenced colonies, was actually a previously isolated sequence from biopanning bulk aluminum: the best performing aluminum binding peptide in that study, "DBAD1" [16]. This was likely present in the manual PA sort due to the use of iron oxide-containing MyOne Streptavidin T1 beads, since DBAD1 could potentially bind to other metal oxides in addition to aluminum oxide. MAN-D36/DBAD1 was a poor PA binder, with nMFI of 1.1. The third repeating sequence from manual MACS, MAN-D09, repeated only twice and was also a poor PA binder, with nMFI of 1.1 as well. Other than MAN-C04, the only candidate examined from the manual sort with nMFI greater than 3, including the upper bound of standard deviation, was MAN-D44. This peptide, which contained two adjacent tryptophan residues, had nMFI of 17.5 for PA488, but a 2-fold higher nMFI of 35.8 for SAPE. This is not surprising since, as noted in Sarkes *et al.*, the double tryptophan motif seems to have lower specificity and affinity than the WXCFTC consensus when the "X" is also an aromatic residue [50]. The PA488 nMFI to SAPE nMFI ratio for all 18 manual MACS isolates tested was 1.3 or lower, so specificity was problematic. Not one of the sequences obtained from the manual round 4 sort was found to contain the full WXCFTC consensus (Figure 2A and data not shown), or to demonstrate both affinity and specificity for PA (Figure 2A). One sequence, MAN-C05, did contain two phenylalanine residues with the proper separation of two residues, but the nMFI for

PA488 was only 2.0. It was therefore concluded that no promising PA binding candidates were obtained from the manual MACS, even when employing knowledge from previous sorting to select potential binders. It may have been possible to find a good candidate with further sequencing of round 4 candidates, or further rounds of sorting, but the time required to fully characterize all candidates is impractical for routine isolation and study. The characterization bottleneck with manual MACS further necessitates a reliable and reproducible biopanning approach, such as the approach presented herein using the autoMACS' Pro Separator.

Comparison of autoMACS' to (non-commercial) micro-magnetic separation (MMS) platform

To further validate the autoMACS' biopanning approach, the top candidates from round 4 were analyzed by flow cytometry in a manner that more closely resembles the analysis used in Kogot *et al.*, using percentage of cells falling outside a gated population of cells incubated with PBS buffer alone, for candidates isolated using a non-commercial microfluidic system with the same target, PA [19]. This was the standard FACS analysis method used with candidates isolated by bacterial display for many years [17-23], but nMFI is preferred because the percent binding can saturate at 100%, and the nMFI can vary between populations with the exact same percent binding due to the extent of binding [41]. Using percent binding, the affinity and specificity of the autoMACS' candidates for PA binding herein are similar to those discovered by MMS sorting, but were discovered using a widely - available platform. Specifically, the autoMACS' yielded a tighter range of 65.1%-87.4% for PA binding (Supplemental Table 1 and Supplemental Figure 1) when compared to 44.1%-89.8% obtained using the MMS [19]. When comparing the average PA binding for the candidates tested, both approaches yielded similar results: 78.1% and 71.5% for autoMACS' and MMS, respectively. A tighter range of binding to the negative control is also evident upon comparison of the percent binding to streptavidin in this study versus the literature. Specifically, the streptavidin negative control binding range was 0.7%-13.5% for autoMACS' (Supplemental Table 2 and Supplemental Figure 1), while the streptavidin binding range was 0.1%-39.5% for MMS [19]. While these studies were performed at similar (but not exactly the same) concentrations of target and negative control, it is clear that the autoMACS' biopanning method described here yielded more

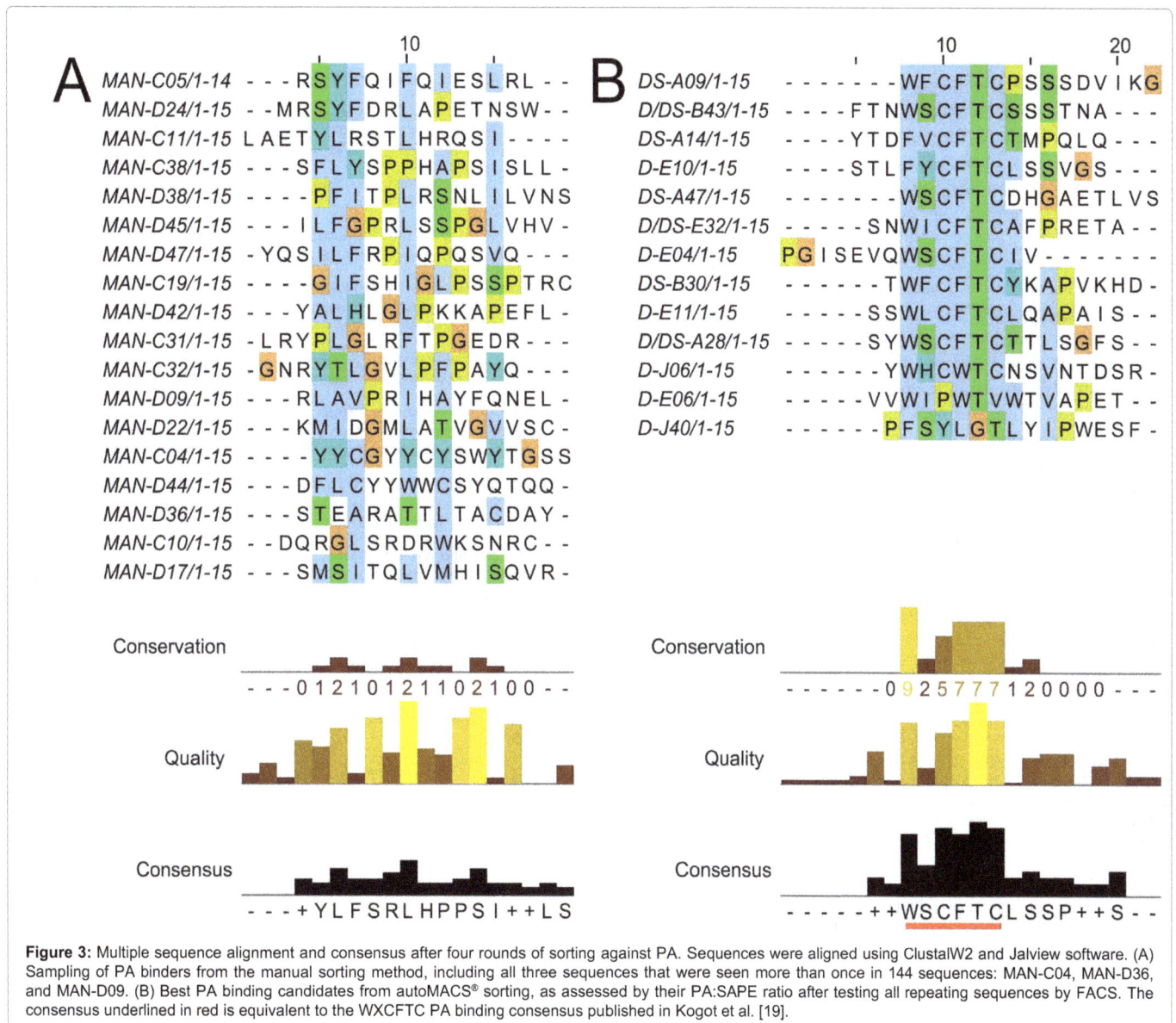

Figure 3: Multiple sequence alignment and consensus after four rounds of sorting against PA. Sequences were aligned using ClustalW2 and Jalview software. (A) Sampling of PA binders from the manual sorting method, including all three sequences that were seen more than once in 144 sequences: MAN-C04, MAN-D36, and MAN-D09. (B) Best PA binding candidates from autoMACS® sorting, as assessed by their PA:SAPE ratio after testing all repeating sequences by FACS. The consensus underlined in red is equivalent to the WXCFTC PA binding consensus published in Kogot et al. [19].

consistent and reliable results. Furthermore, it is important to note that the autoMACS® yields better results than the MACS/FACS standard protocol for bacterial display sorting but in a significantly lower cost and simpler platform. For example, the MACS/FACS sorting method, demonstrated for comparison to MMS in Kogot et al., had a range of 1.2%-62.0% and average of 20.4% for PA binding and range of 0.4%-16.3% and average of 3.5% for streptavidin binding [19]. The range for both target and negative control were tighter using autoMACS®, and the average PA binding was almost 4x greater for autoMACS® than for MACS/FACS, at a fraction of the cost. Overall the commercial availability of the autoMACS® Pro Separator gives it an advantage over the MMS, and when combined with the analysis approach herein is extendable to other cell surface display applications.

Conclusion

In this work, we successfully demonstrated bacterial display sorting using the commercially-available autoMACS® Pro Separator for the first time. Several new PA peptide reagent candidates were discovered as a result, with the same consensus sequence, WXCFTC, as candidates discovered using a non-commercial platform previously tested by our group [19]. Both of these semi-automated platforms are preferred over manual MACS or MACS/FACS sorting due to cost and/or sorting capability, since a FACS capable of sorting cells for downstream use is more expensive, and both MACS/FACS and manual MACS yielded lower affinity binders. Only a single candidate discovered by MACS/FACS yielded the WXCFTC consensus for PA [19], while this consensus was completely absent in the manual MACS study herein. For autoMACS®, whether Program D or Program DS is preferable may depend on the target of interest, the round of sorting (due to sheer number of cells to sort), or how well the peptide is displayed on the cell surface. In this test case, both programs led to PA binders with the same consensus and similar specificity after only three rounds of sorting, although Program D demonstrated more promise after only two rounds of sorting and Program DS yielded more unique sequences containing the consensus. Pushing to a fourth round of sorting led to a high frequency of repeats and is therefore recommended to simplify bioinformatics analysis, and testing these repeat sequences by FACS led to inclusion of candidates that did not quite follow the consensus but that had similar affinity and specificity to candidates containing the consensus. This approach could therefore enable discovery of candidates that bind to different epitopes, and enable rapid discovery of binders in cases where there is no obvious consensus for a particular target, without extensive, brute force characterization of candidates. The specificity obtained without further maturation is notable for a small peptide. The affinity and specificity obtained with minimal effort and relatively low cost using this approach gives this method strong potential for discovery of robust peptide capture candidates and peptide therapeutic agents that can be further matured by incorporation into protein catalyzed capture agents, for instance [49]. We are confident in this instrument's ability as a semi-automated tool for bacterial display sorting due to the strong enrichment and consensus, and future work will include continued use of this platform to screen for peptide capture candidates against existing and future bio-threat agents, and further automation of the biopanning protocol, which could become the new workhorse for bacterial display. This method should be extendable to use with other aerobic bacteria and the autoMACS' Pro Separator is small enough that it may be used inside of a large anaerobe chamber if needed for a specific application. Although the current study was for biopanning of rare cells expressing peptides which bind to a biothreat agent, the platform could be used to isolate bacteria from a variety of mixtures, such as whole blood. The key requirement for extending this protocol to other applications with bacteria is obtaining a binding partner for the target of interest that can be attached covalently or indirectly to a magnetic bead, and that specifically recognizes the cell surface of the bacteria directly, or through a signaling molecule or displayed peptide.

Acknowledgement

The authors would like to thank Dr. Bryn Adams and the Daugherty Lab (UCSB) for their help with cloning the YPet Mona reagent. This research was primarily supported by the U.S. Army Research Laboratory. The content does not necessarily reflect the position or the policy of the United States Government, and no official endorsement should be inferred.

References

1. Bessette PH, Rice JJ, Daugherty PS (2004) Rapid isolation of high-affinity protein binding peptides using bacterial display. Protein Eng Des Sel 17: 731-739.

2. Georgiou G, Stathopoulos C, Daugherty PS, Nayak AR, Iverson BL, et al. (1997) Display of heterologous proteins on the surface of microorganisms: from the screening of combinatorial libraries to live recombinant vaccines. Nat Biotechnol 15: 29-34.

3. Stratis-Cullum D, Kogot JM, Sarkes DA, Val-Addo I, Pellegrino PM (2011) Bacterial display peptides for use in biosensing applications. In: Pramatarova L (ed) On Biomimetics. InTech, Washington DC, USA.

4. Uhlig T, Kyprianou T, Martinelli FG, Oppici CA, Heiligers D, et al. (2014) The emergence of peptides in the pharmaceutical business: From exploration to exploitation. EuPA Open Proteom 4: 58-69.

5. Fosgerau K, Hoffmann T (2015) Peptide therapeutics: current status and future directions. Drug Discov Today 20: 122-128.

6. Daugherty PS (2007) Protein engineering with bacterial display. Curr Opin Struct Biol 17: 474-480.

7. Sezonov G, Joseleau-Petit D, D'Ari R (2007) Escherichia coli physiology in Luria-Bertani broth. J Bacteriol 189: 8746-8749.

8. Boder ET, Wittrup KD (1997) Yeast surface display for screening combinatorial polypeptide libraries. Nat Biotechnol 15: 553-557.

9. Jagadish MN, Carter BL (1978) Effects of temperature and nutritional conditions on the mitotic cell cycle of Saccharomyces cerevisiae. J Cell Sci 31: 71-78.

10. Donatan S, Yazici H, Bermek H, Sarikaya M, Tamerler C, et al. (2009) Physical elution in phage display selection of inorganic-binding peptides. Mater Sci Eng C 29: 14-19.

11. Gaskin DJ, Starck K, Turner NA, Vulfson EN (2001) Phage display combinatorial libraries of short peptides: ligand selection for protein purification. Enzyme Microb Technol 28: 766-772.

12. Nixon AE (2002) Phage display as a tool for protease ligand discovery. Curr Pharm Biotechnol 3: 1-12.

13. Goldman ER, Pazirandeh MP, Mauro JM, King KD, Frey JC, et al. (2000) Phage-displayed peptides as biosensor reagents. J Mol Recognit 13: 382-387.

14. New England BioLabs Inc (2014) Ph.D. Phage Display Libraries Instruction Manual, Version 1.2.

15. Puddu V, Perry CC (2012) Peptide adsorption on silica nanoparticles: evidence of hydrophobic interactions. ACS Nano 6: 6356-6363.

16. Adams BL, Finch AS, Hurley MM, Sarkes DA, Stratis-Cullum DN (2013) Genetically Engineered Peptides for Inorganics: Study of an Unconstrained Bacterial Display Technology and Bulk Aluminum Alloy. Adv Mater 25: 4585-4591.

17. Kenrick SA, Daugherty PS (2010) Bacterial display enables efficient and quantitative peptide affinity maturation. Protein Eng Des Sel 23: 9-17.

18. Kogot JM, Pennington JM, Sarkes DA, Kingery DA, Pellegrino PM, et al. (2014) Screening and characterization of anti-SEB peptides using a bacterial display library and microfluidic magnetic sorting. J Mol Recognit 27: 739-745.

19. Kogot JM, Zhang Y, Moore SJ, Pagano P, Stratis-Cullum DN, et al. (2011) Screening of peptide libraries against protective antigen of Bacillus anthracis in a disposable microfluidic cartridge. PLOS ONE 6: e26925.

20. Little LE, Dane KY, Daugherty PS, Healy KE, Schaffer DV (2011) Exploiting bacterial peptide display technology to engineer biomaterials for neural stem cell culture. Biomaterials 32: 1484-1494.

21. Rice JJ, Schohn A, Bessette PH, Boulware KT, Daugherty PS (2006) Bacterial display using circularly permuted outer membrane protein OmpX yields high affinity peptide ligands. Protein Sci 15: 825-836.

22. Thomas JM, Daugherty PS (2009) Proligands with protease-regulated binding activity identified from cell-displayed prodomain libraries. Protein Sci 18: 2053-2059.

23. Dane KY, Chan LA, Rice JJ, Daugherty PS (2006) Isolation of cell specific peptide ligands using fluorescent bacterial display libraries. J Immunol Methods 309: 120-129.

24. Bessette PH, Hu X, Soh HT, Daugherty PS (2007) Microfluidic library screening for mapping antibody epitopes. Anal Chem 79: 2174-2178.

25. Rice JJ, Daugherty PS (2008) Directed evolution of a biterminal bacterial display scaffold enhances the display of diverse peptides. Protein Eng Des Sel 21: 435-442.

26. Kogot JM, Pennington JM, Sarkes DA, Stratis-Cullum DN, Pellegrino PM (2011) Population Enrichment and Isolation with Magnetic Sorting. US Army Research Laboratory, Adelphi, MD Sensors and Electron Devices Directorate.

27. An Z (2011) Therapeutic monoclonal antibodies: from bench to clinic. John Wiley & Sons, New Jersey, USA.

28. Chao G, Lau WL, Hackel BJ, Sazinsky SL, Lippow SM, et al. (2006) Isolating and engineering human antibodies using yeast surface display. Nat Protoc 1: 755-768.

29. Miller KD, Pefaur NB, Baird CL (2008) Construction and screening of antigen targeted immune yeast surface display antibody libraries. Curr Protoc Cytom Chapter 4: Unit4.

30. Schuijt TJ, Narasimhan S, Daffre S, DePonte K, Hovius JW, et al. (2011) Identification and characterization of Ixodes scapularis antigens that elicit tick immunity using yeast surface display. PLoS One 6: e15926.

31. Puri V, Streaker E, Prabakaran P, Zhu Z, Dimitrov DS (2013) Highly efficient selection of epitope specific antibody through competitive yeast display library sorting. MAbs 5: 533-539.

32. Monaci P, Luzzago A, Santini C, De Pra A, Arcuri M, et al. (2008) Differential screening of phage-ab libraries by oligonucleotide microarray technology. PLoS One 3: e1508.

33. Ahmad MR, Nakajima M, Kojima S, Homma M, Fukuda T (2008) The effects of cell sizes, environmental conditions, and growth phases on the strength of individual W303 yeast cells inside ESEM. IEEE Trans Nanobioscience 7: 185-193.

34. Kubitschek HE, Friske JA (1986) Determination of bacterial cell volume with the Coulter Counter. J Bacteriol 168: 1466-1467.

35. Nelson DE, Young KD (2000) Penicillin binding protein 5 affects cell diameter, contour, and morphology of Escherichia coli. J Bacteriol 182: 1714-1721.

36. Tyson CB, Lord PG, Wheals AE (1979) Dependency of size of Saccharomyces cerevisiae cells on growth rate. J Bacteriol 138: 92-98.

37. Grossman N, Ron EZ, Woldringh CL (1982) Changes in cell dimensions during amino acid starvation of Escherichia coli. J Bacteriol 152: 35-41.

38. Miltenyi Biotec GmbH (2007) autoMACS™ Pro Separator User Manual, Version 1.1.

39. Miles AA, Misra SS, Irwin JO (1938) The estimation of the bactericidal power of the blood. J Hyg (Lond) 38: 732-749.

40. Getz JA, Schoep TD, Daugherty PS (2012) Peptide discovery using bacterial display and flow cytometry. Methods Enzymol 503: 75-97.

41. Chan LY, Yim EK, Choo AB (2012) Normalized Median Fluorescence: An Alternative Flow Cytometry Analysis Method for Tracking Human Embryonic Stem Cell States During Differentiation. Tissue Eng Part C Methods 19: 156-165.

42. Clamp M, Cuff J, Searle SM, Barton GJ (2004) The Jalview Java alignment editor. Bioinformatics 20: 426-427.

43. Thompson JD, Gibson TJ, Plewniak F, Jeanmougin F, Higgins DG (1997) The CLUSTAL_X windows interface: flexible strategies for multiple sequence alignment aided by quality analysis tools. Nucleic Acids Res 25: 4876-4882.

44. Multiple Sequence Alignment.

45. Henriques ST, Thorstholm L, Huang YH, Getz JA, Daugherty PS, et al. (2013) A novel quantitative kinase assay using bacterial surface display and flow cytometry. PLoS One 8: e80474.

46. Stratis-Cullum DN, Finch AS (2013) Current Trends in Ubiquitous Biosensing. J Anal Bioanal Tech S7-009.

47. Christmann A, Walter K, Wentzel A, Kratzner R, Kolmar H (1999) The cystine knot of a squash-type protease inhibitor as a structural scaffold for Escherichia coli cell surface display of conformationally constrained peptides. Protein Eng 12: 797-806.

48. Kogot JM, Sarkes DA, Pennington JM, Pellegrino PM, Stratis-Cullum D (2014) Binding specificity and affinity analysis of an anti-protective antigen peptide reagent using capillary electrophoresis. Adv Biosci Biotechnol 5: 40-45.

49. Farrow B, Hong SA, Romero EC, Lai B, Coppock MB, et al. (2013) A chemically synthesized capture agent enables the selective, sensitive, and robust electrochemical detection of anthrax protective antigen. ACS Nano 7: 9452-9460.

50. Sarkes DA, Dorsey BL, Stratis-Cullum DN (2015) Analysis of protective antigen peptide binding motifs using bacterial display technology. Proceedings of the International Society for Optics and Photonics, SPIE Defense+ Security.

6

A Novel Catechol Electrochemical Sensor Based on Cobalt Hexacyanoferrate/(CoHCF)/Au/SBA-15

Yaqian Yan, Linjing Wu, Qianqiong Guo and Shasheng Huang*

Life and Environmental Science College, Hanghai Normal University, Shanghai, PR China

Abstract

A novel electrochemical sensor for catechol was developed by electrodepositing $HAuCl_4$ and cobalt hexacyanoferrate (CoHCF) on an ordered mesoporous SBA-15 decorated glassy carbon electrode (GCE). The CoHCF/Au/SBA-15 film was characterized by scanning electron microscopy (SEM) and impedance spectra. A mesoporous SBA-15 was used as a platform that enlarged the surface area of the working electrode. The CoHCF/Au/SBA-15 modified electrode showed good electrocatalytic activity to catechol and the electrocatalytic response was measured using cyclic voltammetry and Amperometric *i-t* curve. The electrochemical performance of the sensor for catechol was further enhanced due to the deposition of Au on the electrode surface. Under the optimal conditions, the sensor showed a linear range from 3.0×10^{-7}M to 5.1×10^{-5}M of catechol, with a detection limit of 50 nM (S/N=3). Good reproducibility, stability and good selectivity in the presence of numerous organic phenolics made the CoHCF/Au/SBA-15 modified electrodes applicable to the determination of catechol in the various water samples.

Keywords: Cobalt hexacyanoferrate; Mesoporous molecularsieve; Catechol; Amperometric i-t curve

Introduction

Recent decades, considerable efforts have been invested in the determination of phenolic compounds in environmental, industrial, agricultural, and food fields [1,2]. Phenolic compounds are released into environment by a large number of industries, such as coal mining, oil refinery, paint, polymer and pharmaceutical preparation. Phenolic compounds are secondary metabolites. They are not involved in growth and energy metabolism and are usually generated in response to environmental stress [2]. Some of these phenolic compounds like phenol, hydroquinone, are also harmful to humans and animals [3]. Catechol (CC, 1,2-dihydroxybenzene) is a phenolic derivative with several applications such as an antifungal preservative on potato plantations, a photographic and fur dye developer, and as an antioxidant [4]. Therefore, the determination of catechol is very important in environmental protection. Catechol undergoes oxidation under mild conditions to give benzoquinone. Benzoquinone is said to be antimicrobial, which slows the spoilage of wounded fruits and other plant parts. Catechol is produced by the reversible two-electron, two-proton reduction of 1,2-benzoquinone (Scheme 1) [5].

Catechol contains phenolic hydroxy group and possesses excellent electrochemical activity. Many electrochemical methods using different modified electrodes have been reported for the determination of catechol [6-8].

Transition metal hexacyanoferrates (MHCFs) belong to a class of polynuclear inorganic mixed-valence compounds because of their reversible redox properties and their zeolitic structure [9-12], the color of prussian blue (PB) or iron (III) hexacyanoferrate (II) can be reversibly switched by electrochemical treatments [13]. Among the many transition metal hexacyanoferrates, cobalt hexacyanoferrate (CoHCF) has interesting chemical and electrochemical properties [14-17], which makes it a suitable modifier in many sensing applications. CoHCF exhibits well-defined and reproducible electrochemical responses at not only oxidized but also reduced states [10]. In addition, CoHCF film can also be easily fabricated on various electrode substrates [13] and CoHCF modified electrode shows good electrocatalytic activity toward a variety of substrates [14-17].

Gold nanoparticles (AuNPs) exhibit distinctive physical, chemical and catalytic properties because of their size, shape and high surface to volume ratio in contrast to bulk materials [18]. In particular, Gold nanoparticles have received much attention because of their applications in catalysis, nanoelectronics, drug delivery and chemical sensing [19]. Recently, increasing attention is being given to the fabrication of thin films of metal nanoparticles from both a fundamental and a practical application point of view. Such thin films of metal nanoparticles on solid surfaces have been prepared using a number of strategies including assembly techniques with cross linkers for metal nanoparticles [20].

Mesoporous materials are the focus of research due to their porous structure and high surface area, and the past decade many innovative synthetic methods of the mesoporous materials have been developed employing self-assembled surfactants as structure-directing agents [21-23]. Porous silicon (PS) has emerged as a promising material for sensor applications since it presents a number of advantages like miniaturization, integration of signal processing circuitry, low cost, greater surface area, and greater adsorption capacity [24-26]. They showed remarkable applications such as in molecular sieves, adsorbents, gas sensors, protein immobilization, etc. [27-31].

In this paper, the mesoporous SBA-15 was synthesized and the electrochemical behaviors of the glassy carbon electrode modified with CoHCF-Au nanocomposites based on mesoporous SBA-15 were investigated. The experimental results of the present work indicated that the CoHCF/Au/SBA-15 modified electrode showed an improved electrochemistry response to catechol compared with that without SBA-

*Corresponding author: Shasheng Huang, Life and Environmental Science College, Hanghai Normal University, Shanghai, 200234, PR China
E-mail: sshuang@shnu.edu.cn

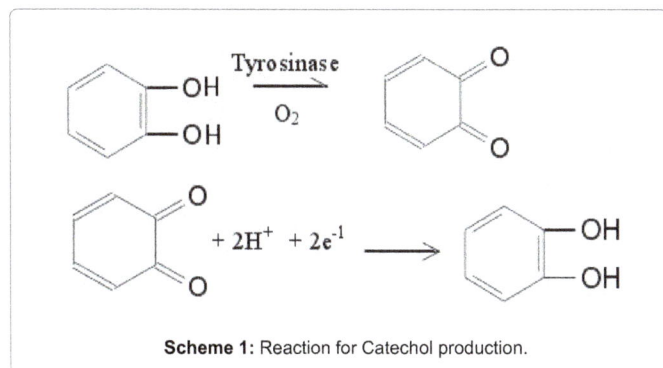

Scheme 1: Reaction for Catechol production.

15 due to the greater surface area of SBA-15 as the substrate material. Electrochemical results showed that the CoHCF-Au nanocomposites in the gaps of SBA-15 could enhance the direct electron transfer. The catechol sensor was successfully used in real waste water sample analysis with a stability and reliable recovery.

Experimental

Apparatus and reagents

All chemicals and reagents used in the study were of A.R. grade and used as received without further purification. Catechol ($C_6H_6O_2$, 99%), tetraethoxysilane (TEOS, 98%), hydrochloric acid (HCl, 37%) and pluronic copolymer P123 (non-ionic triblock copolymer, $EO_{20}PO_{70}EO_{20}$, MW=5800) were purchased from Sigma Chemicals Company. Gold (III) chloride trihydrate ($HAuCl_4 \cdot 3H_2O$, 99.9%), cobalt chloride hexahydrate ($CoCl_2 \cdot 6H_2O$, 99.9%) were obtained from Aldrich. Phosphate buffer solution (PBS, 0.1M) was used to prepare the supporting electrolyte and the pH value was adjusted by mixing the stock solution of NaH_2PO_4 and Na_2HPO_4. Prior to experiments, the solutions were purged with purified nitrogen for at least 15 min to remove oxygen. Milli Q 18.2 MΩ water was used throughout the experiments.

The cyclic voltammetric measurements were carried out on CHI760B electrochemical workstation (Chen Hua Instrument, Shanghai, China). A conventional three electrode system was used in this work consists of CoHCF/Au/SBA-15/GCE as the working electrode, a thin Pt wire as auxiliary electrode and saturated calomel electrode (SCE) as the reference electrode. All the experiments were conducted at room temperature unless otherwise stated. Scanning electron microscopy (SEM) measurements were carried out using a JSM 6360 LV microscope (JEOL Ltd, Japan) operating at 100 kV. The transmission electron micrograph (TEM) was obtained using a JEM-2100 TEM instrument (JEOL). Fourier transformation infrared (FT-IR) spectra (4000-400 cm^{-1}) in KBr were obtained with a Vector 22 FTIR spectrometer (Bruker Optics, Germany). X-ray diffraction (XRD) patterns were collected on a Bruker D8 Advance X-ray diffractometer (Bruker, Germany).

The synthesis of mesoporous SBA-15

The synthesis of SBA-15 was conducted as described in the literature [22,23]. In a typical preparation, 4.0 g P123 was completely dissolved in 130 g ultrapure water, and 20 mL HCl solution (37%) was added with stirring at 35°C. Then 8.5 g TEOS was slowly added to the solution with stirring at 35°C for 24 h. The mixture was then transferred to an autoclave and aged for 24 h at 80°C. The solid product was recovered by filtration, washed with ultrapure water, and air-dried

at room temperature. Calcination was carried out in an air atmosphere at 550°C for 6 h with a heating rate of 1°C/min to remove the template and the final product was denoted as mesoporous SBA-15.

Fabrication of CoHCF/Au/SBA-15/GCE

The GCE was carefully polished with 1.0, 0.3 and 0.05 μm alumina powder successively, followed by rinsing thoroughly with ultrapure water. The polished electrode was sonicated in acetone and water, respectively. Then the cleaned GCE was pretreated by scanning from - 0.2 to 1.2 V in 0.5 M H_2SO_4 and dried at room temperature. The as-prepared SBA-15 (4.0 mg) was dispersed into 2 ml DMF and the mixture was sonicated for 1 h to form a stable white suspension.

The fabrication of CoHCF/Au/SBA-15/GCE was performed as follows: Firstly, 4.0 μL SBA-15 solution was dropped onto the pretreated glassy carbon electrode surface and dried in air. Secondly, Au nanoparticles were immobilized on electrode surface by cyclic voltammetry in 0.1% $HAuCl_4$ between 0 and 1.6 V at scan rate of 50 mV·s^{-1}, then the electrode was washed with water and dried in air (this electrode was denoted as Au/SBA-15/GCE). The Au/SBA-15/GCE was scanned from -0.2 to 1.3 V at 50 mV·s^{-1} in 1.0 mM $CoCl_2$ solution (containing 0.05M KCl), potassium ferrocyanide solution (1.0 mM), with 0.1 M KCl as the background electrolyte to get a thin film of CoHCF on the surface. Unless otherwise stated, the procedure involved 10 full voltammetric cycles (20 segments) in the potential range from -0.2 to 1.3 V. After cycling, the electrode was kept at the potential of 1.3 V for 20 min to get deposition of cobalt hexacyanoferrate (CoHCF) [32]. After being removed from the solution and thoroughly rinsed with water, the electrode was dried in air for later use. The obtained electrode was the CoHCF/Au/SBA-15 /GCE. The preparation process of this sensor is shown in Scheme 2.

Sample preparation and measurement procedures

All experiments were carried out at room temperature and high purity nitrogen was kept flowing over the solution during the experiments. Solutions of catechol (99%, Sigma-Aldrich) were prepared daily. Under the optimal conditions, CV was conducted in 10.0 mL PBS (0.1M, pH=6.98) and CV was carried out between -0.3 to 0.5 V at scan rate of 100 mV·s^{-1}. The determination of catechol in the samples was performed using i-t curve with initial potential of 0.4 V.

Results and Discussion

Characterization of SBA-15

The crystalline structure and the phase composition of SBA-15 were characterized by X-ray diffraction (XRD). The small-angle XRD pattern for as-synthesized SBA-15 prepared with $EO_{20}PO_{70}EO_{20}$ showed three well-resolved diffraction peaks in the 2θ range 0.5-5°,

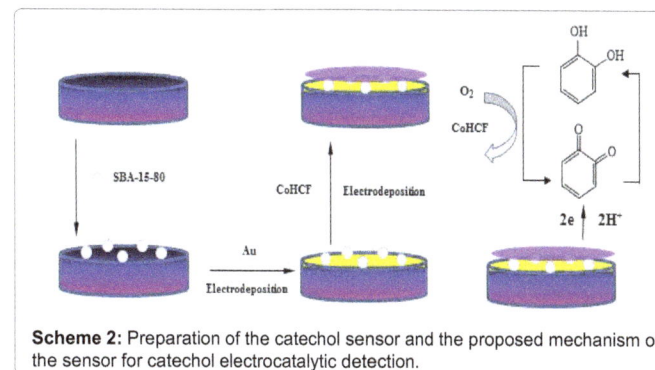

Scheme 2: Preparation of the catechol sensor and the proposed mechanism of the sensor for catechol electrocatalytic detection.

which corresponded to the diffraction of (100), (110), and (200) planes, respectively [33] (Figure 1), indicating that as-synthesized SBA-15 had a high degree of hexagonal mesoscopic organization. These diffraction peaks were characteristic of the ordered structure of SBA-15.

Scanning electron microscopy (SEM) images (Figure 2A) revealed that the as-synthesized SBA-15 sample was consists of many rope-like domains with relatively uniform sizes, which were typical microstructures of the mesoporous SBA-15. Transmission electron microscopy (TEM) images (Figure 2B) of calcined SBA-15 showed well-ordered hexagonal arrays of Mesopores and further confirmed that SBA-15 had a 2D p6mm hexagonal structure.

The N_2-sorption studies of SBA-15-X samples exhibited type IV isotherms according to IUPAC classification with H1 hysteresis loop, which were characteristics of mesoporous materials with one-dimensional cylindrical channels [34]. A pore diameter of 6.24 nm (Figure S1 in the supporting information), and a Brunauer-Emmett-Teller (BET) surface area of 518 $m^2 \cdot g^{-1}$ can be observed. Three well distinguished regions of the adsorption isotherm (Figure S1 in the supporting information) were apparent: (i) monolayer multilayer adsorption, (ii) capillary condensation and (iii) multilayer adsorption on the outer particle surfaces.

Characterization of the modified electrode

The modified CoHCF/Au composite was characterized by scanning electron microscopy (SEM). Figure 3A showed that bare GCE was clean and smooth and the mesoporous SBA-15 had typically rod structures (Figure 2A). After Au nanoparticles and cobalt hexacyanoferrate (CoHCF) were electrodeposited on the Bare GCE, as seen in Figure 3B, the CoHCF/Au composite presented a clear nano-flower structure on the electrode. And these assembled materials were evenly dispersed on glassy carbon electrode.

EIS (Electrochemical impedance spectroscopy) was employed to reveal the impedance changes of the corresponding electrode surface. The frequency varied from 105 to 0.05 Hz and the ac excitation amplitude was 5 mV. Figure 4 shows the Nyquist plots for the electrodes in a 0.1 M KCl solution containing 1 mM $[Fe(CN)_6]^{4-/3-}$ (1:1).

From the Figure 4, it can be seen that the Ret for bare GCE was 600 Ω (curve a GCE). After modification of SBA-15, the value of Ret significantly increased to 4230 Ω (curve b SBA-15/GCE), as a result of the existence of the mesoporous silica skeleton which has poor conductivity. After electropolymerization of Au nanoparticles on SBA-15 surface, the value of Ret immediately increased to 156.4 Ω, indicating that the introduction of AuNPs could improve electron transfer kinetics to a large extent in the self-assembly process of the sensor (curve c Au/SBA-15/GCE). After the introduction of transition metal ferricyanide (CoHCF), the value of Ret of the sensor was larger than that of Au/ SBA-15/GCE maybe the conductivity of CoHCF was weaker than AuNPs (curve d CoHCF/Au/SBA-15/GCE).

Figure 5 showed the cyclic voltammograms obtained for differently modified electrodes in 1mM $[Fe(CN)_6]^{3-/4-}$ aqueous solution containing 0.1M KCl at scan rate of 100 $mV \cdot s^{-1}$. The CVs showed well-defined typical diffusion-limited patterns, The bare glassy carbon electrode (GCE) witnessed a pair of well-defined redox peaks with the anodic (E_{pa}) and cathodic (E_{pc}) peak potential of 0.159 V and 0.233 V, respectively, and a peak potential difference of 74 mV (Figure 5, curve a). These peaks could be definitely attributed to the redox behaviors of $[Fe(CN)_6]^{3-/4-}$. The modification of SBA-15 on electrode surface induced a big decrease of peak current, which was invoked by the

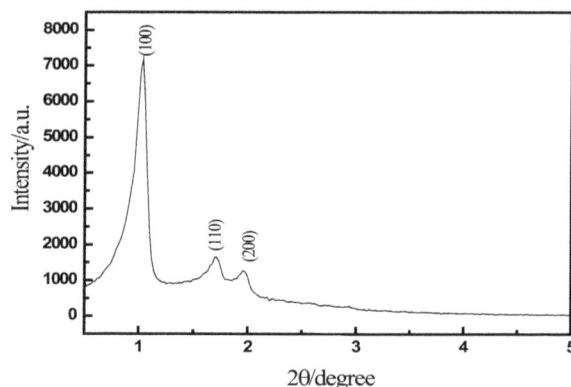

Figure 1: Low-angle powder X-ray diffraction patterns of SBA-15 materials.

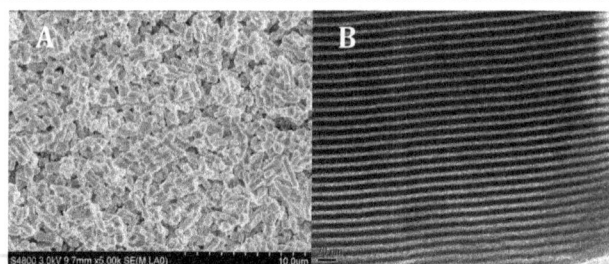

Figure 2: SEM image (A) and TEM image of SBA-15(B).

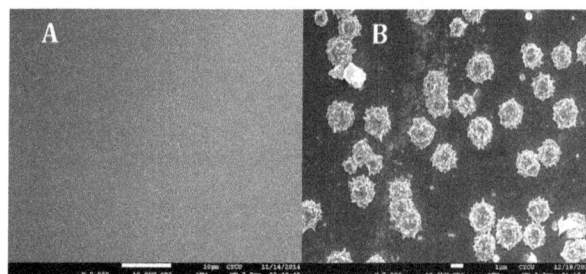

Figure 3: SEM image of bare GCE (A) and CoHCF/Au/ GCE (B).

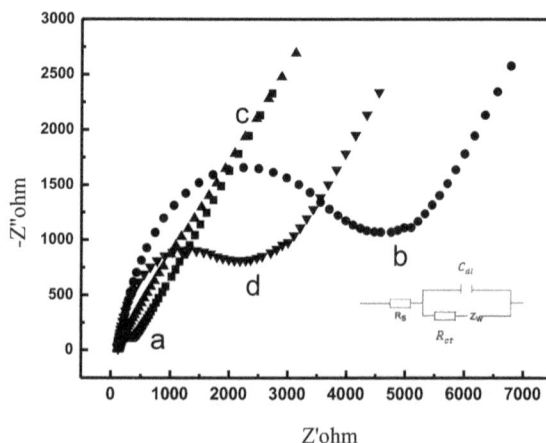

Figure 4: Electrochemical impedance spectra of bare GCE (a), SBA-15/GCE(b), Au/ SBA-15/GCE (c) and CoHCF/Au/SBA-15/GCE (d).

diffusion inhibition of $[Fe(CN)_6]^{3-/4-}$ to the electrode surface (Figure 5, curve b). The Au nanoparticles modification was convinced to increase the effective electrode surface area and the rate of electron transfer at the sensor, which was confirmed by an apparent increase in the voltammetric responses of $[Fe(CN)_6]^{3-/4-}$ than that of SBA-15 modified electrode (Figure 5, curve c). Nevertheless, the further assembly of CoHCF on the Au/SBA-15 modified electrode obviously decreased the peak current of $[Fe(CN)_6]^{3-/4-}$ as observed in curve d, which could be reasonable considering that the electron transfer $[Fe(CN)_6]^{3-/4-}$ to the underlying electrode surface was partially blocked. At the other extreme, indicating that CoHCF has been successfully immobilized onto the GCE modified with SBA-15. The results of the cyclic voltammetric experiments for differently modified electrodes were in good agreement with the results by impedance experiments.

Figure 6 shows the current responses obtained for the different decorated electrodes in o.1 mM catechol solution in a potential range of -0.3-0.5 V. The CVs showed the bare GCE (a) and SBA-15/GCE (b) had small redox current responses to catechol. While the current responses of Au/SBA-15/GEC (c), CoHCF/Au/SBA-15/GCE (d) and CoHCF/Au/GCE (e) were all larger than that of Curve a, b, which indicated that the deposition of Au on the electrode surface could increase the electron transfer rate, enhance electrical conductivity and made the current signal large. As seen in Curve c and d, CoHCF showed good electrocatalytic activity toward catechol. Besides, CVs of CoHCF/Au/SBA-15/GCE (d) and CoHCF/Au/ GCE (e) depicted that the decorated electrode showed larger current response to the catechol.

Electrodeposition processes of Au nanoparticles and cobalt hexacyanoferrate (CoHCF)

The electropolymerization of Au nanoparticles on SBA-15 surface was done by continuous potential cycling for 10 cycles in the range of 0.0 V to 1.6 V. The continuous growth of Au nanoparticles exhibited two redox processs [36]. The results showed CVs obtained at SBA-15/GCE in 0.2% $HAuCl_4$ at scan rate of 120 mV·s^{-1} (Figure S2 in the supporting information). Two quasi-reversible redox waves were observed at SBA-15/GCE in the potential range of 0.0 V and 1.6 V. Both the redox peak currents and the peak-to-peak difference increase with increasing scan rates. It can be seen that the consequential increase of the redox peak currents when the potential scan rate is increased. In order to achieve appropriate AuNPs, scan rate of 50 mVs^{-1} was chosen for the immobilization of Au nanocomposites.

CoHCF was electrodeposited from the solutions containing 1.0 mM $CoCl_2$ solution, 1 mM $K_3[Fe(CN)_6]$ and 0.1M KCl. Before the electrodeposition, CVs were recorded in $K_3[Fe(CN)_6]$ at different scan rates to estimate the effective electrode area. The cyclic voltammograms showed that the deposition of CoHCF nanoparticles started at a potential where the reduction of $[Fe(CN)_6]^{3-}$ to $Fe(CN)_6^{4-}$ occurred, and then Co^{2+} reacted instantaneously with $[Fe(CN)_6]^{4-}$ to form CoHCF nanoparticles on the electrode surface (Figure S3 in the supporting information). Before the electrodeposition, the reversible peaks of the $[Fe(CN)_6]^{3-}/[Fe(CN)_6]^{4-}$ system were recorded. Three well-defined reversible peaks were detected around 0.60 V. CoHCF was discovered to grow with each potential cycle, as revealed by the increasing of peak currents with each cycle. An apparent voltammetric change was observed in the deposition process of CoHCF. The redox peak currents of the electrode increase with increasing the segment number.

Optimization of experimental conditions

Effect of scan rate: The influence of potential scan rate on the

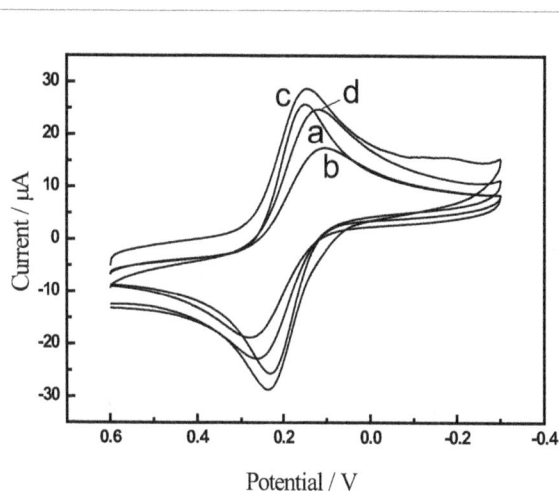

Figure 5: Cyclic voltammograms of the bare electrode (a), SBA-15/GEC (b), Au/SBA-15/GEC (c) and CoHCF/Au/SBA-15/GCE (d).

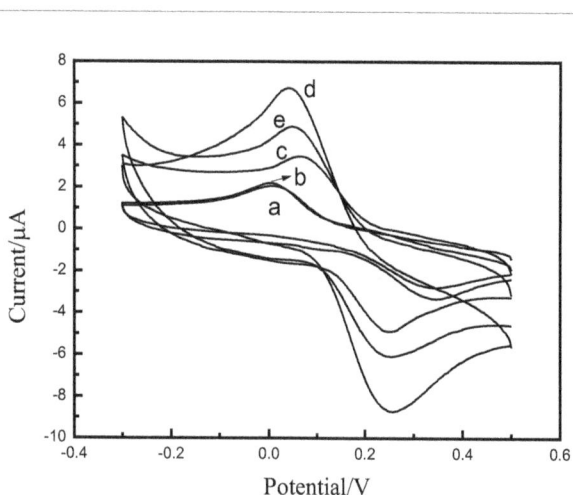

Figure 6: CV curves of the bare electrode (a), SBA-15/GEC (b), Au/SBA-15/GEC (c), CoHCF/Au/SBA-15/GCE (d) and CoHCF/Au/ GCE (e) in 0.1mM catechol (pH=7.0).

electrochemical behavior of catechol at GCE was studied [37] (Figure 6A). The relationship between current (i) versus scan rate at the modified electrode was studied individually in the presence of 0.1 mM catechol. When the sensor was cycled between 20 and 180 mV·s^{-1}, a linear relationship was obtained between the peak intensity I_{pa} and the scan rate v (Figure 6B), indicating that the oxidation of catechol at GCE is a adsorption controlled process. In addition, the anodic peak potential shifted to positive values with the scan rate increased and these results also confirmed that the oxidation reaction was irreversible.

Effect of pH: The effect of pH on the response of catechol at the sensor was investigated by recording the cyclic voltammograms in the range of -0.4-0.6 V at a scan rate of 100 mV·s^{-1}. The anodic peak currents for catechol increased with the increase of pH from 4.5 to 7.0 (Figure 7A) and reached a maximum value at 7.0. When the pH value is higher than 7.0, the peak current decreases rapidly. Within these pH ranges, the relationship between pH and the anodic peak potential was investigated (Figure 7B). It can be seen that the peak potential is shifted to less positive values as the pH of the solution increasing from 4.5 to 8.0 for catechol. These results indicated that protons are participating in the oxidation of catechol. The plot of E_{pa} vs. pH showed a straight line. The equation for peak potential with pH for catechol is expressed as follows:

$$E_{pa}=0.6433\text{-}0.058 \text{ pH } (R=0.999)$$

The slope of $\approx$59 mV·pH^{-1} unit is indicative of single electron transfer processes involving one proton according to the Nernst relationship.

Chronocoulometry studies of the GCE and modified electrodes

The electrochemical effective surface area for the bare GCE and CoHCF/Au/SBA-15/GCE can be calculated by the slope of the plot of Q vs. $t^{1/2}$ obtained by chronocoulometry using [Fe(CN)$_6$]$^{3-}$ as a model complex (the diffusion coefficient D of [Fe(CN)$_6$]$^{3-}$ is 7.6×10^{-6} cm^{2}s^{-1} [38]) The corresponding Q–t curves and Q-$t^{1/2}$ plots were also performed and shown in Figure 8. This was according to the formula given by Anson [38] (1):

$$Q(t)=\frac{2nFAcD^{1/2}t^{1/2}}{\pi^{1/2}}+Q_{dl}+Q_{ads} \qquad (1)$$

where A is the surface area of the working electrode, c is the concentration of substrate, D is the diffusion coefficient, Q_{dl} is double layer charge which could be eliminated by background subtraction, and Q_{ads} is Faradic charge. Based on the slope of the linear relationship between Q and $t^{1/2}$, A can be calculated to be 0.00116 cm^2 and 0.00297 cm^2 for the GCE and CoHCF/Au/SBA-15/GCE, respectively (Figure 8B). The results indicated that the electrode effective surface area was increased obviously after modification of the GCE with CoHCF/Au/SBA-15/GCE, which could enhance the total adsorption capacity of catechol, leading to the increase of current response of catechol, decreasing the limit of detection (Figure 9).

Amperometric determination of CC at CoHCF/Au/SBA-15/GCE

Figure 9 illustrates the real-time amperometric i-t curve of CoHCF/Au/ SBA-15/GCE with successive addition of catechol to a continuously stirred PBS (pH=7.0) solution under optimized experimental conditions. Oxidation current was increased with increasing the catechol concentration. The maximum steady-state current of the sensor was achieved within 4 s. The result indicating that it is a fast

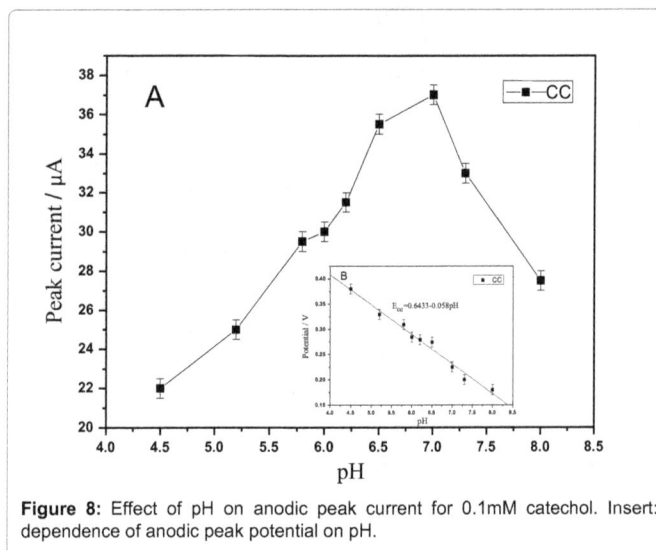

Figure 8: Effect of pH on anodic peak current for 0.1mM catechol. Insert: dependence of anodic peak potential on pH.

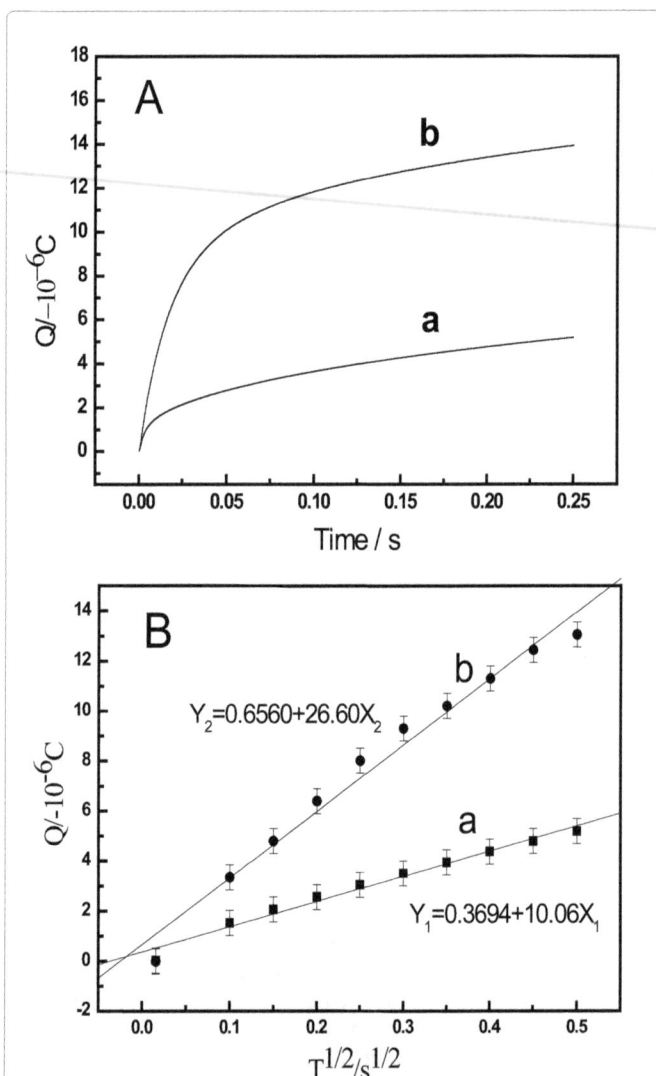

Figure 9: (A) Plot of Q–t curves for the GCE (a) and CoHCF/Au/SBA-15/GCE (b) in 0.1mM K$_3$Fe(CN)$_6$ solution containing 0.1mM KCl. (B)Linear relationship of Q–$t^{1/2}$on the GCE (a) and CoHCF/Au/SBA-15/GCE (b).

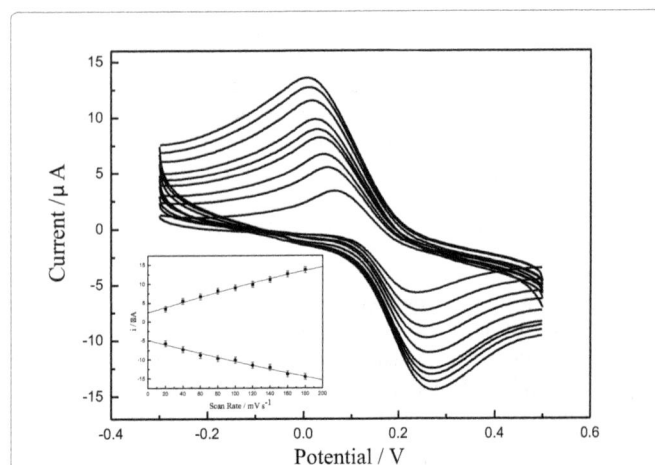

Figure 7: (A) CV curves of 0.1mM catechol in 0.1M PBS (pH=7.0), containing 0.1M KCl at different scan rates: 20, 40, 60, 80, 100, 120, 140, 160, 180 mVs^{-1}. (B) Dependence of peak current on the scan rate.

oxygen reduction reaction. The regression equation of linear current response with catechol concentration is $i(\mu A)=0.44619c+0.00213$ (μM) (R=0.999). The CoHCF/Au/SBA-15/GCE displays a linear response range from 3.0×10^{-7} to 5.1×10^{-5}M of catechol with the detection limit of 5 nM (S/N=3). Conspicuously, the detection limits, sensitivity, and applied potential obtained in the present work exhibited comprehensive superior than other sensors were summarized in Table 1.

Reproducibility durability and selectivity of the modified electrode

Reproducibility of the catechol sensor was investigated with the measurement of the modified electrode in 4.83×10^{-6}M target catechol solution. Three catechol sensors, made independently, showed the response current values of 5.05×10^{-6}A, 4.98×10^{-6}A and 5.17×10^{-6}A with a relative standard deviation of 1.89% (n=3). A RSD 5.19% (n=6) can be obtained in the continuous determination of six times. The catechol in waste water was detemined based on the amperometric i-t curve (Table 2). It indicated that a satisfactory reproducibility could be obtained by this system. The stability of the modified electrode was also tested. No significant changes in cathodic and anodic peaks current were observed for more than 10 complete CV cycles. When a modified electrode was stored in the 0.1M PBS buffer solution (pH=7.0) for at least one week at 4°C, the electrode retained about 95.2% of its initial response.

Interference of coexisting substances and the practical sample analysis

The influences of common interfering species on the determination of catechol were investigated in the presence of 3.0 μM catechol with great details. Some common phenolic complexes and inorganic ions were tested to check their levels of interference in the determination of catechol (Figure 10). Interference was taken as the level causing an error in excess of 5%. It was found that most ions and common substances with high concentration, Na^+, K^+, Fe^{3+}, Zn^{2+}, Cu^{2+}, NO_3^-, Cl^-, SO_4^{2-} and PO_4^{3-}, Ca^{2+}, Zn^{2+}, Mg^{2+}, Pb^{2+} did not interfere on the determination of catechol. 150 fold Lysine, cysteine, glucose citric acid, dopamine, ascorbic acid, uric acid, 100 fold p-aminophenol, phenol, 1-nitroso-2-naphthol and 10 fold ortho-aminophenol produced a negligible change on the response currents of the sensor, indicating that the CoHCF/Au/SBA-15/GCE exhibits good selectivity for catechol detection (Figure 11).

Conclusion

In summary, a novel electrochemical sensor with an excellent current response for the catechol based on CoHCF/Au/SBA-15 film was prepared. The sensor showed a linear relationship of 3.0×10^{-7} to 5.10×10^{-5}M of catechol. Au nanoparticles and cobalt hexacyanoferrate (CoHCF) on the glassy carbon electrode (GCE) obviously improved the response characteristics of the catechol sensor. Results showed that the composite film has promising electrocatalytic activity toward the oxidation of catechol. High sensitivity, good selectivity, low detection limit, good repeatability and anti-interference ability, which have been verified by determination of catechol in waste water, made the proposed sensor show the promising practical application in environmental pollution test.

Acknowledgements

This work was supported by the Project of the National Science Foundation of People's Republic of China (21275100), Shanghai Leading Academic Discipline Project (S30406) and Key Laboratory of Resource Chemistry of Ministry of Education.

Electrode	Methods	Linear range	Detection limit(mol L^{-1})	Reference
SiO$_2$/C/Nb$_2$O$_5$	DPV	3.98×10^{-5}-9.8×10^{-4}	0.8×10^{-6}	[37]
Au/L-lysine/OMC-Au/Tyr/GCE	DPV	4.0×10^{-7}-8.0×10^{-5}	2.5×10^{-8}	[32]
Tyr/CoPc/CGCE	i-t	3.0×10^{-6}-8.63×10^{-4}	4.5×10^{-7}	[38]
CoHCF/Au/SBA-15/GCE	i-t	3.0×10^{-7}-5.10×10^{-5}	5×10^{-8}	This work

Table 1: Comparison of the determination of catechol by different sensing methods.

Sample No	Detection (μmol/L)	Addition (μmol/L)	Found (μmol/L)	Recovery (%)
1	1.0	1.0	1.93	96.5
2	2.0	1.5	3.38	96.6
3	1.5	1.5	2.89	96.3
4	5.0	4.5	9.84	103.6
5	1.7	2.0	3.82	103.2
6	9.0	5.0	13.87	99.1

Table 2: Analysis of catechol in waste water samples (n=6).

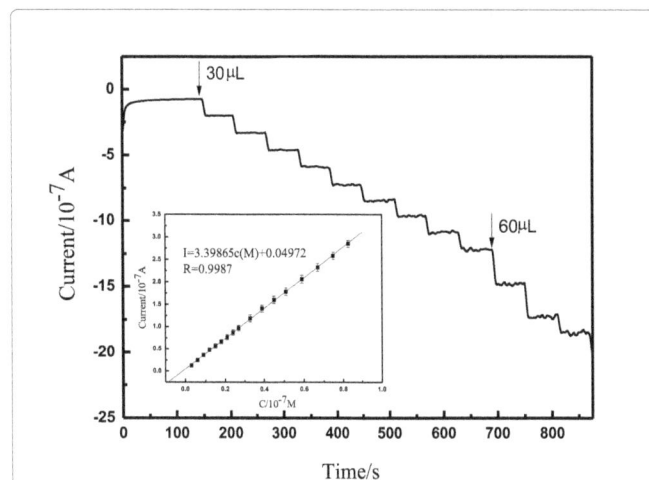

Figure 10: Amperometric response obtained on the CoHCF/Au/SBA-15/GCE upon successive addition of catechol at different concentrations to 0.1M PBS (pH=7.0) with applied potential at 0.15 V. Inset: calibration curves of the catechol sensor.

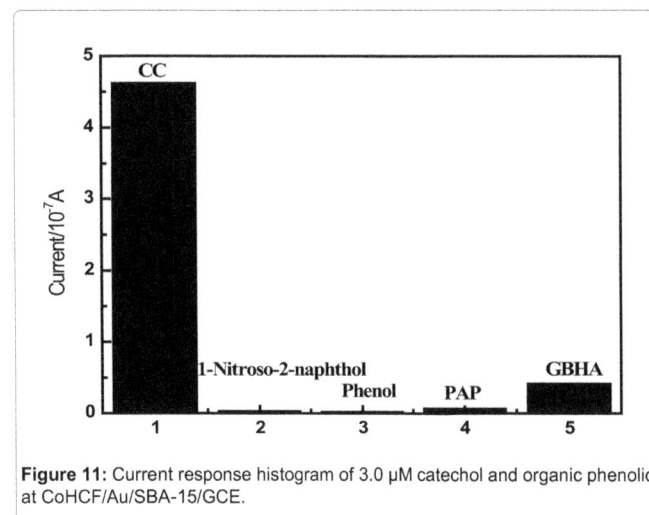

Figure 11: Current response histogram of 3.0 μM catechol and organic phenolics at CoHCF/Au/SBA-15/GCE.

References

1. Yang K, Chen XH, Ni JH, Yao SP, Wang WC, et al. (2010) Palygorskite-expanded graphite electrodes for catalytic electro-oxidation of phenol. Applied Clay Science 49: 64-68.

2. Harnly JM, Bhagwat S, Lin LZ (2007) Profiling methods for the determination of phenolic compounds in foods and dietary supplements. Anal Bioanal Chem 389: 47-61.

3. Irons RD (1985) Quinones as toxic metabolites of benzene. J Toxicol Environ Health 16: 673-678.

4. Bukowska B, Kowalska S (2004) Phenol and catechol induce prehemolytic and hemolytic changes in human erythrocytes. Toxicol Lett 152: 73-84.

5. Du HJ, Ye JS, Zhang JQ, Huang XD, Yu CZ (2011) A voltammetric sensor based on graphene-modified electrode for simultaneous determination of catechol and hydroquinone. Journal of Electroanalytical Chemistry 650: 209-213.

6. Tang L, Zhou Y, Zeng G, Li Z, Liu Y, et al. (2013) A tyrosinase biosensor based on ordered mesoporous carbon-Au/L-lysine/Au nanoparticles for simultaneous determination of hydroquinone and catechol. Analyst 138: 3552-3560.

7. Canevari TC, Arenas LT, Landers R, Custodio R, Gushikem Y (2013) Simultaneous electroanalytical determination of hydroquinone and catechol in the presence of resorcinol at an SiO2/C electrode spin-coated with a thin film of Nb2O5. Analyst 138: 315-324.

8. Carralero V, Mena ML, Gonzalez-cortés A, Pingarron JM (2006) Development of a high analytical performance-tyrosinase biosensor based on a composite graphite-Teflon electrode modified with gold nanoparticles. Biosensors and Bioelectronics 22: 730-736.

9. Liu J, Lin YH, Liang L, Voigt JA, Huber DL, et al. (2003) Templateless Assembly of Molecularly Aligned Conductive Polymer Nanowires: A New Approach for Oriented Nanostructures. Chemistry-A European Journal 9: 604-611.

10. Xun Z, Cai C, Lu T (2004) Effects of a Surfactant on the Electrocatalytic Activity of Cobalt Hexacyanoferrate Modified Glassy Carbon Electrode Towards the Oxidation of Dopamine. Electroanalysis 16: 674-683.

11. Tao WY, Pan DW, Liu YG, Nie LH, Yao SZ (2004) Characterization and electrocatalytic properties of cobalt hexacyanoferrate films immobilized on Au-colloid modified gold electrodes. Journal of Electroanalytical Chemistry 572: 109-117.

12. Haghighi B, Varma S, Alizadeh Sh FM, Yigzaw Y, Gorton L (2004) Prussian blue modified glassy carbon electrodes-study on operational stability and its application as a sucrose biosensor. Talanta 64: 3-12.

13. DeLongchamp DM, Hammond PT (2004) Multiple-Color Electrochromism from Layer-by-Layer-Assembled Polyaniline/Prussian Blue Nanocomposite Thin Films. Chemistry of Materials 16: 4799-4805.

14. Golabi SM, Noor-Mohammadi F (1998) Electrocatalytic oxidation of hydrazine at cobalt hexacyanoferrate- modified glassy carbon, Pt and Au electrodes. Journal of Solid State Electrochemistry 2: 30-37.

15. Cai CX, Xue KH, Xu SM (2000) Electrocatalytic activity of a cobalt hexacyanoferrate modified glassy carbon electrode toward ascorbic acid oxidation. Journal of Electroanalytical Chemistry 486: 111-118.

16. Chen SM (1998) Characterization and electrocatalytic properties of cobalt hexacyanoferrate films. Electrochimica Acta 43: 3359-3369.

17. Cataldi TRI, Benedetto GD, Bianchini A (1999) Enhanced stability and electrocatalytic activity of a ruthenium-modified cobalt–hexacyanoferrate film electrode. Journal of Electroanalytical Chemistry 471: 42-47.

18. Daniel MC, Astruc D (2003) Gold Nanoparticles: Assembly, Supramolecular Chemistry, Quantum-Size-Related Properties, and Applications toward Biology, Catalysis, and Nanotechnology. Chemical Reviews 104: 293-346.

19. Saha K, Agasti SS, Kim C, Li X, Rotello VM (2012) Gold nanoparticles in chemical and biological sensing. Chem Rev 112: 2739-2779.

20. Shipway AN, Willner I (2001) Nanoparticles as structural and functional units in surface-confined architectures. Chem Commun (Camb): 2035-2045.

21. Beck JS, Vartuli JC, Roth WJ, Leonowicz ME, Kresge CT, et al. (1992) A new family of mesoporous molecular sieves prepared with liquid crystal templates. Journal of the American Chemical Society 114: 10834-10843.

22. Zhao D, Feng J, Huo Q, Melosh N, Fredrickson GH, et al. (1998) Triblock copolymer syntheses of mesoporous silica with periodic 50 to 300 angstrom pores. Science 279: 548-552.

23. Zhao DY, Feng JL, Cmelka BF, Stucky GD, Huo QS (1998) Nonionic Triblock and Star Diblock Copolymer and Oligomeric Surfactant Syntheses of Highly Ordered, Hydrothermally Stable, Mesoporous Silica Structures. Journal of the American Chemical Society 120: 6024-6036.

24. Stewart MP, Buriak JM (2000) Chemical and Biological Applications of Porous Silicon Technology. Advanced Materials 12: 859-869.

25. Lillis B, Jungk C (2005) Microporous silicon and biosensor development: structural analysis, electrical characterisation and biocapacity evaluation. Biosensors and Bioelectronics 21: 282-292.

26. Francia GD, Ferrara VL, Manzo S, Chiavarini S (2005) Towards a label-free optical porous silicon DNA sensor. Biosens Bioelectron 21: 661-665.

27. Deere J (2002) Mechanistic and Structural Features of Protein Adsorption onto Mesoporous Silicates. The Journal of Physical Chemistry B 106: 7340-7347.

28. Hartmann M (2005) Ordered Mesoporous Materials for Bioadsorption and Biocatalysis. Chemistry of Materials 17: 4577-4593.

29. Carrott MMLR, Candeias AJE, Carrott PJM, Ravikovitch PI, Neimark AV, et al. (2001) Adsorption of nitrogen, neopentane, n-hexane, benzene and methanol for the evaluation of pore sizes in silica grades of MCM-41. Microporous and Mesoporous Materials 47: 323-337.

30. Sayari A (1996) Catalysis by Crystalline Mesoporous Molecular Sieves. Chemistry of Materials 8: 1840-1852.

31. Yamada T, Zhou HS, Uchida H, Tomita M, Ueno Y, et al. (2002) Surface Photovoltage NO Gas Sensor with Properties Dependent on the Structure of the Self-Ordered Mesoporous Silicate Film. Advanced Materials 14: 812-815.

32. Prabakar SJ, Narayanan SS (2006) Surface modification of amine-functionalised graphite for preparation of cobalt hexacyanoferrate (CoHCF)-modified electrode: an amperometric sensor for determination of butylated hydroxyanisole (BHA). Analytical and Bioanalytical Chemistry 386: 2107-2115.

33. Yang CM, Kalwei M, Schüth F, Chao K (2009) Gold nanoparticles in SBA-15 showing catalytic activity in CO oxidation. Applied catalysis A: General 254: 289-196.

34. Schmidt R, Hansen EW, Stoecker M, Akporiaye D, Ellestad OH (1995) Pore Size Determination of MCM-51 Mesoporous Materials by means of 1H NMR Spectroscopy, N2 adsorption, and HREM. A Preliminary Study. Journal of the American Chemical Society 117: 4049-4056.

35. Du P, Li H, Mei Z, Liu S (2009) Electrochemical DNA biosensor for the detection of DNA hybridization with the amplification of Au nanoparticles and CdS nanoparticles. Bioelectrochemistry 75: 37-43.

36. Abolhasani J, Hosseini H, Khanmiri RH (2014) Electrochemical study and differential pulse voltammetric determination of oxcarbazepine and its main metabolite at a glassy carbon electrode. Analytical Methods 6: 850-856.

37. Heusler KE, Adams RN (1969) Electrochemistry at Solid Electrodes. Erschienen in der Buchreihe "Monographs in Electroanalytical Chemistry and Electrochemistry". Marcel Dekker Inc., New York. 402 Seiten. Berichte der Bunsengesellschaft für physikalische Chemie 73: 1098-1098.

38. Anson FC (1964) Application of Potentiostatic Current Integration to the Study of the Adsorption of Cobalt(III)-(Ethylenedinitrilo(tetraacetate) on Mercury Electrodes. Analytical Chemistry 36: 932-934.

7

Synthesis of Surface Molecularly Imprinting Polymers for Methylphenidate and its Application in Separating Methylphenidate

Mehdi Rajabnia Khansari[1,2], Sara Shahreza[4], Azam Rezvanirad[1], Amin Nikavar[2], Shahrzad Bikloo[3] and Bahareh Sadat Yousefsani[5]

[1]Faculty of Pharmacy, Research Center, Shahid Beheshti University of Medical Sciences, Tehran, Iran

[2]School of Chemical Engineering, Research Center, Iran University of Science and Technology, Tehran, Iran

[3]Lorstan University of Medical Sciences, Research Center, Khoramabad, Iran

[4]Department of Nanobiotechnology, Tarbiat Modates University, Tehran, Iran

[5]Department of Pharmacodynamy and Toxicology, School of Pharmacy, Pharmaceutical Research Center, Mashhad University of Medical Sciences, Mashhad, Iran

Abstract

In this study, a novel approach is proposed for determination of methylphenidate in biological fluids. In this method molecularly imprinted solid-phase extraction (MISPE), as the sample extraction technique, combined with high-performance liquid chromatography (HPLC) is used. The water-compatible molecularly imprinted polymers (MIPs) were prepared using methacrylic acid as functional monomer, ethylene glycol dimethacrylate as cross-linker, Hexane as porogen and methylphenidate as template molecule. Extraction of methylphenidate from human serum was carried out using a novel imprinted polymer as the solid-phase extraction (SPE). Various parameters affecting the extraction efficiency of the polymer were evaluated. Also, the optimal conditions for the MIP cartridges were studied. The limit of detection (LOD) and limit of quantification (LOQ) for methylphenidate in serum samples were 1.3 and 10 ng mL^{-1}, respectively. The recoveries for serum samples were higher than 92%.

Keywords: Molecularly imprinted polymer; Methylphenidate; Pharmaceutical analysis; Solid-phase extraction; Affinity assay; Template polymerization

Introduction

Recently, there have been an increasing interest in potential applications of highly selective molecularly imprinted polymers, MIPs. Especially their applications in analysis of drugs and other compounds in biological and environmental samples. Applicability of Imprinted polymers [1-4] are in various analytical techniques, including liquid chromatography [3,5], capillary electrophoresis, capillary electrochromatography [6], solid-phase extraction [7], and 'immunoassay' [8], have been investigated. An inhere advantage of molecular imprinting, which has extensively been testified by many examples above, is the possibility to synthesize sorbents with selectivity pre-determined for a particular analyte. The fundamental step in this technique is polymerization of functional and cross-linking monomers in the presence of a templating ligand, or imprint species. Subsequent removal of the imprint molecules leaves behind 'memory sites', or imprints, in a solid, highly cross-linked polymer network. The general belief holds that the functional monomers are spatially fixed in the polymer via their interaction with the imprint species during the polymerization reaction.

Attention deficit hyperactivity disorder (ADHD) is a common neurobehavioral disorder in childhood, which is estimated to strike up to 10% of the general population [9,10]. Methylphenidate (MPH) is a psychostimulant drug approved primarily for the treatment of attention deficit hyperactivity disorder (ADHD) and narcolepsy [11]. This drug fits to the piperidine class of compounds and increases the levels of dopamine and noradrenaline in the brain through reuptake inhibition of the monoamine transporters [11]. The main urinary metabolite is a de-esterified product, ritalinic acid (RA), which accounts for 80% of the dosage and has a half-life of about 8 h [11]. Reviews over pharmacy databases and treatment studies have shown that the incidences of medication discontinuation or non-adherence is between 13.2% and 64% [12]. The clinical laboratory has an important role in being able to detect MPH in serum and its metabolite RA in urine.

Various analytical methods such as immunoassay [13], HPLC with UV detection [14] and more recently liquid chromatography–electrospray ionization mass spectrometry [15-17], have been proposed for measuring MPH (in serum).

This study was meant to develop and validate a novel HPLC–SPE method with samples throughput for determinations of MPH in human serum. In these applications solid phase extraction (SPE) sample preparation procedures were used. The method outline with a selective chromatography in combination with specific UV detection fulfills the high quality standard required for accurate determinations in serum samples from human.

Experimental

Materials

MPH (Figure 1) was purchased from Cerillant (Texas, USA) as 1 mg/mL solutions in methanol and RA (Figure 1) from Sigma-Aldrich (Australia). HPLC grade acetonitrile was purchased from Thermo Fisher (Cambridge, UK), ammonium formate and dimethyloctylamine (DMOA) from Sigma-Aldrich (Germany), and HPLC grade methanol and reagent grade 89-91% pure formic acid from BDH (Poole, UK).

*Corresponding author: Mehdi Rajabnia Khansari, Research Center, Faculty of Pharmacy, Shahid Beheshti University of Medical Sciences, Tehran, Iran, E-mail: rajabniamahdi@yahoo.com

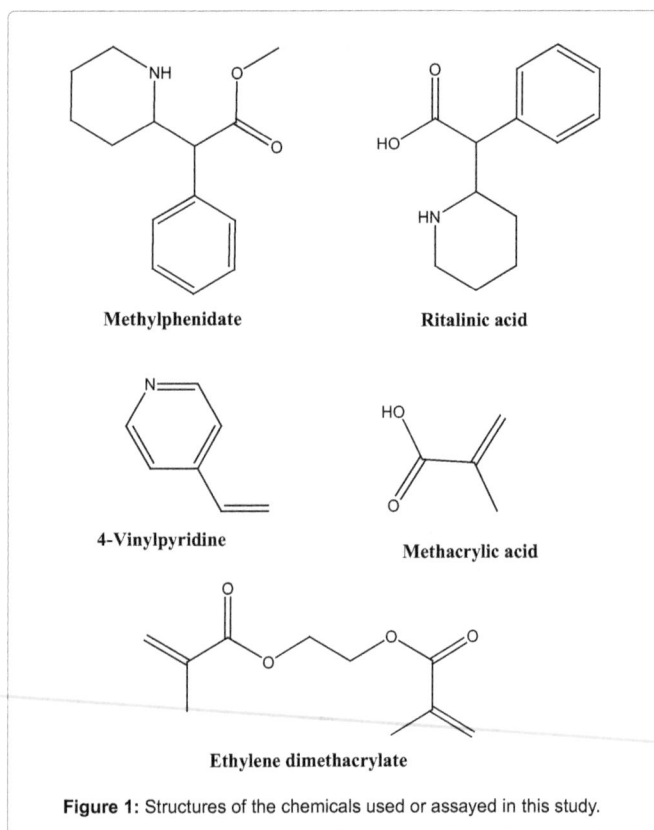

Figure 1: Structures of the chemicals used or assayed in this study.

Methacrylic acid (MAA), 4-vinyl pyridine (4VP) and ethylene glycol dimethacrylate (EDMA) were obtained from Sigma-Aldrich (Milwaukee, USA). 2,2⊠-Azobis-iso-butyronitrile (AIBN) was obtained from Acros (Geel, Belgium). All solvents used [acetonitrile (ACN), tetrahydrofuran (THF), hexane, acetone, methanol, acetic acid and Trifluoroacetic acid (TFA)] were of HPLC grade.

Instrumentation and analytical conditions

The HPLC system consisted of an isocratic HPLC pump (Model 590, Waters), an autoinjector (SIL-10A, Shimadzu) fitted with an injection loop of 200 µl and was directly connected to the chiral AGP column (150 × 4.0 mm id.). Detection of analyte was achieved using a variable wavelength UV detector (Model 481, Waters) set at 220 nm. A reversed-phase mode was used, and the mobile phase consisted of 0.4% acetic acid containing 0.1% DMOA, pH 3.4, which DMOA was applied as an organic modifier. The mobile phase was degassed before use. The flow-rate was 1 ml/min and an ambient column temperature was used. AGP columns are all stable for at least 3 months.

Standards

A standard stock solution of MPH was prepared by dissolving 1 mL of 1 mg/mL commercial standard in 50 mL methanol to give a final concentration of 20 mg/L.

The serum calibration curves for MPH were plotted by spiking drug-free human serum with standard solutions at concentrations of 0.5, 5, 10, 15, 20 µg/mL, giving a calibration range of 0.5-20 µg/mL for both MPH.

Synthesis of polymers

A non-covalent approach was used for preparation of MIPs.

MPH as the template and MAA, 4VP as the functional monomer, were dissolved in 10 mL of organic solvent (hexane or acetone) in a screw-capped glass tube and kept at 4°C for 60 min. Then, EDMA as the cross-linker, and AIBN as the initiator, were added. The mixture was sparged, in an ice bath, with oxygen-free nitrogen for 20 min and heated at 60°C for 22 h to complete radical polymerization. The resultant bulk rigid polymers were crushed, grounded into powder and sieved through a 200-mesh stainless steel sieve (particle size less than 75 µm). The polymer particles were washed with a methanol-acetic acid (60:40, v/v) mixture, centrifuged at 4000 rpm for 5 min and supernatant was analyzed by HPLC. The washing was continued until no MPH or other compound was detected in supernatant. Blank non-imprinted polymers (NIPs) were prepared, in the absence of MPH, under the same condition described above. Imprinted (MIP) and non-imprinted (Blank) polymers prepared and examined in this study are presented in Table 1.

Batch adsorption procedure

The recognition ability of both MIP and Blank polymer was examined by batch rebinding experiments. Dry polymer (10 mg) was incubated in 2 mL ACN with MPH and was shaken for 24 h at room temperature. The solution was centrifuged (3000 rpm for 10 min) and supernatant was analyzed by HPLC. The amount of bound MPH was calculated from the difference between initial and final concentrations in solution. Each test was carried out four times and Mean ± SD was reported.

The imprinting factor (IF) was calculated according to Eq. (1):

Eq. (1): $K = \dfrac{K\,mip}{K\,blank}$
(1)

MISPE procedure

25 mg of polymer (MIP-A1 or NIP), in 3 mL ACN, was slurry packed into an empty polypropylene SPE cartridge. The column was washed and conditioned with 8 mL loading solution. MPH (10 µg) in 200 µL water was loaded into the column. Washing solvent was then passed through the column five times. Finally, 5 mL methanol–TFA (50:50, v/v) was applied to perform complete extraction of MPH. In order to find a solvent with maximum selectivity for MPH, Methanol and Acetic acid were evaluated as washing solvents. The loading, washing and eluting fractions were analyzed by HPLC.

Extraction of MPH from human serum samples

25 mg of MIP-A1 suspended in 3 mL ACN was packed into a polypropylene cartridge. The column was washed and conditioned by 5 mL of methanol-TFA (50:50, v/v) and 4 mL of CAN, respectively. 800 µL of ACN was added to 150 µL of serum in order to precipitate the serum proteins. After centrifugation (3000 rpm for 8 min), 1 mL deionized water was added to 0.5 mL supernatant and the mixture was

Polymer	Template	Functional Monomer	Cross linker	Molar Ratio	Solvent
	MPH	MAA	EDMA	MAA/MPH	Hexane
Blank	-	2.5 mmol	10 mmol	-	5 ml
MIP- A1	0.5 mmol	2.5 mmol	10 mmol	5	5 ml
MIP- A2	0.25 mmol	2.5 mmol	10 mmol	10	5 ml
		4-VP		4-VP/MPH	Acetone
Blank	-	2.5 mmol	10 mmol	-	5 ml
MIP- B1	0.5 mmol	2.5 mmol	10 mmol	5	5 ml
MIP- B2	0.25 mmol	2.5 mmol	10 mmol	10	5 ml

Table 1: MIPs and Blank polymers preparation protocol.

loaded onto the column. Acetic acid (3 mL) was percolated through the column for selective washing and finally MPH was eluted with 8 mL methanol-TFA (50:50, v/v). The solvent was dried under a stream of nitrogen. Later, the residue was dissolved in 100 μL mobile phase and the concentration of MPH was determined by HPLC.

Results and Discussion

Choice of functional monomer and solvent

As molecular recognition of the template molecule by imprinted polymers is based on the intermolecular interactions between the template molecule and functional groups of the polymer [18], choosing a proper functional monomer is of great importance for creating a strong monomer-template comple [19]. The functional monomer used in this study was MAA, 4-VP. However, because of the amine groups in MPH structure, it easily binds to this acidic monomer. The ionic interaction between MAA and MPH makes the MIP suitable for MISPE procedure in aqueous conditions. Solvent has an important effect on conjuncture of functional monomers with the template. Before and during polymerization, the polarity of the solvent affects the extent of non-covalent pre-polymer complex. Less polar solvents such as chloroform or hexane increase complex formation, facilitating polar non-covalent interactions such as hydrogen bonding. On the other hand, more polar solvents tend to dissociate the non-covalent interactions in the pre-polymer complex, especially protic solvents which leads to a high degree of disruption to hydrogen bonds [20-22]. In this work, two organic solvents (hexane and acetic acid) were tested as the polymerization solvents for optimization of molecular imprinting procedure (Figure 2).

Batch adsorption measurements

A best method for evaluating the binding sites in MIPs is batch adsorption test. Batch adsorption involves analysis of an MIP in a solution of substrate [19]. Therefore, in this research binding properties of polymers were studied in a conventional batch adsorption method [9,20]. Two MAA/MPH ratios (5 and 10) in two polymerization solvents

(hexane and acetone) were used for optimization of MPH imprinting condition. The results of binding assay showed that MIP-A1 was the optimized imprinted polymer (Figure 3). The calculated imprinting factor (IF) values of MIPs are shown in Table 2. In comparison with other polymers, the highest affinity of MIP-A1 for MPH and its highest IF, indicates the superiority of binding properties of MIP-A1 for MPH. As the polarity of hexane is less than acetone, more MPH-MAA interaction is possible in hexane compared to acetone. Based on this fact, the optimized MIP was prepared in hexane. The optimized MAA/MPH ratio was 5 and the best polymerization solvent was hexane. It has been proven, in other studies, that the best MIPs, with selective binding properties, have been prepared in less polar organic solvents with a template/monomer ratio of less than 1 [2,6]. Therefore, MIP-A1, as the optimized polymer, was used for Scatchard analysis and selective extraction of MPH from human serum.

Scatchard analysis

Binding of MPH to MIP-A1 and its blank polymer (Blank-A) was studied at different concentrations (Figure 4). The data showed that MPH binding to MIP-A1 was significantly more than Blank-A at all concentrations. From the Scatchard plot (Figure 5) one dissociation constants could be discerned, one was representing high affinity binding sites with a KD of 8 μM, and Bmax of 6.38 μmol g^{-1} polymers. The binding sites of MIPs prepared by non-covalent bulk polymerization are usually heterogeneous. The KD value in other studies ranged from μM to M (lamotrigine KD=16.2 mM [23], sulphametoxazole KD=18.8 mM [24], theophylline KD=1.5 M [25].

MISPE

This study was aimed at determination of MPH in human serum.

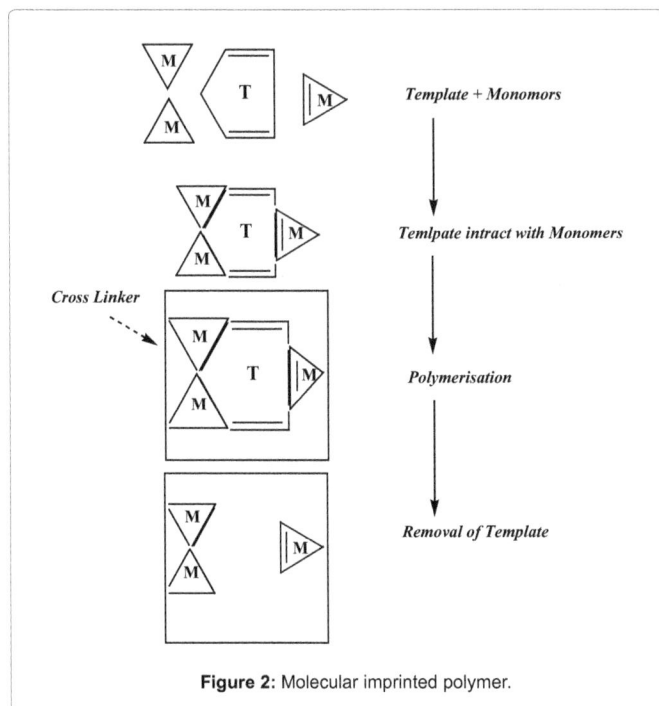

Figure 2: Molecular imprinted polymer.

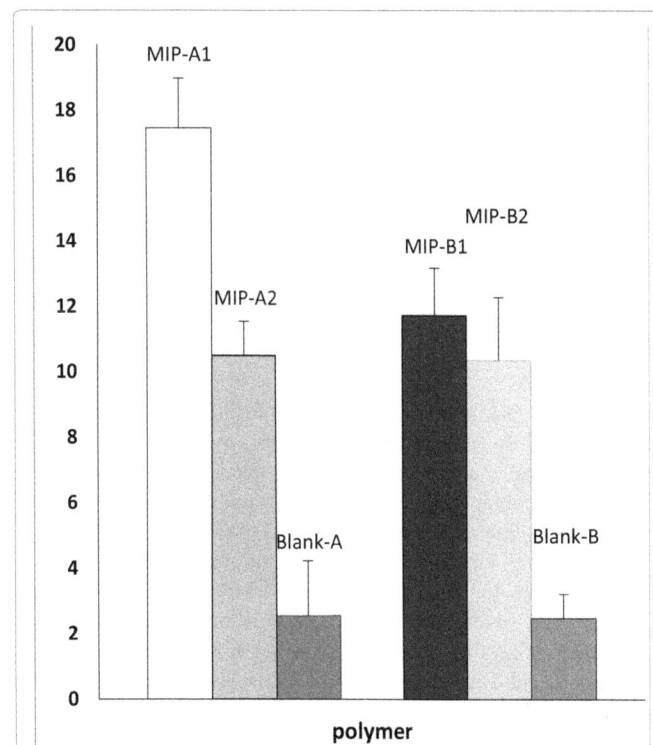

Figure 3: Binding of MPH to 5 mg of the imprinted (MIPs) and non-imprinted (Blank) polymers in ACN (n=4). Each point represents mean ± SEM; MPH concentration in ACN was 20 μg mL^{-1}.

Imprinted polymer	IF=(K_{MIP}/K_{NIP})
MIP-A1	60.3
MIP-A2	18.6
MIP-B1	23.8
MIP-B2	9.1

Table 2: Imprinting factor (IF) of each imprinted polymer for methylphenidate (Ritalin).

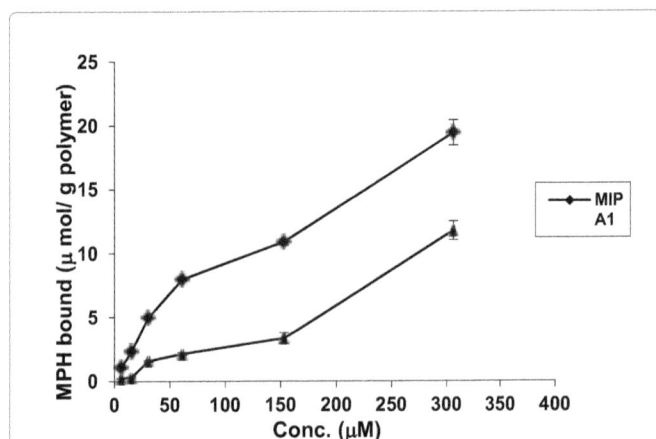

Figure 4: Adsorption isotherm of MIP-A1 and Blank-A using batch adsorption test (n=4). Each point represents mean ± SEM. Experiment conditions: 10 mg of polymer was incubated in 2 mL ACN with different concentrations of MPH for 24 h at room temperature.

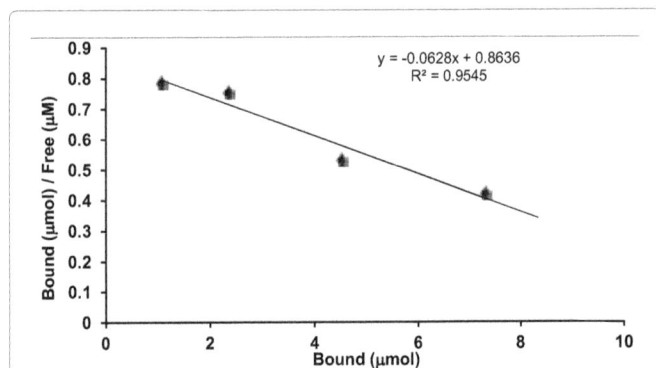

Figure 5: Scatchard plot of the binding of MPH to the imprinted polymer (MIP-A1). Bound is the amount of donepezil bound to MIP-A1 and free is the concentration of free MPH at equilibrium.

To fulfill this goal, water was selected as loading solvent in MISPE procedure [18]. The real challenge in MISPE is to exploit the selectivity of the MIP in aqueous media. The washing step is in fact the key factor for the development of specific interactions with the MIP. The aim of this step is finding a solvent which yields maximum selectivity and recovery of MPH. Two solvents (methanol and acetic acid) were applied in washing step. The percentage of washed MPH relative to the total loaded amount was calculated in each fraction and cumulative recovery was plotted against the volume of washing solvent (Figure 6). Although 5 mL acetic acid could disrupt 80% of non-specific bindings of MPH to Blank-B polymer, it could also wash 6.2% of drug from MIP-A1. Whereas 89% of MPH was removed from Blank-A column which is 14 times more than MIP-A1 polymer. Therefore acetic acid was selected as washing solvent. 3 mL acetic acid could be used to wash the polymer column after loading, without removing considerable amount of MPH

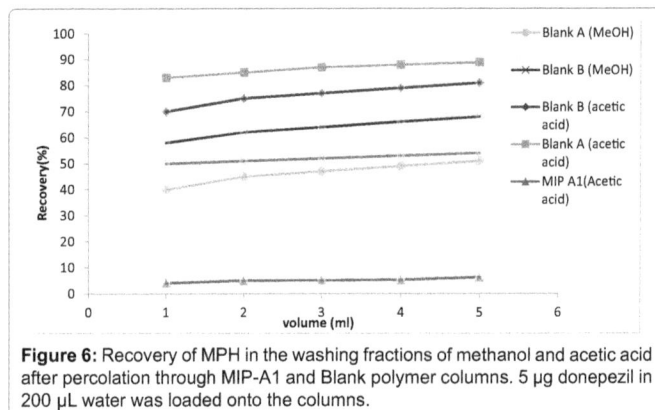

Figure 6: Recovery of MPH in the washing fractions of methanol and acetic acid after percolation through MIP-A1 and Blank polymer columns. 5 μg donepezil in 200 μL water was loaded onto the columns.

from MIP-A1 cartridge. Acidic solutions are usually used for complete washing of template in elution step from the MIP column [23]. In this study, 8 mL of methanol–TFA (50:50, v/v) was used as eluting solvent.

Extraction of MPH from serum samples

Serum samples with or without MPH was loaded onto the MIP-A1 column and MISPE procedure was carried out according to the method described in Section 2.6. The calibration curve of HPLC of spiked serum samples with known concentrations of MPH was established in the range of 0.5-20 μg mL^{-1} (y=98.54x+12.664, R^2=0.9992). The standard curve of MPH was also plotted in the same range (y=106x+17.179, R^2=0.9998) (Figure 7). The recovery of MPH in this MISPE process was calculated as 92%. The limit of detection (LOD) of assay (signal/noise ratio of 3) was found to be 1.3 ng mL^{-1}. The limit of quantification (LOQ) of assay (signal/noise ratio of 20) was 10 ng.mL^{-1} which was much lower than the minimum therapeutic concentration of MPH in patient serum. LOD and LOQ values obtained in other studies (with similar MISPE method and HPLC assay carried out for other templates) ranged from ng mL^{-1} to μg mL^{-1} [14,21]. Also, LOD and LOQ values of HPLC assay of MPH in human serum, determined by other researchers, were less than 1 ng mL^{-1} and 50 ng mL^{-1}, respectively [4]. The intra-day and inter-day precision values for MPH concentration of 0.8 μg mL^{-1} were 1.88 and 3.3%, respectively. In both cases the precision was calculated as the relative standard deviation of the spiked serum samples (n=4) in one day (intra-day precision) and four different days (inter-day precision).

Optimization of MISPE protocol

In the next step, the selectivity of MISPE, in washing step, for methylphenidate was evaluated in presence of other drugs in aqueous solution. After loading 1 mL of an aqueous mixture of sertraline, alprazolam, phenobarbital and methylphenidate onto the MIP-A1 or Blank-A column, the cartridge was washed with 3 mL ACN. The data showed that 90-100% of sertraline, alprazolam and phenobarbital were removed from MIP-A1 and Blank-A polymer. Also, 77% of methylphenidate was washed from Blank-A, while only about 4% of methylphenidate was removed from MIP-A1 column (Figure 8). This indicated a significantly higher affinity of MIP-A1 for methylphenidate in comparison with other drugs that could be present with methylphenidate, simultaneously, in serum of patients. We have also used ACN, in our previous studies, as washing solvent for selective extraction of donepezil in a MISPE method from their aqueous mixture [21]. Thus, optimized MISPE conditions were as follows: washing conditions, 3 mL acetic acid; elution conditions, 8 mL methanol-TFA (50:50, v/v).

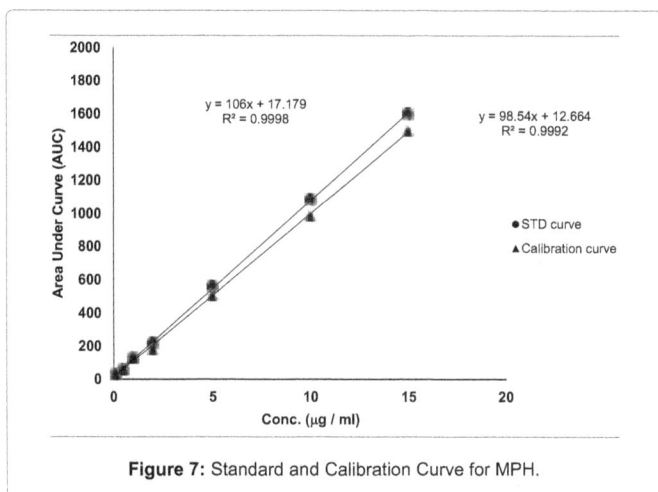

Figure 7: Standard and Calibration Curve for MPH.

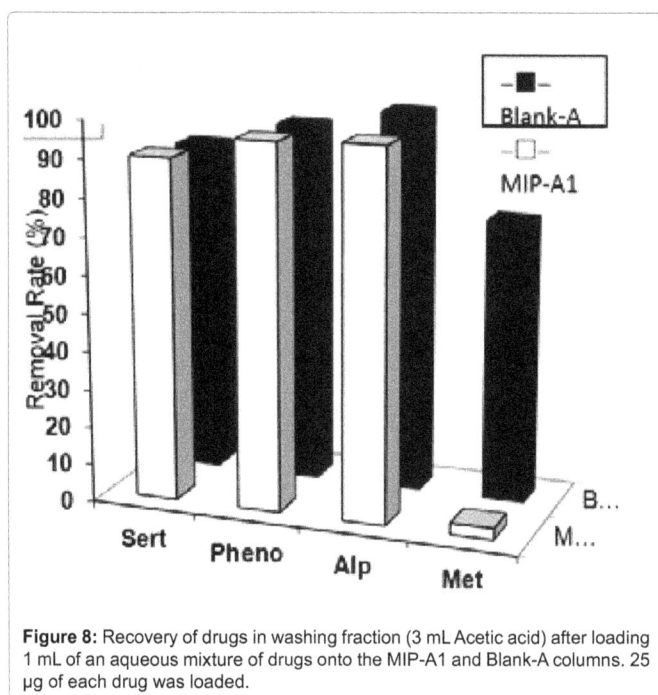

Figure 8: Recovery of drugs in washing fraction (3 mL Acetic acid) after loading 1 mL of an aqueous mixture of drugs onto the MIP-A1 and Blank-A columns. 25 µg of each drug was loaded.

Conclusion

In this paper for the first time, a novel MPH MIP is prepared by bulk polymerization. The MPH MIP shows higher molecular recognition than NIP on chromatographic evaluation. A SPE-HPLC method based on MIP is developed for the extraction of MPH from aqueous solutions. Furthermore, the MIP particles as new sorbents in SPE are successfully investigated for the extraction of human serum samples with an optimized procedure. This efficient method allows extracts with higher purity to be obtained and interfering peaks arising from complicated biologic samples to be suppressed. The method was applied to the trace MPH determination at three levels, and the recoveries for the spiked human serum samples were higher than 92%. It could be concluded that the proposed technique has a great potential in developing selective extraction method for other compounds too.

Acknowledgements

We gratefully acknowledge Research Center, School of Chemical Engineering, Iran University of Science and Technology and Research Center, Faculty of Pharmacy, Shahid Beheshti University of Medical Sciences, Tehran, Iran. The work was financially supported by research grant from the Shahid Beheshti University of Medical Sciences, Deputy of Research.

References

1. Olesen OV, Thomsen K, Jensen PN, Wulff CH, Rasmussen NA, et al. (1995) Clozapine serum levels and side effects during steady state treatment of schizophrenic patients: a cross-sectional study. Psychopharmacology (Berl). J Christensen and R Rosenberg 117: 371-378.

2. Ramstrom O, Yu C, Mosbach K (1996) Chiral recognition in adrenergic receptor binding mimics prepared by molecular imprinting. J Mol Recognit 9: 691-696.

3. Ansell RJO, Ramstrom R, Mosbach K (1996) Towards artificial antibodies prepared by molecular imprinting. Clin Chem 42: 1506-1512.

4. Steinke JHG, Dunkin IR, Sherrington DC (1999) A simple polymerisable carboxylic acid receptor: 2-acrylamido pyridine. TrAC Trends in Analytical Chemistry 18: 159-164.

5. Kempe M, Mosbach K (1995) Separation of amino acids, peptides and proteins on molecularly imprinted stationary phases. J Chromatogr A 691: 317-323.

6. Schweitz L, Andersson LI, Nilsson S (2002) Molecularly imprinted CEC sorbents: investigations into polymer preparation and electrolyte composition. Analyst 127: 22-28.

7. Olsen JP, Martin M, Wilson ID (1998) Molecular imprints as sorbents for solid phase extraction: potential and applications. Analytical Communications 35: 13H-14H.

8. Andersson LI (2000) Molecular imprinting for drug bioanalysis: a review on the application of imprinted polymers to solid-phase extraction and binding assay. Journal of Chromatography B: Biomedical Sciences and Applications 739: 163-173.

9. Cantwell DP (1996) Attention deficit disorder: a review of the past 10 years. Journal of the American Academy of Child & Adolescent Psychiatry 35: 978-987.

10. Rösler M, Casas M, Konofal E, Buitelaar J (2010) Attention deficit hyperactivity disorder in adults. The World Journal of Biological Psychiatry 11: 684-698.

11. Pae CU, Marks DM, Masand PS, Peindl K, Hooper-Wood C, et al. (2009) Methylphenidate extended release (OROS MPH) for the treatment of antidepressant-related sexual dysfunction in patients with treatment-resistant depression: results from a 4-week, double-blind, placebo-controlled trial. Clinical Neuropharmacology 32: 85-88.

12. Adler LD, Nierenberg AA (2010) Review of medication adherence in children and adults with ADHD. Postgraduate Medicine 122: 184-191.

13. Seçilir A, Schrier L, Bijleveld YA, Toersche JH, Jorjani S, et al. (2013) Determination of methylphenidate in plasma and saliva by liquid chromatography/tandem mass spectrometry. Journal of Chromatography B 923: 22-28.

14. Zhang J, Deng Y, Fang J, McKay G (2003) Enantioselective analysis of ritalinic acids in biological samples by using a protein-based chiral stationary phase. Pharmaceutical Research 20: 1881-1884.

15. Marchei EJ, Munoz Garcia-Algar O, Pellegrini M, Vall O, Zuccaro P, at al. (2008) Development and validation of a liquid chromatography–mass spectrometry assay for hair analysis of methylphenidate. Forensic Science International 176: 42-46.

16. Marchei E, Farrè Pellegrini M, Rossi S, García-Algar Ó, Vall O, et al. (2009) Liquid chromatography–electrospray ionization mass spectrometry determination of methylphenidate and ritalinic acid in conventional and non-conventional biological matrices. Journal of Pharmaceutical and Biomedical Analysis 49: 434-439.

17. Letzel MK, Weiss Schüssler W, Sengl M (2010) Occurrence and fate of the human pharmaceutical metabolite ritalinic acid in the aquatic system. Chemosphere 81: 1416-1422.

18. Yang J, Hu Y, Cai JB, Zhu XL, Su QD, et al. (2007) Selective hair analysis of nicotine by molecular imprinted solid-phase extraction: an application for evaluating tobacco smoke exposure. Food Chem Toxicol 45: 896-903.

19. Alizadeh T, Zare M, Ganjali MR, Norouzi P, Tavana B (2010) A new molecularly imprinted polymer (MIP)-based electrochemical sensor for monitoring 2, 4, 6-trinitrotoluene (TNT) in natural waters and soil samples. Biosensors and Bioelectronics 25: 1166-1172.

20. Azodi-Deilami S, Abdouss M, Hasani SA (2010) Preparation and utilization of a molecularly imprinted polymer for solid phase extraction of tramadol. Central European Journal of Chemistry 8: 861-869.

21. Khansari MR, Bikloo S, Shahreza S (2016) Determination of donepezil in serum samples using molecularly imprinted polymer nanoparticles followed by high-performance liquid chromatography with ultraviolet detection. Journal of Separation Science.

22. Javidi JM, Esmaeilpour E, Khansari MR (2015) Synthesis, characterization and application of core-shell magnetic molecularly imprinted polymers for selective recognition of clozapine from human serum. RSC Advances 5: 73268-73278.

23. Mohajeri SA, Ebrahimi SA (2008) Preparation and characterization of a lamotrigine imprinted polymer and its application for drug assay in human serum. J Sep Sci 31: 3595-3602.

24. Huamin Q, Lulu F, Li X, Li L, Min S, et al. (2013) Determination sulfamethoxazole based chemiluminescence and chitosan/graphene oxide-molecularly imprinted polymers. Carbohydr Polym 92: 394-399.

25. Sun HW, Qiao FX, Liu GY (2006) Characteristic of theophylline imprinted monolithic column and its application for determination of xanthine derivatives caffeine and theophylline in green tea. J Chromatogr A 1134: 194-200.

8

LC-MS Method for the Quantitation of Two Monoclonal Antibodies by Multiple Signature Peptides in Monkey Serum

Rita Mastroianni*, Marina Feroggio, Barbara Marsiglia, Clarissa Porzio Vernino, Simona Riva and Luca Barbero*

QPD - NBE Bioanalytics, RBM-Merck Serono, Via Ribes 1, 10010, Colleretto Giacosa (TO), Italy

Abstract

Evaluation of the *in-vivo* concentration of monoclonal antibody (mAb) mixtures is a challenging task. Here we report the application of an LC-MS bioanalytical method to quantify in monkey serum the Sym004, an equimolar mixture of two monoclonal antibodies, 992 mAb and 1024 mAb. This method has been assessed accordingly to industry standards and it is based on the determination of two specific signature peptides that report the single mAbs concentrations and on another one peptide, common to the two mAbs, that measures the total concentration of the two target proteins. It is shown that the total concentration is in agreement with the sum of the two measured single concentrations in spiked monkey serum samples. The consistency of the results will allow monitoring of the metabolic fate of different parts of the mAbs, at least in the central body compartment. This can then help to rationalize the design of the protein therapeutics modulating their stability accordingly.

Keywords: Monoclonal antibody; Quantitation; Bioanalysis; Sym004; LC-MS; EGFR; Catabolism

Abbreviations

EGFR: Epidermal Growth Factor Receptor; sEGFR: Extracellular EGFR; mAb: Monoclonal Antibody; ADA: Anti-Drug Antibodies; LC-MS: Liquid Chromatography Mass Spectrometry; IAM: Iodoacetamide; DTT: D-L Dithiothreitol, BIAS%: Percent Relative Error; CoA: Certificate of Analysis; CV%: Percent Coefficient of Variation; IS: Internal Standard; LLOQ: Lower Limit of Quantitation; QC: Quality Control; RS: Reconstitution Solvent; STD: Calibration Standard; ULOQ: Upper Limit of Quantitation; UPLC-MS/MS: Ultra Performance Liquid Chromatography - Tandem Mass Spectrometry; VS(L,M,H): Validation Sample (Low, Medium, High); S/N: Signal-to-Noise ratio; MRM: Multiple Reaction Monitoring; MCX: Mixed Mode Cation Exchange; SD: Standard Deviation

Introduction

Modification of the epidermal growth factor receptor (EGFR; ErbB1) pathway system has been reported to correlate with human malignancies. Increase in ligand production, receptor over-expression, receptor mutations, and/or cross-talk with other receptor systems are the most frequent modifications involved [1-3].

These changes have been linked to the development and maintenance of a malignant phenotype and correlated to poor clinical prognosis [4]. For this reason, the EGFR is an attractive target for anticancer therapy [5].

To date, four EGFR targeting agents (cetuximab, panitumumab, gefitinib and erlotinib) from two distinct drug classes have received FDA approval [6]. These include mAbs directed against the extracellular ligand-binding domain of EGFR and small molecule tyrosine kinase inhibitors (TKIs) directed against the cytosolic catalytic domain of the EGFR.

Although selected patients receive clear benefit from anti-EGFR mAbs, overall single agent response rates are in the order of 10% [5].

When two mAbs against distinct receptor epitopes are combined, rapid and more efficient receptor internalization is observed, followed by EGFR degradation [7]. The mixed antibody treatment is also more effective than single Abs in inhibiting signaling and tumor growth in tissue culture and animal models [8,9].

Sym004 is a recombinant antibody equimolar mixture of a pair of mouse-human chimeric immunoglobulin G1 monoclonal antibodies, 992 mAb and 1024 mAb. Both antibodies have activity against the epidermal growth factor receptor (EGFR) and bind specifically to two distinct non-overlapping epitopes on the extracellular domain III of the EGFR [10,11].

Unlike other anti-EGFR mAbs, Sym004 induces pronounced internalization and degradation of the EGFR, thereby leading to removal of EGFR from the cell surface. This novel mechanism of action is believed to result in superior anti-cancer activity compared to other anti-EGFR mAbs, specifically if resistance or failure to anti-EGFR mAbs is conferred by the presence of high affinity ligands, receptor cross-talk or constitutively activated EGFR. This has been demonstrated both *in-vitro* using human cancer cell lines and *in-vivo* using EGFR-dependent tumor xenografts [11].

Advanced preclinical development of such mAbs mixtures requires the determination of the single PK profiles of each component independently. From this analysis it should be possible to understand how the relative distribution ratio of mAb992 and mAb1024 in the central compartment is modified or maintained, to correlate it to any safety issues, and finally to grasp further pharmacology insights into the mode of action of mixture itself [12]. It is clear that, given mAbs sequence similarities and any binding partners already present in the

***Corresponding authors:** Rita Mastroianni, QPD-NBE Bioanalytics, RBM-Merck Serono, Via Ribes 1, 10010, Colleretto Giacosa (TO), Italy
E-mail: rita.mastroianni@merckgroup.com

Luca Barbero, QPD-NBE Bioanalytics, RBM-Merck Serono, Via Ribes 1, 10010, Colleretto Giacosa (TO), Italy, E-mail: luca.barbero@merckgroup.com

serum or that may appear (e.g. ADA) during an *in-vivo* assessment, the bioanalytical method needs more stringent selectivity and sensitivity.

LC-MS methods can fulfill such requirements at different levels of specificity and sensitivity depending on their access to more sophisticated MS technology and to their inherent technological limitations. Three main approaches are used to detect and quantify a protein drug in a complex matrix by LC-MS: top-down, middle-down and bottom-up approaches.

In the top-down approach, the intact molecule is detected and quantified as it is. In this case the highest degree of possible specificity is achieved, when a High Resolution High Accuracy MS device is deployed. On the one hand, the direct determination of the MW can immediately determine whether the molecule itself, or the formulation composition in term of active drugs, has been modified. On the other hand, limitations to this approach are intrinsically linked to ionization techniques (ESI remains the most commonly used ionization mode in LC-MS) that produce many charge states from a single molecule, different glycosylation forms or post-translational modifications that further spread the drug(s) signal over a wide range of m/z units. This impacts the overall MS sensitivity and increases the need for MS accuracy and resolution. Finally, the presence of natural ligands has to be taken into account, since they can increase the complexity of the extraction procedure from the matrix and further reduce the overall sensitivity.

The other two approaches, i.e. middle-down and bottom-up based protocols, involve a progressively extensive cleavage step of the drug(s) to be quantified, i.e. the molecular weight of signature peptides will be above 3 kDa or below 3 kDa respectively. They are less demanding in terms of sample preparation and MS sensitivity, since the number of multiple charge states decreases with the dimension of the molecule to be quantified. This explains why they are extensively used in pharmacokinetic assessments of large protein therapeutics. However, their main drawback stems from the fact that these methods rely on the quantification of a surrogate molecule and not of the whole drug. In fact, any metabolic or elimination effects that impact differently on different parts of the drug or on the drugs present in the formulation administered, could jeopardize the assumption of a direct relation between the concentration of the signature peptide(s) and actual concentration of the drug(s).

To mitigate this problem we are presenting a LC-MS bottom-up method that allows quantitation of two signature peptides specific for each mAb and a third signature peptide that is common to both the mAbs but located on a different part of the mAb molecules, in a single run. Comparison of the concentrations of the three peptides further enhance the LC-MS intrinsic selectivity and provide further insight into the degradation status of the two molecules.

Materials and Methods

In-silico analyses

The identification of the signature peptide candidates consisted of a three-step process. In the first step, peptides derived from enzymatic cleavages by different enzymes were obtained by the ExPASy PeptideCutter tool [13] (http://web.expasy.org/peptide_cutter/). In the second step, the peptides were aligned by using the LALIGN algorithm (http://embnet.vital-it.ch/software/LALIGN_form.html), to obtain paired specific sequences from each mAb.

In the third and last step, the peptide pairs were checked for matrix interference by the BLAST® (blastp suite) (http://blast.ncbi.nlm.nih.gov/Blast.cgi?PROGRAM=blastp&BLAST_PROGRAMS=deltaBlast&PROG_DEFAULTS=on&PAGE_TYPE=BlastSearch&LINK_LOC=BlastHomeAd) with Database set to "Non-redundant protein sequences" and Organism set to "*Macaca fascicularis* (taxid:9541)". Only the specific peptide pairs that passed the last two steps were further evaluated by LC-MS to become signature peptides. A further analysis for checking any matrix interference in human serum was performed with the same tool and settings, but with Organism set to "Homo sapiens (taxid: 9606)".

A similar workflow was used in the setup to select common peptides to the two mAbs, but still specific enough to be quantified in monkey serum. These peptides provide the total concentration of the two mAbs.

Average MW and isoelectric points (pI) were calculated by the ProtParam tool (http://web.expasy.org/protparam/) [13].

Proteins, chemicals and reagents

Monoclonal antibody, 992 mAb and 1024 mAb, were provided by CMC Biologics A/S (Denmark). Labeled Peptide Internal Standards - Stable isotope-labeled amino acids, $[^{13}C_6\ ^{15}N]$ Leucine and $[^{13}C_5\ ^{15}N]$ Valine (> 97% by HPLC assay) were purchased from Bachem (Bubendorf, Switzerland). Acetonitrile (LiChrosolv®, Reag. Ph Eur, gradient grade for liquid chromatography), 2-propanol (LiChrosolv®, gradient grade for liquid chromatography), Methanol (LiChrosolv®, Reag. Ph Eur, gradient grade for liquid chromatography), Formic Acid (Emsure® ACS, Reag. Ph Eur, 98-100% for analysis) were purchased from Merck Millipore (Merck KGaA, Darmstadt, Germany). Ammonia Solution 32% (extra pure), Trypsin from porcine pancreas (Type IX-S), Iodoacetamide, IAM (BioUltra, ≥99% NMR), D-L Dithiothreitol, DTT (BioUltra, ≥99.5% RT), Calcium Chloride dihydrate (Reagent Plus® ≥99%), Ammonium bicarbonate (BioUltra, ≥99.5% T), Urea (for electrophoresis gel) and Phosphoric acid (85 wt.% in H_2O, 99.99% trace metals basis) were purchased from Sigma–Aldrich (St. Louis, MO). Ultrapure water was from a Millipore Milli-Q system (Merck Millipore, Billerica, MA). Cynomolgus monkey serum was purchased from R.C. Hartelust BV (Tilburg, Nederland).

LC-MS/MS equipment

The UPLC–MS/MS analyses were performed by an Acquity UPLC® (Waters Corporation Milford, MA) system consisting of Binary Solvent Manager, Sample Manager, and Sample Organizer and Column oven. An Acquity UPLC® CSH (Charged Surface Hybrid) C18 column (2.1 × 150 mm, 1.7 µm particle size; Waters, Milford, MA, USA) was used for the separation. The UPLC system was interfaced with an AB SCIEX TripleQuad® 5500 mass spectrometer (AB SCIEX, Toronto, Canada) used as the detector. Analyst software v.1.5.1 was used for data acquisition and processing.

Monoclonal antibodies-spiked serum samples

Stock solutions of monoclonal antibodies 992 mAb and 1024 mAb were prepared in ammonium bicarbonate buffer 100 mM at a concentration of 1 mg/mL, separately. Protein-spiked serum samples were prepared by diluting protein stock solutions into blank monkey serum followed by further serial dilution in blank monkey serum to obtain the final concentrations desired. 992 mAb and 1024 mAb calibration standard concentrations were 150.0, 250.0, 500.0, 1000.0, 2500.0, 5000.0, 7500.0, 10000.0 ng/ml for the method performance evaluation runs (for the total signature peptide the concentration levels are twice those listed); VS concentrations for all mAb analytes were 150.0, 450.0, 1500.0, 8500.0 and 10000.0 ng/ml. The m992 IS, 1024 mAb IS and total mAbs IS working solutions were prepared at a concentration of 5000.0 ng/ml, 2500.0 ng/ml and 2500.0 ng/ml, respectively.

Serum sample digestion

50 µL of serum samples (STD, VS) were transferred to 96-well polypropylene microplates. 992 mAb, 1024 mAb and total mAbs Internal standards (5 µL) (Table 1), Urea 2M solution (700 µL), DTT 250 mM solution (70 µL) were added to each well followed by vortex mixing at 750 rpm for 30 min. at 60°C. IAM 0.5M solution (45 µL) was added to each well followed by vortex mixing at 750 rpm for 45 min. at RT. Calcium chloride 87.5 mM (10 µL) solution and trypsin solution at 15 mg/mL concentration (85 µL) were added to each well followed by vortex mixing at 750 rpm for 40 min. at 60°C.

Sample clean-up

SPE clean-up: The digests of the serum samples were mixed with 15 µl of phosphoric acid (85%, %W/V) and then loaded onto an Oasis⁺ MCX SPE plate. The samples were washed sequentially with 2 mL each of: 2% formic acid in water; 1 mL of 10% methanol; 1 mL of 5% ammonium hydroxide in water. The analytes were eluted with 1 mL of 5% ammonium hydroxide in 60/40 methanol/water and then dried down. The dried samples were reconstituted into 150 µl of 1% formic acid in water and analyzed by LC-MS/MS.

Chromatographic and mass spectrometric conditions

A gradient solvent system consisting of mobile phase A (0.1% formic acid in water), and mobile phase B (0.1% formic acid in acetonitrile) was used. The column temperature was set at 45°C. The gradient was as follows: 0-0.5 min 10% B; 0.5-5.0 min 10-18% B; 5.0-8.5 min 30% B; 8.6-10.6 min 95%B; 10.7-16.5 min 10%B. The flow rate was 0.3 ml/min, and the injection volume was 20 µl. The mass spectrometer was operated in ESI positive mode. The following parameters were used: curtain gas 30 psi; ion source gas one, 45 psi; ion source gas two, 50 psi; temperature 450°C; ion-spray voltage 5200 V. The MRM channels monitored for the surrogate peptides and their IS are listed in Table 2. The dwell time for each MRM channel was 50 ms.

Results

In-silico analysis

The primary sequences of the light and heavy chain of the two mAbs (Table 3) were analyzed as described in the Materials and Methods section. The enzyme selected for the cleavage was trypsin since it was able to provide peptides with well distributed length and charge, and at least 3 to 4 peptide pairs for heavy as well as light chains specific for each mAb Fab part. From these, the selectivity analysis conducted by the BLAST suite revealed the candidate signature peptides. These findings are reported in Tables 4 and 5 for the mAb Fab parts.

The in-silico analysis regarding the shared mAb Fc parts are reported in Table 6. As described by Furlong et al. [14], these tryptic peptides can be used to obtain the concentration of a humanized mAb in non-human matrices.

Optimization of experimental conditions and selection of signature peptides

The method set-up was thoroughly investigated in order to obtain the most selective peptides, the best SPE conditions and the appropriate chromatographic conditions for peptide separation. During this phase, different SPE cartridges and different solvent mixtures were tested to obtain acceptable results in term of sensitivity, robustness, reproducibility and selectivity. Initially two signature peptides, 992-HC-3 and 1024-HC-3, coming from paired regions were monitored and analyzed using a BEH C18 Column (1 × 100 mm, 1.7 µm, Waters), given their very intense MS signal. Unfortunately the 992-HC-3 peptide from 992 mAb, did not show sufficient selectivity and sensitivity on different MRM transitions when monitored in a biological matrix, while the 1024-HC-3 peptide from 1024 mAb showed acceptable selectivity and sensitivity in a biological matrix (Figure 1). Modifications in chromatographic conditions (including column length) and SPE purification were not able to solve the issue with 992-HC-3 peptide. Of the other 992 mAbs peptides, 992-HC-1 was found to be the second choice in term of sensitivity. Therefore its selectivity was investigated and considered to be sufficient in matrix samples (Figure 2). In order to improve HC-1-992 peptide sensitivity, a CSH C_{18} column and some gradient modifications were introduced (see LC-MS/MS equipment and Chromatographic and mass spectrometric conditions section, respectively). With these modifications, 992-HC-1 could be selected as mAb signature peptide. Additionally, quantitation of a signature peptide coming from the common region of the two monoclonal antibodies was introduced. This strategy was applied in order to confirm the analytical response from HC-1-992 (992 mAb) and HC-3-1024 (1024 mAb) (Figures 3 and 4).

Labeled peptide	Sequence	#of aa	MW (Av.), Da
HC-3-1024 mAb IS	Acetyl-NH-VKQRPGQG-[L($^{13}C_6$;^{15}N)]-EWIGEINPSSGR-COOH	21	2357.6
HC-1-992 mAb IS	NH₂-EVQLQPGSE-[L($^{13}C_6$;^{15}N)]-VRPGASVKLS-CONH₂	21	2228.5
HC-3-total IS	Acetyl-NH-YRVVSVLT-[V($^{13}C_5$;^{15}N)]-LHQDWLNGK-COOH	18	2175.5

Table 1: Primary sequence of the stable isotopes (internal standards).

Peptide	Precursor ion (m/z)	Product ion (m/z)
HC-3-1024 mAb	694.4 [M+3H]³⁺	916.5 y_9^{+1} ion
HC-3-1024 mAb	694.4 [M+3H]³⁺	866.5 b_8^{+1} ion
HC-1-992 mAb	674.7 [M+3H]³⁺	648.8 y_{13}^{+2} ion
HC-1-992 mAb	674.7 [M+3H]³⁺	712.9 y_{14}^{+2} ion
HC-1-992 mAb	674.7 [M+3H]³⁺	777.0 y_{15}^{+2} ion
HC-3-total	603.3 [M+3H]³⁺	805.4 y_{14}^{+2} ion
HC-3-1024 mAb IS	696.7 [M+3H]³⁺	859.4 y_8^{+1} ion
HC-1-992 mAb IS	677.3 [M+3H]³⁺	714.4 y_{14}^{+2} ion
HC-3-total IS	605.3[M+3H]³⁺	807.4 y_{14}^{+2} ion

Table 2: MRM transitions and charge states for surrogate peptides and the stable isotopes (internal standards).

992 mAb, IgG1
Light Chain, 214aa
DIQMTQTTSSLSASLGDRVTISCRTSQDIGNYLNWYQQKPDGTVKLLIYYTSRLHSGVPSRFSGSGSGTDFSLTINNVEQEDVATYFCQHYNTVPPTFGG GTKLEIKRTVAAPSVFIFPPSDEQLKSGTASVVCLLNNFYPREAKVQWKVDNALQSGNSQESVTEQDSKDSTYSLSSTLTLSKADYEKHKVYACEVTHQG LSSPVTKSFNRGEC
Heavy Chain, 452aa
EVQLQQPGSELVRPGASVKLSCKASGYTFTSYWMHWVKQRPGQGLEWIGNIYPGSRSTNYDEKFKSKATLTVDTSSSTAYMQLSSLTSEDSAVYYCTR NGDYYVSSGDAMDYWGQGTSVTVSSASTKGPSVFPLAPSSKSTSGGTAALGCLVKDYFPEPVTVSWNSGALTSGVHTFPAVLQSSGLYSLSSVVTVPS SSLGTQTYICNVNHKPSNTKVDKRVEPKSCDKTHTCPPCPAPELLGGPSVFLFPPKPKDTLMISRTPEVTCVVVDVSHEDPEVKFNWYVDGVEVHNAKT KPREEQYNSTYR**VVSVLTVLHQDWLNGK**EYKCKVSNKALPAPIEKTISKAKGQPREPQVYTLPPSREEMTKNQVSLTCLVKGFYPSDIAVEWESNGQPE NNYKTTPPVLDSDGSFFLYSKLTVDKSRWQQGNVFSCSVMHEALHNHYTQKSLSLSPG
1024 mAb, IgG1
Light Chain, 219aa
DIVMTQAAFSNPVTLGTSASISCRSSKSLLHSNGITYLYWYLQKPGQSPQLLIYQMSNLASGVPDRFSSSGSGTDFTLRISRVEAEDVGVYYCAQNLELPY TFGGGTKLEIKRTVAAPSVFIFPPSDEQLKSGTASVVCLLNNFYPREAKVQWKVDNALQSGNSQESVTEQDSKDSTYSLSSTLTLSKADYEKHKVYACEV THQGLSSPVTKSFNRGEC
Heavy Chain, 448aa
QVQLQQPGAELVEPGGSVKLSCKASGYTFTSHWMHWVK**QRPGQGLEWIGEINPSSGR**NNYNEKFKSKATLTVDKSSSTAYMQFSSLTSEDSAVYYCV RYYGYDEAMDYWGQGTSVTVSSASTKGPSVFPLAPSSKSTSGGTAALGCLVKDYFPEPVTVSWNSGALTSGVHTFPAVLQSSGLYSLSSVVTVPSSSL GTQTYICNVNHKPSNTKVDKRVEPKSCDKTHTCPPCPAPELLGGPSVFLFPPKPKDTLMISRTPEVTCVVVDVSHEDPEVKFNWYVDGVEVHNAKTKPR EEQYNSTYR**VVSVLTVLHQDWLNGK**EYKCKVSNKALPAPIEKTISKAKGQPREPQVYTLPPSREEMTKNQVSLTCLVKGFYPSDIAVEWESNGQPENNY KTTPPVLDSDGSFFLYSKLTVDKSRWQQGNVFSCSVMHEALHNHYTQKSLSLSPG

Table 3: Primary sequences of 992 mAb and 1024 mAb. The signature peptides are reported in bold.

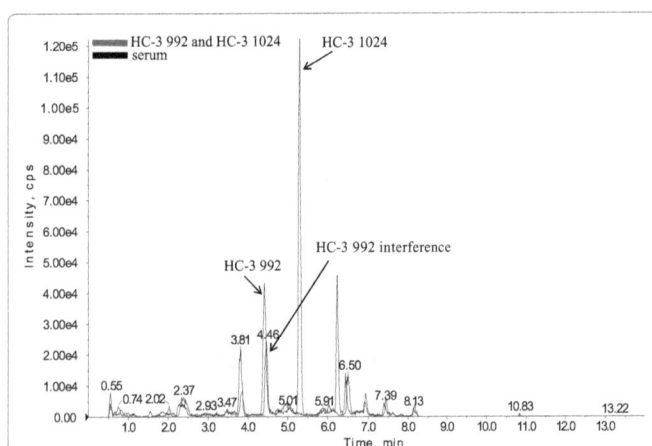

Figure 1: A representative chromatogram showing 992 mAb and 1024 mAb response in spiked monkey serum and ion interference at 992 mAb retention time in monkey serum.

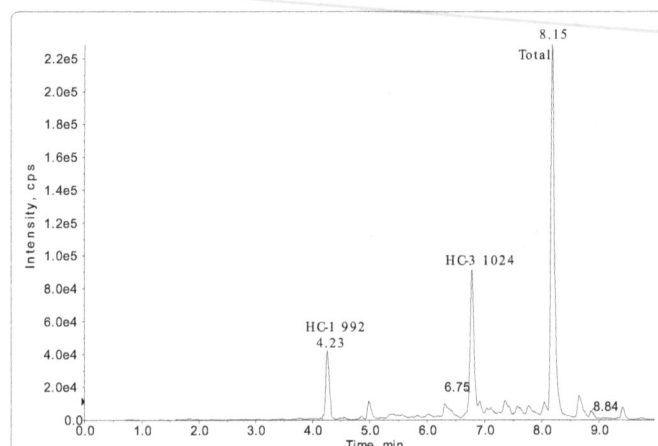

Figure 2: A representative chromatogram showing 992-HC-1 signature peptide response and selectivity.

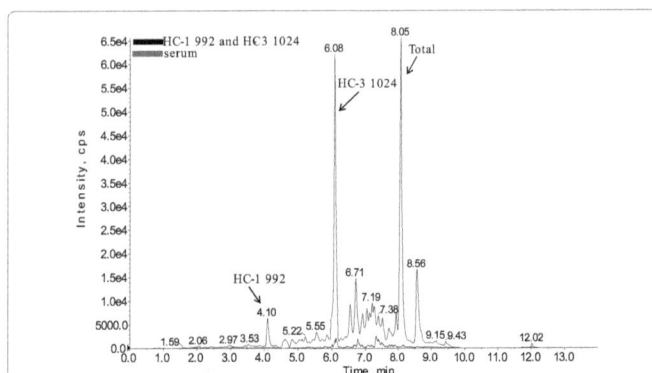

Figure 3: A representative chromatogram showing 992-HC-1 from 992 mAb, 1024-HC-3 from 1024 mAb and total mAbs response in spiked monkey serum.

Method performances

The method was assessed by evaluating linearity, sensitivity, selectivity, accuracy and precision and matrix effect according to the recent industry guidelines and adopting white papers suggestions [15-18] for NBE LC/MS analysis. For both the mAb analytes, a single calibration curve range and a single QC/VS concentration set were analyzed using the same LC method that incorporated all surrogate peptides.

Selectivity, specificity and sensitivity

Selectivity and specificity assessments were performed by evaluating potential analytes traces in different types of serum sample: a) zero sample; b) five different blank matrix sources; c) LLOQ prepared in five different matrix sources in six replicates each; d) 1024

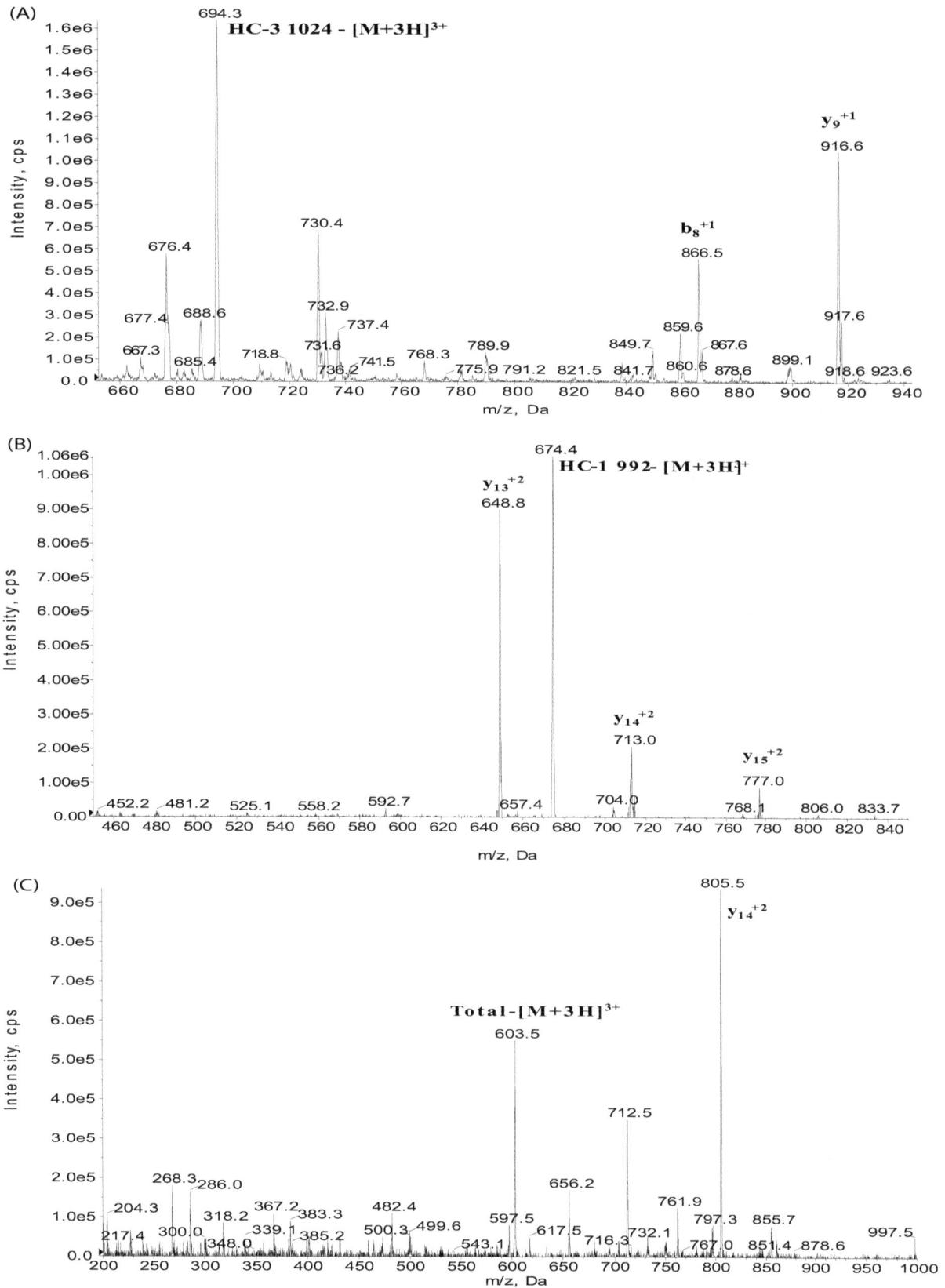

Figure 4: Mass spectra of the detected peptides for (a) 1024-HC-3, (b) 992-HC-1, and (c) total mAbs. Daughter ions that were used to quantify the peptides are shown.

mAb at ULOQ concentration level spiked in the 992 mAb calibration standard curve from STD1 to STD 4 (the most critical concentrations to be determined) and vice-versa. No analyte ions were found in the blank serum samples or in the zero samples, which indicated that serum matrices and the IS do not interfere with the determination of either monoclonal antibodies or total mAbs. The lower limit of detection was 150 ng/mL for both the surrogate peptides (from 992 mAb and 1024 mAb) and 300 ng/mL for the total surrogate peptide (common region of the two mAbs) with a S/N ratio above 5, which represented a concentration of about 1.00 nM, 1.00 nM and 2.00 nM of analytes (about 20 fmol, 20 fmol and 40 fmol) injected on the column for QRPGQGLEWIGEINPSSGR, EVQLQQPGSELVRPGASVK and VVSVLTVLHQDWLNGK peptides, respectively.

Both monoclonal antibodies at LLOQ level met the required accuracy (%BIAS ± 25) and precision (%CV ≤25) criteria in all the sources of matrices tested (Table 7) and in all the signature peptides considered. The total mAb determination was performed at LLOQ+25% only, during preliminary experiments. It was also demonstrated that a high concentration (ULOQ) of one monoclonal antibody does not affect the response of the second monoclonal antibody at low concentrations, and vice-versa (specificity). Moreover, the accuracy (%BIAS ± 25) of the total mAbs quantified (sum of one mAb spiked at ULOQ level and the second one spiked at low levels, and vice-versa) gave confirmation of the method consistency (Table 8).

Assessment of matrix effects

Matrix effect assessment was conducted using the post-extraction spike method. It quantitatively assesses matrix effects by comparing the response of the surrogate peptides in neat solvent to the response of the surrogate peptide spiked into a blank matrix sample that has gone through the sample preparation process. The monoclonal antibodies, at QC-Low and QC-High level concentrations, were spiked in five different monkey serum sources in order to investigate also the matrix effect values among different lots of serum. The results reported in Table 9 show that there is no significant matrix effect on either analyte.

Accuracy and precision

Accuracy and precision were determined in one run of intra-batch and three runs of inter-batch serum samples containing 1024 mAb, 992 mAb and the total mAbs at five concentration levels (LLOQ, VS-Low, VS-Medium, VS-High and ULOQ). The intra-batch relative mean accuracy of back calculated concentrations of the VS compared with theoretical ones ranged from 2.4% to 12.7%, from -6.3% to 16.4% and from -2.2% to 16.4% for QRPGQGLEWIGEINPSSGR peptide (1024 mAb), EVQLQQPGSELVRPGASVK peptide (992 mAb) and for VVSVLTVLHQDWLNGK (total mAbs), respectively. It should be noted that one QC at the medium concentration level was out of the acceptable limit of the assay for both the analytes but was excluded from the calculation since it was an outlier (known sample processing error). The intra-assay precision (%CV) ranged from 2.2 to 4.6, from 4.3 to 11.7 and from 2.0 to 7.4 for QRPGQGLEWIGEINPSSGR peptide (1024 mAb), EVQLQQPGSELVRPGASVK peptide (992 mAb) and VVSVLTVLHQDWLNGK (total), respectively (Table 10).

As shown in Table 11, for QRPGQGLEWIGEINPSSGR peptide (1024 mAb), inter-assay accuracy (% Bias) of less than 10.9% and inter-assay precision (%CV) of less than 9% was achieved. For peptide EVQLQQPGSELVRPGASVK (992 mAb), inter-assay accuracy (%Bias) of less than 9.6% and inter-assay precision (%CV) of less than 11.1% was achieved, while for VVSVLTVLHQDWLNGK (total

mAbs) inter-assay accuracy (%Bias) of less than 11.0% and inter-assay precision (%CV) of less than 12.3% was achieved. The accuracy and precision of all the peptides are well below the 20% (25% for VS at LLOQ level) acceptance criteria typically used in LC–MS/MS applied to large molecule bioanalysis.

Standard curve

The calibration curves of 1024 mAb, 992 mAb and total mAbs in monkey serum (Figure 5) were generated automatically by using the *algorithm classic* of Analyst software. The linear dynamic range evaluated was between 150 ng/mL and 10000 ng/mL for 1024 mAb and 992 mAb, and from 300 ng/mL to 20000 ng/mL for total mAbs. Individual standard curve concentration data in monkey serum are shown in Table 12.

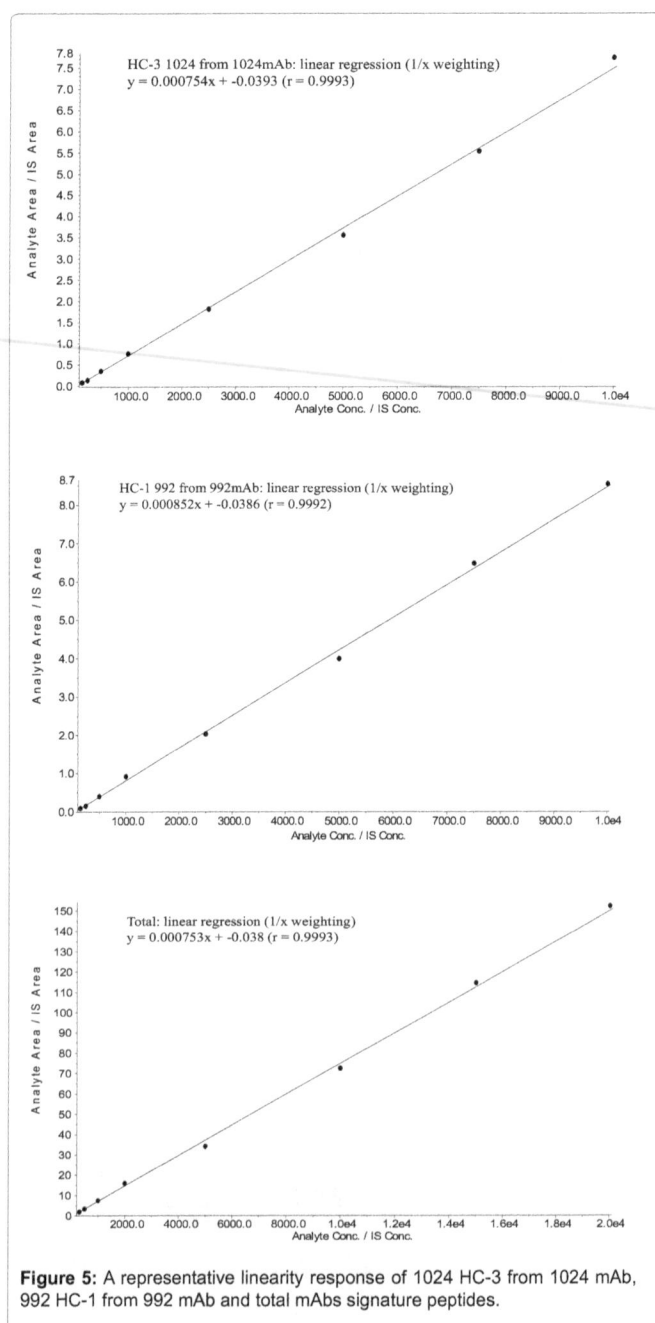

Figure 5: A representative linearity response of 1024 HC-3 from 1024 mAb, 992 HC-1 from 992 mAb and total mAbs signature peptides.

mAb	Sequence	#of aa	MW Da	pI	"Macaca fascicularis BLASTP"	"Homo sapiens BLASTP"	Identity% vs. HC of the other mAb
992	EVQLQQPGSELVRPGASVK	19	2022.2	6.24	OK	OK	78.9
992	ASGYTFTSYWMHWVK	15	1864.1	8.55	OK	"Not OK" Chain B, Mature Metal Chelatase Catalytic Antibody With Hapten pdb|3FCT|B	93.3
992	QRPGQGLEWIGNIYPGSR	18	2028.2	8.75	OK	OK	82.4
992	ATLTVDTSSSTAYMQLSSLTSEDSAVYYCTR	31	3352.6	4.03	OK	OK	90.3
992	NGDYYVSSGDAMDYWGQGTSVTVSSASTK	29	3034.1	3.93	OK	OK	80.0
1024	QVQLQQPGAELVEPGGSVK	19	1964.2	4.53	OK	"Not OK" Chain B, Anti-blood Group A Fv pdb|1JV5|B"	78.9
1024	ASGYTFTSHWMHWVK	15	1838.0	8.65	OK	OK	93.3
1024	QRPGQGLEWIGEINPSSGR	19	2081.2	6.14	OK	OK	82.4
1024	SSSTAYMQFSSLTSEDSAVYYCVR	24	2682.9	4.37	OK	OK	91.7
1024	YYGYDEAMDYWGQGTSVTVSSASTK	25	2766.9	4.03	OK	OK	80.0

Table 4: Result summary of the *in-silico* selection process for Fab Heavy Chains (HC). "OK" and "Not OK" mean "no match" or "match", respectively, for known protein sequences present in monkey or human proteome. When a match is present, some examples of proteins containing that exact sequence are listed. The level of identity with the corresponding peptide in the same region on the other mAb is reported (I to L not considered) as Identity% score. The average MW, and calculated isoelectric point (pI) are also reported for each peptide.

mAb	Sequence	#of aa	MW Da	pI	"Macaca fascicularis BLASTP"	"Homo sapiens BLASTP"	Identity % vs. LC of the other mAb
992	DIQMTQTTSSLSASLGDR	18	1911.0	4.21	OK	"Not OK Chain A, Anti-Blood Group A Fv, pdb|1JV5|A, pdb|1IKF|L, pdb|3U0W|L"	50.0
992	TSQDIGNYLNWYQQKPDGTVK	21	2455.6	5.63	OK	OK	NA
992	LLIYYTSR	8	1028.2	8.59	OK	"Not OK", immunoglobulin VL region=humanized bispecific antibody [human, Peptide Recombinant, 107 aa, gb|AAB24132.1|"	NA
992	LHSGVPSR	8	851.9	9.76	Not OK	Not OK	NA
1024	DIVMTQAAFSNPVTLGTSASISCR	24	2469.8	5.83	OK	"Not OK, anti-GlcNAc antibody variable region:SUBUNIT=light chain prf||1911357B"	50.0
1024	FSSSGSGTDFTLR	13	1361.4	5.84	OK	OK	NA
1024	VEAEDVGVYYCAQNLELPYTFGGGTK	26	2824.1	4.00	OK	OK	NA

Table 5: Result summary of the *in-silico* selection process for Fab Light Chains (LC). "OK" and "Not OK" mean no match or match, respectively, for known protein sequences present in monkey or human proteome. When a match is present, some examples of proteins containing that exact sequence are listed. The level of identity with the corresponding peptide in the same region on the other mAb is reported (I to L not considered) as Identity% score. The average MW, and calculated isoelectric point (pI) are also reported for each peptide.

Peptide	Sequence	Human chain subclass	# of aa	MW (Av.), Da	pI
LC-1	TVAAPSVFIFPPSDEQLK	NA	18	1946.2	4.37
LC-2	SGTASVVCLLNNFYPR	NA	16	1740.9	7.94
LC-3	VDNALQSGNSQESVTEQDSK	NA	20	2136.1	3.92
LC-4	DSTYSLSSTLTLSK	NA	14	1502.6	5.83
HC-1	TPEVTCVVVDVSHEDPEVK	IgG1	19	2082.3	4.17
HC-2	FNWYVDGVEVHNAK	IgG1	14	1677.8	5.32
HC-3	VVSVLTVLHQDWLNGK	IgG1, IgG4	16	1808.1	6.71
HC-4	GFYPSDIAVEWESNGQPENNYK	IgG1, IgG2, IgG4	22	2544.6	4.00
HC-5	TTPPVLDSDGSFFLYSK	IgG1	17	1874.0	4.21

Table 6: List of common tryptic Fc peptides from 992 and 1024 mAbs [14]. None of them is present in monkey serum.

1024 Monoclonal antibody					
Parameters	Matrix 1	Matrix 2	Matrix 3	Matrix 4	Matrix 4
Mean Conc. (ng/mL)	157.4	161.1	163.6	142.0	167.8
BIAS%	4.9	7.4	9.1	-5.3	11.9
SD	11.7	21.0	27.4	18.7	28.4
CV%	7.4	13.0	16.7	13.2	16.9
992 Monoclonal antibody					
Parameters	Matrix 1	Matrix 2	Matrix 3	Matrix 4	Matrix 4
Mean Conc. (ng/mL)	169.9	150.0	169.2	135.8	172.9
BIAS%	13.3	0.0	12.8	-9.5	15.3
SD	33.6	27.7	26.4	9.7	10.5
CV%	19.8	18.5	15.6	7.1	6.1

Table 7: Selectivity of the LC-MS/MS analysis of 1024 mAb and 992 mAb spiked at LLOQ in five different monkey serum lots.

	Sample	Nominal Conc. (ng/mL)	992 mAb Calculated Conc. (ng/mL)	992 mAb %BIAS	1024 mAb Calculated Conc. (ng/mL)	1024 mAb %BIAS	1024 mAb at ULOQ	992 mAb at ULOQ	Sum of 992 mAb and 1024 mAb at ULOQ	Sum of 1024 mAb and 992 mAb at ULOQ	Total calculated conc. (ng/mL)	%BIAS
Set 1	Std 1	150.0	158.0	105.3	-	-	11051.6	-	11209.6	-	11120.5	99.2
	Std 2	250.0	263.0	105.2	-	-	10426.9	-	10689.9	-	11502.6	107.6
	Std 3	500.0	590.0	118.0	-	-	10491.9	-	11081.9	-	12180.1	109.9
	Std 4	1000.0	1146.6	114.7	-	-	10154.0	-	11300.6	-	12687.6	112.3
Set 2	Std 1	150.0	-	-	161.6	107.8	-	11730.0	-	11891.6	12320.1	105.0
	Std 2	250.0	-	-	239.0	95.6	-	13011.3	-	13250.3	12249.6	94.1
	Std 3	500.0	-	-	434.3	86.9	-	14181.3	-	14615.6	9934.21	**70.1**
	Std 4	1000.0	-	-	997.4	99.7	-	10596.9	-	11594.3	12312.7	116.2

Sample	992 and 1024 mAb nominal conc. (ng/mL)	1024 mAb calculated conc. (ng/mL)	992 mAb calculated conc. (ng/mL)	Sum of 1024 mAb and 992 mAb calculated conc. (ng/mL)	Total mAbs calculated conc. (ng/mL)	%Bias
VS-LLOQ_1	150.0	165.8	141.5	307.2	301.6	98.2
VS-LLOQ_2	150.0	167.1	144.6	311.7	288.6	92.6
VS-LLOQ_3	150.0	173.1	162.1	335.3	290.1	86.5
VS-LLOQ_4	150.0	164.6	117.2	281.8	288.8	102.5
VS-LLOQ_5	150.0	152.4	152.2	304.6	297.7	97.7
VS-LLOQ_6	150.0	125.5	160.0	285.5	310.9	108.9
VS-LLOQ_7	150.0	141.8	148.9	290.7	263.1	90.5
VS-LLOQ_8	150.0	162.0	153.6	315.6	264.0	83.6
VS-LLOQ_9	150.0	162.5	156.7	319.2	313.8	98.3
VS-LLOQ_10	150.0	150.7	164.1	314.7	266.1	84.5
VS-LLOQ_11	150.0	152.6	158.2	310.7	267.0	85.9
VS-Low_1	450.0	498.5	460.1	958.5	942.4	98.3
VS-Low_2	450.0	509.0	490.1	999.1	1074.1	107.5
VS-Low_3	450.0	479.4	469.8	949.2	1016.4	107.1
VS-Low_4	450.0	530.8	477.9	1008.7	956.2	94.8
VS-Low_5	450.0	511.5	513.8	1025.3	1120.3	109.3
VS-Low_6	450.0	447.1	479.5	926.7	1007.3	108.7
VS-Low_7	450.0	414.8	450.2	865.0	783.9	90.6
VS-Low_8	450.0	462.6	458.4	921.0	873.7	94.9
VS-Low_9	450.0	393.5	424.1	817.6	1019.4	**124.7**
VS-Low_10	450.0	446.2	523.4	969.6	1124.5	116.0
VS-Low_11	450.0	461.5	377.6	839.1	1072.3	**127.8**
VS-Medium_1	1500.0	DEV	DEV	DEV	DEV	DEV
VS-Medium_2	1500.0	1701.1	1837.1	3538.3	3181.9	89.9
VS-Medium_3	1500.0	1634.9	1750.2	3385.1	3569.9	105.5
VS-Medium_4	1500.0	1717.1	1782.8	3499.9	3593.0	102.7
VS-Medium_5	1500.0	1706.7	1614.5	3321.1	3235.2	97.4
VS-Medium_6	1500.0	1475.4	1632.8	3108.2	2586.1	83.2
VS-Medium_7	1500.0	1524.1	1389.1	2913.1	2789.6	95.8

VS-Medium_8	1500.0	1443.1	1424.3	2867.4	2832.3	98.8
VS-Medium_9	1500.0	1697.3	1497.9	3195.1	3387.2	106.0
VS-Medium_10	1500.0	1698.5	1722.1	3420.6	3285.3	96.0
VS-Medium_11	1500.0	1875.6	1792.7	3668.3	3591.1	97.9
VS-High_1	8500.0	8401.0	8187.5	16588.5	15863.2	95.6
VS-High_2	8500.0	8948.2	9003.0	17951.2	16840.3	93.8
Sample	992 and 1024 mAb nominal conc. (ng/mL)	1024 mAb calculated conc. (ng/mL)	992 mAb calculated conc. (ng/mL)	Sum of 1024 mAb and 992 mAb calculated conc. (ng/mL)	Total mAbs calculated conc. (ng/mL)	%Bias
VS-High_3	8500.0	8501.8	8986.3	17488.1	17506.2	100.1
VS-High_4	8500.0	8797.9	8114.8	16912.7	19210.4	113.6
VS-High_5	8500.0	8881.8	8691.8	17573.6	18023.8	102.6
VS-High_6	8500.0	8191.7	8791.02	16982.8	14244.0	83.9
VS-High_7	8500.0	8800.2	8909.84	17710.0	15840.6	89.4
VS-High_8	8500.0	8421.5	9575.83	17997.4	14445.5	80.3
VS-High_9	8500.0	8429.3	8714.6	17143.8	17328.8	101.1
VS-High_10	8500.0	8183.6	9253.3	17436.9	16197.1	92.9
VS-High_11	8500.0	8406.3	8774.3	17180.6	16795.2	97.8
VS-ULOQ_1	10000.0	11091.2	13600.9	24692.1	22622.1	91.6
VS-ULOQ_2	10000.0	11414.5	11984.6	23399.1	24885.9	106.4
VS-ULOQ_3	10000.0	10760.8	11311.2	22072.0	21742.9	98.5
VS-ULOQ_4	10000.0	11190.5	12435.0	23625.5	22692.7	96.1
VS-ULOQ_5	10000.0	11018.2	14001.5	25019.7	24498.1	97.9
VS-ULOQ_6	10000.0	10099.1	11867.0	21966.1	16253.0	**74.0**
VS-ULOQ_7	10000.0	11463.1	11306.4	22769.5	20383.4	89.5
VS-ULOQ_8	10000.0	11299.2	10712.1	22011.3	17992.2	81.7
VS-ULOQ_9	10000.0	11412.3	11301.9	22714.2	23017.8	101.3
VS-ULOQ_10	10000.0	11870.9	11608.8	23479.7	24067.1	102.5
VS-ULOQ_11	10000.0	10427.4	11056.1	21483.5	21557.2	100.3

Table 8: Evaluation of total mAbs accuracies when compared to the sum of individual mAbs measured concentrations. DEV stands for deviation from the sample preparation procedure. In Set1 and Set2 are also reported the specificity experiment results where one mAb at ULOQ is spiked over a concentration curve of the other mAb. The last column reports the accuracy of the calculated total mAb concentrations (%BIAS) where the nominal concentration was assumed to be the sum of the individual mAbs calculated concentrations. Out of accuracy acceptance criteria, i.e. ±20% BIAS% (±25%BIAS % at LLOQ) are shown in bold.

1024 Monoclonal antibody				
	AN/IS Area ratio in samples spiked after extraction		AN/IS Area ratio in neat solvent	
Parameters	VS Low 450.0 ng/mL	VS High 8500.0 ng/mL	VS Low 450.0 ng/mL	VS High 8500.0 ng/mL
Mean area ratio	0.658	9.66	0.501	9.39
Normalized matrix effect	1.3	1.0		
n	6	6	6	6
992 Monoclonal antibody				
	AN/IS Area ratio in samples spiked after extraction		AN/IS Area ratio in neat solvent	
Parameters	VS Low 450.0 ng/mL	VS High 8500.0 ng/mL	VS Low 450.0 ng/mL	VS High 8500.0 ng/mL
Mean area ratio	0.334	5.13	0.275	5.92
Normalized matrix effect	1.2	0.9		
n	6	6	6	6

Table 9: Matrix effect of the LC-MS/MS analysis of 1024 mAb and 992 mAb in monkey serum.

1024 Monoclonal antibody					
Parameters	VS LLOQ 150.0 ng/mL	VS Low 450.0 ng/mL	VS Medium 1500.0 ng/mL	VS High 8500.0 ng/mL	VS ULOQ 10000.0 ng/mL
Mean Conc. (ng/mL)	164.6	505.8	1690.0	8706.2	11095.0
Accuracy (%BIAS)	9.7	12.4	12.7	2.4	10.9
SD	7.6	18.8	37.3	241.2	239.2
Precision (%CV)	4.6	3.7	2.2	2.8	2.2
n	5	5	4*	5	5
992 Monoclonal antibody					

Parameters	VS LLOQ 150.0 ng/mL	VS Low 450.0 ng/mL	VS Medium 1500.0 ng/mL	VS High 8500.0 ng/mL	VS ULOQ 10000.0 ng/mL
Mean Conc. (ng/mL)	143.5	482.3	1746.1	8596.7	9372.9
Accuracy (%BIAS)	-4.3	7.2	16.4	1.1	-6.3
SD	16.7	20.7	94.8	425.9	771.4
Precision (%CV)	11.7	4.3	5.4	5.0	8.2
n	5	5	4*	5	5
Total					
Parameters	VS LLOQ 300.0 ng/mL	VS Low 900.0 ng/mL	VS Medium 3000.0 ng/mL	VS High 17000.0 ng/mL	VS ULOQ 20000.0 ng/mL
Mean Conc. (ng/mL)	293.4	1021.9	3395.0	17488.8	23288.3
Accuracy (%BIAS)	-2.2	13.5	13.2	2.9	16.4
SD	5.9	76.0	216.6	1256.3	1341.9
Precision (%CV)	2.0	7.4	6.4	7.2	5.8
n	5	5	4*	5	5

*One outlier of a set of five values, excluded from the statistic calculation

Table 10: Intra-assay accuracy and precision of the LC-MS/MS analysis of 1024 mAb, 992 mAb and total mAbs in monkey serum.

1024 Monoclonal antibody					
Parameters	VS LLOQ 150.0 ng/mL	VS Low 450.0 ng/mL	VS Medium 1500.0 ng/mL	VS High 8500.0 ng/mL	VS ULOQ 10000.0 ng/mL
Mean Conc. (ng/mL)	156.2	468.6	1647.4	8542.1	11095.2
Accuracy (%BIAS)	4.1	4.2	9.8	0.5	10.9
SD	13.6	42.4	131.3	270.3	504.6
Precision (%CV)	8.7	9.0	8.0	3.2	4.5
n	11	11	10*	11	11
992 Monoclonal antibody					
Parameters	VS LLOQ 150.0 ng/mL	VS Low 450.0 ng/mL	VS Medium 1500.0 ng/mL	VS High 8500.0 ng/mL	VS ULOQ 10000.0 ng/mL
Mean Conc. (ng/mL)	152.8	465.9	1644.3	8818.4	10428.8
Accuracy (%BIAS)	0.6	3.5	9.6	3.7	4.3
SD	13.2	40.5	160.4	419.1	1158.5
Precision (%CV)	8.8	8.7	9.8	4.8	11.1
n	11	11	10*	11	11
Total					
Parameters	VS LLOQ 300.0 ng/mL	VS Low 900.0 ng/mL	VS Medium 3000.0 ng/mL	VS High 17000.0 ng/mL	VS ULOQ 20000.0 ng/mL
Mean Conc. (ng/mL)	286.5	999.1	3205.2	16572.3	21792.0
Accuracy (%BIAS)	-4.5	11.0	6.8	-2.5	9.0
SD	18.9	104.5	360.3	1475.0	2687.7
Precision (%CV)	6.6	10.85	11.2	8.9	12.3
n	11	11	10*	11	11

*One outlier of a set of five values, excluded from the statistic calculation

Table 11: Inter-assay accuracy and precision of the LC-MS/MS analysis of 1024 mAb, 992 mAb and total mAbs in monkey serum.

1024 Monoclonal antibody					
Nominal conc. (ng/mL)	Calculated conc. (ng/mL)	Accuracy (%Bias)	Calculated conc. (ng/mL)	Accuracy (%Bias)	Calculated conc. (ng/mL)	Accuracy (%Bias)
150.0	136.4	-9.0	155.4	3.6	146.8	-2.1
250.0	271.0	8.4	228.9	-8.5	151.4	*Dev
500.0	506.5	1.3	516.7	3.3	549.2	9.8
1000.0	1064.2	6.4	1060.6	6.1	980.0	-2.0
2500.0	2302.6	-7.9	2457.1	-1.7	2421.1	-3.2
5000.0	5098.9	2.0	4769.3	-4.6	4582.0	-8.4
7500.0	7092.6	-5.4	7390.1	-1.5	7835.3	4.5
10000.0	10427.8	4.3	10321.9	3.2	10135.6	1.4
992 Monoclonal antibody					
Nominal conc. (ng/mL)	Calculated conc. (ng/mL)	Accuracy (%Bias)	Calculated conc. (ng/mL)	Accuracy (%Bias)	Calculated conc. (ng/mL)	Accuracy (%Bias)
150.0	141.1	-5.9	152.6	1.7	128.8	-14.1
250.0	265.5	6.2	222.7	-10.9	222.4	-11.0
500.0	436.1	-12.8	513.3	2.7	584.6	16.9

1000.0	1147.0	14.7	1119.4	11.9	1122.2	12.2
2500.0	2721.4	8.9	2428.7	-2.9	2466.2	-1.4
5000.0	4489.4	-10.2	4730.4	-5.4	4855.1	-2.9
7500.0	6643.5	-11.4	7646.6	2.0	7514.5	0.2
10000.0	11056.0	10.6	10086.2	0.9	10006.2	0.1
Total MABS						
Nominal conc. (ng/mL)	Calculated conc. (ng/mL)	Accuracy (%Bias)	Calculated conc. (ng/mL)	Accuracy (%Bias)	Calculated conc. (ng/mL)	Accuracy (%Bias)
300.0	290.1	-3.3	297.9	-0.7	240.1	-20.0
500.0	527.1	5.4	486.3	-2.7	349.4	'Dev
1000.0	935.2	-6.5	1039.0	3.9	1101.4	10.1
2000.0	2140.5	7.0	2156.1	7.8	2250.6	12.5
5000.0	4707.9	-5.8	4589.8	-8.2	4977.7	-0.4
10000.0	10526.8	5.3	9673.2	-3.3	9922.9	-0.8
15000.0	14727.5	-1.8	15267.2	1.8	14693.3	-2.0
20000.0	19944.9	-0.3	20290.7	1.5	20114.1	0.6
'Dev: Deviation from nominal concentration						

Table 12: Individual standard curve concentration data of the LC-MS/MS analysis of 1024 mAb, 992 mAb and total mAbs in monkey serum.

	Measured signature peptide concentrations			
Condition	992 mAb	1024 mAb	Tot mAbs	Degradation inference
1	C992 = C	C1024 = C	=2'C	= at HC N-term and at C-term Identical degradation
2	C992 = C	C1024 = C	≠ 2'C	= at HC N-term ≠ C-term. Different degradation at C-term
3	C992	C1024	= C992+C1024	≠ at HC N-term = at C-term, but the same on individual mAb at HC N-term and C-term. Different catabolism of one mAb
4	C992	C1024	≠ C992+C1024	Different catabolism, ≠ at HC N-term ≠ at C-term

Table 13: The four possible situations that can be reported from the measurement of *in-vivo* samples with the multi-analyte method, with some potential implications described. C: concentration, C992 specific 992mAb concentration, C1024 specific 1024mAb concentration, HC heavy chain. Two values are considered as identical if the differences are within 20% of BIAS% (25% at LLOQ).

With the exception of a single 1024 mAb calibration point, the deviations of the back-calculated concentrations from their nominal values were all within ±20.0% (within ±25.0% at LLOQ) for all the surrogate peptides. The correlation coefficients (r) from three batches of calibration samples were 0.999, between 0.993 and 0.999 and 0.999 for QRPGQGLEWIGEINPSSGR peptide (1024 mAb), EVQLQQPGSELVRPGASVK (992 mAb) and VVSVLTVLHQDWLNGK (total mAbs), respectively.

These results indicated that this LC-MS method can provide sensitive, specific, selective, precise and accurate analysis of 1024 mAb, 992 mAb and total mAbs in monkey serum.

Therefore it met the validation criteria set by Health Authorities' guidelines, industry best practice and opinion leaders' white papers in bioanalytical method validation.

Discussion

In-silico analysis led to the selection of 3-4 candidate signature peptides specific for each mAb, that were then tested using LC-MS to assess sensitivity and selectivity in the real matrix. An additional criterion that guided the selection was to select, as far as possible, tryptic peptide pairs deriving from homologous mAb regions. This was taken to allow for the possibility of differential *in-vivo* degradation on different parts of the mAbs. Selecting peptides from paired regions would normalize the measured concentration by leveling off possible differential catabolic effects.

This was not entirely possible, since the sensitivity and selectivity for the N-terminus 1024 mAbs candidate were not ideal. Nevertheless, the two selected specific peptides derive from Heavy Chain N-terminus regions of the mAbs.

Since the quantitative method had to be developed in a non-human matrix and the mAbs belong to the h-IgG1 class, it was hypothesized that a further improvement could be the quantitation of peptide(s) not present in monkey matrix but common to the two mAbs. This had already been reported by Furlong *et al.* [14], but in our work the investigation was extended. We used them for the total quantitation of the mAbs, but also attempted a comparison between the concentrations from the mAbs specific signature peptides and the concentration obtained from non-specific peptides. The added value lies not only in the consistency of the concentrations obtained, but in the help it can provide in understanding if and how the two mAbs degrades *in-vivo*.

To this end, first of all we had to prove that the method (for each of the three signature peptides independently) was robust, sensitive and specific enough to meet design requirements. This was successfully demonstrated by assessing linearity, accuracy and precision, sensitivity, selectivity, specificity, and matrix effect according to GLP industry guidelines for each of the three signature peptides.

Upon this solid basis, we were then able to confirm, that the total mAbs concentration in spiked monkey serum samples was consistent with the sum of the specific single concentrations from the two mAbs within the same accuracy criteria that were applied for method performance assessment, i.e. 20% BIAS% (25% for LLOQ) (Table 12): 94% of the results met the acceptance criteria. Of course this should be further verified doing incurred samples reanalysis, and in spiked incurred samples to have a more complete picture of the applicability of the principle.

As a future perspective, this method represents a seed of moving the LC-MS bottom-up approach to more reliable and informative quantification procedure.

In real samples, due to catabolic processing, we could hypothesize four different situations (Table 13). In the first case the sum of 992 mAb and 1024 mAb is identical to the concentration determined for the total mAbs: we could therefore infer that there is not any difference in the catabolic fate of the different HC regions. In the second situation, the HC N-terminus degrades differently with respect to the HC C-terminus, but identical degradation has occurred on the two HC N-termini of the two mAbs. Situation 3 is the opposite case from situation 2: there is a differential degradation between the two HC N-termini of the two mAbs, but no differences on the HC C-termini. Finally, as in case 4, the two HC chains are subject to different degradation processes. For the first time, these kinds of measurements will allow identification of which part of the molecule is impacted by degradation and connect this parameter directly to their relative body exposure. This information is of paramount importance in molecule design, and can be used to fine-tune the molecular structure in order to maximize its efficacy.

References

1. Peghini PL, Iwamoto M, Raffeld M, Chen YJ, Goebel SU, et al. (2002) Overexpression of epidermal growth factor and hepatocyte growth factor receptors in a proportion of gastrinomas correlates with aggressive growth and lower curability. Clin Cancer Res 8: 2273-2285

2. Damstrup L, Kuwada SK, Dempsey PJ, Brown CL, Hawkey CJ, et al. (1999) Amphiregulin acts as an autocrine growth factor in two human polarizing colon cancer lines that exhibit domain selective EGF receptor mitogenesis. Br J Cancer 80: 1012-1019.

3. Wong AJ, Ruppert JM, Bigner SH, Grzeschik CH, Humphrey PA, et al. (1992) Structural alterations of the epidermal growth factor receptor gene in human gliomas. Proc Natl Acad Sci U S A 89: 2965-2969.

4. Arteaga CL (2001) The epidermal growth factor receptor: from mutant oncogene in nonhuman cancers to therapeutic target in human neoplasia. J Clin Oncol 19: 32S-40S.

5. Mendelsohn J, Baselga J (2006) Epidermal growth factor receptor targeting in cancer. Semin Oncol 33: 369-385.

6. Wheeler DL, Dunn EF, Harari PM (2010) Understanding resistance to EGFR inhibitors-impact on future treatment strategies. Nat Rev Clin Oncol 7: 493-507.

7. Friedman LM, Rinon A, Schechter B, Lyass L, Lavi S, et al. (2005) Synergistic down-regulation of receptor tyrosine kinases by combinations of mAbs: implications for cancer immunotherapy. Proc Natl Acad Sci U S A 102: 1915-1920.

8. Ben-Kasus T, Schechter B, Lavi S, Yarden Y, Sela M (2009) Persistent elimination of ErbB-2/HER2-overexpressing tumors using combinations of monoclonal antibodies: relevance of receptor endocytosis. Proc Natl Acad Sci U S A 106: 3294-3299.

9. Spangler JB, Neil JR, Abramovitch S, Yarden Y, White FM, et al. (2010) Combination antibody treatment down-regulates epidermal growth factor receptor by inhibiting endosomal recycling. Proc Natl Acad Sci U S A 107: 13252-13257.

10. Pedersen MW, Jacobsen HJ, Koefoed K, Hey A, Pyke C, et al. (2010) Sym004: a novel synergistic anti-epidermal growth factor receptor antibody mixture with superior anticancer efficacy. Cancer Res 70: 588-597.

11. Koefoed K, Steinaa L, Søderberg JN, Kjær I, Jacobsen HJ, et al. (2011) Rational identification of an optimal antibody mixture for targeting the epidermal growth factor receptor. MAbs 3: 584-595.

12. Skartved NJ, Jacobsen HJ, Pedersen MW, Jensen PF, Sen JW, et al. (2011) Preclinical pharmacokinetics and safety of Sym004: a synergistic antibody mixture directed against epidermal growth factor receptor. Clin Cancer Res 17: 5962-5972.

13. Gasteiger E, Hoogland C, Gattiker A, Duvaud S, Wilkins MR, et al. (2005) Protein Identification and Analysis Tools on the ExPASy Server. In: Walker JM (ed) The Proteomics Protocols Handbook. Humana Press, pp. 571-607.

14. Furlong MT, Zhao S, Mylott W, Jenkins R, Gao M, et al. (2013) Dual universal peptide approach to bioanalysis of human monoclonal antibody protein drug candidates in animal studies. Bioanalysis 5: 1363-1376.

15. DeSilva B, Garofolo F, Rocci M, Martinez S, Dumont I, et al. (2012) 2012 white paper on recent issues in bioanalysis and alignment of multiple guidelines. Bioanalysis 4: 2213-2226.

16. Stevenson L, Garofolo F, DeSilva B, Dumont I, Martinez S, et al. (2013) 2013 White Paper on recent issues in bioanalysis: 'hybrid'--the best of LBA and LCMS. Bioanalysis 5: 2903-2918.

17. European Medicines Agency (2011) Guideline on bioanalytical method validation, pp. 1-23.

18. Food and Drug Administration (2001) Guidance for Industry: Bioanalytical Method Validation. U.S. Department of Health and Human Services, pp. 1-25.

9

Nicotine as Corrosion Inhibitor for 1018 Steel in 1M HCl under Turbulent Conditions

Araceli Espinoza-Vázquez*, Sergio Garcia-Galan and Francisco Javier Rodríguez-Gómez

Faculty of Chemistry, Department of Metallurgical Engineering, Universidad Nacional Autónoma de México, C.U., Distrito Federal, 04510, Mexico

Abstract

An electrochemical impedance technique for determining corrosion inhibition of nicotine in HCl on AISI 1018 steel under concentrations from 0 to 50 ppm found that the organic compound is a better corrosion inhibitor under static conditions. The inhibition efficiency (IE) increased with inhibitor concentration reaching an IE>90%, at 10 ppm. For [nicotine] ≤ 50 ppm, the IE value reached 71%, at 40 rpm, but diminished then to 33% upon changing the working electrode rotation speed to 500 rpm. The thermodynamic analysis showed a process ruled by physisorption according to the Langmuir adsorption model mechanism. Furthermore, the inhibition kinetics study showed that nicotine gave good protection against corrosion up to 72 hours of immersion with IE ≤ 87%. Finally, with increased temperature the IE values diminished from 90% at 25°C to 57% at 70°C, concluding that at high temperatures nicotine is ineffective at inhibition, because the temperature decrease, persistence layer easily desorbs.

Keywords: AISI 1018 steel; Nicotine; Corrosion inhibition

Introduction

The transport of hydrocarbons in the oil industry depends on the use of pipelines that can be damaged by corrosion, causing large impacts on production, significant damage to property, as well as pollution, and risk to human lives [1].

Corrosion inhibitors [2-5], such as molybdates, phosphates, and ethanolamines, are effective, but they are very toxic. The development of corrosion inhibitors, non-toxic and compatible with the environment, is an area of great importance in the science and technology of corrosion [6-8]. Inhibitor substances extracted from plants offer environmental and cost advantages; for example alkaloid extracts from *Oxandra asbeckii* plant [9], *Hibiscus sabdariffa* [10], Geissospermum leave [11] *Euphorbia falcate* [12] show 89% inhibition efficiency, *Morinda tinctoria* has a 70% at 30% v/v [13]; they have been tested, but there is a lack in assessing and identyfing the active substance.

Nicotine ((S)-3-(1-methylpirrolidin-2-il) pyridine) is an organic compound belonging to the alkaloids: a liquid, oily, and colorless derivative of the ortinina, Figure 1, synthesized in the areas of high activity in the roots of tobacco plants, and transported by the sap to the greens. Structurally, this compound is formed by a pyridine and a pyrrole that could have bifunctional activity from nitrogen atoms. Given its chemical structure, this organic compound is a candidate for the protection of petroleum pipeline systems, since it is of natural origin, readily found in tobacco plants (nicotiana tabacum), in which it is the major active chemical component. Futhermore, the nicotiana tabacum extract has been reported by Njokua et al. and they demonstrated that the best concentration was 1200 mg/L with 89% IE [14] and Bhawsar et al. [15] found to have inhibitory corrosion properties for mild steel in acidic medium, attaining 94.13% IE at an optimum concentration of 10 g/L under static conditions for 6 hours at 303 K.

In the oil industry the hydrocarbon flow is typically turbulent, when the Reynolds number (Re) is greater than 1000. Laboratory standard testing (ASTM G170 and ASTM G-185) of corrosion inhibitors using rotating cylinder electrodes recommend flow rates from 100 to 500 rpm corresponding to hydrocarbon transport of 194 and 970 L/min, respectively.

Many substances, able to be adsorbed, have been introduced as corrosion inhibitors, for example azol derivatives [16-20], so it is important to evaluate candidates to determine the conditions under which they might be effective.

Consequently, our interest in evaluating this organic molecule as pure compound has been demonstrated inhibition properties against corrosion thus making it an eco-friendly material equally proper for engineering uses under different hydrodynamic conditions and temperatures, which simulates the transport of hydrocarbons.

The inhibition efficiency of nicotine, IE, can be calculated as [21]:

$$IE\ /\ \% = \frac{\left(\frac{1}{Rp}\right)blank - \left(\frac{1}{Rp}\right)inhibitor}{\left(\frac{1}{Rp}\right)blank} \times 100 \quad (1)$$

where R_p is the polarization resistance with ("inhibitor") or without inhibitor ("blank").

The value of polarization resistance is an important parameter that can be obtained from electrochemical impedance spectroscopy to estimate the rate of corrosion, by using small polarization, i.e., no damage on the electrode is caused due to experiment.

Materials and Methods

Inhibitor solution 0.01M nicotine (Aldrich 97% purity) was prepared in water, with different aliquots taken from this solution, with concentrations ranging from 5 to 50 ppm added to the 1M HCl

*Corresponding author: Araceli Espinoza-Vázquez, Faculty of Chemistry, Department of Metallurgical Engineering, Universidad Nacional Autónoma de México, C.U., Distrito Federal, 04510, Mexico
E-mail: arasv_21@yahoo.com.mx

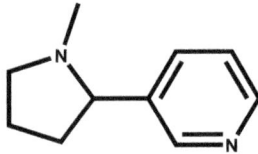

Figure 1: Chemical structure of nicotine.

electrolyte, prepared through dilution of the HCl analytical grade reagent at 37% with doubly distilled water.

The electrochemical impedance study was performed at room temperature using the Gill AC workstation, applying a sinusoidal ± 10 mV perturbation, within the 10^{-1} Hz to 10^{4} Hz frequency range to an electrochemical cell with a three-electrode setup. A saturated Ag / AgCl electrode was used as reference, with a graphite rod as counter electrode, while the working electrode (rotating cylinder electrode, RCE) was the AISI 1018 steel sample with ~ 3.92 cm^2 exposed area, duly prepared through standard metallographic procedures. The working electrode was first immersed into the test solution for 30 min to establish steady-state open-circuit potential.

The study under hydrodynamic conditions was carried out at 0, 40, 100 and 500 rpm. In addition, a study of kinetic inhibition of nicotine (50 ppm) for immersion times of 30 days maximum was performed, observing the effect of temperature at the same concentration from 30 to 70°C, with ± 2°C error.

Results and Discussion

Effect of concentration

Figure 2a shows impedance diagrams for the immersed metal, with an increase in the value of Z_{re} at higher rotation rates in 1M HCl due to the increase of mass transfer and localized attack [22].

Figure 2b-2e show the results for different rotation rates; the diameters of semicircles in Nyquist plots increas with increasing the inhibitor concentration. Analysis of the shape of the Nyquist plots revealed the curves were approximated by two semicircles: one time-constant at high frequency associated with a corrosion process, while the second time-constant at low frequency would likely be associated with inhibitor adsorption [23,24].

When the RCE was at static condition (0 rpm) and at 10 ppm a Z_{re} of 900 Ω·cm^2 was found, slightly increasing to 1000 Ω·cm^2 at 50 ppm. On the other hand, at 100 rpm, Figure 2d, Z_{re} decreases to 350 Ω·cm^2 at 50 ppm, attributable to a desorption of organic molecules on the electrode surface [25].

To determine electrochemical parameters of nicotine for different rotation rates (static and turbulent flows), simulation using electrical circuits was carried out using the Zview program, which are shown in Figure 3. Electrical equivalent-circuit diagram in Figure 3a corresponds to the metal/solution interface (double layer between the metallic surface and acid solution). Figure 3b describes a two-time-constant model, corresponding to the metal/solution interface and the inhibitor adsorption. R_s is the solution resistance, R_{ct} the charge transfer resistance, R_{mol} the molecular resistance, CPE$_1$ and CPE$_2$ the constant phase elements corresponding to the R_{mol} and R_{ct}, respectively.

The value of polarization resistance is calculated as:

$$Z_{CPE} = Y_0^{-1}(j\omega)^{-n} \qquad (2)$$

For the description of a frequency-independent phase shift between an applied AC potential and its current response, a constant phase element (*CPE*) is used, which is defined in impedance representation as:

$$Z_{CPE} = Y_0^{-1}(j\omega)^{-n} \qquad (3)$$

where Y_0 is the *CPE$_1$* or *CPE$_2$*, n is the *CPE* exponent that can be used as a gauge of the heterogeneity or roughness of the surface, j the unit imaginary number, and ω is the angular frequency in rad/s. Depending on n, *CPE* can represent a resistance ($Z_{CPE}=R$, $n=0$); capacitance ($Z_{CPE}=C$, $n=1$); a Warburg impedance ($Z_{CPE}=W$, $n=0.5$); or inductance ($Z_{CPE}=L$, $n=-1$). The equation to convert Y_0 into double layer capacitance (C_{dl}) is given by ref. [26-28]:

$$Cdl = Y_0(\omega_{max})^{n-1} \qquad (4)$$

Tables 1-4 show the corresponding values from the simulation of the experimental data, demonstrating that in all cases the concentration has greater polarization resistance compared with non-inhibitor.

Adequate corrosion protection was found when nicotine was assessed under static conditions, since at 10 ppm it reached 90%, attributable to the organic compound being adsorbed more easily under these conditions.

However, for rotation rates greater than 40 rpm (turbulent flow) the polarization resistance and the inhibition efficiency were reduced. This effect could be due to the high Reynolds number regime causing the partial desorption of inhibitor at higher wall shear stresses [29,30].

Thus, Table 1 indicates that C_{dl} decreases at greater inhibitor concentration, because increase the thickness of the inhibitor layer or decrease in local dielectric constant [31,32]. Also, inhibition efficiency is higher than for turbulent flow (Tables 2-4). Therefore, with agitation the inhibitor-adsorbed layer is thinner than in static conditions, and even be heterogeneous and non-persistent. It can be assumed that the decrease of C_{dl} is caused by the gradual replacement of water molecules by adsorption of inhibitor molecules on the mild steel surface [33,34].

Figure 4 shows the IE at different rotation rates, noting that good protection is achieved in the static condition, reaching 90% IE at 10 ppm, attributed to the presence of electrodonor groups (pairs of free electrons in atoms in nitrogen) that favor the adsorption of nicotine on the metal surface.

Adsorption process

The adsorption of organic compounds can be described by two types of interaction: physisorption and chemisorption [35]. To determine the value of the standard Gibbs energy ($\Delta G°_{ads}$) we used Equations 5 and 6, where K_{ads} is the adsorption equilibrium constant and θ is surface coverage, R is the universal gas constant, C is inhibitor molar concentration and T is the absolute temperature:

$$\frac{C}{\theta} = \frac{1}{K_{ads}} + C \qquad (5)$$

$$\Delta G°_{ads} = -RT \ln K_{ads} \qquad (6)$$

The results in Table 5 indicate that the process is physisorption (electrostatic interaction), since the value of $\Delta G°_{ads}$ obtained from the linear fit shown in Figure 5 is less than -20 kJ/mol for the four-rotation rates tested [36-39].

Strong correlation coefficient ($R^2 \geq 0.9979$) of the Langmuir adsorption isotherm for nicotine was observed. The Langmuir

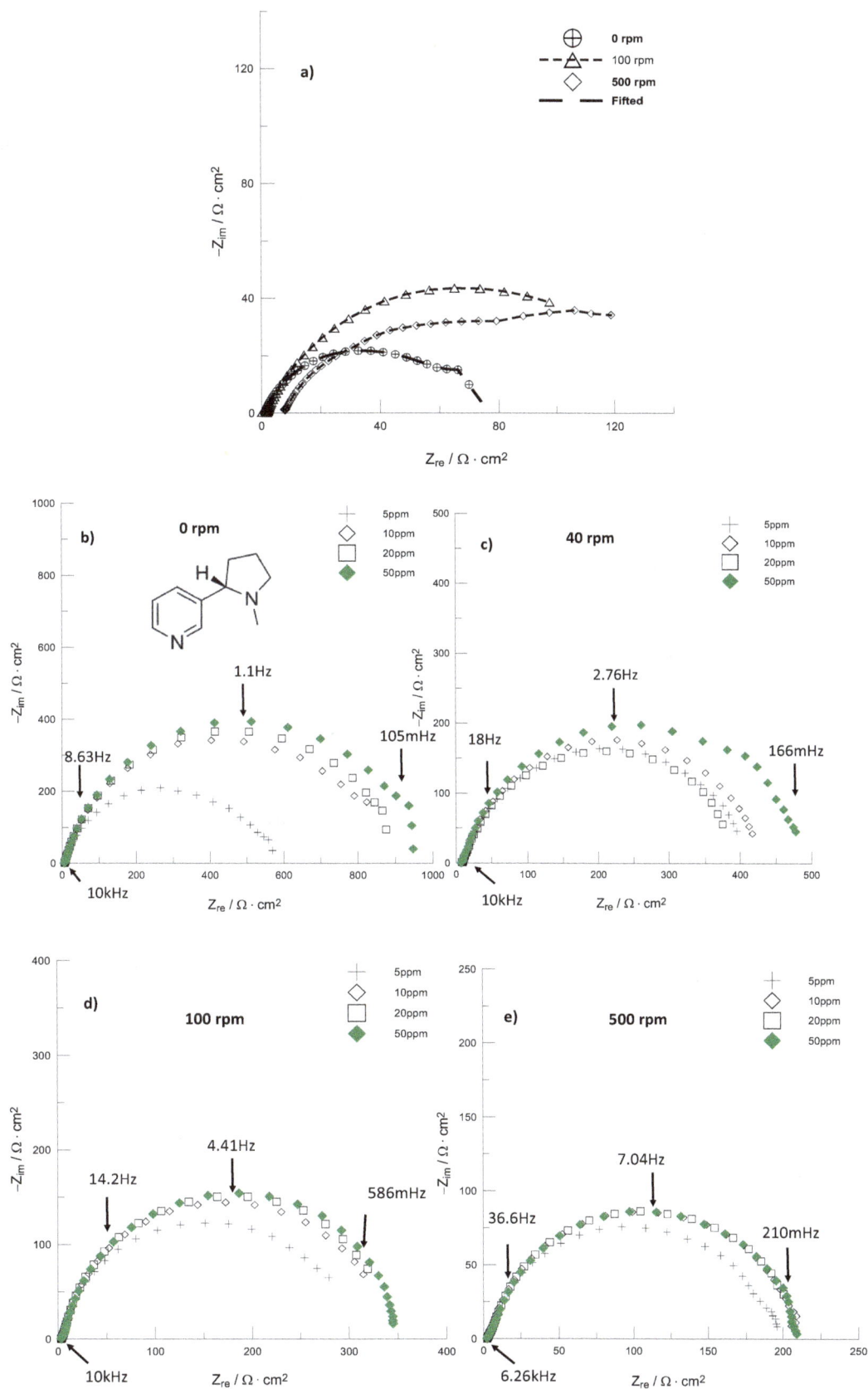

Figure 2: Nyquist diagrams at different concentrations of nicotine as a corrosion inhibitor at different rotation rates.

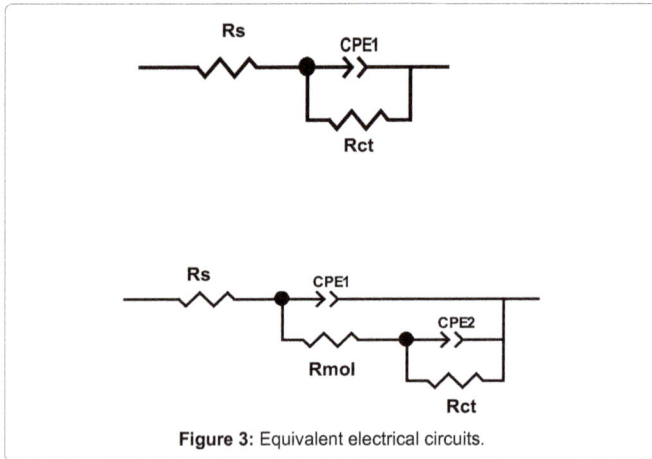

Figure 3: Equivalent electrical circuits.

C/ppm	Rs/Ω cm²	n	Cdl/μF cm⁻²	Rp/Ω cm²	IE/%
0	3.4	0.8	351.0	83.0	-
5	6.4	0.9	81.6	552.7	85.0
10	6.2	0.9	71.4	878.1	90.5
20	6.2	0.9	69.2	907.1	90.8
50	6.1	0.9	72.4	974.1	91.5

Table 1: Electrochemical parameters of nicotine (0 rpm).

C/ppm	Rs/Ω cm²	n	Cdl/μF cm⁻²	Rp/Ω cm²	IE/%
0	8.5	1.0	400.3	141.1	0.0
5	10.9	0.9	392.2	403.8	65.0
10	9.5	0.9	432.5	431.2	67.3
20	8.9	0.8	441.0	394.7	64.3
50	7.6	0.8	429.9	490.8	71.2

Table 2: Electrochemical parameters of nicotine/solution interface (40 rpm).

C/ppm	Rs/Ω cm²	n	Cdl/μF cm⁻²	Rp/Ω cm²	IE/%
0	3.3	0.8	435.3	117.6	0.0
5	3.4	0.9	374.1	308.4	61.9
10	3.4	0.9	396.8	337.1	65.1
20	3.5	0.9	431.5	349.2	66.3
50	3.5	0.9	489.2	355.9	67.0

Table 3: Electrochemical parameters of nicotine/solution interface (100 rpm).

C/ppm	Rs/Ω cm²	n	Cdl/μF cm⁻²	Rp/Ω cm²	IE/%
0	8.5	1.0	760.0	141.1	0.0
5	2.9	0.9	314.9	191.1	26.1
10	2.8	0.9	372.9	206.9	31.8
20	2.6	0.9	462.9	210.5	33.0
50	2.6	0.9	515.8	210.7	33.0

Table 4: Electrochemical parameters of nicotine/solution interface (500 rpm).

adsorption isotherm assumes that the adsorption of organic molecules on the adsorbent is a monolayer. The high values of the adsorption equilibrium constant reflect the high adsorption ability of the nicotine molecules on the steel surface.

Effect of immersion time

Corrosion inhibitors are often evaluated in order to determine their performance under different conditions, such as the concentration, but few evaluated for persistence of the layer inhibitor. It is important to study the effect of immersion time to optimize concentration [40,41].

In the Nyquist plot of Figure 6, some examples of impedance curves at three different immersion times in the presence of 50 ppm of nicotine in 1M HCl are shown, with 24 hours of immersion producing a maximum Z_{re} value of 854 Ω·cm². After that time, this value decreases; at 240 hours it is remarkable to observe two time-constants, one attributed to the charge transfer resistance (corrosion process) and a second time-constant corresponding to an adsorbed molecular layer [42]. In Table 6, the values of corresponding resistances for each immersion time are summarized.

In order to clarify the behavior of nicotine (50 ppm) as a function of immersion time, Figure 7, a maximum inhibition period of 72 hours

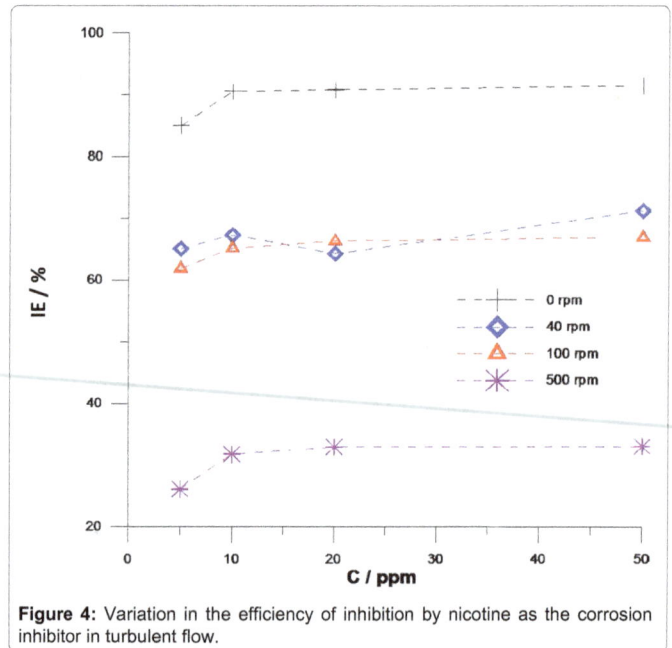

Figure 4: Variation in the efficiency of inhibition by nicotine as the corrosion inhibitor in turbulent flow.

Rotation rate/rpm	Ln k_ads	ΔG°_ads/KJ mol⁻¹	Regression Equation Lineal/mM	R² (correlation coefficient)
0	5.77	-13.11	C/θ=1.0864 C+0.0031	0.9999
40	4.16	-9.44	C/θ=1.3839 C+0.0156	0.9979
100	5.13	-11.65	C/θ=1.4812 C+0.0059	1
500	3.61	-8.21	C/θ=2.9628 C+0.0269	0.9795

Table 5: Thermodynamic analysis of nicotine at different rotation rate by the Langmuir adsorption model.

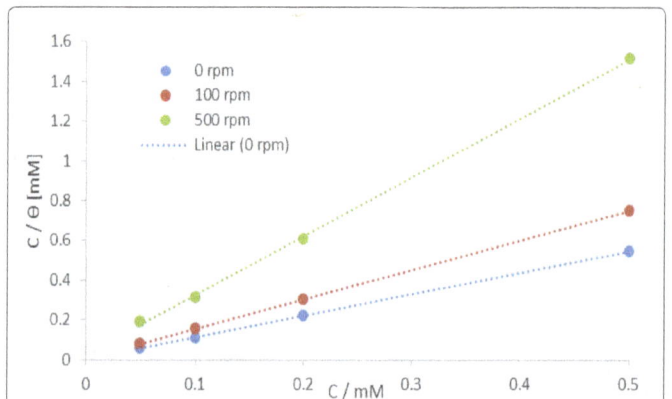

Figure 5: Adsorption isotherm of nicotine in 1M HCl on immersed 1018 steel at different rotation rates.

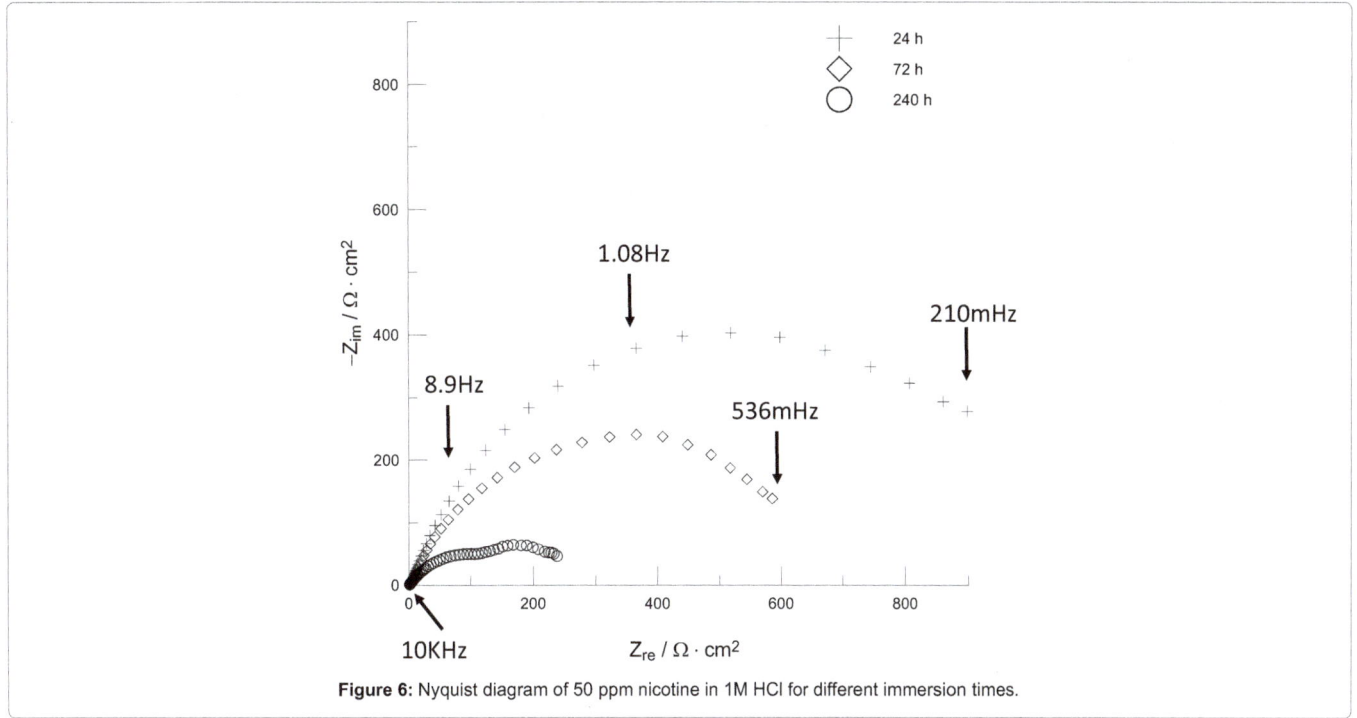

Figure 6: Nyquist diagram of 50 ppm nicotine in 1M HCl for different immersion times.

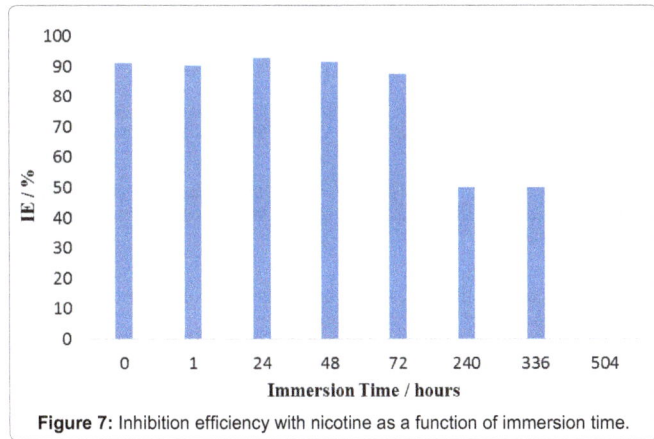

Figure 7: Inhibition efficiency with nicotine as a function of immersion time.

t/h	$R_{ct}/\Omega \cdot cm^2$	$R_{mol}/\Omega \cdot cm^2$	IE/%
1	854.0	5.9	90.3
24	1145.0	1.9	92.8
48	999.2	6.5	91.7
72	661.7	4.4	87.5
240	166.5	106.4	50.2
336	166.5	106.3	50.2
504	77.5	52.6	-

Table 6: Performance parameters of nicotine as a function of immersion time at 50 ppm.

T/C	$Rct_{blank}/\Omega\,cm^2$	$Rct_{inh}/\Omega\,cm^2$	IE/%
30	188.0	510.0	63.1
40	141.2	525.3	73.1
50	71.91	371.7	80.7
60	23.7	105.0	77.4
70	45.0	105.1	57.2

Table 7: Effect of temperature in the presence of 50 ppm of 1M nicotine in HCl.

provides adequate protection (87%), after which this value decreases considerably to ~50%, due to a degradation of the nicotine adsorbed during the corrosion process.

Effect of Temperature

It has been observed that temperature can modify the interaction between the metal and the inhibitor [43]. In Figure 8, Nyquist plots are shown corresponding to: a) steel immersed in 1M HCl, and b) steel immersed in 50 ppm nicotine in the corrosive medium, at different temperatures.

Figure 8a shows that the value of Z_{re} decreases with increasing temperature, attributable to the 1M HCl solution causing material (steel) loss due to a faster corrosion process. When the inhibitor is added (50 ppm at Figure 8b), the previously mentioned effect in immersion time with temperature is also shown, with low charge transfer resistance at high temperature, as can be seen in Table 7. This fact is attributed to the inhibitor formed on the metal surface, thus a decrease in strength of the adsorption process at elevated temperature, suggesting physical adsorption [44] or desorption of the organic molecules (nicotine) at higher temperatures [45].

Conclusion

It was demonstrated that nicotine is an efficient corrosion inhibitor in HCl under static conditions, but only up to 10 ppm with efficiencies around ~ 90%. However, turbulent flow conditions (>40 rpm) affected the corrosion inhibition with IE less than 70%.

The nicotine is adsorbed on the metallic surface consistent with the Langmuir isotherm model.

The inhibitors behavior is affected with increase in temperature decreasing the persistence of the layer due to desorption process; according to the ASTM G 170 standard, this inhibitor is not suitable for high temperatures.

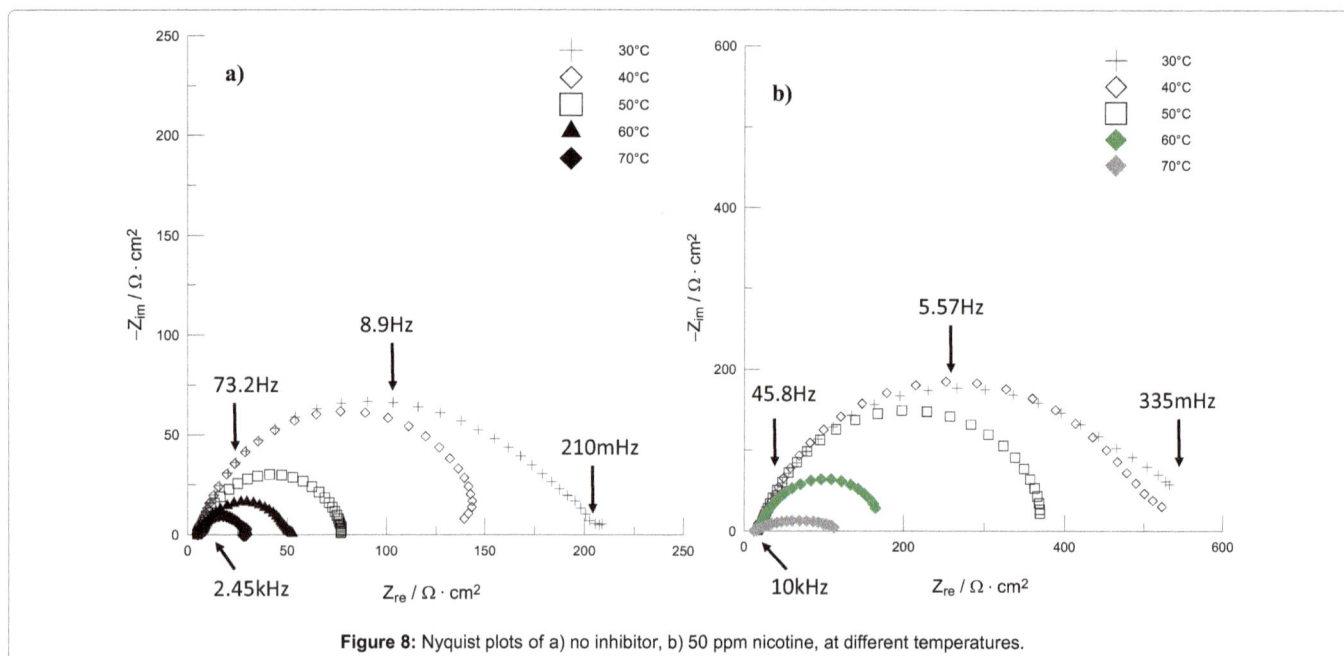

Figure 8: Nyquist plots of a) no inhibitor, b) 50 ppm nicotine, at different temperatures.

According to the ASTM G48 or G78 standard, the inhibitor is accepted when there is persistence of a test layer after 72 hours immersion. The inhibitors for this research meet that criterion.

Acknowledgements

The authors are grateful to the Faculty of Chemistry (UNAM), Department of Metallurgy and DGAPA for a postdoctoral fellowship for the corresponding author. AEV and FJRG wish to acknowledge the SNI (Sistema Nacional de Investigadores) for the honor of membership and the stipends received.

References

1. Espinoza A, Negrón GE, González R, Angeles D, Herrera H, et al. (2014) Mild steel corrosion inhibition in HCl by di-alkyl and di-1H-1,2,3-triazole derivatives of uracil and thymine. Mat Chem Phys 145: 407-417.

2. Yurt A, Bereket G, Kivrak A, Balaban A, Erk B (2005) Effect of Schiff bases containing pyridyl group as corrosion inhibitors for low carbon steel in 0.1 M HCl. J App Electrochem 35: 1025-1032.

3. Awad MI (2006) Eco-friendly corrosion inhibitors: Inhibitive action of quinine for corrosion of low carbon steel in 1M HCl. J of Appl Electrochem 36: 1163-1168.

4. Thiraviyam P, Kannan K (2013) Inhibition of Aminocyclohexane Derivative on Mild Steel Corrosion in 1N HCl. Arab J Sci Eng 38: 1757-1767.

5. Tansug G, Tüken T, Sigircik G, Findikkiran G, Giray ES, et al. (2015) Methyl 3-((2-mercaptophenyl) imino) butanoate as an effective inhibitor against steel corrosion in HCl solution. Ionics 21: 1461-1475.

6. Khadraoui A, Khelifa A, Hamitouche H, Mehdaoui R (2014) Inhibitive effect by extract of *Mentha rotundifolia* leaves on the corrosion of steel in 1M HCl solution. Res Chem Intermed 40: 961-972.

7. El Ouariachi E, Bouyanzer A, Salghi R, Hammouti B, Desjobert J-M, et al. (2015) Inhibition of corrosion of mild steel in 1 M HCl by the essential oil or solvent extracts of *Ptychotis verticillata*. Res Chem Intermed 41: 935-946.

8. Raja PB, Kaleem A, Abdul A, Awang K, Ropi M, et al. (2013) Indole Alkaloids of *Alstonia angustifolia* var. latifolia as Green Inhibitor for Mild Steel Corrosion in 1M HCl Media. JMEPEG 22: 1072-1078.

9. Lebrini M, Robert F, LecanteA, Roos C (2011) Corrosion inhibition of C38 steel in 1 M hydrochloric acid medium by alkaloids extract from *Oxandra asbeckii* plant. Corros Sci 53: 687-695.

10. Oguzie E (2008) Corrosion inhibitive effect and adsorption behaviour of *Hibiscus sabdariffa* extract on mild steel in acidic media. Portug Electro Acta 26: 303-314.

11. Faustin M, Maciuk A, Salvin P, Roos C, Lebrini M (2015) Corrosion inhibition of C38 steel by alkaloids extract of *Geissospermum laeve* in 1M hydrochloric acid: Electrochemical and phytochemical studies. Corros Sci 92: 287-300.

12. El Bribri A, Tabyaoui M, Tabyaoui B, El Attari H, Bentiss F (2013) The use of *Euphorbia falcata* extract as eco-friendly corrosion inhibitor of carbon steel in hydrochloric acid solution. Mat Chem Phys 141: 240-247.

13. Krishnaveni K, Ravichandran J (2014) Influence of aqueous extract of leaves of *Morinda tinctoria* on copper corrosion in HCl medium. J Electroanal Chem 735: 24-31.

14. Bhawsar J, Jain PK, Jain P (2015) Experimental and computational studies of *Nicotiana tabacum* leaves extract as green corrosion inhibitor for mild steel in acidic médium. Alex Eng J 54: 769-775.

15. Njokua DI, Chidiebere MA, Oguzie KL, Ogukwe CE, Oguzie EE (2013) Corrosion inhibition of mild steel in hydrochloric acid solution by the leaf extract of *Nicotiana tabacum*. Adv Mater Corros 1: 54-61.

16. Palomar MP, Romo MR, Hernández HH, Abreu-Quijano MA, Likhanova NV, et al. (2012) Influence of the alkyl chain length of 2 amino alkyl 1,3,4, thiadiazole compounds on the corrosion inhibition of Steel immersed in sulfuric acid solutions. Corros Sci 54: 231-243.

17. Popova A, Christov M, Zwetanova A (2007) Effect of the molecular structure on the inhibitor properties of azoles on mild steel corrosion in 1M hydrochloric acid. Corros Sci 49: 2131-2143.

18. Elayyachy M, Elkodadi M, Aouniti A, Ramdani A, Hammouti B, et al. (2015) New bipyrazole derivatives as corrosion inhibitors for steel in hydrochloric acid solutions. Mat Chem Phys 93: 281-285.

19. Danaee I, Gholami M, RashvandAvei M, Maddahy MH (2015) Quantum chemical and experimental investigations on inhibitory behavior of amino-imino tautomeric equilibrium of 2-aminobenzothiazole on steel corrosion in H2SO4 solution. J Ind Eng Chem 26: 81-94.

20. Bouklah M, Attayibat A, Kertit S, Ramdani A, Hammouti B (2005) A pyrazine derivative as corrosion inhibitor for steel in sulphuric acid solution. App Surf Sci 242: 399-406.

21. Ameh PO, Eddy NO (2014) *Commiphora pedunculata* gum as a green inhibitor for the corrosion of aluminium alloy in 0.1 M HCl. Res Chem Intermed 40: 2641-2649.

22. Khan PF, Shanthi V, Babu RK, Muralidharan S, Barik RC (2015) Effect of benzotriazole on corrosion inhibition of copper under flow conditions. J Environ Chem Eng 3: 10-19.

23. Khaled KF (2008) Guanidine derivative as a new corrosion inhibitor for copper in 3% NaCl solution. Mat Chem Phys 112: 104-111.

24. Mahdavian M, Ashhari S (2010) Corrosion inhibition performance of 2-ercaptobenzimidazole and 2-mercaptobenzoxazole compounds for protection of mild steel in hydrochloric acid solution. Electrochimica Acta 55: 1720-1724.

25. Wang X (2012) The Inhibition Effect of Bis-Benzimidazole Compound For Mild Steel in 0.5M HCl Solution. Int J Electro Sci 71: 11149-11160.

26. Ouici H, Benali O, Harek Y, Larabi L, Hammouti B, et al. (2013) The effect of some triazole derivatives as inhibitors for the corrosion of mild steel in 5% hydrochloric acid. Res Chem Intermed 39: 3089-3103.

27. Lopez DA, Simison SN, de Sánchez SR (2003) The influence of steel microstructure on CO2 corrosion. EIS studies on the inhibition effciency of benzimidazole. Electrochimica Acta 48: 845-854.

28. Hassan HH (2007) Inhibition of mild steel corrosion in hydrochloric acid solution by triazole derivatives: Part II: Time and temperature effects and thermodynamic treatments. Electrochimica Acta 53: 1722-1730.

29. Barmatov E, Hughes T, Nagl M (2015) Efficiency of film-forming corrosion inhibitors in strong hydrochloric acid under laminar and turbulent flow conditions. Corros Sci 92: 85-94.

30. Gopiraman M, Sakunthala P, Kesavan D, Alexramani V, Kim IS, et al. (2012) An investigation of mild carbon steel corrosion inhibition in hydrochloric acid medium by environment friendly green inhibitors. J Coat Technol Res 9: 15-26.

31. Joseph A, Mohan R (2014) Electroanalytical and computational studies on the corrosion inhibition behavior of ethyl (2-methylbenzimidazolyl) acetate (EMBA) on mild steel in hydrochloric acid. Res Chem Intermed 41: 4795-4823.

32. Yadav M, Sharma D, Kumar S (2015) Thiazole derivatives as efficient corrosion inhibitor for oil-well tubular steel in hydrochloric acid solution. Korean J Chem Eng 32: 993-1000.

33. Ramesh S, Rajeswari S (2004) Corrosion inhibition of mild steel in neutral aqueous solution by new triazole derivatives. Electrochimica Acta 49: 811-820.

34. Jafari H, Danaee I, Eskandari H (2014) Inhibitive Action of Novel Schiff Base Towards Corrosion of API 5L Carbon Steel in 1M Hydrochloric Acid Solutions. Trans Indian Inst Met 68: 729-739.

35. Chauhan LR, Gunasekaran G (2007) Corrosion inhibition of mild steel by plant extract in dilute HCl médium. Corros Sci 49: 1143-1161.

36. Djamel D, Tahar D, Saifi I, Salah C (2014) Adsorption and corrosion inhibition of new synthesized thiophene Schiff base on mild steel X52 in HCl and H2SO4 solutions. Corros Sci 79: 50-58.

37. Torres V, Rayol V, Magalhães M, Viana G, Aguiar L, et al. (2014) Study of thioureas derivatives synthesized from a green route as corrosion inhibitors for mild steel in HCl solution. Corros Sci 79: 108-118.

38. Golestani G, Shahidi M, Ghazanfari D (2014) Electrochemical evaluation of antibacterial drugs asenvironment-friendly inhibitors for corrosion of carbon steel in HCl solution. Appl Sur Sci 308: 347-362.

39. Lozano I, Mazario E, Olivares C, Likhanova N, Herrasti P (2014) Corrosion behaviour of API 5LX52 steel in HCl and H2SO4 media in the presence of 1,3-dibencilimidazolio acetate and 1,3-dibencilimidazolio dodecanoate ionic liquds as inhibitors. Mat Chem Phys 147: 191-197.

40. Olvera M, MendozaJ, Genesca J (2015) CO2 corrosion control in steel pipelines. Influence of turbulent flow on the performance of corrosion inhibitors. J ofoss Prev in the Process Industries 35: 19-28.

41. Yadav M, Sharma U, Yadav PN (2013) Isatin compounds as corrosion inhibitors for N80 steel in 15% HCl. Egyptian Journal of Petroleum 22: 335-344.

42. Yildiz R (2015) An electrochemical and theoretical evaluation of 4,6-diamino-2-pyrimidinethiol as a corrosion inhibitor for mild steel in HCl solutions. Corros Sci 90: 544-553.

43. Tourabi M, Nohair K, Nyassi A, Hammouti B, Jama C, et al. (2014) Thermodynamic characterization of metal dissolution and inhibitor adsorption processes in mild steel/3,5-bis(3,4-dimethoxyphenyl)-4-amino- 1,2,4-triazole/ hydrochloric acid system. J Mater Environ Sci 5: 1133-1143.

44. Missoum N, Guendouz A, Boussalah N, Hammouti B, Chetouani A, et al. (2013) Synthesis and evaluation of bipyrazolic derivatives as inhibitors of corrosion of C38 steel in molar hydrochloric acid. Res Chem Intermed 39: 3441-3461.

45. Karthik G, Sundaravadivelu M, Rajkumar P (2015) Corrosion inhibition and adsorption properties of pharmaceutically active compound esomeprazole on mild steel in hydrochloric acid solution. Res Chem Intermed 41: 1543-1558.

10

Determination of Epinephrine at a Screen-Printed Composite Electrode Based on Graphite and Polyurethane

Dias IARB, Saciloto TR, Cervini P and Cavalheiro ETG*

Departamento de Química e Física Molecular, Instituto de Química de São Carlos, Av. Trabalhador São Carlense, 400, Centro, São Carlos, São Paulo CEP 13566-590, Brazil

Abstract

A screen-printed electrode based on a graphite and polyurethane composite (SPGPU) was used in the determination of epinephrine (EP) in cerebral synthetic fluid (CSF) sample. Both Differential Pulse Voltammetry (DPV) and Square Wave Voltammetry (SWV) were used to investigate the suitability of sensor for determination of EP. Under the optimum conditions, the analyte oxidation signal was observed at 0.17 and 0.080 V for DPV and SWV, respectively (*vs. pseudo*-Ag|AgCl) in phosphate buffer (pH=7.4). A linear region between 0.10 and 1.0 µmol L^{-1} was observed in DPV, with detection limit of 6.2×10^{-7} mol L^{-1} (R=0.997). In SWV two linear ranges were observed, the first one between 0.10 and 0.80 µmol L^{-1} and the second from 1.0 to 8.0 µmol L^{-1} with limit of detection 9.5×10^{-8} and 6.0×10^{-7} mol L^{-1} (R=0.998) respectively. Recoveries of 99 to 100% were observed using the sensor for determination of epinephrine in the CSF. Interference tests showed that uric and ascorbic acids as well as dopamine increase the current of epinephrine, with acceptable levels for UA. The use of a standard addition of EP in the CSF solution containing the ascorbic acid allowed minimizing such interference.

Keywords: Epinephrine; Screen printed electrode; Graphite-polyurethane composite; Dopamine

Introduction

Catecholamines control many processes in the body. The interest in the study of such compounds is due to fact that they are connected to the neurotransmitting process and associated with the Parkinson and Alzheimer diseases [1]. Thus, studies regarding determination of this analyte are of utmost importance.

Epinephrine (EP, Figure 1) is one of the members of this group present in sympathetic nervous system and synthesized by adrenal glandule, acting when the individual is subjected to strong emotions and whose main function is reducing the amount of blood in peripheral circulation, transferring the blood flow for other organs as heart, brain, liver and kidneys [2].

Despite their importance as neurotransmitter, the amount of catecholamines present in organism can induce a series of diseases, for example, in high concentrations can cause diabetes, trauma in central nervous system and arterial hypertension, in low concentrations can induce cognition and memorization problems and even schizophrenia [1].

Several analytical methods have been developed and employed in the determination of EP. Examples are those involving high performance liquid chromatography (HPLC) [3,4], extraction prior of fluorometric detection [5], solid-phase extraction for liquid chromatography [6], solid-phase extraction combined with LC-MS/MS [7], chemiluminescence [8], solid phase reactor based on molecularly imprinted polymer (MIP) [9] and others. Usually these techniques result in low analytical frequencies requiring complex

sample preparation and generation of relatively large amounts of waste.

In a general sense, electroanalytical methods present many advantages in the determination of electroactive substances such as analyzing small volumes of samples, faster analysis time, among others. They are also used for the determination of EP with hundreds of proposed procedures and strategies of electrode modifications. Some representative examples of these determinations are summarized in Table 1.

On top of these advantages, screen-printed electrodes (SPE) present, suitability for automation procedures associated to higher reliability and repeatability [20,21] with a so low cost that they are disposable. Furthermore the preparation and modification of the ink with others materials such as metals, enzymes, polymers, nanomaterials and complexing agents increase the selectivity and sensibility of sensor in the determination of several analytes.

In this work the use of a screen-printed electrode based on a graphite and polyurethane composite (SPGPU) is presented as an electroanalytical sensor in the determination of EP in cerebrospinal synthetic fluid. The same sensor has been earlier evaluated with satisfactory results in electroanalytical determination of acetaminophen [22], simultaneous determination of acetaminophen and caffeine [23] and simultaneous determination of Zn(II), Pb(II), Cu(II), and Hg(II) in ethanol fuel using the SPGPU modified with organofunctionalized SBA-15 silica [24].

The ink used in this work is based on the same graphite-

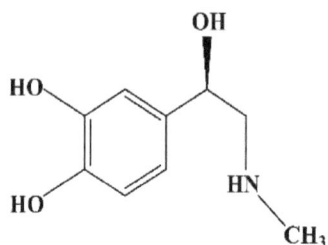

Figure 1: Structural formula of Epinephrine.

*Corresponding author: Cavalheiro ETG, Departamento de Química e Física Molecular, Instituto de Química de São Carlos, Av. Trabalhador São Carlense, 400, Centro, São Carlos, São Paulo CEP 13566-590, Brazil
E-mail: cavalheiro@iqsc.usp.br

Electrode	Modification	Linear Range (μmol L^{-1})	LOD (mol L^{-1})	Reference
Paste Carbon	Pre-anodized	0.2-400	6.2×10^{-8}	[10]
Nanotube Carbon Paste	Au	10-150	2.8×10^{-6}	[11]
Carbon Paste	Pristine Multi-Walled Carbon Nanotubes	0.1-1.0 and 1.0- 100	4.5×10^{-8}	[12]
Nanotubes Carbon Paste	Poly(Serine)/Multi-Walled	1-220	6.0×10^{-7}	[13]
Glassy Carbon	Composite Film of MnO$_2$ and Nafion	0.03-10 and 10-100	5.0×10^{-9}	[14]
Glassy Carbon	Au-nanoparticle poly-fuchsine acid film	0.5-792.7	1.0×10^{-8}	[15]
Screen Printed Electrodes	Oxidized Single-Wall Carbon Nanohorns (o-SWCNHs)	2-2500	1.0×10^{-7}	[16]
Carbon Nanotube	Multiwalled Carbon Nanotubes (MWCNTs) in a Chitosan Matrix	10-100	9.0×10^{-7}	[17]
Carbon Paste	Multi-walled carbon nanotube	10-100 and 0.5-10	2.9×10^{-8}	[18]
Glassy Carbon	Polypyrrole/multi-walled carbon nanotube	0.1-8.0 and 10-100	4.0×10^{-8}	[19]
Screen-Printed Electrodes (SPE)	Graphite and Polyurethane Composite (SPGPU)	0.1-1.0 and 0.1-0.8	6.23×10^{-7} and 9.51×10^{-8}	This work

Table 1: Comparison of some analytical characteristics of the proposed method with other examples of electrochemical procedures found in the literature.

polyurethane composite previously used as conventional disk working electrode in the determination of many analytes, as recently reviewed [25] while here the material was used to construct the printed device as well as the electrical contacts of the SPE.

Experimental

Reagents and solutions

All solutions were prepared with water purified in a Barnstead™ EasyPure® RoDi (Thermoscientific, model D13321) system with resistivity $\geq$ than 18 MΩ cm. The EP used was of analytical grade (Sigma-Aldrich, Germany). A stock 1.0×10^{-4} mol L^{-1} solution was prepared daily by dissolving the analyte in 0.1 mol L^{-1} phosphate buffer pH 7.4 and protect of light.

Apparatus

Voltammetric experiments were performed in an AUTO-LAB PGSTAT-30 (Ecochemie, The Netherlands) potentiostat/galvanostat coupled to a personal computer and controlled with a GPES 4.9 software.

Preparation of the screen-printed electrodes

The SPGPU was prepared as previously described Saciloto et al. [26]. The graphite-polyurethane composite (60%, graphite m/m) was used to prepare the components of the screen-printed electrodes as working and auxiliary electrodes as well as in the electrical contacts of the printed device. As already discussed [27], the composite does not respond in graphite contents lower than 50% (m/m) and it is not physically resistant in graphite contents higher than 70% (m/m). The best signals can be found using 60% (graphite, m/m). Briefly, the process consists in forcing the ink composed by composite GPU and solvent to pass through a mask to be deposited on a PVC plate (0.5 mm thickness). So, this set is partly covered by a layer of pure PU resin, used as insulation to define the area of electrical contact at one end. At the other end, there was another uncoated area to define the active area allowing the electrodes to be exposed. To one of the imprints it was attached a strip of silver epoxy (Conductive Silver Epoxy Kit, Electron Microscopy Sciences, USA) to serve as a *pseudo*-reference electrode (Figure 2).

After curing of the polymers and concluded the assembling step

the electrodes were activated by cycling between -0.8 to +1.2 V (*vs* Ag | AgCl), in 0.10 mol L^{-1} buffer phosphate pH 7.4, with 150 cycles at 200 mVs^{-1} scan rate.

Procedures for preparation of the cerebrospinal synthetic fluid

For determination of EP, a 250.0 mL solution of the cerebrospinal synthetic fluid (CSF) was prepared according to Oser [28] and Zhang et al. [29] as described in Table 2, containing the major constituents of the biologic fluid. This solution was spiked with EP and the pH adjusted for 7.4. The solution was immediately used after preparation to avoid the hydrolysis of the urea.

Optimization of the DPV and SWV parameters

The experimental parameters for DPV and SWV were optimized in 0.1 mol L^{-1} phosphate buffer pH 7.4, containing 5.0×10^{-5} mol L^{-1} de EP.

For the DPV, the scan rates ranged from 10 to 50 mV s^{-1} and pulse amplitude between 10.0 and 100 mV in a potential window -0.25 to +0.20 V (*vs. pseudo*-Ag | AgCl). In SWV the potential window used was -0.12 to +0.40 V (*vs. pseudo*-Ag | AgCl), scan rate ranged from 10 to 50 mV s^{-1}, frequency 8.0 to 50 s^{-1}, scan increment 2.0 to 20 mV and amplitude between 10.0 to 100 mV.

Effect of the pH in redox process of EP

The experiments were realized using cyclic voltammetry in 0.1 mol L^{-1} phosphate buffer with pH varying of 0.5 unit between 5.0 to 8.0 in the presence of 1.00×10^{-4} mol L^{-1} epinephrine.

Interference studies

Interference studies were performed using substances with biological relevance as uric acid, ascorbic acid and dopamine at SPGPU in 0.1 mol L^{-1} phosphate buffer pH 7.4. In these studies, the peak current of 4.0×10^{-7} mol L^{-1} of epinephrine was compared with those obtained in the presence of 2.0, 4.0 and 8.0×10^{-7} mol L^{-1} of each potential interfering specie.

Results and Discussion

Electrochemical behavior of epinephrine at SPGPU

The morphology of the imprinted layer of the polymer can be

Substance	Amount/g
NaCl	2.10
KCl	0.07
CaCl$_2$	0.08
Glucose	0.20
NaHCO$_3$	0.40
Urea	0.002

Table 2: Composition of the CSF for the biologic analysis, amount of each component in 250.0 mL [28,29].

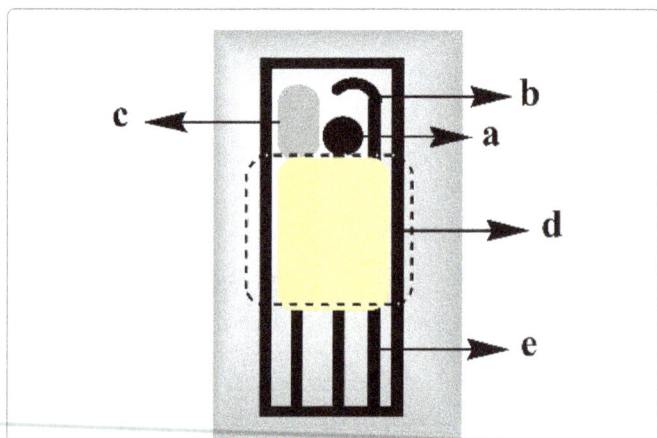

Figure 2: Schematic representation of the SPGPU with all the components: working electrode (a), auxiliary electrode (b), silver glue (c), insulating (d) and electrical contact (e).

observed in the SEM image in Figure 3. In which a homogeneous distribution of the particles of the composite can be noticed with average size in the 10 μm order as well as several porous inherent from the imprinting process and solvent evaporation. The diameter of the working electrode is presented in experimental section as 3 mm and the thickness of the composite layer is defined by the imprinting mask as 100 μm.

Figure 4 presents the cyclic voltammograms at SPGPU in 0.1 mol L^{-1} phosphate buffer pH 7.4 after the activation of electrode in potential window -0.80 to +1.2 V, at 200 mV s^{-1} scan rate during 150 cycles, in the presence of 1 × 10^{-4} mol L^{-1} of EP. The first peak in 0.15 V generates a product that presents a reversible redox couple of peaks II/III in -0.25/-0.33 V (vs. pseudo-Ag | AgCl) associated to the presence of epinephrinechrome and leucoepinephrinechrome [30,31].

The mechanism of EP oxidation was proposed by Hawley and Kim et al. as represented in Scheme 1 [32,33].

According to the literature [34] the peak I represents the oxidation of EP, which involves two electrons and two protons producing epinephrinequinone (peak II). The oxidation-reduction involving the couple of peaks II/III corresponds to the electron transferring process that occurs via cyclization of leucoepinephrinechrome to epinephrinechrome. Because of signal intensity the peak I in 0.15 V (vs pseudo-Ag | AgCl) was chosen for further detailed studies of electrochemical determination of EP, being the potential window set at -0.25 to 0.20 V and -0.12 to 0.40 V (vs pseudo-Ag | AgCl) in DPV and SWV respectively.

The same process was observed in other studies for determination of the EP using sensor based on carbon nanotube film modified carbon electrodes [17] and electrochemical studies of oxidation of catecholamines [32].

Effect of the pH in redox process of EP

The effect of pH in redox process of the EP at SPGPU was investigated by studying the dependence of EP oxidation peak potential with the pH. The experiments were performed by cyclic voltammetric with the values of pH varying of 0.5 unit between 5.0 to 8.0 in the presence of 1.0 × 10^{-4} mol L^{-1} EP solutions. Figure 5 shows the dependence of the anodic peak current and the peak potential with of pH.

The curve of I_{pa} vs. pH revealed that the peak of anodic current have a maximum value in pH=7.5, decreasing in higher pH values. This is explained considering the oxidation-reduction mechanism of the EP, presented in Scheme 1. The oxidation process involves the liberation of 2H$^+$/molecule that is injured in acid medium. On the other hand EP presents a pK$_a$=8.88 [12], and in higher pH values the increasing concentration of the non-electroactive deprotonated form, causes a decrease in peak current. The physiologic pH of the cerebrospinal fluid for healthy individuals is approximated of 7.4, that corresponding to the region of maximum current in Figure 5, thus this pH 7.4 was chosen for the further experiments.

Analytical curve for epinephrine

After optimizing of the experimental conditions for DPV the potential interval of -0.25 to +0.20 V (vs. pseudo-Ag | AgCl) was chosen due to the presence of an irreversible and more intense peak near to 0.17 V (vs. pseudo-Ag | AgCl) relative to epinephrinequinone oxidation.

An analytical curve was thus obtained using 50 mV pulse amplitude and 10 mV s^{-1} scan rate for concentrations between 0.10 to 8.0 μmol L^{-1} as presented in Figure 6, with linear response between 0.10 to 1.0 μmol L^{-1}, represented by Equation 1:

$$\Delta I = 5.377 \times 10^{-8}(\mu A) + 0.313(\mu A\,\mu mol^{-1}L) \times C_{Epineprine} \quad (1)$$

The correlation coefficient obtained was 0.997 and a detection limit of 6.23 × 10^{-7} mol L^{-1} calculated as three times the standard deviation of the background divided by the slope of the linear curve [35] and a limit of quantification 2.07 × 10^{-6} mol L^{-1} calculated as 10 times the background standard deviation divided by the slope of the straight line.

It is important to note that there is a displacement in the peak potential (Figure 5) while increasing the analyte concentration, so currents were taken at the higher values, in the respective peak potential. In addition for concentrations higher than 1.0 μmol L^{-1} saturation of electrode surface, resulting in non-linear responses were observed.

In SWV, an analytical curve was obtained in the potential interval of -0.12 a +0.40 V (vs pseudo-Ag | AgCl), 35 mV s^{-1} scan rate; 5 mV step potential, 8 s^{-1} frequency and 50 mV amplitude. The analytical curve presented two linear regions, being most sensitive the first one between 0.10 μmol L^{-1} to 0.80 μmol L^{-1} as shown in Figure 7, represented by the Equation 2:

$$\Delta I = 2.276 \times 10^{-7}(\mu A) + 0.871(\mu A\,\mu mol^{-1}L) \times C_{Epineprine} \quad (2)$$

The first region presented a correlation coefficient of 0.998, with detection limit of 9.51 × 10^{-8} mol L^{-1} and a limit of quantification 3.17 × 10^{-7} mol L^{-1}. The second one presented a correlation coefficient of 0.998, with a detection limit of 6.00×10^{-7} mol L^{-1}. Thus for analytical purposes the first linear range will be considered.

The presence of two linear regions in the analytical curves from both techniques suggests a competition by the active sites of the electrode above a certain concentration, probably provoked by the

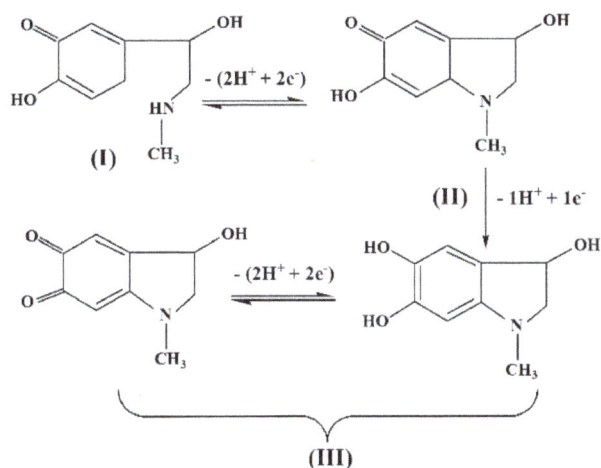

Scheme 1: Electrochemical oxidation mechanism of EP at SPGPU. (I) Epinephrine; (II) Reduction of Epinephrinequinone to Leucoepinephrinechrome and (III) Oxidation of Leucoepinephrinechrome to Epinephrinechrome [33].

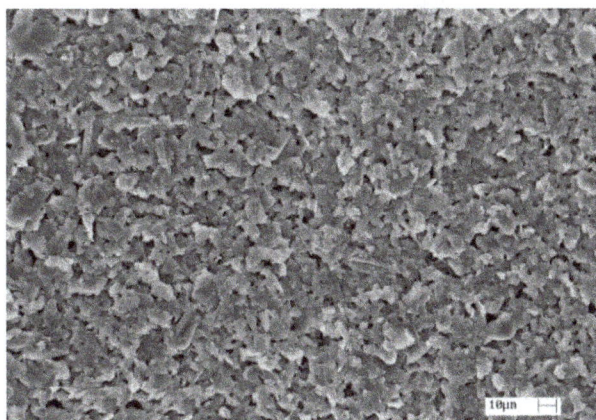

Figure 5: Dependence of peak potential and anodic peak current against pH for EP oxidation in 0.1 mol L^{-1} phosphate buffer pH 7.4, in the presence of 1.0×10^{-4} mol L^{-1} EP.

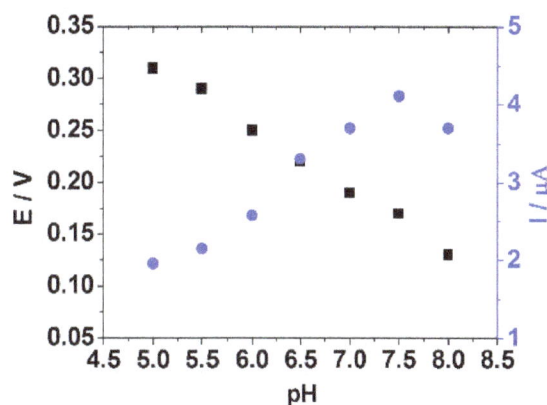

Figure 3: Scanning electron microscopy picture of the graphite-polyurethane composite imprinted on the PVC support.

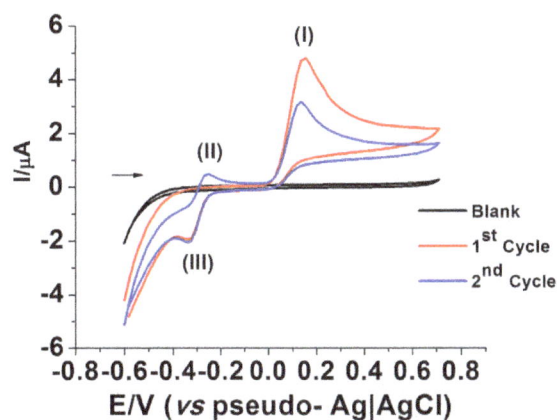

Figure 6: Differential pulse voltammograms obtained with the SPGPU for concentrations (1) 1×10^{-7}; (2) 2×10^{-7}; (3) 4×10^{-7}; (4) 6×10^{-7}; (5) 8×10^{-7}; (6) 1×10^{-6}; (7) 2×10^{-6}; (8) 4×10^{-6}; (9) 6×10^{-6}; (10) 8×10^{-6} mol L^{-1} EP. In detail: dependency of peak current with concentration of EP.

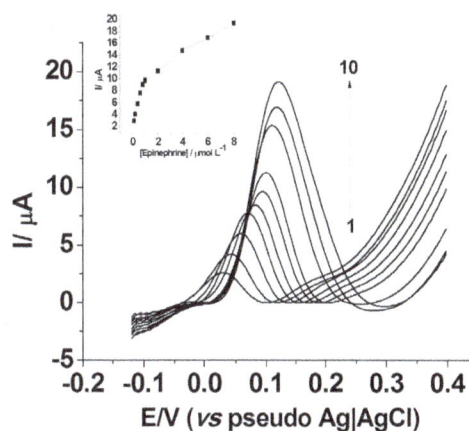

Figure 4: Cyclic voltammograms at SPGPU, showing two successive cycles in presence of de 1×10^{-4} mol L^{-1} of EP in 0.1 mol L^{-1} phosphate buffer pH 7.4, v=25 mV s^{-1}.

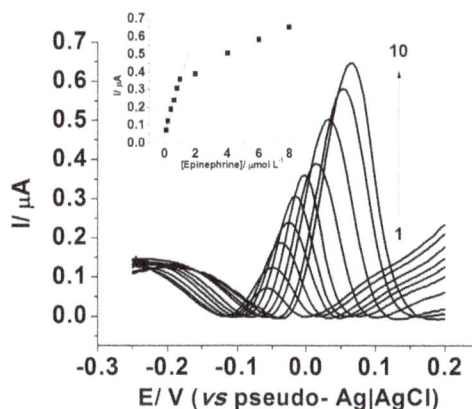

Figure 7: Square wave voltammograms obtained with the SPGPU for concentrations: (1) 1×10^{-7}; (2) 2×10^{-7}; (3) 4×10^{-7}; (4) 6×10^{-7}; (5) 8×10^{-7}; (6) 1×10^{-6}; (7) 2×10^{-6}; (8) 4×10^{-6}; (9) 6×10^{-6}; (10) 8×10^{-6} mol L^{-1} EP. In detail: linear dependency of peak current with concentration of EP.

Technique	Interferant/µmol L⁻¹		Epinephrine signal/µA	Interference/%
DPV	AA	0.20	0.142	350
		0.40	0.152	317
		0.80	0.152	394
	DA	0.20	0.450	220
		0.40	0.675	380
		0.80	0.932	562
	UA	0.20	0.113	1.49
		0.40	0.124	8.20
		0.80	0.111	2.54
SWV	AA	0.20	0.0296	40.2
		0.40	0.193	291
		0.80	0.160	222
	DA	0.20	0.737	362
		0.40	1.39	729
		0.80	1.54	868
	UA	0.20	0.171	3.60
		0.40	0.160	9.84
		0.80	0.164	7.65

Table 3: Effect of interfering in the presence of 0.40 µmol L⁻¹ of Epinephrine.

Technique	Addition	Ep/µmol L⁻¹		Recovery/%
		Added	Found	
DPV	1	0.100	0.103	103
	2	0.150	0.149	99.3
	3	0.200	0.198	99.0
			Average	100 ± 2*
SWV	1	0.150	0.145	97.3
	2	0.200	0.198	99.3
	3	0.250	0.253	101
			Average	99.3 ± 4*

*Average ± standard deviation, n=3

Table 4: Recovery test for Epinephrine in CSF under optimized conditions.

turnover on these active sites occupied by the electroactive molecules, once the kinetics of electron transfer is relatively slow, in an irreversible process.

Interferences study

To investigate the chemical interference in the EP analytical signal the effect of substances usually present in biologic fluids was evaluated using a fixed 4×10^{-7} mol L⁻¹ EP solution in the presence of varying concentrations of uric (UA) and ascorbic (AA) acids and dopamine (DA), as detailed in Table 3.

Severe interference from AA and DA while acceptable interference levels were observed for UA was observed in both techniques. The increase in peak current of the EP in presence of these interferants suggests a superposition of redox process involving both the analyte and the interfering species at similar potentials. The interference from AA was overcome by standard addition of the EP in CSF solutions containing the ascorbic acid.

Determination of epinephrine in sample of cerebrospinal synthetic fluid (CSF)

Table 4 presents the results of recoveries of EP in CSF using the SPGPU.

These results are comparable to that using edge plane pyrolytic graphite electrode modified with nanotubes for determination of EP in samples of blood and urine. This study showed recoveries of 97-104%

[36]. The results obtained showed that the EIGPU can be used as sensor electroanalytical in the determination of EP in biologic samples with good results regarding recoveries.

These devices are supposed to be disposable due to its relatively low cost and have demonstrated to be stable for a relatively large period of time, at least 2 years, after production. After electrochemical pretreatment the devices are being used in our lab for more than 50 measurements, at least.

Conclusions

The composite SPGPU presented a satisfactory limit of detection and recoveries when compared to the literature and the possibility of determining EP in cerebral synthetic fluid using DPV and SWV in a simple, low cost and waste generation procedure in a disposable device even in the presence of AA and interference from UA and DA.

Acknowledgements

The authors are indebted to the Brazilian agencies CNPq and FAPESP (Grants 2010/05913-6 and 2010/11080-7).

References

1. Grossman M, Glosser G, Kalmanson J, Morris J, Stern MB, et al. (2001) Dopamine supports sentence comprehension in Parkinson's Disease. J Neurol Sci 184: 123-130.

2. Hoffman BB (2004) Cathecolamines: Encyclopedia of Endocrine Diseases. Amsterdam: Academic Press, p: 480.

3. Silva DAF, Menezes ML, Kempinas WG (2007) A new method for simultaneous

determination of catecholamines in reproductive organs from rats by high performance liquid chromatography with electrochemical detection. Ecletica Quim 32: 35-42.

4. Carrera V, Sabater E, Vilanova E, Sogorb MA (2007) A simple and rapid HPLC–MS method for the simultaneous determination of epinephrine, norepinephrine, dopamine and 5-hydroxytryptamine: Application to the secretion of bovine chromaffin cell cultures. J Cromatography B 847: 88-94.

5. Davletbaeva P, Falkova M, Safonova E, Moskvin L, Bulatov A (2016) Flow method based on cloud point extraction for fluorometric determination of epinephrine in human urine. Anal Chim Acta 911: 69-74.

6. Woo HI, Yang JS, Oh HJ, Cho YY, Kimd JH, et al. (2016) A simple and rapid analytical method based on solid-phase extraction and liquid chromatography-tandem mass spectrometry for the simultaneous determination of free catecholamines and metanephrines in urine and its application to routine clinical analysis. Clin Biochem 49: 573-579.

7. Li XS, Li S, Kellermann G (2016) Pre-analytical and analytical validations and clinical applications of a miniaturized, simple and cost-effective solid phase extraction combined with LC-MS/MS for the simultaneous determination of catecholamines and metanephrines in spot urine samples. Talanta 159: 238-247.

8. Li T, Wang Z, Xie H, Fu Z (2012) Highly sensitive trivalent copper chelate-luminol chemiluminescence system for capillary electrophoresis detection of epinephrine in the urine of smoker. J Chromatogr B Analyt Technol Biomed Life Sci 911: 1-5.

9. Sartori LR, Santos WJR, Kubota LT, Segatelli MG, Tarley CRT (2011) Flow-based method for epinephrine determination using a solid reactor based on molecularly imprinted poly (FePP–MAA–EGDMA). Mat Sci and Eng C 31: 114-119.

10. Li J, Shangguan E, Guo D, Gao F, Li Q, et al. (2015) Influence of acidity and auxiliary electrode reaction on the oxidation of Epinephrine on the pre-anodized carbon paste electrode. Electrochim Acta 186: 209-215.

11. Wierzbicka E, Szultka-Młynska M, Buszewskib B, Sulka GD (2016) Epinephrine sensing at nanostructured Au electrode and determination its oxidative metabolism. Sens and Act B 237: 206-215.

12. Thomas T, Mascarenhas RJ, D'Souza OJ, Detriche S, Mekhalif Z, et al. (2014) Pristine multi-walled carbon nanotubes/SDS modified carbon paste electrode as an amperometric sensor for epinephrine. Talanta 125: 352-360.

13. Narayana PV, Reddy TM, Gopal P, Reddy MM, Naidu GR (2015) Electrocatalytic boost up of epinephrine and its simultaneous resolution in the presence of serotonin and folic acid at poly(serine)/multi-walled carbon nanotubes composite modified electrode: A voltammetric study. Mat Sci and Eng C 56: 57-65.

14. Liu X, Ye D, Luo L, Ding Y, Wanga Y, et al. (2012) Highly sensitive determination of epinephrine by a MnO$_2$/Nafion modified glassy carbon electrode. J Electroanal Chem 665: 1-5.

15. Taei M, Hasanpour F, Tavakkoli N, Bahrameian M (2015) Electrochemical characterization of poly(fuchsine acid) modified glassy carbon electrode and its application for simultaneous determination of ascorbic acid, epinephrine and uric acid. J Mol Liq 211: 353-362.

16. Valentini F, Ciambella E, Conte V, Sabatini L, Ditaranto N, et al. (2014) Highly selective detection of Epinephrine at oxidized Single-Wall Carbon Nanohorns modified Screen Printed Electrodes (SPEs). Biosens and Bioelectron 59: 94-98.

17. Ghica ME, Brett CMA (2013) Simple and efficient Epinephrine sensor based on carbon nanotube modified carbon film electrodes. Anal Lett 46: 1379-1393.

18. Thomas T, Mascarenhas RJ, Martis P, Mekhalif Z, Swamy BEK (2013) Multi-walled carbon nanotube modified carbon paste electrode as an electrochemical sensor for the determination of epinephrine in the presence of ascorbic acid and uric acid. Mat Sci and Eng C 33: 3294-3302.

19. Shahrokhiana S, Saberib RS (2011) Electrochemical preparation of over-oxidized polypyrrole/multi-walled carbon nanotube composite on glassy carbon electrode and its application in epinephrine determination. Electrochim Acta 57: 132-138.

20. Alonso-Lomillo MA, Domínguez-Renedo O, Arcos-Martínez MJ (2010) Screen-Printed biosensors in microbiology: A Review. Talanta 82: 1629-1636.

21. Barry RC, Lin Y, Wang J, Liu G, Timchalk CA (2009) Nanotechnology based electrochemical sensors for biomonitoring chemical exposures. J Expo Sci Environ Epidemiol 19: 1-18.

22. Saciloto TR, Cervini P, Cavalheiro ETG (2013) New Screen printed electrode based on graphite and polyurethane composite for the determination of acetaminophen. Anal Lett 46: 312-322.

23. Saciloto TR, Cervini P, Cavalheiro ETG (2013) Simultaneous voltammetric determination of acetaminophen and caffeine at a graphite and polyurethane screen-printed composite electrode. J Braz Chem Soc 24: 1461-1468.

24. Saciloto TR, Cervini P, Cavalheiro ETG (2014) Simultaneous voltammetric determination of Zn (II), Pb (II), Cu (II) and Hg (II) in ethanol fuel using an organofunctionalized modified graphite-polyurethane composite disposable screen-printed device. Electroanal 26: 1-14.

25. Cavalheiro ETG, Brett CMA, Oliveira-Brett AM, Fatibello-Filho O (2012) Bioelectroanalysis of pharmaceutical compounds. Bioanal Rev 4: 31-53.

26. Saciloto TR, Cervini P, Cavalheiro ETG (2012) Tinta e processo para preparação de eletrodos impressos descartáveis a base de um compósito de grafite e poliuretana eletrodo obtido pelo referido processo. Br PI 104: 355.

27. Mendes RK, Claro-Neto S, Cavalheiro ETG (2002) Evaluation of new rigid carbon-castor oil polyurethane composite as an electrode material. Talanta 57: 909-917.

28. Oser BL (1952) Hawk's Physiological Chemistry. TATA McGraw-Hill, New Delhi, p: 1054.

29. Zhang F, Yang L, Shuping B, Liu J, Liu F, et al. (2001) Neurotransmitter dopamine applied in electrochemical determination of aluminum in drinking waters and biological samples. J Inorg Biochem 87: 105-115.

30. Luczak T (2009) Epinephrine oxidation in the presence of interfering molecules on gold and gold electrodes modified with gold nanoparticles and thiodipropionic acid in aqueous solution. A comparative study. Electroanalysis 21: 2557-2562.

31. Mang D, Yueming Z, Xizhen L, Hongbin Z, Zhenzhen W, et al. (2016) An electrochemical sensor based on graphene/poly (brilliant cresyl blue) nanocomposite for determination of epinephrine. J Electrochem Soc 763: 25-31.

32. Hawley MD, Tatawadi SV, Piekarski S, Adams RN (1967) Electrochemical studies of oxidation pathways of cathecolamines. J Am Chem Soc 89: 447-450.

33. Kim SH, Lee JW, Yeo IH (2000) Spectroelectrochemical and electrochemical behavior of epinephrine at a gold electrode. Electrochim Acta 45: 2889-2895.

34. Wierzbicka E, Szultka-Młynska M, Buszewskib B, Sulka GD (2016) Epinephrine sensing at nanostructured Au electrode and determination its oxidative metabolism. Sens and Act B 237: 206-215.

35. Long GL, Winefordner JD (1983) Limit of detection. Anal Lett 55: 712-724.

36. Goyal RN, Bishnoi S (2011) Simultaneous determination of epinephrine and norepinephrine in human blood plasma and urine samples using nanotubes modified edge plane pyrolytic graphite electrode. Talanta 84: 78-83.

11

Analytical Method for Transdermal Delivery of the Anti-angiogenic Compound TNP-470

Eva Abramov, Ouri Schwob and Ofra Benny*

The Institute for Drug Research, The School of Pharmacy, Faculty of Medicine, The Hebrew University of Jerusalem, Jerusalem, Israel

Abstract

Pathological angiogenesis is a critical component in cancer, in chronic systemic inflammatory diseases such as psoriasis and rheumatoid arthritis, and in ocular diseases. Anti-angiogenic drugs have the ability to prevent, inhibit, and regress newly formed blood vessels. The activity of TNP-470 (chloro acetylcarbamoylfumagillol), a potent anti-angiogenic drug, has been demonstrated in numerous preclinical studies and in eight clinical studies involving more than three hundred patients. Despite its encouraging efficacy, TNP-470 is unstable compound with short plasma half-life, and, as was found clinically it can cause neurotoxicity side-effects at high doses. In light of these limitations, developing a transdermal drug delivery for TNP-470, can offer a novel and promising clinical usage for this drug by improving its bioavailability, controlled dosage and safety profile. In this work, we developed a reliable method for skin permeation studies of TNP-470, using the pig skin in Franz diffusion cells and High-Performance Liquid Chromatography (HPLC) analysis. Additionally, we performed a broad stability and degradation studies of TNP-470 in different mediums and identify optimal stabilizing conditions in acetate buffer pH-4.5, which can be used for transdermal formulation. Our results demonstrated excellent permeability properties of TNP-470 through the pig skin, where 25% from the initial amount was crossed through the skin membrane after 72 hours. Our results are suggesting that TNP-470 is a good candidate for transdermal drug delivery, whereas, an optimal dermal formulation would improve drug's pharmacokinetic properties and toxicity profile by introducing it in a slow release system.

Keywords: Transdermal delivery; TNP-470; Franz cell; HPLC; Anti-angiogenic therapy

Background

Angiogenesis, the formation of new blood vessels, has an essential role in development and reproduction [1]. However, pathological angiogenesis is a critical component in neovascular associated diseases such as cancer, infantile haemangiomas, psoriasis, rheumatoid arthritis and ocular diseases such as age-related macular degeneration [2]. Since angiogenesis is a mutli-step process, several strategies are being developed to target distinct steps which would ultimately interfere in capillary formation. Such anti-angiogenic drugs have the ability to prevent, inhibit, and regress newly formed blood vessels [1,3].

TNP-470 is a synthetic analogue of fumagillin, a natural product secreted by the fungus Aspergillus fumigatus Fresenius [4] which was found by Folkman lab to impair endothelial tube formation [4]. From over 100 synthetic derivatives, TNP-470 was identified as the most potent angiogenic inhibitor, which was 50 times more active than fumagillin and less cytotoxic [5]. The molecular target of TNP-470 was found to be the intracellular enzyme, methionine aminopeptidase 2 (MetAp2) an enzyme responsible for the removal of methionine from newly synthesized proteins [6,7]. MetAp2 is overexpressed in proliferating endothelial cells, and therefore inhibition of MetAp2 leads to selective inhibition of angiogenesis. TNP-470 showed a broad spectrum anti-cancer activity in numerous preclinical studies and in clinical trials, performed in more than 300 patients, it showed indication of disease stabilization and a few cases of complete remission [8-16].

However, despite its encouraging efficacy, TNP-470 has a several serious limitations for clinical uses. Extremely rapid degradation of TNP-470 in plasma, half-life about 7-8 minutes [14,16] leading to poor pharmacokinetics. TNP-470's poor oral availability reinforced administration regime of frequent continuous intravenous infusion. Additionally, the dose limiting toxicity involved neurological side effects such as dizziness, decreased concentration, short-term memory loss, confusion and depression [17]. To overcome these limitations

pharmacological approaches such as advanced systemic nano-formulations [18] local delivery, or slow release can be used. Transdermal drug delivery can potentially offer a novel and promising clinical usage for TNP-470 for improving its pharmacological properties and safety profile [19].

Transdermal drug delivery is the transport of drugs across the skin into systemic circulation [19]. It has significant advantages over more traditional delivery dosages including controlled drug release and consistent drug levels in the plasma [20]; reduced toxicity due to reduction of drug peak concentrations; improved stability of drugs by avoiding the gastrointestinal tract and first pass metabolism. Additionally, transdermal drug delivery provides an alternative mode of administration to accommodate patients who cannot tolerate oral dosage forms or those who are unconscious. Human skin is the largest organ of the human body and most readily accessible surface for drug delivery which receives about one-third of the blood circulating through the body. However, transdermal drug delivery has also several significant limitations: one of the greatest disadvantages of transdermal drug delivery is the skin's low permeability. The primary function of the stratum corneum, the keratinised outmost layer of the skin, is to provide an outstanding transdermal barrier against the absorption of chemical and biological toxins [21]. This physiological function of the

*Corresponding author: Ofra Benny, Institute for Drug Research, The School of Pharmacy, Faculty of Medicine, Campus Ein Karem, The Hebrew University, Jerusalem 91120, Israel, E-mail: ofrab@ekmd.huji.ac.il

skin substantially limits the ability of drugs to penetrate in this manner through the skin [21-23].

In this work, we demonstrate that TNP-470 can be delivered transdermally and we present the development of an *in vitro* skin permeation analysis using the pig skin in Franz diffusion cells method which enables the detection of drug transport through the skin membrane [24,25]. We performed a straight forward and robust analytical HPLC method [26] for assay, dissolution and impurity studies of TNP-470. The system suitability parameters of our method meet common U.S. Pharmacopeial Convention (USP) requirements. Additionally, in this work we performed a broad stability and degradation study of TNP-470 in different mediums and several conditions, and identify such that inhibit the degradation of TNP-470 and stabilize it. This broad investigation is critical for selecting suitable dissolution medium for *in vitro* studies and for stable transdermal formulation development in future. The degradation product of TNP-470 was detected in HPLC followed by high resolution mass spectrometry.

Our results demonstrated excellent permeability properties of TNP-470 through the pig skin and suggested its suitability for use in transdermal drug delivery system which can improve its pharmacokinetic properties, safety profile and introduce a slow release long-term maintenance anti-angiogenic therapy.

Methods and Materials

HPLC detection method

HPLC analytical method for TNP-470 detection, dissolution and impurity studies was based on the previously published data [26,27] using Shimadzu model LC-20 instrument with Photodiode Array (PDA) detector at 202 nm. The mobile phase consisted of organic phase, acetonitrile (HPLC grade, Baker) and water (HPLC grade, J. Baker). The method robustness was tested on a wide range of acetonitrile:water ratio; different injection volume parameters and flow rate conditions. The chromatographic conditions were robust for 2 different column types: Gemini-NX 5u, C18 110A, 250 × 4.6 mm and XBridge BEH C18, 250 × 5 µm. Initially the method was developed using XBridge BEH column and later was adjusted to Gemini-NX column. The following parameters remained constant during the development: sample temperature -5°C, column temperature -30°C and the wavelength of UV detector -202 nm. The stock standard solution was prepared by weighing about 30 mg of TNP-470 (MedChem partners, MA, USA) and dissolving with acetonitrile in 10 ml volumetric flask. The working standards solutions were prepared by further dilution of the stock solution with acetonitrile for calibration curve linearity test.

Medium solution preparation

The stability and degradation rates of TNP-470 were investigated in 10 different mediums: sodium lauryl sulfate (SLS) 2% solution was prepared by dilution of SLS 20% solution (Innotrain) 10 times; SLS 4% solution was prepared by dilution of SLS 20% solution 5 times; Tween 80 solution 4% was prepared by dissolving of 2 g of Tween 80 (Fisher) in 50 ml of water (HPLC grade, Baker); medium-chain triglyceride(MCT) emulsion 4% was prepared by dispersing of 2 g of MCT (Gattefosse) in 50 ml water, ethanol solution 30% prepared by mixing 9 ml of absolute ethanol (HPLC grade, Baker) with 21 ml water, acetic acid 0.1M was prepared by dilution of 0.3 g of glacial acetic acid (Biolab) with 50 ml water, the checked pH was 2.8; acetate buffer pH-4.5 solution was prepared according to USP procedure by dissolving 0.15 g of sodium acetate (Baker) in 50 ml water, adjusted to pH-4.5 by glacial acetic acid; buffer acetate pH-4.5 in ethanol 70% solution was

prepared by mixing 21 ml of acetate buffer pH-4.5 with 9 ml ethanol; ehylenediaminetetraacetic acid (EDTA) solution 2% was prepared by dissolving of 1 g of EDTA disodium salt dehydrate (Baker) in 50 ml water; citric acid solution 0.5% was prepared by dissolving of 0.25 g of citric acid (Sigma) in 50 ml water.

Stability and degradation study procedure

The tested sample solutions for degradation studies were prepared in duplicate by dilution the TNP-470 stock solution (1 mg/ml) in each medium to obtain the final concentrations of 50 mg/ml in working sample. The working solutions of TNP-470 in different medium were incubated during the experiment in water bath preheated to 37°C and aliquots samples were tested at each time point of 1, 2, 24, 96 and 168 hours, diluted with acetonitrile, injected and detected by HPLC.

Force degradation procedure

In order to detect degradation products of TNP-470 we performed forced TNP-470 decomposition by incubation of TNP-470 solution in acetonitrile: water [50:50] in 50°C for 3 days. After complete decomposition of TNP-470, the isolated degradant solution was lyophilized for 24 hours and stored at 4°C. The obtained white powder was used for identification in mass spectrometric analysis.

Mass spectrometric analysis

Mass spectrometric analysis of TNP-470 degradant was carried out with an LTQ Orbitrap XL mass analyzer (ThermoFisher Scientific) employing positive-mode electrospray ionization. The instrument was fully calibrated prior to all measurements according to manufacturer's instructions. The measurements were performed at a resolution of 30,000. All FT scans consisted of 15 microscans. The acquired spectra was evaluated using Xcalibur 2.1 software.

Franz cell method

In vitro and *ex vivo* studies enable the detection of drug concentration in the skin, penetration, rate of drug transport across the skin and permeation. *Ex vivo* skin permeation studies are commonly performed using the Franz diffusion cells method [25]. We used a customized setting of multi Franz stirred chambers in order to perform experiments in parallel. The Franz cell apparatus consists of two primary units, the donor and receptor chambers, separated by a membrane. We used standard O-ring Franz cells (PermeGear), with receptor volume of 5 ml and surface area of orifice area of 0.2 cm². The tested solution was added to the upper chamber, the bottom chamber contains dissolution medium from which samples were withdrawn at different time points, diluted with acetonitrile, injected and detected by HPLC. The amount of drug that crossed the membrane was determined at each time point according to calibration curve of TNP-470. The membrane we used was pig's ears skin dermatome with thickness of 500-700 µm purchased from the Institute of Animal Research (Kibbutz Lahav).

Results

HPLC method development

The robustness of following parameters of the method was tested during the development: injection volume, mobile phase, flow rate, column manufactures. Mobile phase robustness was tested on different ratios of acetonitrile: water- 50:50, 60:40, 70:30, and 80:20. As expected, the retention time (Rt) of TNP-470 varied according to mobile phase: from Rt - 2.7 min, in case of high content of organic phase acetonitrile in mobile phase to Rt - 6.3 min in case of low content of acetonitrile in

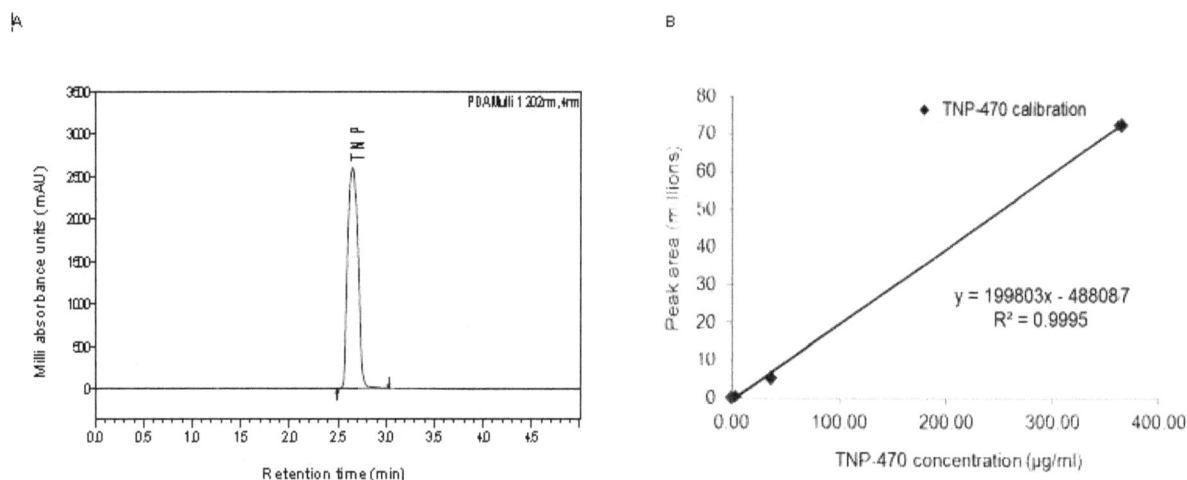

Figure 1: HPLC detection method for TNP-470 assay, dissolution and degradation studies. **A:** TNP-470 peak in HPLC chromatogram: retention time (Rt)-2.7 min, symmetry of peak (tailing factor) - 1.1, number of theoretical plates - 2687, **B:** The calibration curve is linear (R²=0.9995) for a wide range of TNP-470 standard solution concentrations (0.04-360 µg/ml).

mobile phase as we can see in Figure 1A. The symmetry factor value, tailing factor, of the TNP-470 peak was around 1.0 and the number of theoretical plates was around 2500 in different conditions. The flow rate of the mobile phase was changed from 1 to 2 ml/min. The robustness of different injection volume parameters was measured in 10, 20, 30 µl. Different sets of standards solutions were prepared for linearity determination of different chromatographic conditions. For example, Figure 1B show the calibration curve for a wide range of TNP-470 concentrations, from 0.04 to 360 µg/ml. The calibration curve linearity factor - R^2 was 0.9995 for injection volume 30 µl and 0.9967 for injection volume 10 µl. For most tested chromatographic conditions the calibration curve linearity factor- R^2 was more than 0.99. The repeatability test was performed for 5 injections and relative standard deviation (RSD) was found 0.4%. The stability of the working standard solutions was verified for 5 days at cooling conditions 5°C, the RSD for 5 injections was found around 1.5%.

Stability and degradation study of TNP-470 in different mediums

In order to assess the stability of TNP-470 in different conditions, the degradation rate of TNP-470 was investigated in 10 different mediums: sodium lauryl sulfate (SLS) 2%, SLS solution 4%, Tween 80 solution 4%, MCT emulsion 4%, ethanol solution 30%, acetic acid 0.1M, buffer acetate pH-4.5 in ethanol 70%, EDTA solution 2%, citric acid solution 0.5%, acetate buffer pH-4.5 solution. The specific mediums were selected to find conformed dissolution medium for skin penetration studies and to find stabilizing and destabilizing factors for future formulation development of TNP-470. According to TNP-470 stability results we can sort the mediums to two main categories, stabilizing and destabilizing mediums.

The first category of destabilizing mediums, which included SLS 2%, SLS 4%, Tween 80 4%, acetic acid 1M and citric acid solution 0.5%, characterized by intensive degradation of TNP-470, see Figure 2A-2E. The assay of TNP-470 decreased to 70% in SLS 2%, 45% in SLS 4%, 62% in Tween 80 4%, 69% in acetic acid 0.1M and 13% in citric acid after 24 hours, about 10% of TNP-470 remained after 96 hours and after 168 hours the TNP-470 peak was not detected in all mediums of destabilizing category. Additionally, we observed that TNP-470 peak

decrease was followed by unknown peak formation which was detected at Relative Retention Time, RRT-1.5 related to TNP-470. The intensive degradation and decrease in TNP-470 assay resulted in rapid growing of its degradation product peak, degradant RRT-1.5, see Figure 2A-2E. After 168 hours when the assay of TNP-470 peak dropped from 100 to 0% the assay of its degradant RRT-1.5 peak increased in average from 0 to 35% in the destabilizing category mediums.

The second category of stabilizing mediums, which included MCT 4%, ethanol 30%, buffer acetate pH-4.5 ethanolic solution 70%, EDTA 2%, acetate buffer pH-4.5, characterized by moderate degradation of TNP-470, see Figure 3A-3E. The assay of TNP-470 after 24 hours was found 99% in MCT 4%, 53% in ethanol 30%, 110% in buffer acetate pH-4.5 in ethanol 70%, 107% in EDTA 2%, and 104% in acetate buffer pH-4.5. Around 80% of TNP-470 remained after 96 hours in all stabilizing mediums except ethanol 30% solution where assay of TNP-470 was found around 50%. Finally, after 168 hours the assay of TNP-470 remained about 50% in stabilizing category mediums. The degradant RRT-1.5 peak formation was detected also in stabilizing category mediums however its increase was moderate, see Figure 3A-3E. After 168 hours when the assay of TNP-470 peak decreased from 100 to 50% the assay of its degradant RRT-1.5 peak increased in average from 0 to 5%, except ethanol 30% and MCT 4% solutions, where the raise was more significant, from 0 to 35%.

TNP-470 degradation product identification

The degradation of TNP -470 was associated with formation of TNP-470 degradation product. The rate of TNP-470 degradation and its degradation product formation varied in different buffers. Figure 4A shows a single peak in chromatogram of TNP-470 when sample is fresh in SLS 2% without detectable degradation peak. In Figure 4B after 1 day at 37°C the height of degradant peak was equal to TNP-470 peak. After 5 days, as shown in Figure 4C all TNP-470 was degraded with a clear peak of degradation product. The overlaid chromatograms of described samples solutions are shown in Figure 4D. The RRT value of TNP-470 degradation product varied from 1.2 to 1.5 related to TNP-470 according to chromatographic conditions. With the decrease of organic part, acetonitrile to 50% in mobile phase, the RRT value increased to 1.5, and with increase of organic part to 80% the distance between two peaks reduced to RRT-1.2.

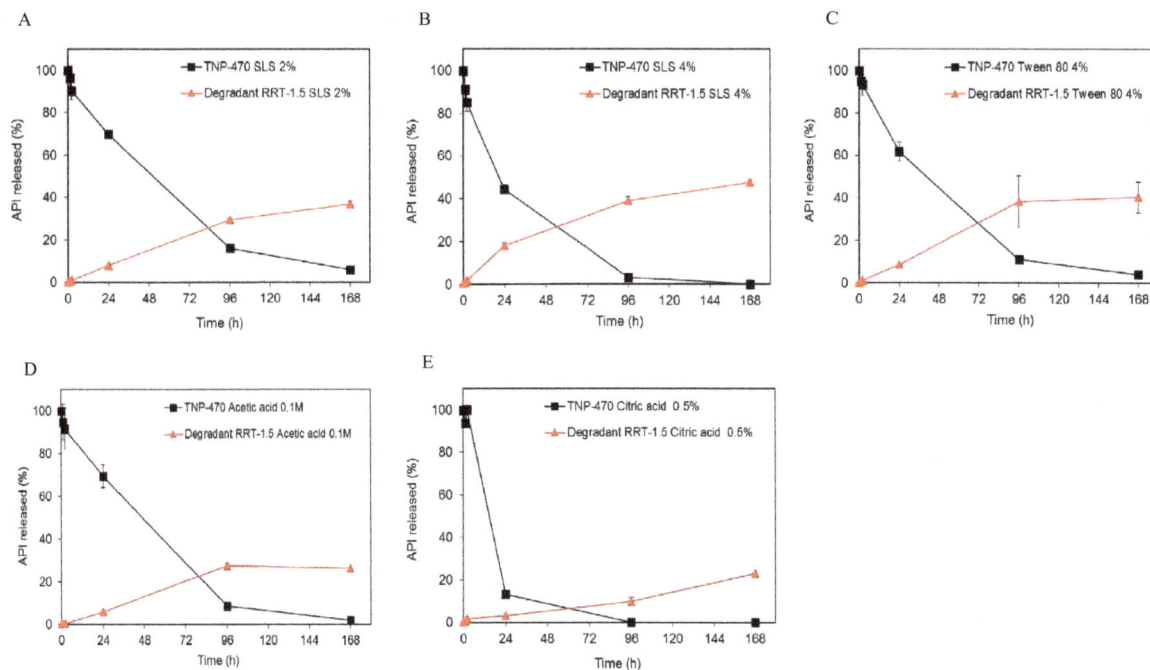

Figure 2: Intense degradation of TNP-470 and rapid formation of degradant peak RRT-1.5 in destabilizing mediums. **A:** SLS 2%, **B:** SLS 4%, **C:** Tween 80.4%, **D:** Acetic acid 0.1 M, **E:** Citric acid 0.5%.

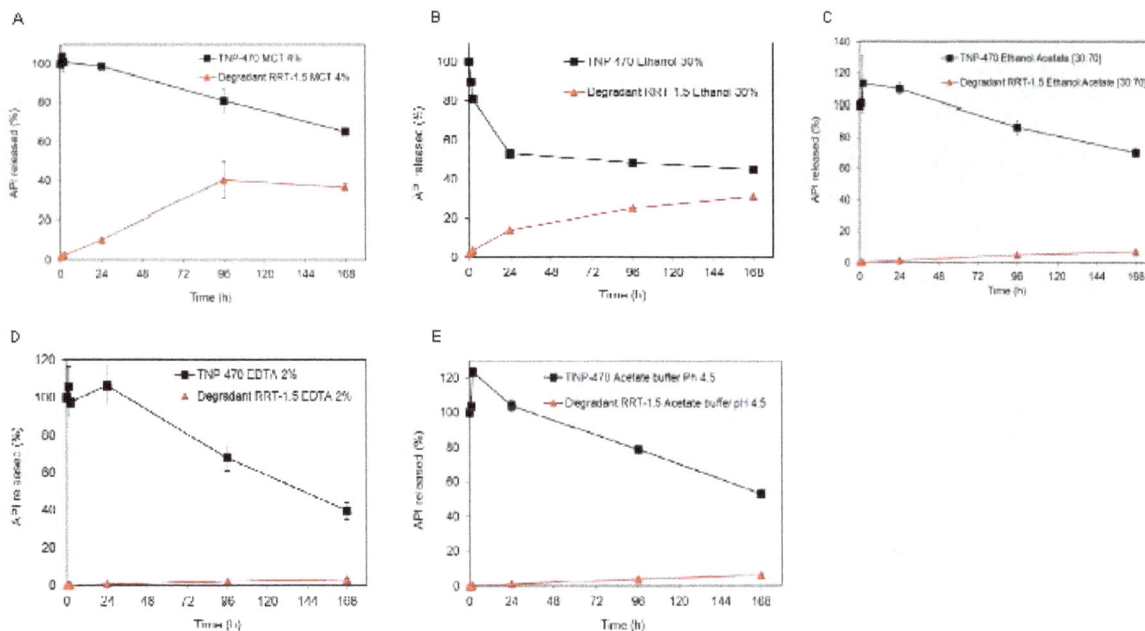

Figure 3: Moderate degradation of TNP-470 and inhibited formation of degradant peak RRT-1.5 in stabilizing mediums. **A:** MCT 4%, **B:** Ethanol 30%, **C:** Ethanol: Acetate buffer pH 4.5 [30:70], **D:** EDTA 2%, **E:** Acetate buffer pH 4.5.

To further identify the main degradation product of TNP-470 we performed mass spectrometry analysis. Mass spectrogram of the isolated and purified TNP-470 degradation product is shown in Figure 4E. According to high resolution mass spectrum the TNP-470 degradation product was identified as one of its main metabolites of TNP-470, AGM-1883 [27,28]. The metabolic pathway of TNP-470 and formation of AGM-1883 following by chloroacetyl group cleavage is shown in Figure 4F.

TNP-470 skin permeation study

In vitro skin permeation study was performed using Franz cell method [25]. The customized setting of multi Franz stirred chambers is shown at Figure 5A. The Franz cell apparatus consists of two primary units, the donor and receptor chambers, separated by a membrane. As a membrane, we used pig skin dermatome with uniform thickness of 500-700 micron, as shown in Figure 5B and 5D. The tested solution

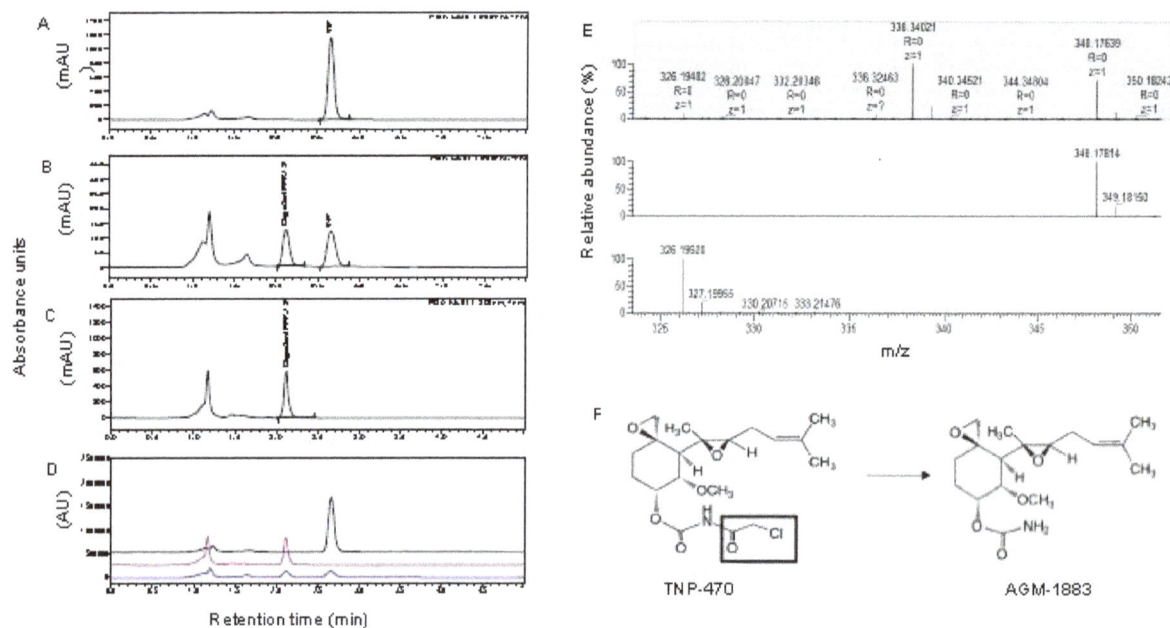

Figure 4: Degradation product RRT-1.2 identification. **A:** HPLC chromatogram of TNP-470 solution 50 mg/ml in SLS 2% medium at time 0, **B:** HPLC chromatogram of TNP-470 solution 50 mg/ml in SLS 2% medium after 1 day at 37°C, **C:** HPLC chromatogram of TNP-470 solution 50 mg/ml in SLS 2% medium after 5 day at 37°C, **D:** Overlay HPLC chromatograms at time 0, 1 day and 5 days show TNP-470 degradation and the degradation product RRT-1.2 formation, **E:** Mass spectrum of the degradation product RRT-1.2 identify TNP-470 main metabolite AGM-1883, **F:** Metabolic pathway, formation of AGM-1883 following by chloroacetyl group cleavage.

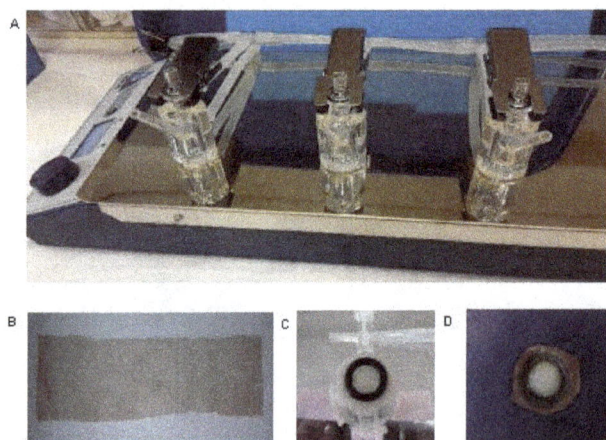

Figure 5: Franz cell diffusion method for TNP-470 skin permeation studies. **A:** Customized setting of multi Franz stirred chambers, **B:** Pig skin dermatome sample (dimensions 12 × 4 cm, thickness 500-700 μm), **C:** Pig skin sample with tested composition before the diffusion test, **D:** Pig skin sample after the diffusion test.

was added to the upper chamber as shown in Figure 5C and the bottom chamber contained dissolution medium from which samples were taken at different time points for analysis. After the stability and degradation study, we perform the Franz permeability assay using acetate buffer pH 4.5 stabilizing medium and compare the results to SLS 2%, destabilizing method. The amount of drug permeated the skin was sampled and determined in four time points by HPLC detection method. The released amount of TNP-470 in acetate buffer pH-4.5 medium is shown in Figure 6A. The amount of TNP-470 which crossed through the pig skin membrane is presented in μg per skin area (cm^2) plotted in each time point (hours). A maximum value of 1250 μg/cm^2

TNP-470 was obtained after 72 hours which was 25% from the initial amount. In Figure 6B the released amount of TNP-470 in acetate buffer pH-4.5 medium is presented in % from the initial amount and compared to results in SLS 2% medium in the same conditions. The skin permeability values of TNP-470 sampled in acetated buffer after 24 hours was 14% compared to permeability value in SLS 2% which was only 1%.

Discussion

Anti-angiogenic drugs have the ability to modify or regress newly formed blood vessels. Several forms of angiogenesis inhibitors exist, ranging from endogenous proteins and small molecule cytokine antagonists, to antibodies and tyrosine kinase inhibitors [3,29-31]. Historically, the main target of anti-angiogenic drugs in clinical development has been blocking the Vascular Endothelial Growth Factor (VEGF) pathways. However, other angiogenic factors also play critical roles in neovascularization; in addition, most neovascular diseases are accompanied by inflammation and involve several pro-inflammatory cytokines with VEGF independent mechanisms (such as MCP-1, IL-6, TNFα). These alternative pathways substantially limit the therapeutic index of solely anti-VEGF drugs. Therefore, using strategies to administer broad spectrum angiostatic agents, such as TNP-470, that have molecular target downstream to VEGF may have an important therapeutic advantage. Moreover, small molecules can potentially penetrate the skin, unlike macromolecules [21,32].

In this work, we introduced for the first time the transdermal delivery of TNP-470. We systematically studied the potential of using MetAp2 inhibitor in a transdermal formulation and established a reliable skin permeation assay using Franz diffusion cells method, which enables the detection of drug transport through the skin and quantitation the permeated amount of the drug. For this reason, we

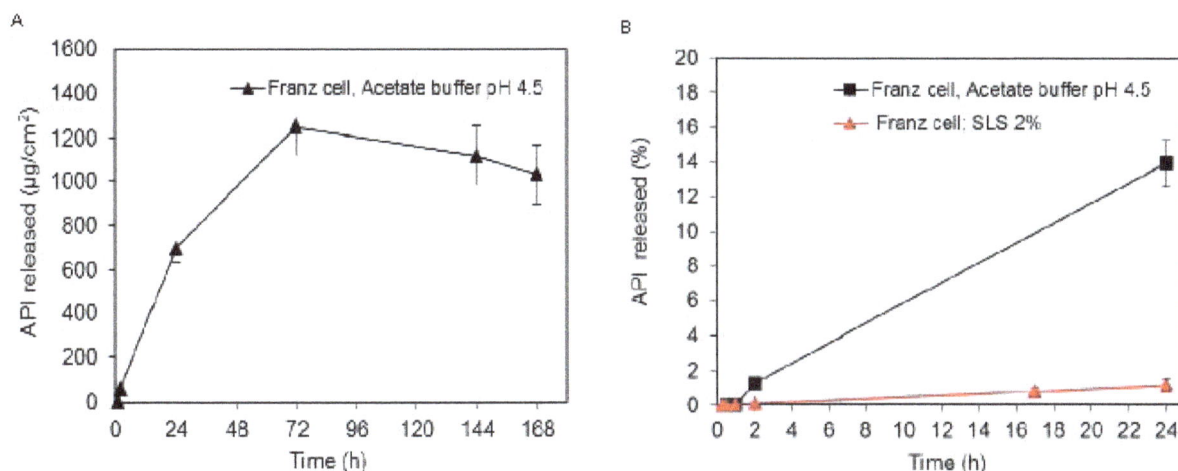

Figure 6: TNP-470 permeability results through the pig skin. **A:** TNP-470 solution release amount through the pig skin during the diffusion test in buffer acetate pH-4.5 medium, in µg/cm²; **B:** TNP-470 solution release amount through the pig skin during the diffusion test in buffer acetate pH-4.5 (stabilizing medium) compared to SLS 2% (destabilizing medium), in % from the initial content of TNP-470.

developed a simple analytical HPLC method for assay, dissolution and impurity studies of TNP-470. The TNP-470 peak parameters, symmetry factor and number of theoretical plates, conformed a common USP requirements (Figure 1A). According to the calibration curve linearity factor, R^2=0.9995, the developed method was linear for a wide range of TNP-470 concentrations (Figure 1B). The repeatability test, the RSD for 5 injections was found less than 2% which also conformed to common USP system suitability test requirements. Additionally, the method was found robust for a wide range of chromatographic parameters: injection volume, mobile phase, flow rate and different types of column, which enables its wide and universal application.

In this study, we investigated the stability and degradation of TNP-470 in 10 different mediums. The specific list of mediums was selected both for analytical purposes, to find suitable dissolution medium for *in vitro* skin penetration studies and for future formulation development, to find stabilizing and destabilizing factors for TNP-470. According to TNP-470 stability results we can sort the mediums to two main categories, stabilizing and destabilizing mediums (Figures 2 and 3). The first group of destabilizing mediums, which included SLS 2%, SLS 4%, Tween 80 4%, acetic acid 0.1M and citric acid solution 0.5%, characterized by intensive decomposition of TNP-470 followed by degradation product peak formation which was detected at relative retention time RRT-1.5 related to TNP-470 (Figure 2).

The second group of stabilizing mediums, which included MCT 4%, ethanol 30%, buffer acetate pH-4.5 ethanolic solution 70%, EDTA 2%, acetate buffer pH-4.5, characterized by moderate degradation of TNP-470 followed by insignificant degradant peak formation (Figure 3). It is known that TNP-470 is stable in organic solvents and undergoes rapid hydrolysis in the presence of water [26,27,33] which explains its rapid degradation in aqueous solution of SLS 2%, SLS 4% and Tween 80 4% and relative stability in ethanol 30%. The stability of TNP-470 in MCT could be explain by solubility of highly lipophilic drug TNP-470 in MCT, lipid compound and it is supposed that water could not so easily access the TNP-470 surrounded by fat and oil molecules. Our stability results of TNP-470 in MCT 4% emulsion supports previous microspheres formulation development containing TNP-470 with MCT performed by Kakinoki et al. [34].

Relative stability of TNP-470 in buffer acetate pH-4.5, EDTA 2% and buffer acetate pH-4.5 ethanolic solution 70% led us to believe that we found stabilizing agent of TNP-470 acetate ion, CH_3COO-, which presented in all 3 solutions. Unfortunately, this hypothesis disproved by rapid degradation of TNP-470 in acetic acid 0.1M, pH-2.8. This fact led us to understanding that the main stabilizing factors of TNP-470 in aqueous solutions is pH. Our findings correlated to previous study performed by Figg et al. suggested that TNP-470 would be most stable in a slightly acidic medium, pH 4-5 [27]. Although the stability of TNP-470 was almost equivalent in MCT 4%, buffer acetate pH-4.5 and buffer acetate pH-4.5 ethanolic solution 70%, we selected buffer acetate pH-4.5 as dissolution medium for *in vitro* skin permeation studies because it is common buffer solution accepted by US pharmacopeia and most approximate to physiological environment conditions.

We performed TNP-470 *in vitro* skin permeation study using Franz cell method in selected dissolution medium, stabilizing buffer acetate pH-4.5 and compared the results to the destabilizing medium, SLS 2%. Skin's protective function provides an outstanding transdermal barrier which limits the penetration of drugs through the skin and usually most transdermal patches after removal contain at least 95% of the total amount of drug initially in the patch, which means that only 5% from the initial amount permeate the skin [19,20]. In light of this facts, the TNP-470 permeability results through the pig skin, 1250 µg/cm² after 72 hours which is 25% from the initial amount (Figure 6A), demonstrate excellent delivery of TNP-470 via transdermal route. The great difference between permeability results in two different dissolution mediums, stabilizing buffer acetate pH-4.5 and destabilizing SLS 2%, highlights the significance of TNP-470 stability studies and the importance of appropriate dissolution medium selection. In the acetate buffer pH-4.5 the permeated amount of TNP-470 after 24 hours was 14% compared to only 1% in SLS 2% (Figure 6B). It is obvious that the negligible amount of TNP-470 detected in SLS 2% was associated with degradability of TNP-470 in this medium and well correlated to stability studies we performed where destabilizing mediums group which included SLS 2% characterized by intensive decomposition of TNP-470.

Conclusion

This study demonstrates a simple reliable method for skin permeation studies of TNP-470, a potent anti-angiogenic drug, using Franz diffusion cells and HPLC detection methods. Our results introduce TNP-470 excellent candidature to transdermal drug delivery, which can offer a novel and promising clinical usage for potent anti-angiogenic drug to improve its poor bioavailability and safety profile, caused by peak concentration of the drug in plasma. Additionally, our study provides a broad data of TNP-470 stability in different mediums, which can lead to stable and improved formulation development in future. Taking together our findings can lead to the development of slow release transdermal delivery system, which may be used as a long-term maintenance therapy for angiogenesis-dependent diseases such as cancer, psoriasis, rheumatoid arthritis and age-related macular degeneration.

Acknowledgement

This research was partially supported by Israel Science Foundation (ISF), no. 0394883.

References

1. Folkman J (2007) Angiogenesis: an organizing principle for drug discovery. Nat Rev Drug Discov 6: 273-286.

2. Miller JW, Adamis AP, Shima DT, D'Amore PA, Moulton RS, et al. (1994) Vascular endothelial growth factor/vascular permeability factor is temporally and spatially correlated with ocular angiogenesis in a primate model. Am J Pathol 145: 574-584.

3. Folkman J (1997) Angiogenesis and angiogenesis inhibition: An overview. Exs 79: 1-8.

4. Ingber D, Fujita T, Kishimoto S, Sudo K, Kanamaru T, et al. (1990) Synthetic analogues of fumagillin that inhibit angiogenesis and suppress tumour growth. Nature 348: 555-557.

5. Figg WD, Pluda JM, Lush RM, Saville MW, Wyvill K, et al. (1997) The pharmacokinetics of TNP-470, a new angiogenesis inhibitor. Pharmacotherapy 17: 91-97.

6. Chun E, Han CK, Yoon JH, Sim TB, Kim YK, et al. (2005) Novel inhibitors targeted to methionine aminopeptidase 2 (MetAP2) strongly inhibit the growth of cancers in xenografted nude model. Int J Cancer 114: 124-130.

7. Griffith EC, Su Z, Niwayama S, Ramsay CA, Chang YH, et al. (1998) Molecular recognition of angiogenesis inhibitors fumagillin and ovalicin by methionine aminopeptidase 2. Proc Natl Acad Sci USA 95: 15183-15188.

8. Bhargava P, Marshall J, Rizvi N, Dahut W, Yoe J, et al. (1999) A phase I and pharmacokinetic study of TNP-470 administered weekly to patients with advanced cancer. Clin Cancer Res 5: 1989-1995.

9. Herbst R, Madden T, Tran H, Blumenschein GJ, Meyers C, et al. (2002) Safety and pharmacokinetic effects of TNP-470, an angiogenesis inhibitor, combined with paclitaxel in patients with solid tumors: evidence for activity in non-small-cell lung cancer. J Clin Oncol 20: 4440-4447.

10. Kruger E, Figg WD (2000) TNP-470: an angiogenesis inhibitor in clinical development for cancer. Expert Opin Investig Drugs 9: 1383-1396.

11. Kudelka A, Levy T, Verschraegen C, Edwards C, Piamsomboon S, et al. (1997) A phase I study of TNP-470 administered to patients with advanced squamous cell cancer of the cervix. Clin Cancer Res 3: 1501-1505.

12. Kudelka A, Verschraegen C, Loyer E (1998) Complete remission of metastatic cervical cancer with the angiogenesis inhibitor TNP-470. N Engl J Med 338: 991-992.

13. Logothetis C, Wu K, Finn L, Daliani D, Figg W, et al. (2001) Phase I trial of the angiogenesis inhibitor TNP-470 for progressive androgen-independent prostate cancer. Clin Cancer Res 7: 1198-1203.

14. Moore JD, Dezube BJ, Gill P, Zhou XJ, Acosta EP, et al. (2000) Phase I dose escalation pharmacokinetics of O-(chloroacetylcarbamoyl) fumagillol (TNP-470) and its metabolites in AIDS patients with Kaposi's sarcoma. Cancer Chemother Pharmacol 46: 173-179.

15. Stadler W, Kuzel T, Shapiro C, Sosman J, Clark J, et al. (1999) Multi-institutional study of the angiogenesis inhibitor TNP-470 in metastatic renal carcinoma. J Clin Oncol 17: 2541-2545.

16. Tran H, Blumenschein GJ, Lu C, Meyers C, Papadimitrakopoulou V, et al. (2004) Clinical and pharmacokinetic study of TNP-470, an angiogenesis inhibitor, in combination with paclitaxel and carboplatin in patients with solid tumors. Cancer Chemother Pharmacol 54: 308-314.

17. Kruger EA, Figg WD (2000) TNP-470: an angiogenesis inhibitor in clinical development for cancer. Expert Opin Investig Drugs 9: 1383-1396.

18. Benny O, Fainaru O, Adini A, Cassiola F, Bazinet L, et al. (2008) An orally delivered small-molecule formulation with antiangiogenic and anticancer activity. Nature Biotechnology 26: 799-807.

19. Schoellhammer CM, Blankschtein D, Langer R (2014) Skin permeabilization for transdermal drug delivery: recent advances and future prospects. Expert Opin Drug Deliv 11: 393-407.

20. Gupta H, Babu RJ (2013) Transdermal delivery: product and patent update. Recent Pat Drug Deliv Formul 7: 184-205.

21. McLafferty E, Hendry C, Alistair F (2012) The integumentary system: anatomy, physiology and function of skin. Nurs Stand 27: 35-42.

22. Ita KB (2015) Chemical penetration enhancers for transdermal drug delivery-success and challenge. Curr Drug Deliv.

23. Jain A, Jain P, Kurmi J, Jain D, Jain R, et al. (2014) Novel strategies for effective transdermal drug delivery: a review. Crit Rev Ther Drug Carrier Syst 31: 219-272.

24. Lehman PA, Raney SG, Franz TJ (2011) Percutaneous absorption in man: in vitro-in vivo correlation. Skin Pharmacol Physiol 24: 224-230.

25. Bartosova L, Bajgar J (2012) Transdermal drug delivery in vitro using diffusion cells. Curr Med Chem 19: 4671-4677.

26. Whalen CT, Hanson GD, Putzer KJ, Mayer MD, Mulford DJ (2002) Assay of TNP-470 and its two major metabolites in human plasma by high-performance liquid chromatography-mass spectrometry. J Chromatogr Sci 40: 214-218.

27. Figg WD, Yeh HJ, Thibault A, Pluda JM, Itoh F, et al. (1994) Assay of the antiangiogenic compound TNP-470, and one of its metabolites, AGM-1883, by reversed-phase high-performance liquid chromatography in plasma. J Chromatogr 652: 187-194.

28. Ong VS, Stamm GE, Menacherry S, Chu S (1998) Quantitation of TNP-470 and its metabolites in human plasma: sample handling, assay performance and stability. J Chromatogr B Biomed Sci Appl 710: 173-182.

29. Folkman J (1996) Fighting cancer by attacking its blood supply. Scientific American 275: 150.

30. Folkman J (1996) Tumor angiogenesis and tissue factor. Nature Medicine 2: 167-168.

31. Folkman J (1995) Angiogenesis inhibitors generated by tumors. Molecular Medicine 1: 120-122.

32. Wiedersberg S, Guy R (2014) Transdermal drug delivery: 30+ years of war and still fighting. Journal of Controlled Release 190: 150-156.

33. Ong V, Stamm G, Menacherry S, Chu S (1998) Quantitation of TNP-470 and its metabolites in human plasma: sample handling, assay performance and stability. Journal of Chromatography B 710: 173-182.

34. Kakinoki S, Yasuda C, Kaetsu I, Uchida K, Yukutake K, et al. (2003) Preparation of poly-lactic acid microspheres containing the angiogenesis inhibitor TNP-470 with medium-chain triglyceride and the in vitro evaluation of release profiles. Eur J Pharm Biopharm 55: 155-160.

12

Microemulsification-Based Method: Coupling with Separation Technique

Gabriela Furlan Giordano[1,2], Karen Mayumi Higa[1,2], Adriana Santinom[1,2], Angelo Luiz Gobbi[1], Lauro Tatsuo Kubota[2,3] and Renato Sousa Lima[1,2*]

[1]Laboratory of Microfabrication , National Nanotechnology Laboratory , National Center for Research in Energy and Materials, Campinas , Sao Paulo, Brasil

[2]Instituto of Chemistry , State University of Campinas, Campinas, Sao Paulo, Brasil

[3]Instituto National Science and Technology Bioanalytics, Campinas, Sao Paulo, Brasil

Abstract

The outcomes described herein outline the potentiality of the microemulsification-based method (MEC) for development of rapid testing (point-of-use) technologies. MEC was recently proposed by these authors for analytical determinations wherein the detection is conducted in solution with naked eyes. It relies on effect of analyte over the colloid thermodynamics by changing the minimum volume fraction of amphiphile needed to generate microemulsions (MEs) (Φ_{ME}), which represents the analytical response of the method. We report in this paper the successfully coupling of MEC-based detection with gas diffusion separation. Such result extends the field of application of MEC in analytical sciences by improving its selectivity. One custom-designed module was constructed on PTFE for the separation measurements. It was utilized in combination with MEC for determining water in ethanol fuels using water/n-propanol/oleic acid MEs and water-rich compositions. In this situation, accurate direct determinations by MEC are not possible. In addition, further studies on analytical performance and robustness of MEC by using n-propanol amphiphile are described. The method was robust as regards to deviations in dispersion preparing and changes in temperature. Concerning the analytical performance, the analytical curves presented wide linear range with limits of linearity of up to 70.00% v/v ethanol to water (Φ_E). The limits of detection (S/N=3) were of 1.03%, 7.21%, and 0.68% v/v Φ_E for compositions with water- (region A) and oil-rich (region C) domains as well as equal volumes of water and oil phases (region B), respectively. With respect to the regions A and B, the analytical performance stressed herein exhibited best linearity and comparable sensitivities when compared to these levels reached with ethanol amphiphile (our first publication on MEC) rather than n-propanol.

Keywords: Point-of-care; Rapid test; Colloid; Interfacial tension; *In situ* analysis

Abbreviations: ME: Microemulsion; AP: Amphiphile; W: Hydrophilic phase of dispersions; O: Hydrophobic phase of dispersions; Φ_{ME}: Minimum volume fraction of amphiphile needed to get ME; Φ_O: Volume fraction of oil to water; Φ_E: Volume fraction of ethanol to water; Φ_W: Volume fraction of water to ethanol; $\Delta\Phi$: Absolute error determined for Φ_E

Point-of-use devices represent currently a key field in quantitative analytical sciences. These platforms are low-cost, fast, portable, and simple to use eliminating the necessity for qualified operators [1]. Rapid tests enable in-situ measurements presenting substantial social and economic implications at industry, environment, and medicine [2-9]. One potential output to perform point-of-use analyses is the accomplishment of the tests in solution with naked eye detection using disposable systems. It allows the determination of different analytes from the use of modified nanomaterial [10]. Naked eye methods bypasses the use of instrumental readers, an essential feature for in-situ technologies. Furthermore, the analyses in solution surpass precision-related downsides when making the tests on substrates such as paper [6,9]. In this case, the diverse paper substrates that are employed to fabricate the devices affect the flow rates and interactions with analytes [10].

This paper reports further investigations and application of the microemulsification-based method (MEC), a point-of-use platform that was recently proposed by these authors [11]. It relies on solution-based-detection with naked eyes. In contrast with colorimetric tools [10], MEC response depends on colloid thermodynamics by relying on effect of analyte on the entropy of emulsions or Winsor systems. It changes the formation of thermodynamically stable dispersions, the microemulsions (MEs). The minimum volume fraction of amphiphile

(AP) needed to get MEs (Φ_{ME}) for a fixed water-oil ratio expressed the analytical signal of the method. The generation of nanodroplets in MEs (transparent) allows the naked eye detection of Φ_{ME} by monitoring the change of turbidity from the emulsions or Winsor systems (cloudy) as shown in Figure 1a. This cloudy-to-transparent conversion acts like a turning point in titrations, ensuring the visual measurement of Φ_{ME} and, therefore, not only screening analyses (positive/negative data) as the most of naked eye colorimetry platforms [10] as well as precise quantitative analyses [11]. The response in colorimetry changes with the intensity of colour or tonality. Herein, subjective uncertainties by personal and surrounding conditions are observed [12].

MEC presents powerful aspects concerning the deployment of point-of-use tools. Such a method is straightforward, cheap, fast, portable, and provides precise analytical determinations with satisfactory precision, linearity, robustness, and accuracy. Lastly, volumes of approximately 20 μL for dispersions assure the visual measurement of Φ_{ME} [11]. It contributes for a low sample consuming.

The first outcomes achieved by MEC were promising with respect to its analytical performance [11]. Direct analyses based on analytical

***Corresponding authors:** Renato Sousa Lima, Laboratory of Microfabrication , National Nanotechnology Laboratory, National Center for Research in Energy and Materials, Campinas , Sao Paulo 13083-970, Brasil
E-mail: renato.lima@lnnano.cnpem.br

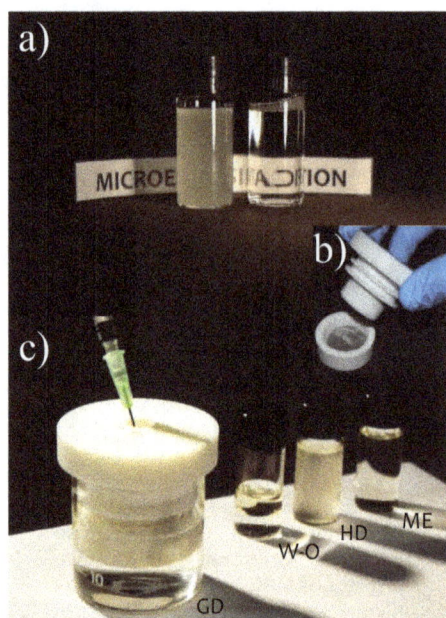

Figure 1: Microemulsification-based method. Photos displaying the transition from cloudy (left) to transparent (right) that allows the naked eye measurement of Φ_{ME} (a), the components of separator (b), and the sequence of analysis in gas diffusion (GD) separation (c) The top and bottom pieces in (b) have receptor media and membrane of PTFE, respectively. After separation (separator on a bottle with sample), one aliquot of receptor was mixed to oil (oleic acid) to get W-O mixtures. Next, n-propanol amphiphile was added generating heterogeneous dispersions (HD) and, finally, the microemulsions (ME).

curves were conducted for determining monoethylene glycol in samples associated to the processing of liquefied natural gas and water in ethanol fuels. For this latter case, accurate direct determinations were obtained by employing oil-rich MEs and dispersions containing similar volumes of the water (W) and oil (O) phases. Nonetheless, accurate direct determinations were not verified when using water-rich MEs. Considering the excess of water in this composition, we supposed these results were because the ionic strength of the samples of ethanol fuel. Studies addressed in literature revealed the presence of NO_3^-, K^+, Ca^{2+} (0.5 to 3.5 mg L^{-1}) [13], Cu^{2+}, Zn^{2+}, Ni^{2+}, and Fe^{3+} (8 to 57 mg L^{-1}) in these samples [14].

We describe herein accurate determinations of water in ethanol fuels by MEC utilizing water-rich MEs. For this, the method was coupled with gas diffusion separation. It represents an important breakthrough by improving the reliability and selectivity of MEC, enlarging its range of application. In addition, further studies into analytical performance and robustness of MEC by using n-propanol AP are described. The robustness level was investigated as a function of deviations in dispersion preparing and changes in temperature.

Materials and Methods

Chemicals

Ethanol and n-propanol alcohols were supplied from Merck (Whitehouse Station, NJ) whereas oleic acid was obtained from Labsynth (São Paulo, Brazil). Deionized water (Milli-Q, Millipore Corp., Bedford, MA) was attained with resistivity no less than 18 MΩ cm.

Microemulsification

MEC routine relies on nature of sample. The analyte can be added in W, AP, or O phase for generation of MEs according to the sample polarity. As a consequence, MEC is applicable to polar, nonpolar, and amphiphile media. When the analyte is dipped into W phase, for instance, W phase solutions are initially obtained changing the concentration of analyte that is added in polar solvent. These solutions are used to attain W-O mixtures for a fixed volume ratio. Following, the microemulsification is made by adding pure AP until cloudy-to-transparent transition. The measured values of Φ_{ME} are, then, employed to construct the analytical curve. For application, the sample acts directly like W phase with the intent to stabilization of the dispersions. Finally, the analyte content is obtained through the linear regression line equation.

The detection of Φ_{ME} was conducted with naked eyes. Dispersions were prepared in Eppendorf tubes with the aid of micropipette at room temperature (23°C). AP-added W-O mixtures were vigorously shaken for microemulsifications. The values of Φ_{ME} were detected by gradually transferring the amphiphile in a unique tube with the W-O mixture. The first attempt in finding Φ_{ME} was intended only to obtain an approximate analytical response. This step took approximately 5 min whereas the other attempts lasted less than 2 min.

Dispersions were composed of water (W), oleic acid (O), and n-propanol hydrotrope (AP phase). Ethanol analyte was added in W phase to get the analytical curves as the previously described experimental procedure. Such a parameter does not take into account the AP volume. The fraction of analyte was expressed by the volume fraction of ethanol to water (Φ_E).

Phase behavior

Φ_{ME} was obtained in diverse compositions of water/n-propanol/oleic acid MEs to attain a ternary phase diagram and, thus, to evaluate the phase behavior of dispersions. For this, the W-O mixtures were prepared using different volume fractions of oil to water (Φ_O). W-O mixtures had a volume of 600 μL in all of the measurements addressed in this paper.

Analytical performance and robustness level

Analytical curves were obtained to compare the n-propanol-observed results with those achieved by ethanol amphiphile (data in our first publication) [11].

The curves ensured the (i) calculation of merit figures (correlation factor, R^2, analytical sensitivity, and limit of detection, LOD) in determination of water in ethanol fuels and the (ii) evaluation of robustness level as discussed below by establishing a relationship between Φ_{ME} signal and Φ_E content. Confidence intervals for each concentration level were calculated for $\alpha=0.05$ and $n=4$ in all of the cases. In both tests of analytical performance and robustness, analyses were conducted for three compositions of W-O mixture, namely: water- (region A) and oil-rich (region C) domains and dispersions based on equal volumes of W and O phases (region B). W-O mixtures were prepared with 5.00% (A), 50.00% (B), and 95.00% v/v Φ_O (C).

The robustness was mathematically expressed by absolute errors calculated for Φ_E ($\Delta\Phi$, % v/v) in determination of ethanol in water. These errors were because deviations in preparation of W-O mixtures and alterations in temperature.

Analytical curves were obtained to calculate $\Delta\Phi$ by taking up (i) relative standard deviations (RSD) of 5.00% and 10.00% v/v in Φ_O and (ii) diverse temperatures, namely: 20, 23, 26, and 29°C. $\Delta\Phi$ values were related to 7.00%, 30.00%, and 60.00% v/v Φ_E at 23°C without alterations in Φ_O (reference values). Φ_{ME} signals related to the reference values were employed in the linear regression equations of all of the analytical curves in order to calculate Φ_E-considering deviations in dispersion preparing and changes in temperature for regions A, B, and C.

Application

MEC was coupled to gas diffusion separation seeking to improve its accuracy. This separation consists of separating volatile species from sample (donor media) to a receptor solution through gas permeable membranes [15]. Usually adopted membranes are hydrophobic and microporous. During the volatilization, a gas thin layer is stored in the micropores. Afterwards, the analyte diffuses through this layer that separates the donor and receptor solutions.

One custom-designed separation module was constructed on PTFE as approached in Figure 1b. It was utilized for determination of water in ethanol fuels using water-rich MEs (region A: 5.00% v/v Φ_O). The routine of analysis is generically shown in Figure 1c. To construct the analytical curve, deionized water (receptor) and PTFE membrane were placed inside the separation module. We inserted such a module manually on an external glass bottle containing ethanol standard (donor medium). The hydrophobic membranes were in contact with the donor and receptor solutions. The volumes of donor and receptor were 17 and 2 mL, respectively. Separations were based on diffusion of ethanol gas through membrane from donor to receptor [15]. The tests were made at room temperature (23°C) under two conditions of time: 5 and 20 min. To accelerate the homogenization of volatized ethanol in receptor medium, a magnetic bar was inserted inside the extraction module. The stirrer was placed under the bottle with sample whereas the stirring was conducted with 500 rpm rotation speed. Subsequently, MEC was accomplished (protocol addressed above) using receptor water, oleic acid, and n-propanol as W, O, and AP phases of the dispersions, respectively, with 5.00% v/v Φ_O (570 μL of receptor and 30 μL of oleic acid to get W-O mixtures). Lastly, the microemulsification processes were conducted by adding n-propanol. Analytical curve was expressed in terms of Φ_E. This parameter allowed us to indirectly calculate the volume fractions of water to ethanol fuel (Φ_W). For application, the samples were dipped into the glass bottles (donor media) acting as W phase.

The conductivity of samples was measured by AJ Micronal AJX-522 (Sao Paulo, Brazil). Statistical evaluation between the data attained by MEC and Karl Fischer titration (Metrohm, Titrando 890, Herisau, Switzerland) was made by Student's t-tests at 95% confidence level to assess the accuracy.

Results and Discussions

Phase behaviour

Phase diagram of dispersions composed of water/n-propanol/ oleic acid at 23°C is exhibited in Figure 2. The regions above and below the binodal curve correspond to MEs and unstable dispersions, respectively. Average values of Φ_{ME} ($n=4$) were used to plot the diagram with confidence intervals of 0.18 up to 0.39% v/v.

Ethanol hydrotrope was employed as AP phase in ours previous investigations [11]. Taking up the phase behavior for water/ethanol/ oleic acid at 23°C, n-propanol AP generated an efficiency higher (it requires less Φ_{ME} for microemulsification) than that obtained with

ethanol in the region of water-rich dispersions. In this region, the global average of Φ_{ME} for n-propanol was approximately 80% of the value attained with ethanol. It is in accordance with Traube's rule that defines an increase in surface activity with the size of AP hydrophobic chain, reducing the interfacial tension values [16]. Accordingly, the decrease observed in Φ_{ME} for n-propanol was expected. For MEs based on oil-excess and similar volumes of W and O phases, the efficiency was almost the same for ethanol and n-propanol amphiphiles.

The efficiency results achieved for water-excess dispersions outline an advantage as regards to the employment of n-propanol as AP phase rather than ethanol. In addition, the wide availability and low cost of this solvent in countries like United States represent an important feature regarding the successfully use of n-propanol in MEC.

Analytical performance

Figure 3 shows analytical curves for standards of ethanol in W phase at 23°C in the A, B, and C regions. The curves presented wide linear range with limits of linearity of 70.00% v/v Φ_E for B and C and 60.00% v/v Φ_E for A.

The regions A and B presented the best sensitivities and

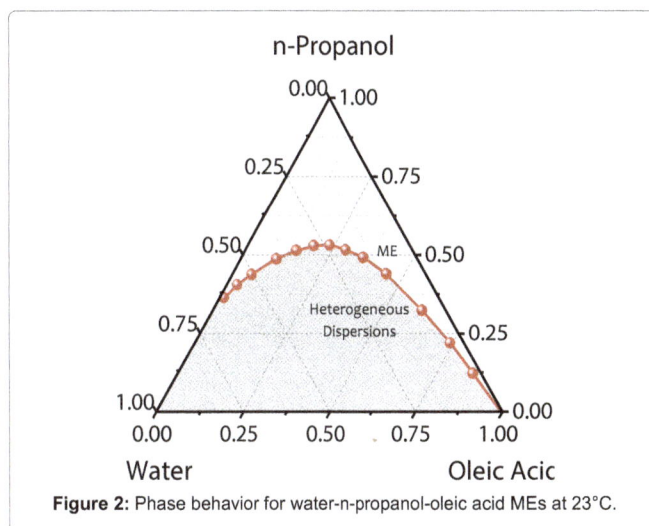

Figure 2: Phase behavior for water-n-propanol-oleic acid MEs at 23°C.

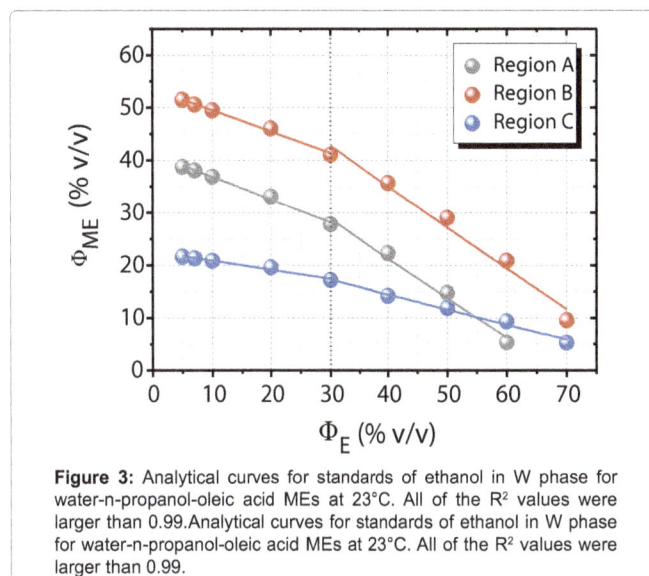

Figure 3: Analytical curves for standards of ethanol in W phase for water-n-propanol-oleic acid MEs at 23°C. All of the R² values were larger than 0.99. Analytical curves for standards of ethanol in W phase for water-n-propanol-oleic acid MEs at 23°C. All of the R² values were larger than 0.99.

detectabilities. All of the curves had an increase in analytical sensitivities upon 30.00% v/v Φ_E. Their values were: -0.43 and -0.75 for A, -0.41 and -0.78 for B, and -0.17 and -0.29 for region C. Lastly, the LODs (S/N=3) were of 1.03%, 0.68%, and 7.21% v/v Φ_E for A, B, and C regions, respectively. Confidence intervals ranged from 0.09 to 0.40% v/v.

It is important to highlight the differences between the tests for determining water in ethanol fuel using ethanol (our previous publication) [11] and n-propanol AP (this paper). In first case, the analyte in analytical curves was water that was added in AP instead W phase. The curves correlated Φ_{ME} and Φ_W. For application, the samples acted as AP. In addition, the O phase was chlorobenzene, a much more hazardous solvent than oleic acid as regards to safety and environment risks.

With respect to the regions A and B, the analytical performance stressed herein exhibited best linearity and comparable sensitivities when compared to these levels reached with ethanol AP [11]. In contrast, meanwhile, the data attained with ethanol were best for region C, presenting limit of linearity of 70.00% v/v Φ_W, analytical sensitivity of 2.04, and LOD equal to 0.32% v/v Φ_W.

The negative deviations in Φ_{ME} shown in Figure 3 are due to progressive addition of ethanol in W phase, requiring decreasingly AP volumes for microemulsifications. Such a decrease in Φ_{ME} by building up ethanol in W phase relates to the increase in surface activity phenomenon. It favours the thermodynamic stabilization of the dispersions by diminishing the interfacial tension [17].

Robustness

This parameter is essentially crucial for the development of point-of-use technologies by considering the accomplishment of in-situ assays wherein the effect of interferents on analytical response is critical. MEC was satisfactorily robust regarding deviations in preparing of W-O mixtures and changes in temperature.

Assuming the theory of dispersions with non-ionic surfactants, the surface activity depends mainly on temperature. Such a parameter reduces the polar group solvation and increases the number of nonpolar chain conformations of the amphiphile, diminishing and raising the surface activity, respectively [17]. Besides, this phenomenon changes with temperature owing to deviations in AP monomeric solubility. It alters the surface activity by modifying the fraction of amphiphile adsorbed at W-O interfaces. Therefore, Φ_{ME} do not show a simple and generic relationship with the temperature. In relation to deviation in W-O ratio, it affects the Φ_{ME} because the changes in surface pressures which are responsible for decreasing the surface activity and, thus, the interfacial tension.

For the procedure of W-O mixture preparation in which Φ_O was increased by 5.00% and 10.00% v/v, the values of $\Delta\Phi$ were positive for A and B and negative for C, with exception of the errors obtained in relation to 60.00% v/v Φ_E (reference value). The absolute error values changed among -0.31% and +5.39% v/v with a global average (in module) of $1.00 \pm 0.94\%$ v/v (n=18) and averages for each region (n=6) equal to $1.22 \pm 0.24\%$ (A), $2.85 \pm 1.28\%$ (B), and $1.39 \pm 0.76\%$ v/v (region C). Figure 4 illustrates the curves related to this robustness investigation as well as the calculated values of $\Delta\Phi$ for A, B, and C regions. The confidence intervals ranged from 0.08 to 0.42% v/v Φ_{ME}.

Resulting data showing the temperature-function robustness, in turn, are depicted in Figure 5. Confidence intervals were of 0.09 to 0.56% v/v Φ_{ME} whereas $\Delta\Phi$ ranged from +0.05% up to -9.46% v/v. Its global average (in module) was $1.69 \pm 0.84\%$ v/v (n=27) with averages

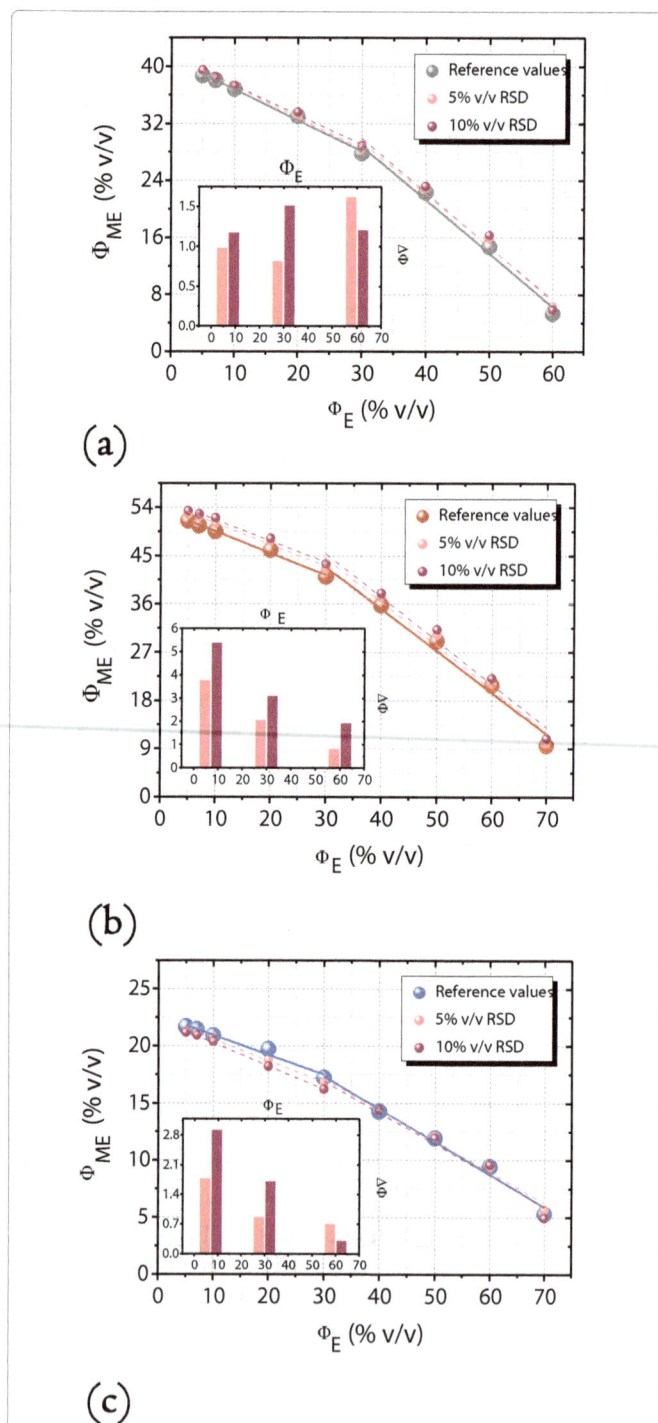

(a)

(b)

(c)

Figure 4: Analytical curves in A (a), B (b), and C (c) regions for ethanol standards in W phase with deviations in Φ_O utilizing water-n-propanol-oleic acid MEs at 23°C. Inset: values of $\Delta\Phi$ as a function of Φ_E for 7.00%, 30.00%, and 60.00% v/v Φ_E taking up 5.00% and 10.00% v/v RSD in Φ_O. All of the R^2 values were larger than 0.99. Analytical sensitivities were: -0.44/-0.73 (5.00%) and -0.41/-0.76 (10.00%) in A; -0.40/-0.81 (5.00%) and -0.39/-0.81 (10.00%) in B; and -0.19/-0.27 (5.00%) and -0.20/-0.27 (10.00% v/v RSD) in C. $\Delta\Phi$ parameter is given in module.

for each investigated region (n=9) of $0.92 \pm 0.61\%$ (A), $0.59 \pm 0.47\%$ (B), and $3.56 \pm 1.91\%$ v/v (C). Herein, $\Delta\Phi$ had unsystematically diverse positive and negative values.

In both of the cases of robustness assessing, the analytical

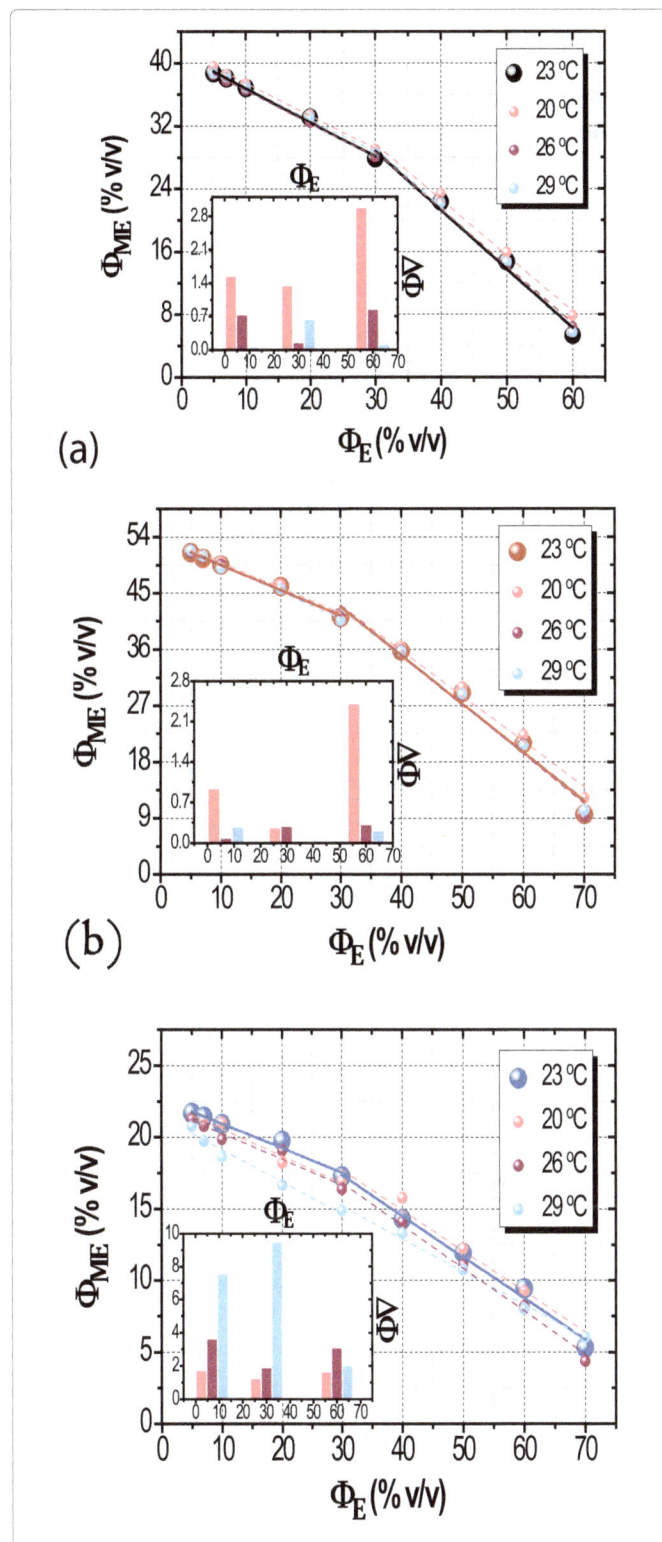

Figure 5: Analytical curves in A (a), B (b), and C (c) regions for ethanol in W phase with alterations in temperature using water-n-propanol-oleic acid MEs at 23°C. Inset: ΔΦ as a function of Φ_E for 7.00%, 30.00%, and 60.00% v/v Φ_E taking the changes of 23 to 20°C, 23 to 26°C, and 23 to 29°C. R^2 values were larger than 0.99. Analytical sensitivities were: -0.41/-0.71 (20), -0.42/-0.73 (26), and -0.40/-0.76 (29°C) in A; -0.40/-0.73 (20), -0.40/-0.79 (26), and -0.44/-0.76 (29°C) in B; and -0.19/-0.28 (20), -0.18/-0.30 (26), and -0.22/-0.23 (29°C) in C. ΔΦ is given in module.

sensitivities remained almost constant when compared to the curve at 23°C without deviations in Φ_O. Indeed, the robustness of MEC was outstanding. All of the values of analytical sensitivity are exhibited in legends of Figures 4 and 5.

Application

One potential factor for changing the signal of MEC is chemical interfering through modification of surface activity phenomenon. For example, a few samples associated to natural gas processing that presented diethylene and triethylene glycol in excess were not successful analyzed by MEC in our group to determine ethylene glycol. Hence, the development of alternatives for improving the MEC accuracy as regards to the presence of chemical interfering is necessary. In this case, a platform integrating gas diffusion separation and MEC detection was applied to test the adulteration of ethanol fuel by water utilizing analytical curve at 23°C and water-rich MEs. According to our preceding publication, direct determinations based on this composition did not generate accurate analyses as aforementioned [11].

The most usual adulteration of ethanol fuel is the excessive addition of water because the attained mixtures are colorless and do not present a distinctive smell. It produces loss of power and raise in fuel consumption rate [18]. Different methods were proposed for determination of water in ethanol fuels such as: i) near infrared spectrometry [19], ii) conductometry [20], iii) enthalpimetry [21], iv) cyclic voltammetry [22], v) photothermal detector [23], vi) ultrasonic propagation velocity [24], vii) evanescent field absorption spectroscopy [25], and viii) colorimetry [18].

The samples of ethanol fuel had 2.9 ± 0.7 µS cm^{-1}. Considering this low conductivity, we suppose a decrease in amount of diffused ethanol vapour due to tonometry was not observed.

High concentrations of ethanol in water (donor media) were used to construct the analytical curve by taking two aspects: the low analyte content that is volatilized in gas diffusion separation and the poor detectabilidade of MEC. The resulting analytical curve at 23°C for a separation time of 5 min is portrayed in Figure 6. For Φ_E bigger than 40.00% v/v, the limit of linearity was of 70.00% v/v. Thus, the samples were previously diluted 50.00% v/v to water for application once ethanol is found in fuel samples on fractions of around 95.00% v/v. Negative deviations in Φ_{ME} are because the progressive increase in ethanol fraction as well as happens in Figure 3. Confidence intervals were of 0.08% to 0.21% v/v Φ_{ME}.

According to Figure 6, the analytical sensitivity was only -0.06. Ways to improve such a poor sensitivity were studied, namely, the increase in the time and temperature of separation. For the warming of donor solutions, the separator was immersed in thermostatted bath that was placed on the stirrer. Measurements at 40°C during 5 min produced non-linear data (not shown). Conversely, assays at 23°C for 20 min generated linear data with enhancement in analytical sensitivity, -0.12. The analytical curved for 20 min is shown in Figure 7. The values of Φ_{ME} reduced with the time of separation because the increase of volatilized ethanol in receptor media that is employed for the subsequent performing of MEC.

For application, gas diffusion separations at 23°C for 5 min generated accurate results at 95% confidence level despite the low sensitivity. This time was adopted by considering the gain in analytical frequency. The data recorded by Karl Fischer and MEC are demonstrated in Table 1 after correcting Φ_E through dilution factor and its conversion in Φ_W.

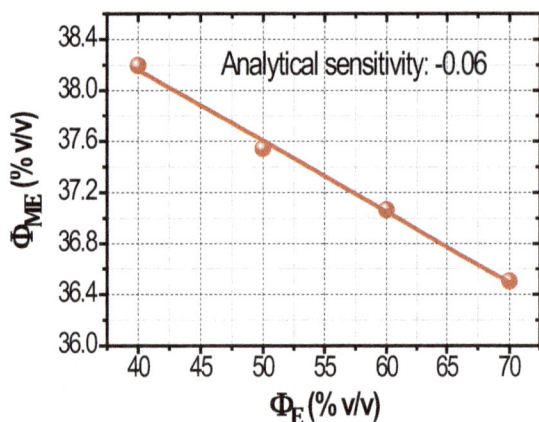

Figure 6: Analytical curve in region A for ethanol added in donor solutions considering separations at 23°C for 5 min and MEC using receptor water-n-propanol-oleic acid MEs. R^2 was 0.9950.

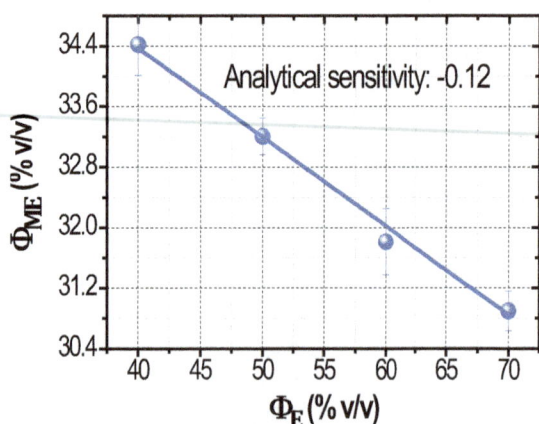

Figure 7: Analytical curve in region A for ethanol added in donor solutions taking up separations at 23°C for 20 min and MEC in receptor water-n-propanol-oleic acid MEs. R^2 was 0.9941.

Samples	Karl Fischer (% v/v, $n = 3$)	MEC (% v/v) (% v/v, $n = 4$)
F_1	5.1 ± 0.1	5.1 ± 0.1
F_2	5.1 ± 0.1	5.2 ± 0.1
F_3	5.0 ± 0.1	5.0 ± 0.2

Table 1: Fractions of H_2O (Φ_w) in ethanol fuel (F_1-F_3) determined by Karl Fischer and MEC with gas diffusion separation.

Conclusion

In summary, the results stressed in this paper outline the potentiality of MEC for the deployment of point-of-use analytical technologies. The method was robust as regards to deviations in preparation of W-O mixtures and changes in temperature for the determination of ethanol in water. Furthermore, the successful coupling with gas diffusion separation extends the field of application of MEC in analytical sciences by improving its selectivity. Herein, other sample preparation methods could be used such as solid phase extraction. For the platforms integrating gas diffusion separation and MEC-based detection, the time showed to be a satisfactory parameter with the intent to increase the sensitivity. In this case, the microemulsification could be also made with greater volumes of dispersion. This fact helps for precision and

accuracy by enhancing the gap among the values of Φ_{ME} for diverse concentrations of analyte.

Acknowledgment

Centro Nacional de Pesquisa em Energia e Materiais is recognized for its facilities. Financial support from the *Fundação de Amparo à Pesquisa do Estado de São Paulo* (FAPESP, Grant No. 2014/24126-6) is gratefully acknowledged.

References

1. Yetisen AK, Akram MS, Lowe CR (2013) Paper-based microfluidic point-of-care diagnostic devices. Lab Chip 13: 2210-2251.

2. Bissonnette L, Bergeron MG (2010) Diagnosing infections--current and anticipated technologies for point-of-care diagnostics and home-based testing. Clin Microbiol Infect 16: 1044-1053.

3. Gubala V, Harris LF, Ricco AJ, Tan MX, Williams DE (2012) Point of care diagnostics: status and future. Anal Chem 84: 487-515.

4. Chin CD, Linder V, Sia SK (2012) Commercialization of microfluidic point-of-care diagnostic devices. Lab Chip 12: 2118-2134.

5. Hartman MR, Ruiz RC, Hamada S, Xu C, Yancey KG, et al. (2013) Point-of-care nucleic acid detection using nanotechnology. Nanoscale 5: 10141-10154.

6. Hu J, Wang S2, Wang L, Li F3, Pingguan-Murphy B4, et al. (2014) Advances in paper-based point-of-care diagnostics. Biosens Bioelectron 54: 585-597.

7. Song Y, Huang YY2, Liu X, Zhang X2, Ferrari M, et al. (2014) Point-of-care technologies for molecular diagnostics using a drop of blood. Trends Biotechnol 32: 132-139.

8. Drain PK, Hyle EP2, Noubary F3, Freedberg KA2, Wilson D4, et al. (2014) Diagnostic point-of-care tests in resource-limited settings. Lancet Infect Dis 14: 239-249.

9. Weaver W, Kittur H, Dhar M, Di-Carlo D (2014) Research highlights: microfluidic point-of-care diagnostics. Lab Chip 14: 1962-1965.

10. Paterson S, de la Rica R (2015) Solution-based nanosensors for in-field detection with the naked eye. Analyst 140: 3308-3317.

11. Lima RS, Shiroma LY, Teixeira AV, de Toledo JR, do Couto BC, et al. (2014) Microemulsification: an approach for analytical determinations. Anal Chem 86: 9082-9090.

12. Hong JI, Chang BY (2014) Development of the smartphone-based colorimetry for multi-analyte sensing arrays. Lab Chip 14: 1725-1732.

13. Munoz RAA, Richter EM, De-Jesus DP, Do-Lago CL, Angnes L (2004) Determination of inorganic ions in ethanol fuel by capillary electrophoresis. J Brazil Chem Soc 15: 523-526.

14. Vieira EG, Soares IV, Dias Filho NL, da Silva NC, Garcia EF, et al. (2013) Preconcentration and determination of metal ions from fuel ethanol with a new 2,2'-dipyridylamine bonded silica. J Colloid Interface Sci 391: 116-124.

15. Giordano GF, Vieira LC2, Gobbi AL2, Lima RS, Kubota LT3 (2015) An integrated platform for gas-diffusion separation and electrochemical determination of ethanol on fermentation broths. Anal Chim Acta 875: 33-40.

16. Shaw DJ (1992) Introduction to colloid and surface chemistry. Elsevier Science, Butterworth Heinemann pp 48-53.

17. Stubenrauch C (2009) Microemulsions: background, new concepts, applications, perspectives. John Wiley & Sons pp 1-30.

18. Giordano GF, Ferreira DCM, Carvalho TR, Vieira LCS, Piazzetta MHO, et al. (2014) Portable platform for rapid and indirect photometric determination of water in ethanol fuel samples. Anal Methods 6: 9497-9502.

19. Silva AC, Pontes LF, Pimentel MF, Pontes MJ (2012) Detection of adulteration in hydrated ethyl alcohol fuel using infrared spectroscopy and supervised pattern recognition methods. Talanta 93: 129-134.

20. Ribeiro MS, Angnes L, Rocha FRP (2013) A simple and fast procedure for in situ determination of water in ethanol fuel. J Braz Chem Soc 24: 418-422.

21. De Oliveira WA, Pasquini C (1984) Determination of water in ethanol and acetone by direct injection enthalpimetry based on the heat of dilution. Talanta 31: 82-84.

22. Pereira PF, Sousa RMF, Munoz RAA, Richter EM (2013) Simultaneous determination of ethanol and methanol in fuel ethanol using cyclic voltammetry. Fuel 103: 725-729.

23. Omido CR, Oliveira SL, Shiraishi RS, Magalhaes KF, Ferreira VS, et al. (2013) Quantification of water in ethanol using a photothermal transparent transducer. Sensor Actuat B Chem 178: 581-585.

24. Figueiredo MKK, Costa-Felix RPB, Maggi LE, Alvarenga AV, Romeiro GA (2012) Biofuel ethanol adulteration detection using a ultrasonic measurement method. Fuel 91: 209-212.

25. Xiong FB, Sisler D (2010) Determination of low-level water content in ethanol by fiber-optic evanescent absorption sensor. Opt Commun 283: 1326-1330.

13

Effect of Somatic Cell Count on Bovine Milk Protein Fractions

Ramos TM[1]*, Costa FF[2], Pinto ISB[3], Pinto SM[4] and Abreu LR[4]

[1]Department of Animal and Food Sciences, University of Delaware, Christina Mill Drive, Newark, Delaware, United States of America
[2]Department of Food Science, Federal University, Juiz de Fora, Brazil
[3]Department Animal and Food Sciences, Embrapa Gado de Leite, Juiz de Fora, Brazil
[4]Department of Food Science, Federal University of Lavras, Lavras, Brazil

Abstract

The objective of this study was to evaluate the influence of somatic cell count (SCC) on the physicochemical properties and protein fractions of milk. Milk was collected and analyzed for somatic cell count, fat, lactose, acidity, total solids, ash, total nitrogen, soluble nitrogen at pH 4.6, and soluble nitrogen in trichloroacetic acid (TCA) 12%. Milk was divided into four groups according to the value of SCC, each constituting a treatment, as follows: Treatment 1 (<300,000 cells/ml), Treatment 2 (300,000 to 750,000 cells/ml), Treatment 3 (750,000-1,000,000 cells/ml), and Treatment 4 (>1 million cells/ml). The electrophoretic profile of milk was also evaluated using microfluidic electrophoresis for separation and quantification of milk proteins. An increase in the concentration of SCC resulted in a significant increase in the amount of fat, soluble nitrogen and soluble protein (casein) fractions, and a reduction of α-casein, β-casein, and κ-casein. There was a higher proteolytic activity associated with high SCC. Changes in protein fractions of milk caused by high SCC had strong implications regarding the potential of milk as raw material for manufacturing products as the industrial yield of milk is mainly associated with the casein fraction.

Keywords: Casein; SCC; Mastitis; Electrophoresis

Abbreviations

SCC: Somatic Cell Count; TN: Total Nitrogen; NCN: Noncasein Nitrogen Content; CP: Crude Protein; SN: Soluble Nitrogen; NPN: Non-Protein Nitrogen; CN: Casein Nitrogen; TP: True Protein; SP: Soluble Protein; β-CN: β-Casein; α-CN: α-Casein; κ-CN: κ-Casein; TA: Titratable Acidity; La: α-Lactalbumin; β-Lg: β-Lactoglobulin

Introduction

Caseins are milk proteins secreted by cells of the mammary gland. They constitute approximately 78-82% of bovine milk proteins and are divided into four main groups: α_{S1}-casein, α_{S2}-casein, β-casein and κ-casein, forming supramolecular structures known as micelles [1,2]. The protein composition of cow's milk is an important factor for the profitability of the dairy industry. An increase in the proportion of casein, in particular α- and β-CN, results in better product yield, especially in cheese [3]. The caseins are phosphoproteins containing a variable number of phosphate radicals linked to serine (P-Se) and are concentrated in different regions of polypeptide chains. Based on the location of these phosphate radicals, the resulting molecule regions are more hydrophilic or more hydrophobic, and consequently, the caseins are more susceptible to proteolysis.

Proteolysis in milk is an important quality criterion that can have beneficial or detrimental effects, depending on processing. Milk protein proteolysis can be attributed to both indigenous proteases and also proteases produced by psychrotrophic bacteria during the cold storage of milk [4]. Proteolytic bacterial enzymes act mostly on the κ-casein, resulting in the destabilization of the casein micelles and coagulation of milk in a manner analogous to chymosin [5,6]. Proteolysis of bovine milk can also occur naturally [7]. During this process, native thermostable protease activity is mainly related to the activity of plasmin, a serine protease derived from its inactive precursor plasminogen [8].

Plasmin is an alkaline proteolytic enzyme which participates in the hydrolysis of casein. It is of great importance in natural milk proteolysis [7]. Plasmin is the active form which is produced from the zymogen called plasminogen. The conversion of plasminogen to plasmin occurs by the action of specific plasminogen activators, which are also proteases [9]. The increase in plasmin activity is caused by somatic cells from its inactive precursor plasminogen, which is converted into plasmin, in a process initiating in the mammary gland and continuing throughout the storage period [7,8,10,11]. Increased SCC in milk results in elevated activation of plaminogen into plasmin, that in turn leads to high breakdown of some proteins chains, primarily β-casein, because protein fraction partially diffuses into solution at low temperature, which facilitates enzyme attack, producing small fragments, such as γ-caseins and other small peptides that diffuse to the aqueous phase of the milk [12,13]. This protease has specificity for Lys-x and Arg-x bonds [13-15].

The level and activity of plasmin in milk can vary and depends on biological factors, such as stage of lactation and somatic cell count [16]. The milk somatic cells, mainly composed of neutrophils and macrophages, have a wide range of proteolytic and lipolytic enzymes, which are released during the intracellular mechanism, killing microorganisms in subclinical mastitis, and may significantly contribute to proteolysis and lipolysis of the milk constituents [17,18]. Therefore, concentrations of many enzymes or their activity in the milk are increased during mastitis [19-21]. The enzymes of primary concern for the dairy industry are those with proteolytic activities, because the increase of proteolysis in milk and milk products has a negative impact on the quality and technological properties.

Proteolysis associated with increased somatic cell count in milk promotes the breakdown of casein micelles [22], one of these is the

*Corresponding author: Ramos TM, Department Animal and Food Sciences, University of Delaware, 1045, Christina Mill Drive, Newark, Delaware, 19711, United States of America, E-mail: thaisramos85@yahoo.com.br

indigenous milk proteinase plasmin, which is associated primarily with the casein micelles [23], where it is capable of hydrolysing all caseins except κ-casein [24-26], in which contributes to increased susceptibility to defects in dairy products such as technological problems related to proteolytic enzymes include the gelling of UHT milk (Ultra High Temperature) [27,28], generation of free amino acids during cheese ripening and development of undesirable flavors and a bitter taste in milk and dairy products [29,30]. Even ultrahigh temperature (UHT) treatment of milk is insufficient to inactivate plasmin completely, but typical retort sterilisation does inactivate plasmin completely [25]. The use of milk with elevated SCC has detrimental technological implications, such as low yield, and decreased shelf life of products, changes in the characteristics of milk, and milk products, and interference in manufacturing technologies, especially in cheese.

Cooling is important and a way of improving milk quality. However, extended refrigeration time leads to modifications in composition and physical properties of milk. Among the many changes that occur during the cooling process, includes the dissociation of caseins, specifically the β-casein, which can solubilize up to 18% of its total fraction, solubilization of colloidal calcium phosphate, and as a consequence decrease in size of the micelles.

The separation and quantification of major milk proteins are fundamental in dairy research. Therefore, accurate and rapid methods are profoundly important. The microfluidic chip technique is faster, and uses considerably fewer chemicals and materials traditional techniques [31].

The aim of this study was to elucidate the behavior of protein fractions of milk with different somatic cell counts; specifically β-casein, it can be broken by plasmin with potentially bitter peptide formation and reducing the total solids.

Materials and Methods

Milk and milk proteins

Sample collection: Raw milk samples were collected from isothermal stirred bulk tanks with an internal temperature no more than 5°C. The samples were collected by specially educated technicians from the bulk tank milk of raw milk suppliers. The samples were labeled and transported according to the procedures established by the laboratory responsible for testing. The samples milk was collected in a dairy located in the city of Lavras, MG, Brazil.

Analysis of milk: The analyses were developed into different steps. Milk Quality Analyze Laboratory (LABUFMG) at Federal University of Minas Gerais (UFMG), Belo Horizonte MG, Brazil, developed Somatic Cells Counting (SSC) analysis. Both physicochemical and microbiological analyses were performed at Federal University of Lavras (UFLA), Lavras MG, Brazil. Thereafter, the frozen samples were transported to the Brazilian Agricultural Research Corporation (EMBRAPA) Laboratory, Juiz de Fora MG, Brazil, to do the electrophoresis profile of the proteins analysis.

Milk proteins: For analysis of proteins were used purified α-lactoalbumina (α-La), β-lactoglobulin (β-Lg), $α_s$-casein ($α_s$-CN), β-casein (β-CN) and κ-casein (κ-CN) were obtained from Sigma-Aldrich (USA). Solutions (10 mg mL^{-1}) of each individual protein were prepared by adding each individual protein purified water (Ultrapure Milli-Q; Millipore Corp., USA) and stirring until dissolved. Mixed protein standards were prepared by combining each of the individual protein solutions (1 mL) and making the final volume up to 10 mL to give a mixed protein standard with an individual protein concentration of 1 mg mL^{-1}.

Microbiological examination

Mesophilic bacteria count: decimal dilutions of raw milk samples were taken and plated on Plate Count Agar - PCA mesophilic bacteria to viable counts after incubation at 32°C for 48 hours. Count of psychrotrophic and proteolytic psychrotrophic: Dilutions decimal of raw milk samples were plated on agar Calcium Caseinate (Merck°) for the bacterial count of psychrotrophic and proteolytic psychrotrophic viable, with incubation at 7°C ± 0.5°C for 10 days.

Analysis of chemical composition and SCC

Fat, lactose, total solids and somatic cell counts were determined by infrared absorption (Bentley CombSystem 2300).

Physical-chemical analysis

Total nitrogen (TN), noncasein nitrogen content (NCN) corresponding to the milk soluble fraction at pH 4,6, and NPN content corresponding to the non-precipitated fraction with 12% trichloroacetic acid were determined by the Kjeldahl method following the AOAC [32]. Nitrogen was then multiplied by a standard factor (6.38) so that the results are expressed as total protein. Ash was determined by an AOAC (Association of the Official Analytical Chemists) technique using carbonization of the samples in a direct flame and subsequent calcination in a muffle at 550°C for 4-6 hours.

Titratable acidity

The acidity was determined by titration with a 0.1N NaOH solution using phenolphthalein as an indicator, and the result was expressed in grams of lactic acid or percentage of compounds having acidic character [32].

Microfluidic chip electrophoresis

Milk samples were subjected to ultracentrifugation in triplicate (40,000 × g) at 4°C for 60 min using a CR21 Himac ultracentrifuge (Hitachi, Japan). After centrifugation, the supernatant (soluble phase) was separated for analysis of protein profiles. Separation of individual milk proteins was performed using the microfluidic chip electrophoresis system (Agilent 2100 Bioanalyser - Technologies GmbH, Waldbronn, Germany) and the associated Protein 80 kit (Agilent Technologies, Germany). These kits contain the chips and proprietary reagents such as the gel matrix solution, proteins in a concentrated solution, a marker protein buffer solution and a protein molecular mass ladder solution to perform the electrophoresis [31,33-35].

The TPS buffer consisted of 0.1 mol L^{-1} tris chloride acid (Amresco, USA), pH 8.8, containing 2 mol L^{-1} urea (USB, Germany), 15% glycerol (Invitrogency, New Zealand) and 0.1 mol L^{-1} dithiothreitol (DTT) (Bioangency, Brazil). It was prepared according to the SOP (Standard Operating Procedure) available

from the Food Standards Agency (FSA) of the United Kingdom [31,35]. The SEP buffer solution, pH 3.0, used to separate the proteins consisted of 6.0 mol L1 urea (USB, Germany), 20 mmol L1 trisodium citrate dehydrate (Synth, Brazil), 0.1 mol L^{-1} citric acid (Merck, Brazil) and 0.05% (w/w) hydroxypropylmethyl cellulose (Sigma-Aldrich, USA) [31,36].

Segundo Costa et al. [31], milk was diluted in a 1 : 4 ratio with TPS buffer, SEP buffer and pure water (Ultrapure Milli-Q; Millipore Corp., USA) to compare and select the more efficient diluting agent. Samples were allowed for at least 2 h at 4°C for protein solubilization before application in microfluidic chip electrophoresis which was performed using an Agilent 2100 Bioanalyzer system (Agilent Technologies, Germany). The gel matrix, solutions and samples for electrophoresis were prepared according to the Bioanalyser protocols (Agilent Technologies, Germany). In Eppendorf tubes (0.5 mL total volume) 4 μL of samples (milk; milk+TPS buffer; milk+SEP buffer; milk+pure water; and milk added with each individual protein+SEP buffer) were mixed with 2 μL of 2-mercaptoethanol (Sigma-Aldrich, USA), heated (95°C, 5 min), cooled in an ice bath, briefly spun in a centrifuge (3000 g) and then 84 μL of Milli-Q water was added to give a total volume of 90 μL. All chips were loaded with ten samples with three replicates each

Quantification was carried out considering the area under the electropherogram using the Agilent 2100 Expert software associated with the instrument. The results were expressed as percentages (%) according to all the proteins identified in the electropherograms.

Experimental design

Bovine milk with mesophilic bacterial counts below 40,000 cfu/ml and psychrotrophic counts below 2000 cfu/mL were collected and analyzed for somatic cell count. Milk samples were grouped according to SCC in four groups, each representing one treatment as follows:

Treatment 1: (<400.000 cells/mL);

Treatment 2: (400.000-750.000 cells/mL);

Treatment 3: (750.000-1.000.000 cells/mL);

Treatment 4: (>1.000.000 cells/mL).

Statistical analysis

Results were analyzed by ANOVA and the Tukey test at 5% probability using the R statistical package (R Development Core Team, Vienna, Austria).

Results and Discussion

Composition of milk with different somatic cell counts

The chemical composition of milk with different somatic cell counts is shown in Table 1. There were no significant differences ($p>0.05$) in concentrations of total solids, solids-not-fat, ash, acidity, lactose and total protein among treatments. The concentrations of total solids, although not statistically significant, presented tendence to increase as SCC increased. Research conducted by Fernandes [37] found an elevation in total solids with higher SCC. However, Marques [38] and Klei et al. [39] reported that the total solids content of milk with high SCC did not change. Moslehishad et al. [40], Lee et al. [41], and Salah El-Tahawy [42] showed higher percentage of total solids with an increase in SCC. Theses and other reports indicate inconsistent variation for this attribute in relation to SCC, once some compounds have their values increasing whereas others have theirs decreasing.

Protein results, despite the non-significant differences ($p>0.05$), had a slightly increment with higher concentrations of SCC. However, the experimental results do not agree with the effects of high SCC milk on the total protein of milk, measured by the concentration of total nitrogen, as reported by several studies. The effect of mastitis on the total concentration of milk protein is variable [43]. Research [39,40] has shown that a higher milk somatic cell count results in higher levels of total protein. On the other hand, Verdi et al. [44], Rogers [45], and Albenzio et al. [10] reported no change in the total protein content of the milk, which possessed high SCC compared to milk with lower

values, while Lee et al. [41] stated that total protein was lower in milk from cows with high SCC. Overall, total protein in milk with high SCC can remain unchanged or undergo small changes, because the content of casein decrease is accompanied by an increase in whey proteins, resulting in a negligible change in total milk protein.

There is an inverse relationship between the values of lactose and SCC but with no significant difference ($p>0.05$). Some authors [10,46] agree that there is a reduction in the concentration of lactose in milk with high SCC. Inflammation of the mammary gland results in lesser synthesis of lactose [46,47]. During mastitis, the NaCl concentration in milk more elevated, resulting in an augment in its osmotic potential, making the milk in the lumen hyper-osmotic relative to the surrounding blood. Because these two mediums must be iso-osmotic for the synthesis of milk, there is a physiological compensation by reducing the lactose content of the milk [48], which explains the results obtained.

Higher fat content ($p<0.05$) was observed between groups of SCC (<300, 300-750, and 750-1000) when associated with higher values of SCC. Similar results were reported by Miller et al. [49], Mitchel et al. [50], Marques et al. [38]. However, Munro et al. [51] and Moslehishad et al. [40] found no significant difference ($p>0.05$) for the values of milk with different fat content, and Najafi et al. [52] observed an inverse relationship between fat content and SCC values. These results indicate that there can be no standard established relative to fat content and SCC.

Although Nafaji et al. [52] reported that high SCC milk reduces the acidity by reducing its solid content, the mean values of acidity did not differ ($p>0.05$). The average milk composition related to crude protein (CP), total nitrogen (TN), soluble nitrogen (SN), non-protein nitrogen (NPN), casein nitrogen (CN), true protein (TP), soluble protein (PS), casein, and the ratio of CN/TP are reported in Table 2. The content of total protein (TP), total nitrogen (TN), true protein (TP), and the relationship between CN/TP was not affected by the milk SCC ($p>0.05$). Santos et al. [53] found similar results regarding the content of total protein (TP), non-protein nitrogen (NPN) and true protein. In contrast, Ma et al. [54] demonstrated that milk with a lower (45,000 cells/ml) SCC concentration had lower CP than that observed in milk with increased SCC (849,000 cells/mL).

Occurred significant difference ($p<0.05$) in the contents of soluble nitrogen (SN), non-protein nitrogen (NPN), soluble protein (PS), and casein. The levels of casein were reduced ($p<0.05$). It is well known that during mastitis, casein synthesis is usually reduced, similar to results found by Santos et al. [53] and O'Connell et al. [55]. Nevertheless, some authors found no significant reduction in casein when correlated with high SCC [10,40,56-59].

Reports [39,44] have previously described that the CN/TP is reduced with lower SCC. This finding accounts for the reduction of casein without changing the total protein (Table 2).

TP concentrations were not significantly different ($p>0.5$) between treatments, while the levels of casein showed differences ($p<0.05$) between milk below 750,000 SCC (treatments 1 and 2) and milk above 750,000 SCC (treatments 3 and 4). The reduction of casein and CN/TP probably occurs by partial degradation of casein, particularly of β-casein by more intense proteolytic activity of plasmin in high SCC milk. The values of soluble nitrogen (SN) and soluble protein (PS) increased ($p<0.05$) with increasing SCC. The higher values of the soluble fractions seem to be a clear indication of a intense proteolytic activity of plasmin, coupled with a low integrity of the casein micelle, due to the solubility of β-casein in cold milk.

Treatments[1]	Analysis[2]							
	SCC	TS	SNF	Ash	Fat	TA	Lactose	CP
<300	271.5 ± 33.09[a]	12.27 ± 0.85[a]	8.99 ± 0.69[a]	0.68 ± 0.06[a]	3.28 ± 0.27[a]	0.15 ± 0.86[a]	4.58 ± 0.12[a]	3.16 ± 0.62[a]
300-750	528.7 ± 241.69[b]	12.56 ± 0.63[a]	9.02 ± 0.40[a]	0.70 ± 0.04[a]	3.54 ± 0.30[b]	0.15 ± 0.35[a]	4.63 ± 0.06[a]	3.16 ± 0.43[a]
750-1000	796.33 ± 49.52[c]	12.71 ± 0.30[a]	9.17 ± 0.23[a]	0.71 ± 0.03[a]	3.54 ± 0.12[b]	0.15 ± 0.74[a]	4.49 ± 0.16[a]	3.18 ± 0.14[a]
>1000	1145 ± 95.9[d]	12.80 ± 0.30[a]	9.22 ± 0.32[a]	0.72 ± 0.01[a]	3.58 ± 0.12[b]	0.14 ± 0.59[a]	4.51 ± 0.20[a]	3.10 ± 0.44[a]

a-dMeans within a columns with different superscripts differ (P<0.05)
Values are given as means ± standard deviation
[1]Treatments: <300=Tank milk with somatic cell count of <300000 cells/ mL; 300-750=Tank milk with somatic cell count of 300000 to 750000 cells/mL; 750-1000=Tank milk with somatic cell count of 750000 to 1000000 cells/ mL; >1000=Tank milk with somatic cell count of >1000000 cells/mL
[2]Analysis: SCC: Somatic cell count; TS: Total solid (%); SNF: Solid-non-fat (%); Ash (%); Fat (%); TA=Total Acidity (grams of lactic acid /100 grams of sample); CP: Crude protein (%)

Table 1: Chemical composition of bulk tank milk with different somatic cell counts.

Treatments[1]	Analysis[2]								
	CP	TN	SN	NPN	CN	TP	SP	Casein	Casein/TP
<300	3.16 ± 0.62[a]	0.49 ± 0.09[a]	0.09 ± 0.1[a]	0.023 ± 0.01[a]	0.39 ± 0.09[a]	3.13 ± 0.62[a]	0.63 ± 0.06[a]	2.49 ± 0.57[a]	0.26 ± 0.08[a]
300-750	3.16 ± 0.43[a]	0.49 ± 0.06[a]	0.10 ± 0.01[b]	0.026 ± 0.01[b]	0.38 ± 0.07[a]	3.13 ± 0.39[a]	0.69 ± 0.09[b]	2.42 ± 0.44[a]	0.25 ± 0.04[a]
750-1000	3.18 ± 0.14[a]	0.50 ± 0.02[a]	0.11 ± 0.05[c]	0.028 ± 0.01[c]	0.37 ± 0.07[b]	3.15 ± 0.15[a]	0.76 ± 0.03[c]	2.36 ± 0.44[b]	0.24 ± 0.01[a]
>1000	3.10 ± 0.44[a]	0.48 ± 0.04[a]	0.12 ± 0.04[c]	0.029 ± 0.01[d]	0.36 ± 0.06[b]	3.07 ± 0.30[a]	0.74 ± 0.02[c]	2.29 ± 0.38[b]	0.25 ± 0.02[a]

a-dMeans within a column with different superscripts differ (P<0.05)
Values are given as means ± standard deviation
[1]Treatments: <300=Tank milk with somatic cell count of <300000 cells/ mL; 300-750=Tank milk with somatic cell count of 300000 to 750000 cells/mL; 750-1000=Tank milk with somatic cell count of 750000 to 1000000 cells/ mL; >1000=Tank milk with somatic cell count of >1000000 cells/mL
[2]Analysis: CP=Crude protein (%); TN=Total nitrogen (%); SN=Soluble nitrogen (%); NPN=Non-protein nitrogen (%); CN=Casein nitrogen (%); TP=True protein (%); SP=Soluble protein (%)

Table 2: Nitrogen components of tank milk with different somatic cell counts.

Moslehishad et al. [40] found no significant difference in the content of total nitrogen (TN) and casein (CN) at three levels of SCC (<200, 200-800 and >800).

Separation and identification of major milk proteins by microfluidic chip electrophoresis

As a starting point, the analysis of the milk proteins of raw bovine milk was carried out using deionized water and two different buffers for the treatment of milk samples before the standard procedure recommended by the manufacturer of the electrophoresis equipment microfluidics. The two buffers compared were a total protein solubilization buffer (TPS buffer) and a separating milk protein buffer (SEP buffer). The first one is recommended for the preparation of milk samples before application in microfluidic chip electrophoresis [31,35] while the latter is commonly used for the separation of protein fractions of milk during the sample preparation for analysis by CE [60].

In order to identify the peaks corresponding to each of the protein fractions, the addition of individual protein standards to the sample of milk was carried out. The identification was confirmed by the observation of an increased signal of each one of the individual proteins added (Figure 1). Thus, the Figure 1, presented here in only for illustrative purposes, shows results from the percentage of total protein fractions of the samples with the highest and the lowest SCC respectively. The elution order is: α-lactalbumin (peak 1), β-lactoglobulin (peak 2), β-casein (peak 3), αs-casein (peak 4) and κ-casein (peak 5). The electropherograms are presented as fluorescence units (FU), the molecular weight (kDa) and migration time (FU × kDa; FU × Time). By comparing the signals detected in milk samples submitted at low and high SCC (Figure 1A and 1B, respectively), variations in the quantification of protein fractions are observed.

The literature Costa et al. [31] showed the addition of both the SEP and TPS buffers in the treatment of milk samples made it possible to separate different peaks corresponding to the major milk proteins with

a good resolution. These results are explained because the milk caseins are dissociated by the addition of urea [61] and both buffers contained urea, the TPS buffer had a concentration of 2 mol L^{-1} and the SEP buffer had a concentration of 6 mol L^{-1} of urea, respectively [31].

Data of the average percentage of individual protein fractions of milk are displayed in Table 3. The microfluidic electrophoresis revealed that the greatest number of somatic cells significantly elevated the products generated by casein hydrolysis in milk. Figure 1 presents, in descending order, significant reductions ($p<0.05$) of percentage of β-casein (peak 3), α-casein (peak 4), and κ-casein (peak 5) of milk associated with SCC, which in turn produced higher concentration of the soluble fractions of milk (Table 2).

Quantitative determination of major milk proteins by microfluidic chip electrophoresis

Approximately 80% of total nitrogen in bovine milk consists of casein. The bovine casein can be classified into four types of proteins with different properties: α_{s1}, α_{s2}, β and κ, comprising 38%, 10%, 34% and 15% of total casein, respectively [48]. It may be observed (Figure 1) a reduction in the β-casein fraction (the fraction most affected by the enzymatic action of plasmin), in the order of 15%, 18%, 26%, and 30% due to the elevation of SCC, it can be noted as well a total variation (between treatment 1 and 4) of approximately 48%. The migration of the β-casein from the aggregate micellar form to dispersed molecules in the soluble phase of milk is more intense at lower temperatures [33,62], becoming in turn, more susceptible to the enzymatic action of plasmin, decreasing its concentration and increasing the concentration of lower molecular weights peptides. As these peptides are of high solubility, they are carried by the whey during the cheese making process, significantly reducing milk yield. In addition, the texture of the cheese changes, because the reduction in the concentration of the β-casein in the micelle causes changes in the physicochemical properties of the cheese mass. High SCC causes serious technological problems in manufacturing dairy products. For example, in cheese manufacturing process, changes in the CN/SP of the milk (Table 2) due to the elevation of SCC, increases the clotting time, particularly by affecting the access of the enzyme to the κ-casein, and reducing the development of the proper pH. Moreover, the time to reach the draining point is lengthened, because the soluble components have higher water holding capacity, and high SCC refrain development of acidity, facts that reduce syneresis. These characteristics affect not only the manufacturing process but also significantly impair the standardization of each type of cheese.

The high proteolytic activity in milk from diseased udders likely leads to a reduction in the concentration of both α-CN and β-CN, with a simultaneous elevation of γ-CN concentration, with evidence that the hydrolysis of casein occurs within the udder previous to the milking process [63]. With respect to α-CN and κ-CN fractions, concentrations did not suffer significant interference, except in SSC over 1,000,000 (Table 3). However, Moslehishad et al. [40] studied the influence of three levels (<200 to >800) of somatic cells to examine the electrophoresis profile of milk using polyacrylamide gel electrophoresis (SDS-PAGE) and achieved significant reductions ($p<0.05$) of α-casein and β-casein fractions, with higher SCC.

Casein is considered to be the more important protein, as far as economical issue is concerned, due to its relation to the production of milk products. Mastitis can significantly affect the quality of dairy products. Rogers and Mitchell [64] reported that an increase in SCC impaired the sensory characteristics of nonfat yogurt. Munro et al. [51] found that yogurt obtained from milk with high SCC showed a color change, characterized as slightly yellow. Also, Oliveira et al. [65] showed a decrease in sensory quality of yogurt after 20 days of cold storage, especially in consistency and flavor attributes when milk with >800,000 cells/mL was used. Fernandes [37] observed elevated viscosity of yogurt obtained from milk with SCC >800,000 cells/mL after 10 days of storage. High SCC can also be correlated to a reduced quality of butter, and Auldist and Hubble [43] reported that SCC alter the composition of butter and elongate churning time. Sensory properties are also affected and butter deteriorates faster during storage. In the production of milk powder, Auldist et al. [46] reported that milk powder with high SCC has lower heat stability and that, other properties deteriorate more rapidly in comparison to milk powder with low SCC, which is highly probably due to more intense lipolysis and proteolysis [66-68].

High somatic cell counting significantly affect the protein fractions, particularly β-CN, with remarkable reduction of protein values, that directly affect the dairy industry, by causing economical losses, decreasing stability of fluid milk that leads to low thermal stability, lower yield and poorer sensory properties of milk products [69,70].

Conclusions

Milk with high somatic cell count undergoes several chemical changes. In general, total solids and solids not fat had a slight increase, whereas fat content had a significant increment. Lactose was reduced. The more intense changes occurred in the proteins. Crude protein had a small elevation in milk with SCC around 700,000 and decreasing with SCC above 1,000,000. Percentage of casein reduced and that of soluble proteins decreased which led to a considerable reduction of the Ratio casein/soluble proteins. The percentage of true proteins was lower and NPN had a remarkably increment in milk with higher SCC. Regarding the casein fractions, high SCC caused reduction in β and α and κ in descendent order. In the particular case of the β-casein, it reduced approximately 48% from milk with SCC lower than 300,000 to milk above 1,000,000. One may conclude with high certainty that the quality of milk is directly and negatively affected by high somatic cell counting.

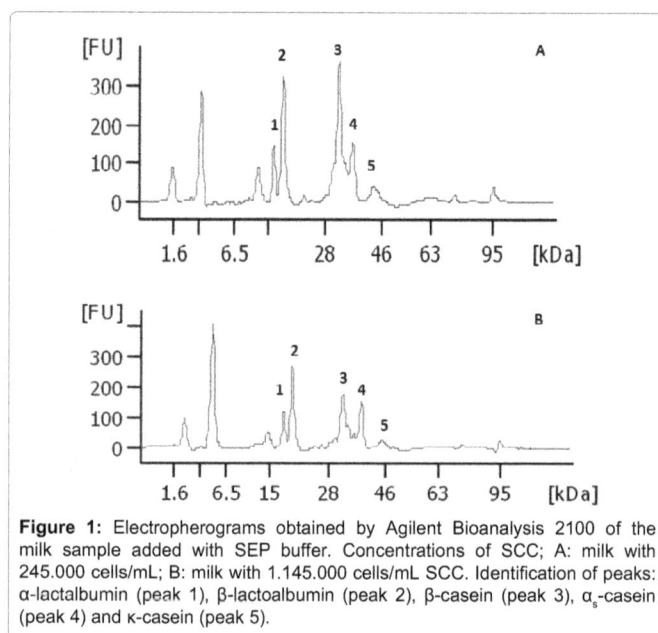

Figure 1: Electropherograms obtained by Agilent Bioanalysis 2100 of the milk sample added with SEP buffer. Concentrations of SCC; A: milk with 245.000 cells/mL; B: milk with 1.145.000 cells/mL SCC. Identification of peaks: α-lactalbumin (peak 1), β-lactoalbumin (peak 2), β-casein (peak 3), α$_s$-casein (peak 4) and κ-casein (peak 5).

Protein Fraction	Treatments[1]			
	<300	300-750	750-1000	>1000
α-lactalbumin	7.88 ± 1.46[a]	7.26 ± 0.57[a]	7.89 ± 0.65[a]	6.97 ± 0.80[a]
β-lactoglobulin	17.64 ± 1.39[a]	17.34 ± 1.29[a]	16.77 ± 0.46[b]	15.58 ± 1.55[a]
β-casein	31.85 ± 1.37[a]	27.08 ± 1.68[a]	22.1 ± 1.30[c]	16.35 ± 2.03[d]
α$_s$-casein	19.00 ± 1.57[a]	18.08 ± 1.63[a]	18.58 ± 0.90[a]	13.00 ± 1.25[b]
κ-casein	5.54 ± 1.52[a]	5.19 ± 0.71[a]	5.27 ± 1.27[a]	3.47 ± 1.22[b]

[a-d]Means within a row with different superscripts differ ($P<0.05$)

Values are given as means ± standard deviation

[1]Treatments: <300=Tank milk with somatic cell count of <300000 cells/ mL; 300-750=Tank milk with somatic cell count of 300000 to 750000 cells/mL; 750-1000=Tank milk with somatic cell count of 750000 to 1000000 cells/mL; >1000=Tank milk with somatic cell count of >1000000 cells/mL

Table 3: Distribution of protein fraction from the soluble phase of tank milk with different levels of somatic cells after ultracentrifugation count.

References

1. Eigel WN, Butler JE, Ernstrom CA, Farrell HM, Harwalker VR, et al. (1984) Nomenclature of proteins of cow's milk: fifth revision. J Dairy Sci 76: 1599-1631.

2. Dalgleish DG, Morris ER (1988) Interactions between carrageenans and casein micelles: electrophoretic and hydrodynamic properties of the particles. Food Hydrocoll 2: 311-320.

3. Pabst K (1994) Organic milk, is the change worth while. Tierzuchter 46: 22-25.

4. Ismail B, Nielsen SS (2011) Plasmin System in Milk. In: Fuquay JW, Fox PF, McSweeney PLH (Eds) Encyclopedia of Dairy Sciences. pp. 929-934.

5. Fairbairn DJ, Law BA (1986) Proteinases of psychrotrophic bacteria: their production, properties, effects and control. J Dairy Res 53: 139-177.

6. Recio I, Frutos M, Olano A, Ramos M (1996) Protein changes in stored ultra-high temperature-treated milks studied by capillary electrophoresis and high-performance liquid chromatography. J Agric Food Chem 44: 3955-3959.

7. Fox PF, McSweeney PLH (2003) Advanced Dairy Chemistry: Volume 1: Proteins Parts A&B. Springer US, USA.

8. Bastian ED, Brown RJ (1996) Plasmin in milk and dairy products: an update. Int Dairy J 6: 435-457.

9. Lahteenmaki K, Kuusela P, Korhonen TK (2001) Bacterial plasminogen activators and receptors. FEMS Microbiol Rev 25: 531-552.

10. Albenzio M, Caroprese M, Santillo A, Marino R, Muscio A, et al. (2005) Proteolytic patterns and plasmin activity in ewes' milk as affected by somatic cell count and stage of lactation. J Dairy Res 72: 86-92.

11. Fox PF, Mcsweeney PLH (1996) Proteolysis in cheese during ripening. Food Reviews International 12: 457-509.

12. Crudden A, Fox PF, Kelly AL (2005) Factors affecting the hydrolytic action of plasmin in milk. Int Dairy J 15: 305-313.

13. Fox PF, Kelly AL (2006) Indigenous enzymes in milk: Overview and historical aspects - Part 2. Int Dairy J 16: 517-532.

14. Sgarbieri VC (2005) Revisão: Propriedades estruturais e físico-químicas das proteínas do leite. Brasilian Journal of Food Technology. Campinas. 8: 43-56.

15. Souza MJ, Ardo Y, Mcsweeney PLH (2001) Advances in the study of proteolysis during cheese ripening. Int Dairy J 11: 327-345.

16. Larsen LB, Hinz K, Jorgensen AL, Moller HS, Wellnitz O, et al. (2010) Proteomic and peptidomic study of proteolysis in quarter milk after infusion with lipoteichoic acid from Staphylococcus aureus. J Dairy Sci 93: 5613-5626.

17. Santos MV, Ma Y, Barbano DM (2003) Effect of somatic cell count on proteolysis and lipolysis in pasteurized fluid milk during shelf-life storage. J Dairy Sci 86: 2491-2503.

18. Philpot NW, Nickerson SC (1991) Mastitis: counter attack. Babson Bro, USA.

19. Kitchen BJ (1981) Review of the progress of dairy science: bovine mastitis: milk compositional changes and related diagnostic tests. J Dairy Res 48: 167-188.

20. Fox PF, Morrissey PA (1981) Enzymes and food processing. In: Birch GC, Blakeborough N, Parker KJ (Eds) Enzymes and Food Processing. Applied Science Publishers, London, UK, 213-238.

21. Andrews AT, Olivercrona T, Bengtssonolivercrona G, Fox PF, Bjorck L, et al. (1991) Indigenous enzymes in milk. In: Fox PF (Ed) Food Enzymology. Elsevier Applied Science, New York, USA, 1: 53-129.

22. Datta N, Deeth HC (2001) Age gelation of UHT milk-a review. Food and Bioproducts Processing 79: 197-210.

23. Politis I, Barbano DM, Gorewit RC (1992) Distribution of plasminogen and plasmin in fractions of bovine milk. J Dairy Sci 75: 1402-1410.

24. Eigel WN (1977) Effect of bovine plasmin on alpha-S1-B and kappa-A caseins. J Dairy Sci 60: 1399-1403.

25. Grufferty MB, Fox PF (1988) Milk alkaline proteinase. J Dairy Res 55: 609-630.

26. Le Bars D, Gripon JC (1993) Hydrolysis of as 1-casein by bovine plasmin. Lait 73: 337-344.

27. Rauh VM, Anja S, Mette B, Richard I, Marie P, et al. (2014) Plasmin activity as a possible cause for age gelation in UHT milk produced by direct steam infusion. Dairy J 38: 199-207.

28. Rauh VM, Johansen LB, Ipsen R, Paulsson M, Larsen LB, et al. (2014) Plasmin activity in UHT milk: relationship between proteolysis, age gelation, and bitterness. J Agric Food Chem 62: 6852-6860.

29. Fernandes AM, Moretti TS, Bovo F, Lima CG, Oliveira CAF (2008) Effect of somatic cell counts on lipolysis, proteolysis and apparent viscosity of UHT milk during storage. Int J Dairy Technol 61: 327-332.

30. Lemieux L, Simard RE (1992) Bitter flavour in dairy products. II. A review of bitter peptides from caseins: their formation, isolation and identification, structure masking and inhibition. Lait 72: 335-385.

31. Costa FF, Brito MAVP, Furtado MAM, Martins MF, de Oliveira MAL, et al. (2014) Microfluidic chip electrophoresis investigation of major milk proteins: study of buffer effects and quantitative approaching. Analytical Methods 6: 1666-1673.

32. AOAC (2005) Official methods of analysis of the Association of the Official Analytical Chemists. AOAC, Gaithersburg.

33. Costa FF, Resende JV, Abreu LR, Goff HD (2008) Effect of calcium chloride addition on ice cream structure and quality. J Dairy Sci 91: 2165-2174.

34. Costa FF, Brito MAVP, Guimaraes MFM, Furtado MAM, de Oliveira MAL, et al. (2010) Protein distribution in a supernatant of milk ultra-centrifuged using lab-on-a-chip microfluid electrophoresis. In: 16 Latin-American Symposium LACE 2010, 2010, Florianópolis.

35. Dooley J, Brown H, Wellum S, Burch B, Jasionowicz P (2010) Determining the milk content of milk-based food products. Food Standards Agency-FSA Final Report Q01117.

36. Anema SG (2009) The use of lab-on-a-chip microfluid SDS electrophoresis technology for the separation an quantification of milk protein. Inter Dairy J 19: 198-204.

37. Fernandes AM (2003) Avaliação do iogurte produzidos com leite contendo diferentes níveis de células somáticas. Tese (Mestrado em Zootecnia) - Faculdade de Zootecnia e Engenharia de Alimentos, Universidade de São Paulo, Pirassununga.

38. Marques LT, Balbinoti M, Ficher V (2002) Variations in the Milk chemical compisition according to somatic cell count. Panamerican congress on milk quality and mastitis control, 2, Ribeirão Preto, Brazil.

39. Klei L, Yun J, Sapru A, Lynch J, Barbano D, et al. (1998) Effects of milk somatic cell count on cottage cheese yield and quality. J Dairy Sci 81: 1205-1213.

40. Moslehishad M, Hamid E, Mehdi A (2010) Chemical and electrophoretic properties of Holstein cow milk as affected by somatic cell count. Inter J Dairy Technol 63: 512-515.

41. Lee SC, Yu JH, Jeong CL, Back YJ, Yoon YC (1991) The influence of mastitis on the quality of raw milk and cheese. Korean Journal of Dairy Science 13: 217-223.

42. El-Tahawy AS, El-Far AH (2010) Influences of somatic cell count on milk composition and dairy farm profitability. Int J Dairy Technology 3: 463-469.

43. Auldist MJ, Hubble IB (1998) Effects of mastitis on raw milk and dairy products. Aust J Dairy Technol 53: 28-36.

44. Verdi RJ, Barbano DM, Dellavalle ME, Senyk GF (1987) Variability in true protein, casein, nonprotein nitrogen, and proteolysis in high and low somatic cell milks. J Dairy Sci 70: 230-242.

45. Rogers SA, Slattery SL, Mitchell GE, Hirst PA, Grieve PA (1989) The relationship between somatic cell count, composition and manufacturing properties of bulk milk III. Individual proteins. Aust J Dairy Technol 44: 49-52.

46. Auldist MJ, Coats S, Sutherland BJ, Mayes JJ, McDowell GH, et al. (1996) Effects of somatic cell count and stage of lactation on raw milk composition and the yield and quality of Cheddar cheese. J Dairy Res 63: 269-280.

47. Silva PLF, Pereira AR, Machado PF, Sarries GA (2000) Effects of somatic cell levels on milk components II-lactose and total solids. Braz J Vet Res Anim Sci 37: 1678-4456.

48. Fox PF, Guinee TP, Cogan, TM, McSweeney PLH (2000) Fundamentals of cheese science. Springer, New York.

49. Miller RH, Emanuelsson U, Persson E, Brolund L, Philipsson J, et al. (1983) Relationships of milk somatic cell counts to daily milk yield and composition. Acta Agric Scand 33: 209-223.

50. Mitchell GE, Fedrick IA, Rogers SA (1986) The relationship between somatic cell count, composition and manufacturing properties of bulk milk. 2. Cheddar cheese from farm bulk milk. Aust J Dairy Technol 41: 12-14.

51. Munro GL, Rieve GPA, Kitchen BJ (1984) Effects of mastitis on milk yield, milk composition, processing properties and yield and quality of milk products. Aust J Dairy Technol 39: 7-16.

52. Najafi MN, Mortaza SA, Koocheki A, Khorami J, Rekik B (2008) Fat and protein contents, acidity and somatic cell counts in bulk milk of Holstein cows in the Khorasan Razavi Province, Iran. Int Dairy J 62: 19-26.

53. Santos MV, Oliveira CAF, Lima YVR, Botaro BG (2006) Somatic cell removal by microfiltration does not affect composition and proteolysis of milk. Ciência Rural 36: 1486-1493.

54. Ma Y, Ryan C, Barbano DM, Galton DM, Rudan MA, et al. (2000) Effects of somatic cell count on quality and shelf-life of pasteurized fluid milk. J Dairy Sci 83: 264-274.

55. O'Connell JE, Grinberg VY, de Kruif CG (2003) Association behavior of beta-casein. J Colloid Interface Sci 258: 33-39.

56. Pirisi A, Piredda G, Podda F, Pintus S (1996) Effect of somatic cell count on sheep milk composition and cheesemaking properties. Somatic Cells and Milk of Small Ruminants. EAAP Publication, Wageningen Pers, Wageningen, The Netherlands 77: 245-251.

57. Pellegrini O, Remeuf F, Rivemale M, Barillet F (1997) Renneting properties of milk from individual ewes: influence of genetic and non-genetic variables, and relationship with physicochemical characteristics. J Dairy Res 64: 355-366.

58. Nudda A, Feligini M, Battacone G, Macciotta NPP, Pulina G (2003) Effects of lactation stage, parity, ß-lactoglobulin genotype and milk SCC on whey protein composition in Sarda dairy ewes. Italian Journal of Animal Science 2: 29-39.

59. Albenzio M, Caroprese M, Santillo A, Marino R, Taibi L, et al. (2004) Effects of somatic cell count and stage of lactation on the plasmin activity and cheese-making properties of ewe milk. J Dairy Sci 87: 533-542.

60. Gouldsworthy AM, Banks JM, Law AJR, Leaver J (1990) Casein degradation in Cheddar cheese monitored by capillary electrophoresis. Milk Science International 54: 620-623.

61. Hames BD, Rickwood D (1998) In: 4th edn, Gel Electrophoresis of Proteins: A Practical Approach, Oxford University Press, pp. 98-145.

62. Trejo R, Harte F (2010) The effect of ethanol and heat on the functional hydrophobicity of casein micelles. J Dairy Sci 93: 2338-2343.

63. Zafalon LF, Nader FA, Carvalho MRB, Lima TMA (2008) Influence of bovine subclinical mastitis on milk protein fractions. Arq Inst Biol São Paulo 75: 135-140.

64. Rogers SA, Mitchell GE (1994) The relationship between somatic cell count, composition and manufacturing properties of bulk milk. 6. Cheddar cheese and skim milk yoghurt. Aust J Dairy Technol 49: 70-74.

65. Oliveira CAF, Fernades AM, Neto COC, Fonseca LFL, Silva EOT, et al. (2002) Composition and sensory evaluation of whole yogurt produced from milk with different somatic cell counts. Aust J Dairy Technol 57: 192-196.

66. Aslam M, Hurley WL (1997) Proteolysis of milk proteins during involution of the bovine mammary gland. J Dairy Sci 80: 2004-2010.

67. Barbano DM, Clark JL (1990) Kjeldahl method for determination of total nitrogen content of milk: collaborative study. Journal AOAC International 73: 849-859.

68. Considine T, Healy A, Kelly AL, Mcsweeney PLH (2002) Proteolytic specificity of cathepsin G on bovine α- and ß-caseins. Food Chemistry 79: 59-67.

69. Lima MCG, Sena MJ, Mota RA, Mendes ES, Almeida CC, et al. (2006) Somatic cell count and physicochemical and microbiological analyzes of raw milk produced in the wild type c region of Pernambuco state. Arq Inst Biol 73: 89-95.

70. Walstra P, Vliet VT (1986) The physical chemistry of curd making. Netherlands Milk and Dairy Journal 40: 241-259.

14

Characterization of Extra Virgin Olive Oil Obtained from Whole and Destoned Fruits and Optimization of Oil Extraction with a Physical Coadjuvant (Talc) Using Surface Methodology

Zina Guermazi[1], Mariem Gharsallaoui[2], Enzo Perri[3], Slimane Gabsi[1] and Cinzia Benincasa[3]*

[1]National School of Engineering, University of Sfax, Route de Soukra, 3038 Sfax, Tunisia
[2]Olive Tree Institute, University of Sfax, 3000 Sfax, Tunisia
[3]Consiglio per la ricerca in agricoltura e l'analisi dell'economia agraria, Centro di ricerca per l'olivicoltura e l'analisi dell'economia agraria, Italy

Abstract

The separation of the two components pulp-stone has allowed the development of a more scented product that respects the environment. According to a preliminary study, it was noticed that the destoning process produced a paste containing 67% moisture without resorting to the addition of water. The destoning process presents an ecologically sensitive system as the amount of waste water generated was reduced to 0.46 m³/ton of olives. Experimental studies indicate that stone removal before crushing affects to a greater extent the quality of the oil. Indeed the use of stoned olives decreases the acidity by 20%. The specific results of extinctions K232 and K270 confirm that using stoned olives improves the stability of oil since less oxidized compounds (K232 and K270 values 1.69 and 0.11 respectively) were detected. This study also shows that stoned olive oil had a better sensory note than the conventional one and consequently a higher commercial value in the market. Enhanced extraction yield of oil from stoned olives was attained using response surface methodology. Different olive oil extractions were investigated considering three variables: temperature of malaxation, time of malaxation and dose of adjuvant using Central Composite Design. Highest extraction yields were obtained when the temperature and the time of malaxation were respectively 37°C and 35 min. For the proportion of coadjuvant, the optimum dose was found at 1.5%.

Olives   Stones   Talc

37°C for 35 min

More Aromatic Olive Oil   Olive Oil with Lower Acidity

Keywords: Extra virgin olive oil; Stoned olive oil; Sensorial analysis; Quality; Oxidative stability; Coadjuvant

Introduction

Olive oil constitutes one of the most appreciated and consumed vegetable oils, whose benefits for health have been known for millennia [1-4]. Currently, Tunisia is a country that has experienced a significant evolution in the olive sector through the plantations of 80 million olive trees that account for 20% of the world olive-growing area. Tunisia, according to the magazine Olive Oil Times, is the 2nd largest producer and exporter of the southern Mediterranean. Indeed, the harvest of the 2014/2015 season was quadrupled, rising from 70,000 to 285,000

*Corresponding author: Cinzia Benincasa, Management for Research in Agriculture and Agricultural Economy Analysis, Research Centre for Olive Growing and Olive Oil Industry, Italy, E-mail: cinzia.benincasa@entecra.it

tons of oil. Olive oil is, according to the Ministry of Agriculture, the main agricultural export, with nearly 40% of total national exports. The increase in olive sales prices pushed growers to minimize the loss of oil in the by-product such as olive pomace and waste water. To minimize this loss, an ironing pomace was performed. This procedure involves subjecting the wet pomace, after pulp/stone separation, to a second centrifugation to remove between 40 and 60% of waste oils. This technique has known a large spread in the Sfax region, a city of 6 million olive feet, where the production of olive oil represents 40% of total national production. The drawback of this action is that it leads to the production of oils that do not achieve the standard requirements established by the COI.

With an aim of preserving olive oil's best physicochemical and organoleptic characteristics and in order to make the reconciliation between olive-growing industries and the environment, a technological improvement in the industrial production of olive oil seems to be represented by the manufacturing procedure which considers the use of stoned olives [5,6].

The benefits of the destoning process on the phenolic compounds have been shown in some studies [7]. Many enzymes of degradation that are present in the stones, in fact, once removed, do not affect the phenolic heritage of the oil during the process of extraction, preserving the antioxidant profile.

Also, higher amounts of volatile compounds were observed for oil produced from stoned olives [8].

The use of de-stoner produces improved sensory characteristic oil, greater impact and volatile content of phenolic compounds and consequently a high stability and long shelf life but has some disadvantages as is the performance problem: the yield of the olive oil has decreased 15% from the whole fruit [9]. Due to the growing interest

stoned olive oil testing is performed to solve this major problem [7,8] as the use of the heat exchanger coupled to the stoner improving olive oil yields and makes the more efficient mixing and design of a specific coring device: the spring pitting apparatus (SPIA) [7]. The use of these techniques remain limited given the large investments that is why we thought was a simpler and less expensive technique that is based on the addition of talc which is a natural coadjutant recognized as a food additive by the European Regulations (E-553B). Talc does not change the oil composition, the positive response is presented by these lipophilic characteristics that can destabilize the oil / water emulsion and increase the rate of oil extraction. On the basis of these observations, the objective of this work was to study the influence of this technology on the quality parameters as well as the optimization of parameters for stoned olive oil extraction.

Materials and Methods

Preparation of samples

Seventeen pairs of olive oils, obtained from whole and stoned olives, were studied. Each pair was obtained from the same batch of olives. Oil was produced from only one cultivar, Chemlali, collected in February 2014. The traditional extraction of the oil has been conducted by using an oleodosor, while, the destoning procedures has been achieved by a reorganization of the extraction steps of the traditional system (Figure 1). In both extraction system, olives were washed and leaves removed. The olives were crushed in a metal crusher and the paste mixed at different temperatures and times with different doses of coadjuvant (talc). The paste was then centrifuged at a speed of 3500 rpm for 1 min. The oils obtained were stored hermetically in 90 mL glass bottles at 4°C. In particular, in the traditional system the paste was malaxed for 45 min at 40°C and the final products were olive oil, vegetation water and pomace. The pomace was then destoned and treated in order to obtain

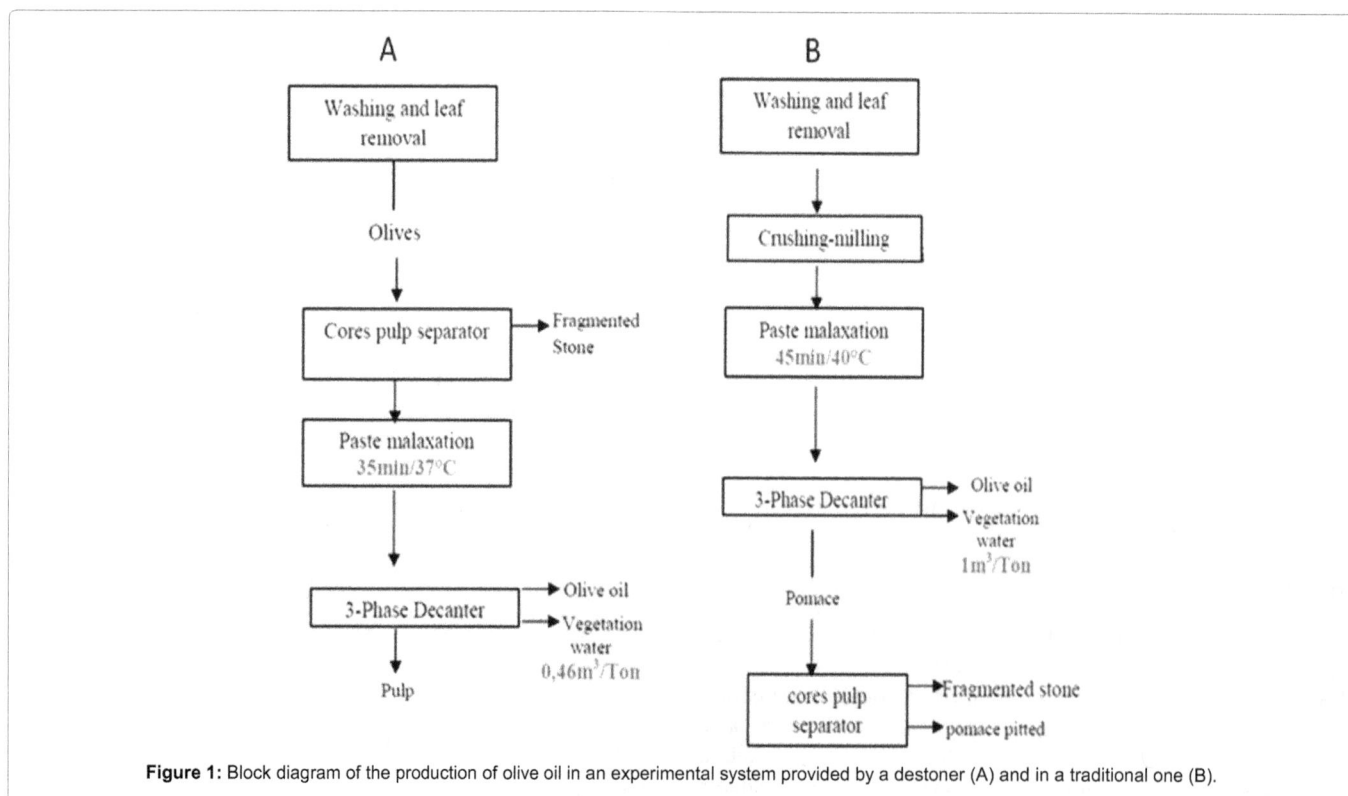

Figure 1: Block diagram of the production of olive oil in an experimental system provided by a destoner (A) and in a traditional one (B).

extra olive oil. In the new experimental system olives were destoned by means of a separating pulp cores directly after been washed and then crushed. The paste was then malaxed for 35 min at 37°C. In this system, instead of pomace, an olive pulp is produced.

Quality indices

Free fatty acidity, peroxide value (PV), conjugated dienes (K232) and trienes (K270) were determined according to the methods reported in the Regulation EEC/2568/91 [10] and subsequent amendments and additions of the European Union Commission.

Phenolic compounds

Quantitative analysis: Folin Ciocolteau Test: Total phenolic compounds were determined as reported in the literature [11]. Briefly, 1 g of olive oil was dissolved in 5 mL of hexane; the solution was loaded on a C18 cartridge (1 g; 6 mL) and washed twice with 5 mL of hexane. Phenols were eluted with 10 mL of MeOH. Final solution (1 mL) was submitted to the Folin Ciocolteau assay.

Quantitative analysis by HPLC/MS/MS

Extraction of phenols: According to the method described by COI (COI, 2009), 2 g of olive oil was added to 6 mL methanol/water (80/20 V/V) and shaken for 1 min in order to homogenize the mixture. The sample was placed in an ultrasonic bath for 15 min and then centrifuged at 5000 rpm for 25 min at 4°C. The supernatant was, finally, injected into the HPLC-MS system.

Preparation of standard solutions: The assay of phenols by LC-MS/MS has been carried out by external standard method. Standard stock solutions were prepared by dissolving reference compounds in methanol. Aliquots of these solutions were further diluted with methanol/water (80/20 V/V) to obtain calibration standards at concentrations between 1-200 μg/mL for oleuropein, hydroxytyrosol, tyrosol, luteolin, apigenin, p- and o-coumaric acids, vanillic acid, vanillin, ferulic and caffeic acids.

Calibration curves were constructed using a least-squares linear regression analysis. For each analyte the calibration curves were linear with correlation coefficients ranging between 0.9993 and 0.9998. Recovery tests were carried out injecting three times two standard solutions at different concentration of the eleven active molecules. Results from recovery experiments at levels of 30 and 80 μg/mL gave mean recoveries ranging from 93-114% with satisfactory precision (relative standard deviation (RSD) from 0.2-0.4%). Limits of quantitation (LOQs) from 1.249-2.848 μg/mL.

High performance liquid chromatography (HPLC): HPLC was performed using an Agilent Technologies 1200 series liquid chromatography system equipped with G1379B degasser, G1312A pump, and G1329A autosampler. The analytes were separated on an Eclipse XDB-C8-A HPLC column [5 μm particle size, 150 mm length and 4.6 mm i.d. (Agilent Technologies, Santa Clara, California)] at a flow rate of 300 μl/min and an injection volume of 10 μL. A binary mobile phase made up of 0.1% aqueous formic acid (A) and methanol (B) was programmed to increase B from 10% to 100% B in 10 min, hold for two min and ramp down to original composition (90% A and 10% B) in eight min and hold for 6 min. The total elution time was 26 min per injection.

Mass spectrometry: The ESI-MS/MS analyses were performed using a MSD Sciex Applied Biosystem API 4000 Q-Trap mass spectrometer in negative ion mode using multiple reactions monitoring (MRM). The experimental conditions were set-up as follow: ionspray voltage (IS) -4500 V; curtain gas 25 psi; temperature 400°C; ion source gas(1) 35 psi; ion source gas(2) 45 psi; collision gas thickness (CAD) medium. Entrance potential (EP), declustering potential (DP), collision energy (CE) and collision exit potential (CXP) were optimized for each transition monitored.

Sensory analysis

Sensorial evaluation of the oils was performed according to the panel test method by a fully trained analytical taste panel recognized by the International Olive Oil Council (IOOC). A panel test was established using the IOOC standard profile sheet method (COI, 2011) [12].

Statistical analyses

The results presented are the averages of the analysis carried out in triple exemplary with the corresponding standard deviations. The threshold of significant difference was fixed at 5% in the Students t-test.

Experimental design: Response surface methodology (RSM) comprises a body of methods for exploring for optimum operating conditions through experimental methods that uses quantitative data from appropriate experimental design to determine optimal conditions. Different levels or values of the operating conditions comprise the factors in each experiment. Some may be categorical (e.g., the supplier of raw material) and others may be quantitative (feed rates, temperatures, and such) [13]. When many factors affect a desired response, it can be an exhausting task to optimize a process. Therefore, RSM can be an effective tool for optimizing the response [14].

The first step of the oil extraction procedure has been optimized using RSM (Tables 1 and 2).

A central composite design was chosen to look for the best experimental conditions of three independent factors affecting the extraction process which are: X1, extraction temperature (°C); X2, extraction time (min) and X3, dose of talc (%). For each factor, the experimental range was chosen on the basis of results of preliminary experiments.

In this work, the relationship between the extraction yield and the three selected quantitative variables was approximated by the following polynomial function:

$$Y = b_0 + b_1 \times X_1 + b_2 \times X_2 + b_3 \times X_3 + b_{11} \times (X_1 \times X_1) + b_{22} \times (X_2 \times X_2) + b_{33} \times (X_3 \times X_3) + b_{12} \times (X_1 \times X_2) + b_{13} \times (X_1 \times X_3) + b_{23} \times (X_2 \times X_3)$$

Where Y is the calculated response function, Xj are the coded variables with no dimension related to the natural variables Uj. The

Objective of study	Study in an experimental field: Surface of Answers
Number of variables	3
Number of experiments	20
Number of coefficients	10
Number of answer	1

Table 1: Experimental design: Characteristics of the problem.

	Factor	Unit	Center
U1	Temperature	°C	31.5
U2	Time	Min	45.0
U3	Dose of Talc	%	1.0

Table 2: Experimental design: Experimental field.

obtained response values were used to estimate the model coefficients bj by the least square method using the experimental design software NEMROD-W [15].

IBM SPSS Statistics 20: The analysis of variance (ANOVA) on SPSS is a bi-varied method of analysis which consists of crossing two different variables, one quantitative and the other one qualitative in order to detect the relation of independence between those two variables.

Results and Discussion

Destoning process and its influence on oil quality

Conventional parameters of quality: Table 3 reports the conventional parameters used to characterize the quality of olive oils. Free acidity values remained below the limits reported by Regulation EEC /1989/2003 of the European Union Commission. The destoned olive sample (DO) value was lower than that of the conventional olive sample (CO) (0.6 and 0.8 g of oleic acid per 100 g of oil, respectively). Thus, it has been noticed that the destoning process decreases the acidity by 20%. This is in agreement with the results reported in the literature [16,17]. Also, it has been noticed that acidity increases with the time of malaxation, therefore, it is very important to control this process during all the duration of the oil extraction. The data of specific extinction, K232, showed that oils obtained from stoned olives presented a weakest value (1.8 nm) compared of ordinary oil (2.4 nm). As for K270, the values ranged between 0.11 and 0.16 nm. Another parameter which has an important effect on quality of oil is the time of malaxation; in fact, a positive correlation was noted between this parameter and the state of oxidation, therefore, malaxation time was extended. The stoned samples presented the weakest values of K232 and K270 thus they were less oxidized; this confirmed the assumption of existence of enzymes in stone responsible of degradation of linoleic acid. As regards to the influence of the technology, the difference between WO and DO values of peroxide index was not significant, in agreement with the results of Del Caro [16].

Olive oils are known for their raised capacity of phenolic compounds compared to other refined vegetable oils. These compounds contribute to total flavor complexes in olive oil and provide antioxidant effects [18,19] and are mainly responsible for its shelf life. In agreement with the results found in the literature [7,8], differences in the total phenol content were observed between the oils produced with conventional and the stoned process; two enzymes, polyphenol oxidase (PPO) and peroxidase (POD), are highly concentrated in the destoned olive. PPO and POD can oxidize phenolic compounds resulting in a reduced concentration of phenolic oil. The de-stoning process eliminates a part of peroxidase activity in stone resulting in an increase in the oxidative stability and nutritional value of the olive oil itself [20]. The results of phenolic compounds detected by LC/MS are illustrated in Table 4,

	Stoned olive oil	Conventional olive oil	Signification
Tyrosol	0.46 ± 0.03	0.48 ± 0.04	*
Vanillin	0.26 ± 0.13	0.13 ± 0.12	***
Hydroxytyrosol	4.65 ± 0.55	5.98 ± 0.24	*
P-coumaric acid	0.35 ± 0.11	0.29 ± 0.16	***
O-coumaric acid	0.36 ± 0.03	0.35 ± 0.05	n.s.
Vanillic acid	3.90 ± 0.18	3.13 ± 0.38	***
Cafeic acid	0.03 ± 0.03	0.03 ± 0.02	*
Ferrulic acid	0.33 ± 0.05	0.20 ± 0.03	***
Apigenin	0.33 ± 0.06	0.35 ± 0.05	n.s.
Luteolin	24.95 ± 0.95	27.42 ± 0.77	***
Oleuropein	121.22 ± 0.98	103.54 ± 0.92	***

Table 4: Result of detection of phenolic compounds by LC/MS and ANOVA test. ***Very significant; **Fairly significant; *Little significant.

while in Figure 2 an extracted ion chromatogram for the two different olive oils is showed.

The most important biophenols quantitatively present in the samples analyzed were: tyrosol, vanillin, hydroxytyrosol, p-coumaric and o-coumaric acid, vanillin acid, caffeic acid, ferulic acid, apigenin, luteolin and oleuropein. This detection reveals a similar qualitative composition in individual phenolic compounds, but different from the quantitative stands points that are described in Figure 3.

This is explained by the degradation of some phenolic compounds by existing enzymes in the stone as polyphenol oxidase (PPO); these enzymes can be activated either through H_3O^+ ions or with fatty acids (linolenic acid). In agreement with recent results [7] increased oleuropein content was observed in the olive oil produced from the same drupes on from classics to stoning procedures; this observation helps to add value to the food. The enzymatic activity of stone is indeed a selective activity affecting some phenolic compounds such as oleuropein so it was found a difference between the two types of oil. Indeed, it can be confidently assumed, that the enzymes of the stone give rise to a process of deglycosylation of the oleuropein [21].

Sensory quality: Phenolic compounds, together with the combined bouquet of volatile compounds, modulate the perceived sensory characteristics of VOO, being phenolic molecules responsible of taste (bitter) and chemesthetic (pungency sensations). Also, jointly with other antioxidants such as tocopherols, carotenes and squalene, the phenolic molecules are directly correlated with the shelf-life of virgin olive oil [22]. The process of the elimination of stone before crushing supports the concentration of total polyphenols which plays a significant role in the oxidative stability and the determination of sensory consistencies. Figure 4 shows that for the fruity aspect, conventional oil presented the weakest values (2.6) but this character is very important for oil obtained from stoned fruits (5.1).

The same observation has been noted for the pricking aspect; the values ranged from 2.3 to 5.7. On the other hand, for the bitter aspect it was observed that the stoned samples presented the lowest values (2.1). The destoning process improves the flavor of olive oil since it increases the fruity aspect and the prickly taste in a significant manner whereas it decreases oil bitterness. These results are in agreement with that found in the literature [7,8,23-25] that showing that the oils obtained from destoned pulp had a higher polyphenol content compared to oils obtained from dough together. De-stoned oils had a larger amount of C 5 and C 6 volatile compounds responsible for positive ratings of the flavor. In fact, the increase is due to the removal of seeds that contain 13-metabolizing enzyme activities other than hydro peroxide lyase

Sample	Time of malaxation[a]	Process	Acidity[b]	K232	K270	Peroxide value[c]
1	15	Stoned	0.52	1.72	0.11	5.67
2	15	Conventional	0.71	2.03	0.16	5.56
3	30	Stoned	0.58	1.81	0.13	5.70
4	30	Conventional	0.78	2.11	0.17	5.43
5	45	Stoned	0.61	1.91	0.15	5.64
6	45	Conventional	0.82	2.21	0.18	5.87
7	60	Stoned	0.69	2.11	0.17	5.42
8	60	Conventional	0.91	2.46	0.19	5.53

Table 3: Conventional parameters to assess the quality of olive oils.

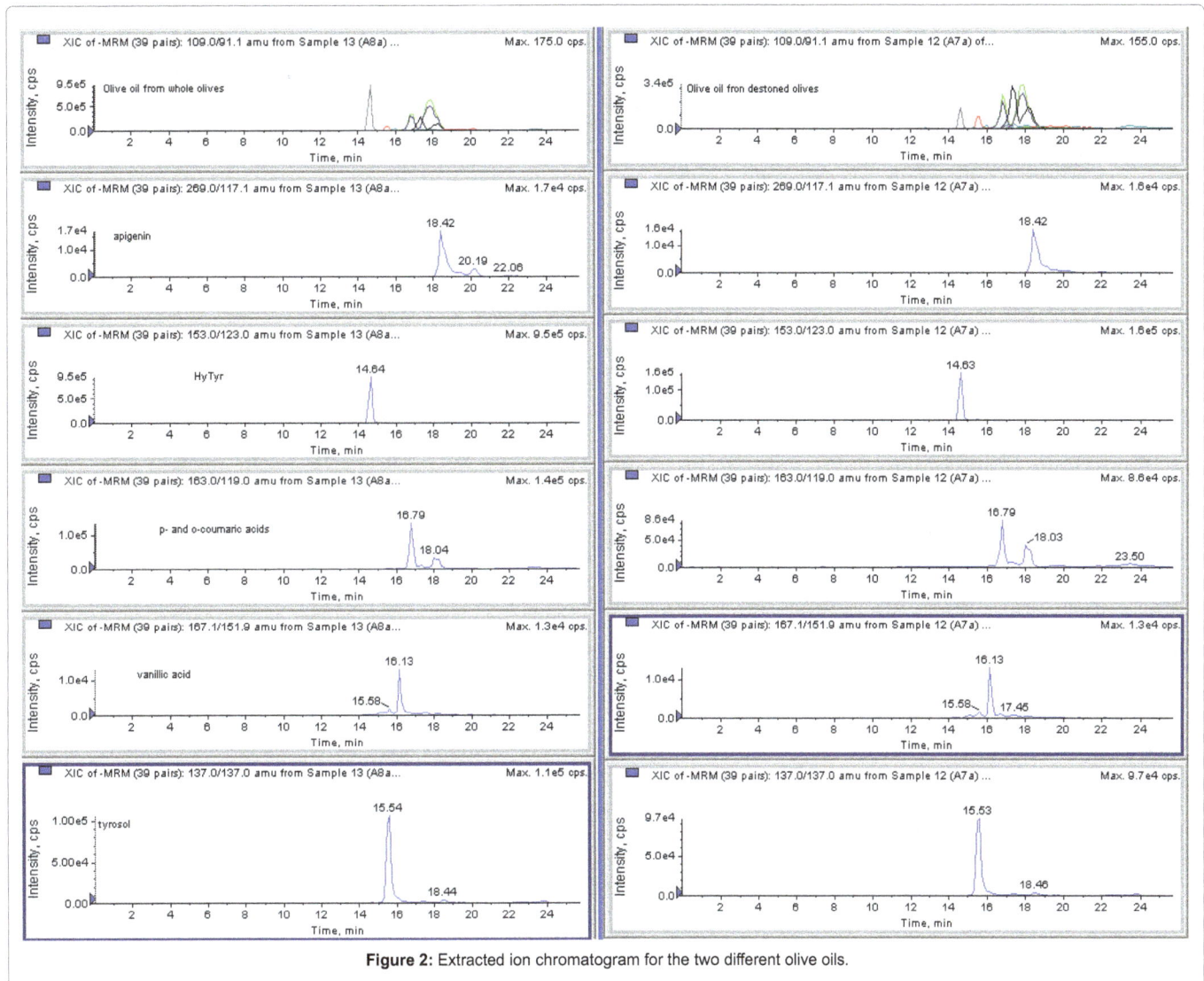

Figure 2: Extracted ion chromatogram for the two different olive oils.

hydro peroxides, resulting in a net decrease in grade C 6 unsaturated aldehydes during the extraction process of oil olive. (C 6 aldehydes, C 6 alcohols, C 5 alcohols and C 5 carbonyls).

Influence of destoning at the environmental and economic level: The extraction of olive oil generates huge quantities of wastes that may have a great impact on land and water environments because of their high phytotoxicity. Several studies have proven the negative effects of these wastes on soil microbial populations [26], on aquatic ecosystems [27] and even in air medium [28]. Therefore, there is a need for guidelines to manage these wastes through technologies that minimize their environmental impact and lead to a sustainable use of resources. Indeed, in Tunisia, the majority of oil mills use the three phase system to process about 100 T of olives per day and there is always the need to add water to achieve the separation. Approximately, about each ton of olives produces 1 m³ of waste water. It has been noted that, when the pulp is separated from the stone before the extraction, the moisture of the paste increases up to 67%, the quantity of water added becomes negligible and thus for each Ton only 0,46 m³ is produced.

On the other hand, considering the calorific value of the pomace and the increase in the price of fossil fuels, it seems appropriate to examine the processing and use this by-product for energy in order to reduce disposal costs of residues from oil mills. The humidity of around 50% present in the pomace obtained from continuous systems does not allow the direct upgrading of these by-products in the energy sector. On the contrary, the pomace obtained after destoning is depleted in water and does not require further treatment, making its utilization in the energy path easier and less expensive.

Optimization of parameters of stoned olive oil extraction

The yields were integrated by the NemrodW software in order to estimate the importance of each parameter (Table 5). In the table it is clear that the temperature of malaxation is a determining factor.

Indeed by increasing the temperature the oil viscosity was decreased and the quantity extracted increased. The time of malaxation and the percentage of additive also have a significant effect. On the other hand, an interaction between the temperature and the time of malaxation improves this phenomenon. An interaction between the temperature and the percentage of talc gives very significant effects. The scale model which connects the yield with the three parameters is presented in the following formula:

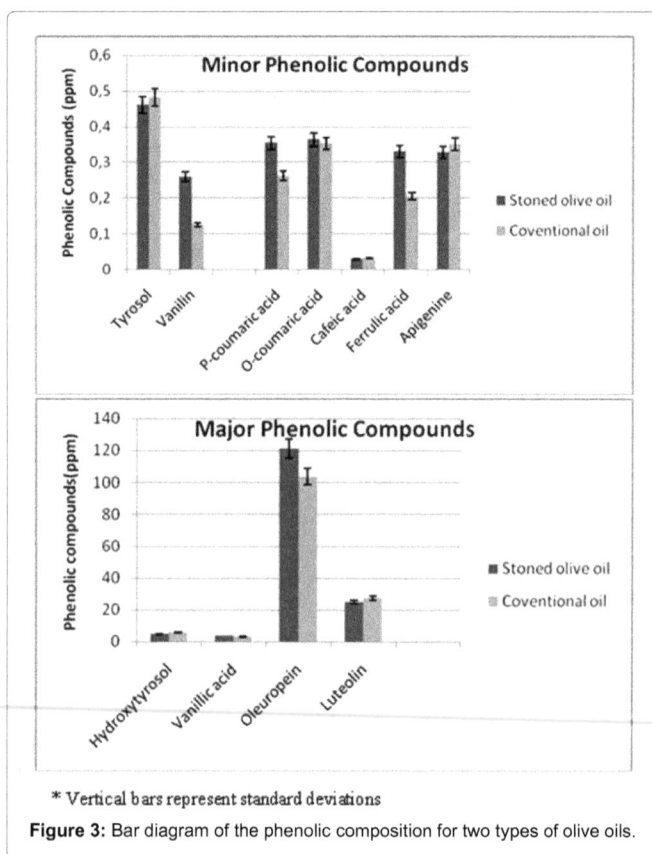

* Vertical bars represent standard deviations

Figure 3: Bar diagram of the phenolic composition for two types of olive oils.

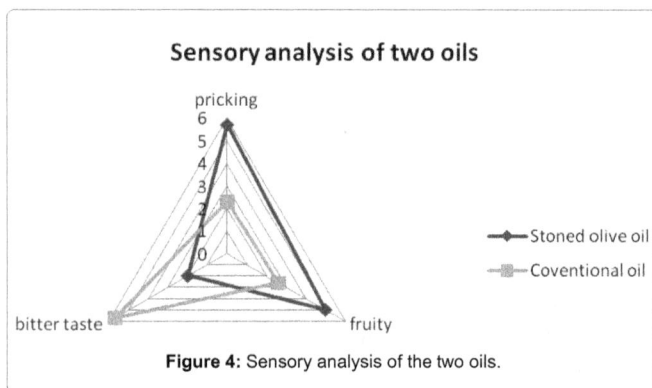

Figure 4: Sensory analysis of the two oils.

$$Y = 16.14 + 4.45 \times X1 + 2.06 \times X2 + 1.12 \times X3 - 1.01 \, (X1 \times X1) + 3.68 \times (X1 \times X2) + 2.96 \times (X1 \times X3)$$

The objective was the optimization of the conditions of olive oil extraction. Table 6 summarizes the levels of the factors which maximize the yield of extraction for each significant interaction. To obtain the best result, the best temperature and time of extraction were found to be at 37°C and 35 min. For the percentage of additive, the optimum was 1.5%.

Influence of the conditions of extraction on sensory quality

Extraction conditions, in particular the time and temperature of malaxation, produce oils with different flavors [29]. Indeed, these two parameters modify the content of aldehydes and esters which are responsible of positive aroma, and are formed by the action of the lipoxygenase enzyme released when the fruit is crushed and mixed

during the stage of malaxation. In this respect, the IBM SPSS software has been used in order to determine the influence of these parameters on the quality. The bar graph of error (Figure 5) shows a reduction in score of the sensory analysis according to temperature and the time of extraction.

The temperature of 23°C and the time of malaxation of 30 min were the critical points beyond which these two parameters negatively affect the sensory score. Otherwise, it was noted that the length of the error bar was variable, which explains the difference between the variances in these groups. It was shown previously how the positive olive oil attributes are correlated positively to the quantity of polyphenols. Figure 6 confirms these results. Indeed the temperature has a positive correlation with the extraction yield.

However, it is negatively correlated with the oil quality. The same results were found as a function of time: long mixing times were associated to an oil rate increase and a concentration of the volatile compounds responsible for the pleasant taste decrease. In other words, increasing the temperature and the mixing time will result in a reduction in viscosity and an improve of the amount of oil extracted, but it may reduce the quality by promoting the oxidation conditions. In conclusion, the optimum conditions to preserve these aromatic compounds and at the same time having a significant performance is presented in Table 7.

Conclusions

In this work we proposed to carry out, before olive oil extraction, the separation of the pulp from the stone. This innovative system brings not only to a reduction of the production cost but also, due to a less water consumption, to a reduction of pollution caused by the exhaust of by-products. In fact, the experimental study has showed how the separation of the stones from the olive reduces the adsorption of water. The absence of the stones in the paste, that are considered water absorbents, preserves the moisture of the mixture and,

Name	Coefficient	Standard deviation	Significance
b0	16.14	5.87	*
b1	4.45	0.27	****
b2	2.06	0.11	*
b3	1.12	1.80	*
b11	-1.01	0.00	*
b22	-0.00	0.00	
b33	-0.64	0.19	
b12	3.68	0.00	****
b13	2.96	0.04	****
b23	0.03	0.02	

Table 5: Estimations and statistics of the coefficients by the NemrodW software. ****Very significant; ***Fairly significant; *Little significant. 1: Temperature (°C); 2: Time of malaxation (min); 3: Percentage of additive (%).

Interactions	Factors			Yield (%)
	T° (°C)	Time (min)	Talc (%)	
T°-Time	37	35		28.7
T°-%	37		1.5	28.9
%-Time		35	1.5	27.9
Optimum conditions	37	35	1.5	28.5

Table 6: Optimum conditions of temperature, time and dose of adjuvant for best extraction.

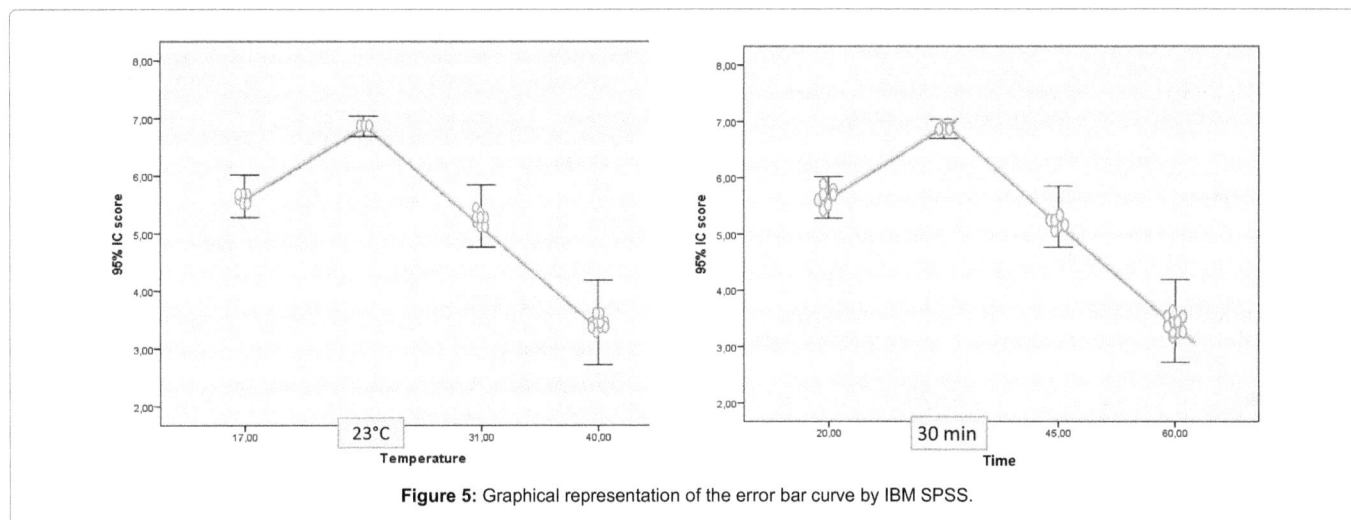

Figure 5: Graphical representation of the error bar curve by IBM SPSS.

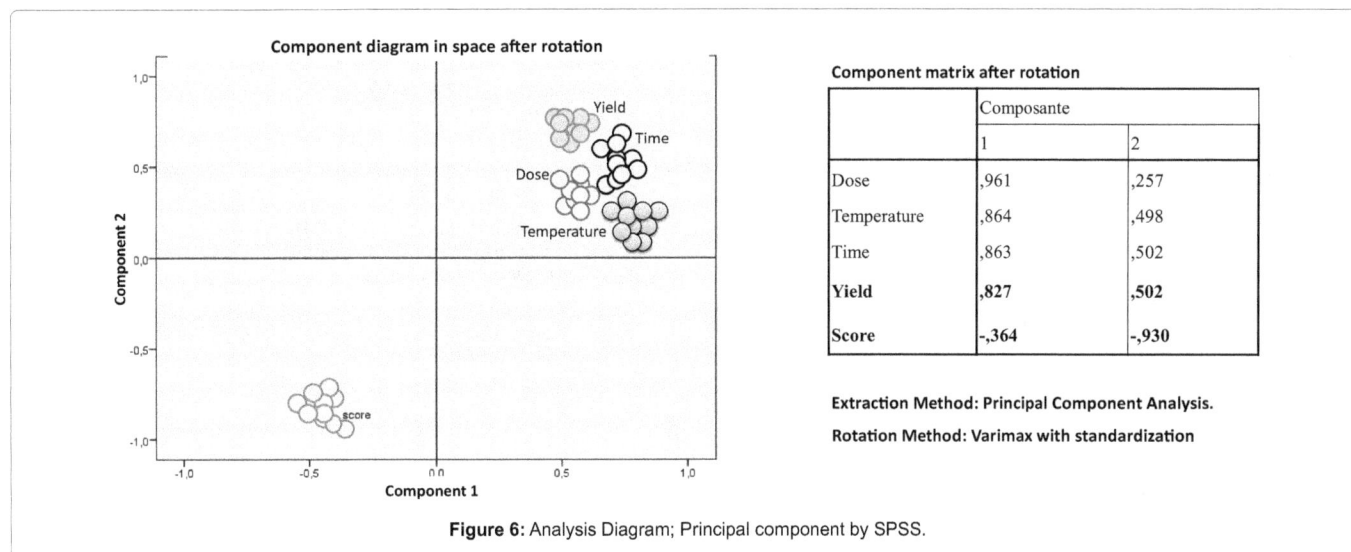

Component matrix after rotation

	Composante	
	1	2
Dose	,961	,257
Temperature	,864	,498
Time	,863	,502
Yield	**,827**	**,502**
Score	**-,364**	**-,930**

Extraction Method: Principal Component Analysis.

Rotation Method: Varimax with standardization

Figure 6: Analysis Diagram; Principal component by SPSS.

Interaction	Factors			Content of total polyphenols (mg d'EAG/kg)	Yield of extraction (%)
	T° (°C)	Time (min)	Talc (%)		
Temp-Time	33	35	2	266	27
T°-%	25		1	280	29
%-Time		27	1	281	27
CO	28	31	1	275	29

Table 7: Optimum condition to achieve the best yield and total polyphenols. [a]Expressed as min, [b]Expressed as % of oleic acid, [c]Expressed as meq of O_2.

therefore, the addition of water during mixing is no longer necessary. Furthermore, as the separation takes place at high temperatures, the paste is instantaneously heated and a mixing of 35 min instead of 45 min is sufficient to squeeze out the oil of higher quality. This system also results in a reduction for the storage period of olives before their treatment, thus a better quality of product. In fact, preliminary results have shown that the extraction of oil after destoning made it possible to decrease acidity, to increase the content of polyphenols (natural antioxidant) and to improve the organoleptic quality of oil.

References

1. Bedbabis S, Rouina BB, Boukhris M (2010) The effect of waste water irrigation on the extra virgin olive oil quality from the Tunisian cultivar Chemlali. Scientia Horiculturae 125: 556-561.

2. Manna C, Migliardi V, Golino P, Scognamiglio A, Galletti P, et al. (2004) Oleuropein prevents oxidative myocardial injury induced by ischemia and reperfusion. J Nutr Biochem 15: 461-466.

3. Owen RW, Giacosa A, Hull WE, Haubner R, Spiegelhalder B, et al. (2000) The antioxidant/anticancer potential of phenolic compounds isolated from olive oil. Eur J Cancer 36: 1235-1247.

4. Somova LI, Shode FO, Ramnanan P, Nadar A (2003) Antihypertensive, and atherosclerotic and antioxidant activity of triterpenoids isolated from Olea europaea, subspecies africana leaves. J Ethopharmacol 84: 299-305.

5. Amirante R, Cini E, Montel GL, Pasqualone A (2001) Influence of mixing and extraction parameters on virgin olive oil quality. Grasas y Aceites 52: 198-201.

6. Frega N, Caglioti L, Mozzon M (1999) Oils extracted from olives pitted: Chemical composition and quality parameters. Olivo & Olio 12: 40-44.

7. De Nino A, Di Donna L, Mazzotti F, Sajjad A, Sindona G, et al. (2008)

Oleuropein expression in olive oils produced from drupes stoned in a spring pitting apparatus (SPIA). Food Chem 106: 677-684.

8. Amirante P, Clodoveo ML, Dugo G, Leone A, Tamborrino A (2006) Advance technology in virgin olive oil production from traditional and de-stoned pastes: Influence of the introduction of a heat exchanger on oil quality. Food Chem 98: 797-805.

9. Amirante P, Baccioni L, Bellomo F, Di Renzo GC (1987 Facilities for olive oil extraction from olive pastes dènoyantèes. Olivae 17: C.O.I. Madrid.

10. EEC (1991) European Union Regulation EEC/2568/91 on the characteristics of olive oil and their analytical methods and later modification. Off J Eur Commun L248: 1-82.

11. Gutierrez F, Albi MA, Palma R, Rios JJ, Olìas JM (1989) Bitter taste of virgin olive oil: Correlation of sensory evaluation and instrumental HPLC analysis. J Food Sci 54: 68-70.

12. Conseil Oleicole International (2011) Sensory analysis of olive oil: organoleptic evaluation method of virgin olive oil.

13. Lenth RV (2009) Response-Surface Methods in R, Using rsm. J Stat Software 32.

14. Masmoudi M, Besbes S, Chaabouni M, Robert C, Paquot M, et al. (2008) Optimization of pectin extraction from lemon by-product with acidified date juice using response surface methodology. Carbohydr Polym 74: 185-192.

15. Mathieu D, Nony J, Phan-Tan-Luu R (2000) NEMROD-W software. Marseille: LPRAI.

16. Caro AD, Vacca V, Poiana M, Fenu P, Piga A (2006) Influence of technology, storage and exposure on components of extra virgin olive oil (Bosana cv) from whole and de-stoned fruits. Food Chem 98: 311-316.

17. Saitta M, Lo Turco V, Pollicino D, Dugo G, Bonaccorsi I, et al. (2003) Olive oil pitted pulp obtained from cv Coratina and Paranzana=Olive oils from stoned paste of Coratina and Paranzana varieties. Rivista Italiana Sostanze Grasse 80: 27-34.

18. Del Carlo M, Sacchetti G, Di Mattia C, Compagnone D, Mastrocola D, et al. (2004) Contribution of the phenolic fraction to the antioxidant activity and oxidative stability of olive oil. J Agric Food Chem 52: 4072-4079.

19. Servili M, Selvaggini R, Taticchi A, Esposto S, Montedoro GF (2003) Volatile compounds and phenolic composition of virgin olive oil: optimization of temperature and time exposure of olive pastes to air contact during the mechanical extraction process. J Agric and Food Chem 51: 7980-7988.

20. Amirante P, Clodoveo ML, Leone A, Tamborrino A, Paice A (2010) Influence of the Crushing System: phenol content in virgin olive oil produced from whole and de-stoned pastes. In: Olives and Olive Oil in Health and Disease Prevention, Academic Press Ltd, Elsevier science Ltd, London, England. pp: 69-76.

21. Wang W, De-Dios-Alche J, Castro AJ, Rodriguez-Garcia MI (2001) Characterization of seed storage proteins and their synthesis during seed development in Olea europaea. Int J Develop Biology 45: S63-S64.

22. Baldioli M, Servili M, Perretti G, Montedoro GF (1996) Antioxidant activity of tocopherols and phenolic compounds of virgin olive oil. J Am Oil Chem Soc 73: 1589-1593.

23. Angerosa F, Basti C, Vito R, Lanza B (1999b) Effect of fruit stone removal on the production of virgin olive oil volatile compounds. Food Chem 67: 295-299.

24. Luaces P, Pérez AG, Sanz C (2003) Role of olive seed in the biogenesis of virgin olive oil aroma. J Agric Food Chem 51: 4741-4745.

25. Runcio A, Sorgona L, Mincione A, Santacaterina S, Poiana M (2008) Volatile, Compounds of virgin olive oil obtained from Italian cultivars grown in Calabria, Effect of processing methods, cultivar, stone removal, and antracnose attack. Food Chem 106: 735-740.

26. Paredes MJ, Moreno E, Ramos-Cormenzana A, Martinez J (1987) Characteristics of soil after pollution with wastewaters from olive oil extraction plants. Chemosphere 16: 1557-1564.

27. Greca MD, Monaco P, Pinto G, Pollio A, Previtera L, et al. (2001) Phytotoxicity of low-molecular-weight phenols from olive mill wastewaters. Bull Environ Contam Toxicol 67: 352-359.

28. Rana G, Rinaldi M, Introna MV (2003) Volatilization of substances after spreading olive oil waste water on the soil in a Mediterranean environment. Agric Ecosystems Envir 96: 49-58.

29. Ranalli, Pollastri L, Contento S, Iannucci E, Lucera L (2003) Effect of olive paste kneading process time on the overall quality of virgin olive oil. Europ J Lipid Sci Tec 105: 57-67.

15

Taxonomic Identification of Hallucinogenic Mushrooms Seized on the Illegal Market Using a DNA-Based Approach and LC/MS-MS Determination of Psilocybin and Psilocin

Veniero Gambaro[1], Gabriella Roda[1*], Giacomo Luca Visconti[1], Sebastiano Arnoldi[1], Eleonora Casagni[1], Caterina Ceravolo[1], Lucia Dell'Acqua[1], Fiorenza Farè[1], Chiara Rusconi[1], Lucia Tamborini[1], Stefania Arioli[2*] and Diego Mora[2]

[1]Department of Pharmaceutical Sciences, University of Milan, Via Mangiagalli 25, Milan, Italy
[2]Department of Food Science and Technology and Microbiology, Via Celoria 2, Milan, Italy

Abstract

The taxonomic identification of mushrooms suspected to contain hallucinogenic active principles was carried out using a DNA-based approach, thus highlighting the usefulness of this approach in the forensic identification of illegal samples also when they are difficult to identify because the morphologic identification is prevented, due to the bad conservation of the vegetable material.

To confirm the presence of the illegal active principles, the optimization of a LC/MS-MS method for the qualitative-quantitative analysis of psilocin and psilocybin in mushroom samples seized by the judicial authority is described. For the quantitative determination it was necessary to identify and synthesize a proper internal standard (IS, i.e., 5-hydroxy-N,N-diethyltryptamine), endowed with chromatographic features suitable for the analysis of the active principles. LC/MS-MS analysis evidenced that the amount of psilocybin ranged from 0.5 to 1.4% while that of psilocin from 1.3 to 2.5% (w/w), confirming literature data. The concentration of psilocin was higher in the cap and in the distal part of the stem (near to the soil) than in the part of the stem proximal to the cap. On the other hand the concentration of psilocybin was higher in the cap and in the proximal part, being lower in the distal part of the stem.

Keywords: Psilocybin; Psilocin; Hallucinogenic mushrooms; DNA-based identification; LC/MS-MS

Introduction

Psychoactive plants or fungi accompanied religious rituals in many ancient civilizations worldwide [1-4]. Hallucinogenic compounds contained in fungi of Psilocybe and Amanita genera have been used as recreational drugs of abuse since the beginning of the 1960s [5,6]. Although hallucinogens are thought to be safe because of relatively low physiologic toxicity and dependence potential [7,8] nowadays the context of hallucinogenic compounds consumption, mainly by young people seeking for unusual experiences, is usually negative [5,6,9]. The psychoactive principles of Psilocybe mushrooms, namely psilocybin and psilocin, were isolated and identified by Hofmann in 1958 [10]. Psilocin and psilocybin are naturally occurring indoles (Figure 1) found in several species of mushrooms [11,12] at concentrations of up to 0.5% m/m and 2% m/m, respectively [13]. Psilocybin is rapidly dephosphorylated to psilocin *in vivo* [14], with the latter compound being structurally related to the neurotransmitter serotonin (Figure 2) which gives rise to its comparable human metabolism [15]. This molecular similarity endows psilocin with high affinity for serotonin receptors, which blocks the release of the neurotransmitter thus giving rise to hallucinogenic effects [16]. Both psilocin and psilocybin are controlled substances in many countries, as intentional intoxication from these compounds continues to be a major problem in USA and Europe [17]. Gas chromatography cannot be recommended since the poorly volatile and heat-labile psilocybin tends to decompose by loss of the phosphate group during injection. All Liquid Chromatography (LC) methods described so far are based on RP LC [18] with conventional C18 columns. Due to the phosphate group the polarity of psilocybin is much higher compared to psilocin.

Hallucinogenic fungi taxonomically fall into several genera, the most well-known being the genus Psilocybe. Certain species of these genera synthesize psilocin and psilocybin. Since not all members of

a given fungal genus contain psilocin or psilocybin, identification of an unknown fungal sample to the species level is required for legal proceedings. The determination of the species requires the analysis of several morphological characteristics. Often the samples seized by the judicial authority are dried, powdered or reduced in small pieces; consequently the morphological identification is difficult. A DNA-based approach ensures the identification of samples in the absence of morphological characteristics [19].

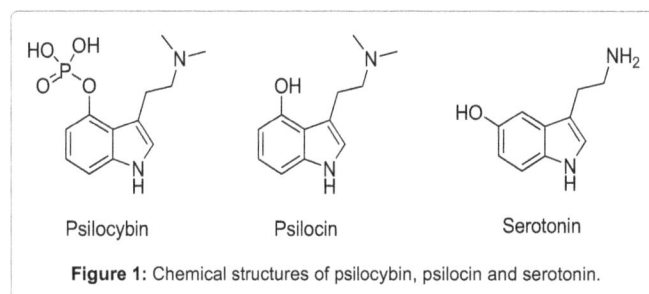

Figure 1: Chemical structures of psilocybin, psilocin and serotonin.

*Corresponding author: Gabriella Roda, Department of Pharmaceutical Sciences, University of Milan, Via Mangiagalli 25, 20133, Milan, Italy E-mail: gabriella.roda@unimi.it

Stefania Arioli, Department of Food Science and Technology and Microbiology, Via Celoria 2, 20133, Milan, Italy, E-mail: stefania.arioli@unimi.it

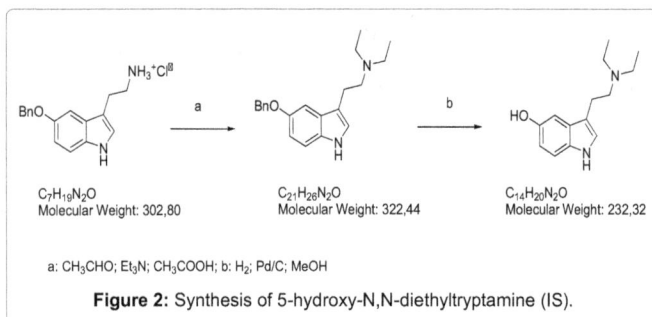

Figure 2: Synthesis of 5-hydroxy-N,N-diethyltryptamine (IS).

In this paper we describe a DNA-based approach for the taxonomic identification of mushrooms species in order to help the forensic identification of illegal samples found on the illicit market, especially when the morphological characteristics cannot be examined. Moreover the optimization of a LC/MS-MS method for the quali-quantitative analysis of psilocybin and psilocin in mushroom samples seized by the judicial authority was carried out. For the quantitative determination it was necessary to identify and synthesize a proper internal standard (IS, i.e., 5-hydroxy-N,N-diethyltryptamine), endowed with chromatographic features suitable for the analysis of the active principles.

Materials and Methods

Vegetable material

Analyses were carried out on four mushrooms seized by the judicial authority at the Malpensa airport in Northern Italy in February 2014. The weights and lengths of the vegetable materials are reported in Table 1.

DNA extraction from growth substrate and fungi

Hundred mg of mushrooms were processed for DNA extraction by means of Ultra Clean™ Microbial DNA Isolation Kit (Mo Bio, Canada), according to manufacturer's instructions. After DNA quantification with a Smart Spec™ Plus Spectrophotometer (Bio-Rad Laboratories, Milan, Italy) and evaluation of 260/280 ratio, 50 ng of DNA were used for species identification based on Internal Transcribed Region located between the 18S- and the 28S rDNA genes as previously well acknowledged for its ability to identify yeast to the species level [20]. The amplification was performed in a final volume of 25 μl, containing 0.5 μM of each primer (NL1: GCATATCAATAAGCGGAGGAAAAG; and NL4: GGTCCGTGTTTCAAGACGG) [20], 1 unit of DreamTaq (Fermentas, Thermo Scientific, Italy) and 50-70 ng of DNA. The reactions were performed in an automatic thermal cycler (EppendorfMastercycler, Italy) under the following conditions: initial denaturation at 94°C for 3 min; 36 cycles of 94°C for 2 min, 52°C for 1 min, 72°C for 2 min; final extension at 72°C for 7 min, holding at 4°C. The amplified products were then purified by means of UltraClean™ PCR Clean-up DNA purification kit (Mo Bio, Canada) and finally sequenced by using primer NL1. The sequences were analyzed by use of the software Chromas2.33 (Technelysium Pty Ltd, South Brisbane, Australia) and compared to the sequences reported in the GenBank using the BLAST algorithm.

Reagents and standards

Psilocybin 1.0 mL (100 μg/mL) in methanol was purchased by Grace (Illinois, USA); Psilocin 99.3%, 1.0 mL (5 mg/mL) in methanol was from Cerilliant, (Texas, USA); 5-benzyloxytryptamine hydrochloride and all the other reagents and solvents of analytical grade were supplied by Sigma Aldrich (St. Louis, USA) and were stored as required by the manufacturer. Water (18.2 MΩ cm) was prepared by a Milly-Q System (Millipore, France).

Synthesis of 5-hydroxy-N,N-diethyltryptamine

a) Triethylamine (68 μL, 0.49 mmol) was added to a suspension of 5-benzyloxytryptamine hydrochloride (150 mg, 0.49 mmol) in anhydrous methanol (7 mL) under an inert atmosphere. The mixture was cooled at 0°C and glacial acetic acid was added (112 μL, 1.96 mmol). A solution of acetaldehyde (66 μL 1.18 mmol) in methanol (7 mL) was slowly dropped (10 min). The mixture was stirred at room temperature for 2 hr until the complete transformation of the starting product (TLC: eluent: dichloromethane/methanol 8:2). The pH was adjusted to 8-9 by a 2M solution of sodium hydrogencarbonate and methanol was evaporated. The aqueous solution was extracted with dichloromethane (3 × 10 mL) and the combined organic phases were washed with water (1 × 10 mL) and brine (1 × 10 mL) and dried with anhydrous sodium sulfate. The solvent was removed under vacuum and the residue purified by a flash silica gel chromatography (eluent: dichloromethane/methanol 9:1), obtaining a pale yellow oil (86 mg, 55% yield).

b) 5-benzyloxy N,N-diethyltriptamine (86 mg, 0.26 mmol) was dissolved in methanol (2 mL) and a catalytic amount of Pd/C 10% (8.7 mg) was added. The mixture was stirred under a hydrogen atmosphere for 24 hr and then filtered on celite. The solvent was evaporated and the residue purified by a flash silica gel chromatography (eluent: dichloromethane/methanol 8:2), obtaining a pale yellow powder (40 mg, 63% yield). In Figure 3 the MS/MS spectrum of the target compound is reported.

Mushroom	Weight (g)	Length (cm)
1	2.8	10.0
2	2.2	9.3
3	2.0	8.2
4	1.3	6.9

Table 1: Characteristics of the vegetable material.

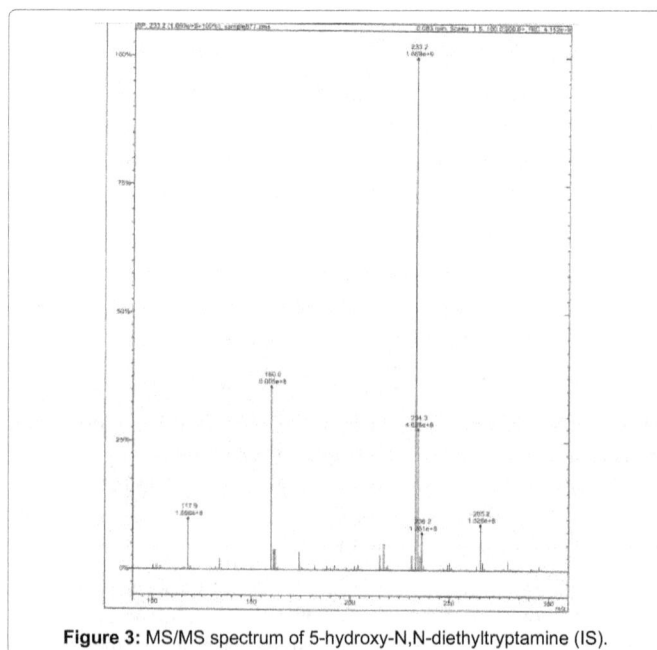

Figure 3: MS/MS spectrum of 5-hydroxy-N,N-diethyltryptamine (IS).

Preparation of the samples

Each mushroom was divided into three parts: cap, proximal part of the stem respect to the cap and distal part of the stem. From each part 100 mg for each mushroom (total 400 mg) were withdrawn and extracted with 4 mL of methanol by maceration overnight. The mixture was sonicated for 10 min and the supernatant filtered on a 0.45 μm membrane. Each extract was analyzed in triplicate adding to 300 μL of the extract 300 μL of a IS solution (10 μg/mL).

LC/MS-MS analyses

Experiments, carried out with a liquid chromatography system coupled to a mass detector, were performed on a LC-320 with an Electron Spray Ionization (ESI) source and a 320-MS triple quadrupole mass spectrometer, equipped with two 212 LC chromatographic pumps and a 410 tray cooled autosampler (Varian, Santa Clara, CA, US). The system was managed by MS Workstation software Version6.9.1 (Varian, Santa Clara, CA, US). The ESI-triple quadrupole mass spectrometer was set to perform collision induced dissociation experiments in positive ionization mode, using argon as collision gas. Particularly, Multiple Reaction Monitoring (MRM) experiments were conducted by the continuous injection at a rate of 20 μL/min of the standards of interest (2 μg/mL) into the mass spectrometer set in positive ionization mode, in methanol solutions. In order to enhance ion formation and to improve conductivity, several proves were performed to optimize ESI parameters and several proves were carried out to improve chromatographic conditions. The method was optimized for the research of the ions of interest, for example, the specific MRM transitions of each standard. In detail, once recognized the molecular ion of each standard $(M + H)^+$, MRM experiments were performed to study the characteristic fragmentation pattern of the analytes (Table 2). ESI source settings and mass spectrometer parameters used for compound identification were Needle Voltage: +5000 V; Shield Voltage: +600 V; Nebulizing Gas (N_2) Pressure: 40.00 psi; Drying Gas (N_2) Pressure: 40.00 psi; Drying Gas (N_2) Temperature: 100°C; Q_0 Offset: +3.0 V; L_4 Offset: +2.000 V; Housing Temperature: 50°C; CID Gas (Ar); Pressure: 2.00 mTorr; Electron multiplier: 1750.0 V; Scan time: 4.0 sec; Dwell time: 0.1 sec.

Chromatographic conditions

Chromatographic column: Kinetex HILIC 150 × 3.0 mm i.d. particle size 2.6 μm; pore dimension 100 A (Phenomenex™, Castel Maggiore, Italy); pre-column: Security Guard Cartridges C18 4 × 2.0 mm (Phenomenex™, Castel Maggiore, Italy); column temperature: 40°C; manifold temperature: 42°C; housing temperature: 50°C; flow rate: 200.0 μL/min; injection volume: 10 μL; injection mode: partial loop fill; solvent A: 0.1% acetic acid adjusted to pH 6 with ammonia; Solvent B: acetonitrile; mobile phase: solvents were filtered under vacuum on 0.45 μm membrane filters and degassed by immersion in ultrasonic bath for 15 minutes before column conditioning; linear gradients: 0.0-4.0 min, 85% B; 4.0-7.0 min, 85-40% B; 7.0-9.3 min 40-85% B; 9.3-12.0 min 85% B. Retention times: psilocin 7.5 ± 0.2 min; psilocybin 9.0 ± 0.2 min.

Linearity of the analytical method

To evaluate the linearity of the analytical method standard solutions of psilocin were analyzed in an interval of concentrations from 10 μg/mL and 100 μg/mL. In this range five solutions of non-sequential concentrations (10, 20, 40, 80, 100 μg/mL) were analyzed (n=5). The linearity equation was y=0.0139x+0.1501 with a good correlation coefficient (R^2=0.9981).

To evaluate the linearity of the analytical method standard solutions of psilocybin were analyzed in an interval of concentrations from 0.625 μg/mL and 10 μg/mL. In this range five solutions of non-sequential concentrations (0.625, 1.25, 2.5, 5.0, 10 μg/mL) were analyzed (n=5). The linearity equation was y=0.0012x+0.0003 with a good correlation coefficient (R^2=0.9958).

Results and Discussion

When dealing with mushrooms it is very useful to assess the genus and species of the biological materials because also the fungi are inserted in the illicit substance list. Sometimes samples seized by the judicial authority are dried or powdered and the morphological characteristics are not recognizable. Therefore an unambiguous identification by a phylogenetic approach could be very helpful, especially in the cases in which active principles could be degraded due to an incorrect storage of the vegetable material. In this context, total DNA directly extracted from small fractions of suspected hallucinogenic mushroom was used in a PCR assay in order to amplify the internal transcribed region between the 18S- and the 28SrDNA gene, and the PCR fragment obtained was subjected to sequence analysis. From all the four samples tested, sequence analysis and comparison allowed to unambiguously identify the biological materials present in the culture medium as belonging to the hallucinogenic species *Psilocybe mexicana* (Accession number LN830951). These results, highlighted that a molecular-based approach, characterized by a total DNA extraction from suspected culture media, amplification and sequence analysis of a discrete region of the 28S rDNA gene, could be useful to disclose the presence of hallucinogenic mushrooms even if the size of the biological material is too small to allow a morphological identification.

A suitable LC/MS-MS method was developed and tested for the analysis of the active principles of mushroom seized on the illicit market, suspected to belong to the genus Psilocybe.

For the quantitative determination it was necessary to identify and synthesize a proper internal standard (IS, i.e., 5-hydroxy-N,N-diethyltryptamine), endowed with chromatographic features suitable for the analysis of the active principles (Figure 4).

LC/MS-MS analysis of the three extracts corresponding to the three different parts of the mushrooms gave the results reported in Table 3. The concentration of psilocybin ranged from 0.5 to 1.4% while that of psilocin from 1.3 to 2.5%, confirming literature data [13]. It is possible to note that the concentration of psilocin is higher in the cap and in the distal part of the stem (near to the soil) than in the part of the stem

		Psilocin		Psilocybin	
	Sample	mg/mL	% (w/w)	mg/mL	% (w/w)
Cap	1	2.14	2.1	0.83	0.8
	2	2.40	2.4	1.44	1.4
	3	2.37	2.3	1.86	2.0
	SD	0.14		0.52	
Proximal stem	1	1.65	1.6	1.40	1.4
	2	1.30	1.3	0.97	1.0
	3	1.36	1.3	1.39	1.3
	SD	0.19		0.25	
Distal stem	1	2.54	2.5	0.50	0.5
	2	2.48	2.4	0.68	0.6
	3	2.57	2.5	0.50	0.5
	SD	0.046		0.10	

Table 3: Concentrations of the active principles.

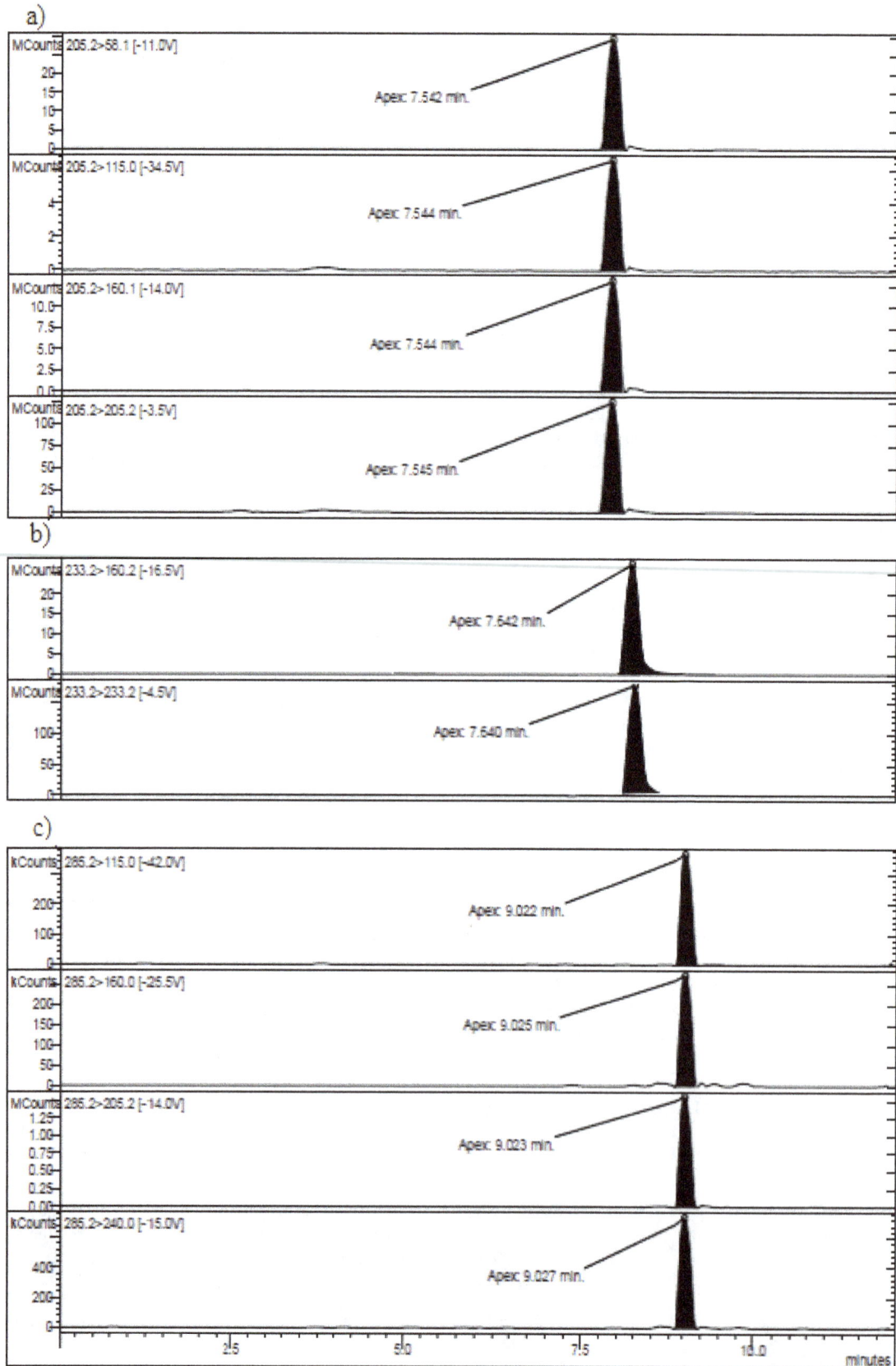

Figure 4: Chromatograms for a) psilocin, b) IS and c) psilocybin.

proximal to the cap. On the other hand the concentration of psilocybin is higher in the cap and in the proximal part, being lower in the distal part of the stem (Figures 5-7). Analyses gave more reproducible results for psilocin as evidenced by the standard deviations.

Conclusion

The taxonomic identification of the biological material contained in the hallucinogenic mushrooms culture media, was carried out using a DNA-based approach, thus highlighting the usefulness of this approach in the forensic identification of illegal samples.

Moreover a LC/MS-MS method for the quali-quantitative analysis of psylocibin and psilocin was proposed and a proper internal

Compound	Q₁ first mass	Q₃ first mass	Capillary (V)	Collision Energy (V)
Psilocin	205.2	58.1	35.209	11.0
	205.2	115.0	35.209	34.5
	205.2	160.1	35.209	14.0
	205.2	205.2	35.209	3.5
Psilocybin	285.2	115.0	60.084	25.5
	285.2	205.2	60.084	14.0
	285.2	240.0	60.084	15.0
	285.2	245.2	60.084	3.5
IS	233.2	160.2	50.046	16.5
	233.2	233.2	50.046	4.5

Table 2: MRM transitions of each analyte and IS.

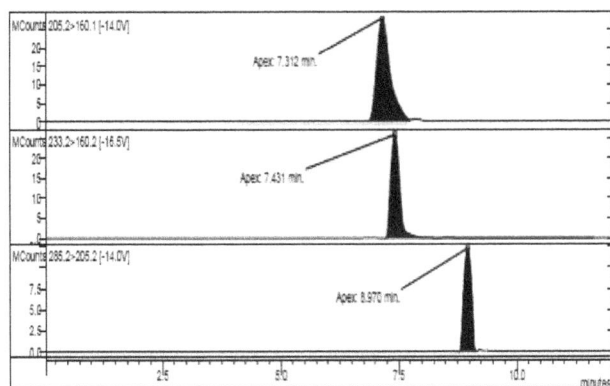

Figure 5: Chromatogram of the active principles contained in the cap of a mushroom (psilocin, IS, psilocybin). The following transitions were used for quantitative analysis. Psilocin: 205→160; IS: 233→160; psilocybin 285→205.

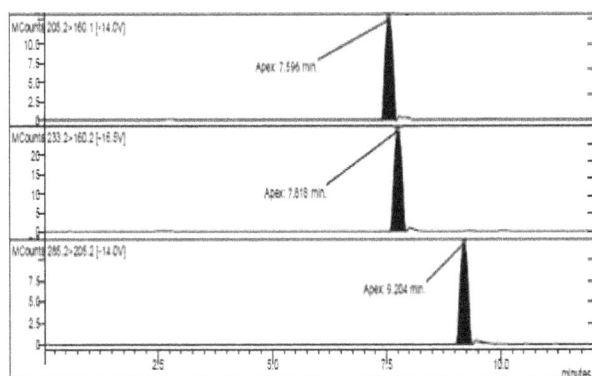

Figure 6: Chromatogram of the proximal part of the stem.

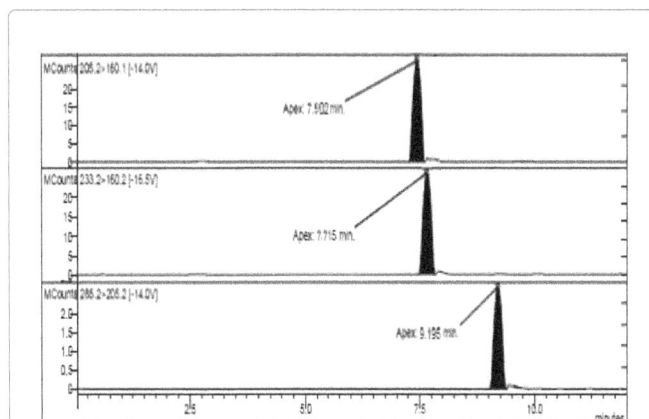

Figure 7: Chromatogram of the distal part of the stem.

standard (5-hydroxy-N,N-diethyltryptamine) was synthesized for the quantitative determination of the analytes in mushroom samples seized by the judicial authority.

References

1. Crundwell E (1987) The unnatural history of the fly agaric. Mycologist 21: 178-181.

2. Furst PT (2004) Visionary plants and ecstatic shamanism. Expedition 46: 26-29.

3. Ott J (1976) Psycho-mycological studies of Amanita-from ancient sacramentto modern phobia. J Psychoactive Drugs 8: 27-35.

4. Sessa BSJ (2008) Are psychedelic drug treatments seeing a comeback in psychiatry? Prog Neurol Psychiatry 12: 5-10.

5. Kuttner RE, Hickey RE (1970) Culture and perception: a note on hallucinogenic drugs. J Natl Med Assoc 62: 25-26.

6. Pavarin RM (2006) Substance use and related problems: a study on the abuse of recreational and not recreational drugs in Northern Italy. Ann Ist Super Sanita 42: 477-484.

7. Johnson M, Richards W, Griffiths R (2008) Human hallucinogen research: guidelines for safety. J Psychopharmacol 22: 603-620.

8. van Amsterdam J, Opperhuizen A, van den Brink W (2011) Harm potential of magic mushroom use: a review. Regul Toxicol Pharmacol 59: 423-429.

9. Riley SC, James C, Gregory D, Dingle H, Cadger M (2001) Patterns of recreational drug use at dance events in Edinburgh, Scotland. Addiction 96: 1035-1047.

10. Hofmann A, Heim R, Brack A, Kobel H (1958) Psilocybin, a psychotropic substance from the Mexican mushroom *Psilicybe mexicana* Heim. Experientia 14: 107-109.

11. Horita A, Weber LJ (1961) The enzymic dephosphorylation and oxidation of psilocybin and psilocin by mammalian tissue homogenates. Biochem Pharmacol 7: 47-54.

12. Horita A, Weber LJ (1962) Dephosphorylation of psilocybin in the intact mouse. Toxicol Appl Pharmacol 4: 730-737.

13. Pedersen-Bjergaard S, Sannes E, Rasmussen KE, Tønnesen F (1997) Determination of psilocybin in *Psilocybe semilanceata* by capillary zone electrophoresis. J Chromatogr B Biomed Sci Appl 694: 375-381.

14. Campbell A (2001) The Australian Illicit Drug Guide. Black Inc., Melbourne, Australia.

15. Helsley S, Fiorella D, Rabin RA, Winter JC (1998) A comparison of N,N-dimethyltryptamine, harmaline, and selected congeners in rats trained with LSD as a discriminative stimulus. Prog Neuropsychopharmacol Biol Psychiatry 22: 649-663.

16. Vollenweider FX, Geyer MA (2001) A systems model of altered consciousness: integrating natural and drug-induced psychoses. Brain Res Bull 56: 495-507.

17. Borowiak KS, Ciechanowski K, Waloszczyk P (1998) Psilocybin mushroom (*Psilocybe semilanceata*) intoxication with myocardial infarction. J Toxicol Clin Toxicol 36: 47-49.

18. Stebelska K (2013) Fungal hallucinogens psilocin, ibotenic acid, and muscimol: analytical methods and biologic activities. Ther Drug Monit 35: 420-442.

19. Nugent KG, Saville BJ (2004) Forensic analysis of hallucinogenic fungi: a DNA-based approach. Forensic Sci Int 140: 147-157.

20. Jespersen L, Nielsen DS, Hønholt S, Jakobsen M (2005) Occurrence and diversity of yeasts involved in fermentation of West African cocoa beans. FEMS Yeast Res 5: 441-453.

16

Classification of Cannabis Cultivars Marketed in Canada for Medical Purposes by Quantification of Cannabinoids and Terpenes Using HPLC-DAD and GC-MS

Dan Jin[1,2], Shengxi Jin[2], Yang Yu[2], Colin Lee[2] and Jie Chen[1,3]*

[1]Biomedical Engineering Department, University of Alberta, Edmonton, Alberta, Canada
[2]Labs-Mart Inc., Edmonton, Alberta, Canada
[3]Electrical and Computer Engineering Department, University of Alberta, Edmonton, Alberta, Canada

Abstract

For over a century, research on cannabis has been hampered by its legal status as a narcotic. The recent legalization of cannabis for medical purposes in North America requires rigorous standardization of its phytochemical composition in the interest of consumer safety and medicinal efficacy. To utilize medicinal cannabis as a predictable medicine, it is crucial to classify hundreds of cultivars with respect to dozens of therapeutic cannabinoids and terpenes, as opposed to the current industrial or forensic classifications that only consider the primary cannabinoids tetrahydrocannabinol (THC) and cannabidiol (CBD). We have recently developed and validated analytical methods using high-pressure liquid chromatography (HPLC-DAD) to quantify cannabinoids and gas chromatography with mass spectroscopy (GC-MS) to quantify terpenes in cannabis raw material currently marketed in Canada. We classified 32 cannabis samples from two licensed producers into four clusters based on the content of 10 cannabinoids and 14 terpenes. The classification results were confirmed by cluster analysis and principal component analysis in tandem, which were distinct from those using only THC and CBD. Cannabis classification using a full spectrum of compounds will more closely meet the practical needs of cannabis applications in clinical research, industrial production, and patients' self-production in Canada. As such, this holistic classification methodology will contribute to the standardization of commercially-available cannabis cultivars in support of a continuously growing market.

Keywords: Cannabis; Classification; Cannabinoids; Terpenes; HPLC-DAD; GC-MS; Cluster analysis; Principal component analysis

Introduction

As of 2014, 545 compounds have been identified in cannabis, among which there are 104 cannabinoids [1]. Cannabinoids have been indicated for sixteen potential therapeutic uses, ranging from pain management to neurological disorders [2]. Although the research focus is mainly on psychoactive THC, other non-psychoactive cannabinoids, such as CBD, cannabigerol (CBG), and cannabichomene (CBC), also have broad therapeutic potential without the negative effects of THC and may enhance the beneficial effects of THC [3]. Apart from cannabinoids, a significant number of compounds produced in cannabis are terpenes, which are responsible for cannabis's distinctive odour [4]. Although clinical studies are still nascent, terpenes are receiving increasing attention for their synergistic interactions with cannabinoids in the treatment of pain, inflammation, depression, anxiety, addiction, epilepsy, cancer, and infections [3]. Cannabis, as an herbal medicine, is suggested to be greater than the sum of its individual components [3].

Cannabis for medical purposes is becoming a global trend and is especially popular in North America. Canada and an increasing number of states in the US have legalized medical cannabis. As of January 2017, 28 states and Washington D.C. have legalized the medical use of cannabis [5] In Canada, 38 licensed producers are authorized to produce and sell dried marijuana, fresh marijuana, cannabis oil, or starting materials to eligible persons in Canada [6]. Effective as of August 2016, the new Access to Cannabis for Medical Purposes Regulations (ACMPR) permits self-production of a limited amount of cannabis for medical purposes as a supplement to purchasing cannabis from licensed producers [7]. However, due to the highly varied and complex composition of active components in cannabis, the suitability of each cultivar for treating particular conditions requires further investigation. In addition, a long history of hybridization has resulted

in hundreds of cannabis cultivars, among which many have similar chemical compositions. In this respect, cannabis cultivar classification is a foundational requirement for standardizing and controlling the quality of cannabis for medical applications.

Currently, there are three classification systems for cannabis. The first, a botanical perspective, attempts to classify cannabis into different species or subspecies based on appearance, THC content, and geographical origins (gene pools) [8-13]. The second, a chemotaxonomic perspective, describes five chemotypes (chemical phenotypes) based on the ratio of two major cannabinoids THC and CBD, which is decided by their corresponding allelic loci [14-21]. Recently, a third perspective seeks to categorize cultivars based on both cannabinoids and terpenes for drug standardization and clinical research purposes [22,23]. However, there is no currently available systematic classification covering the majority of commercially available cultivars. In this project, we aim to classify a portion of cannabis cultivars currently marketed in Canada based on the content of dozens of potentially therapeutic cannabinoids and terpenes. Cannabis cultivar classification will be the foundation for industrial production, clinical research, and informative guidance for individual growers in Canada.

In this work, we used a HPLC method recommended by American

*Corresponding author: Jie Chen, Biomedical Engineering Department, University of Alberta, Edmonton, Alberta, Canada, T6G 2V2
E-mail: jc65@ualberta.ca

Herbal Pharmacopoeia (AHP) [24] to quantify 10 cannabinoids (Cannabidiolic Acid (CBDA), Cannabigerolic Acid (CBGA), CBG, CBD, Tetrahydrocannabivarin (THCV), Cannabinol (CBN), Tetrahydrocannabinolic Acid (THCA), Δ^9-THC, Δ^8-THC, CBC). We also developed a GC-MS method to quantify 14 terpenes (α-Pinene, β-Myrcene, β-Pinene, Δ^3-Carene, Limonene, p-Cymene, Eucalptol, Linalool, Fenchone, Fenchol, Borneol, α-Terpineol, Pulegone, β-Caryophyllene) that have been indicated for pharmacological activities [3,25,26]. We validated these methods for specificity (selectivity), linearity, accuracy, precision (repeatability and intermediate precision), limit of detection (LOD), and limit of quantification (LOQ). We then applied these two quantification methods on 32 medical cannabis samples provided by licensed producers, followed by two classification methods. Hierarchical cluster analysis was first carried out and principal component analysis (PCA) was applied to confirm whether the cultivars in the cluster analysis would also be grouped together by PCA. PCA also revealed the compounds that were responsible in grouping cultivars between clusters. These classification results may have value for clinical researchers in the discrimination and selection of cultivars. They may also assist licensed producers in optimizing cultivar selection with regards to medicinal effects.

Materials and Methods

Sample collection

A total of 32 cannabis samples (dried flower buds) were collected from two licensed producers in Canada. Sample names were provided by the licensed producers and different names may not necessarily represent distinct cultivars. Samples arrived in sealed plastic bags and were stored in a dry and cool storage facility prior to analysis. All samples were pulverized into fine powder. Approximately 1 g of each sample was used for extraction.

Solvents and chemicals

All 10 cannabinoid standards except CBDA were purchased from Cayman Chemical Co. (Ann Arbor, Michigan 48108 USA). CBDA and the internal standard (ISTD) diazepam were purchased from Sigma-Aldrich Company (Oakville, Ontario, Canada). All standards were analytical grade and were provided as 1 mg/mL solution in methanol or acetonitrile.

All 14 terpenes standards and ISTD tridecane were purchased from Sigma-Aldrich Company (Oakville, Ontario, Canada). All standards were analytical grade and came as a pure liquid or white powder.

Hexanes, chloroform, and acetonitrile were purchased from Fisher Scientific Company (Ottawa, Ontario, Canada). Ammonium formate was purchased from Sigma-Aldrich Company (Oakville, Ontario, Canada). Methanol was purchased from EMD Millipore (Etobicoke, Ontario, Canada). Formic acid was purchased from Caledon Laboratory Chemicals (Halton Hills, Ontario, Canada). Water was HPLC grade, produced in-house using a Millipore filtration system which purified water to 18 mΩ resistivity.

HPLC systems and cannabinoids assay

The HPLC-DAD system used in this study was a modular Agilent 1100 Series comprised of the following components: Solvent Degasser (G1322A), Bin Pump (G1312A), WPALS (G1367A), Column. Compartment (G1316A), Hewlett Packard 1100 Series PhotoDiode-Array Detector (DAD) (G1315A), and Agilent Z0RBAX RX-C18 (4.6 mm × 150 mm, 3.5 µm) Column.

	Rate (°C/min)	Value (°C)	Hold Time (min)	Run Time (min)
(Initial)		50	0.1	0.1
Ramp 1	25	100	0.2	2.3
Ramp 2	20	120	0.2	3.5
Ramp 3	5	160	0.2	11.7
Ramp 4	20	200	0.1	13.8
Ramp 5	50	260	5	20.0

Table 1: GC temperature gradient program parameters.

The stock standard solution was prepared by adding 1 mL of 1 mg/mL THCA, CBDA, CBGA, Δ^9-THC, Δ^8-THC, CBD, CBG, CBC, THCV, and CBN standards into a 10 mL volumetric flask. This mixed standard solution was dried under a gentle stream of nitrogen, and then a 1:1 ratio of water and acetonitrile spiked with 20 ppm diazepam as ISTD was added to volume. The resulting concentration of each cannabinoid in the stock solution was 100 ppm (µg/mL). 100 ppm of the mixed standard solution was further diluted to create calibration standard solutions with cannabinoid concentrations of 50 ppm, 25 ppm, 5 ppm, 1 ppm, and 0.5 ppm.

The analytical method was adapted from a published method in AHP monograph (revision 2014) [27] by modifying the dilution factor. In the original method, 200 mg of sample was extracted with methanol/chloroform (9/1, v/v) and the extract was diluted by a factor of 10. In this method, we diluted the extract by a factor of 40 in two steps. A 100 µL aliquot of the filtrate of the extract was first diluted to a volume of 1 mL. Then, a 30 µL aliquot of the diluted extract was evaporated under a gentle stream of nitrogen. The residue was then dissolved in 120 µL of a mixture of water/acetonitrile (5/5, v/v) with 20 ppm ISTD. Finally, 100 µL of the solution was transferred into an amber vial with a spring glass insert for HPLC analysis. Quantifications of cannabinoids were achieved by comparing the ratio of sample/ISTD with the ratio of the external standard (ESTD)/ISTD at the target concentration. Cannabinoid analysis results were reported as mass fraction (w/w %).

GC-MS systems and terpenes assay

The GC-MS system used in this study was an Agilent 7890A GC system comprised of the following components: Agilent 7890A GC (G3440A), Agilent 5975C inert MSD with Triple-Axis Detector, K`Prime GC Sample Injector (MXY 02-01B), and a GC Column (Phenomenex, Zebron, ZB-624 30 m × 0.25 mm ID, 1.40 µm film thickness). A temperature gradient program was used for the separation of terpenes (Table 1). The injector temperature was 250°C. Injection volume was 2 µL. Split ratio is 20:1. The carrier gas (helium) flow rate was 1.2 mL/min. Run time was 20 minutes. SIM was carried out to quantify terpenes.

ISTD was prepared by weighing 216.8 mg of tridecane and dissolving it into 1 L of extraction solvent, resulting in an ISTD concentration of 216.8 ppm. A stock standard solution of each terpene was prepared separately by weighing approximately 200 mg of each terpene and dissolving it into 10 mL of ISTD-spiked extraction solvent. A 500 ppm mixed working standard was prepared by taking a calculated volume of each of the stock standards and adding extraction solvent to a final volume of 100 mL. The 500 ppm standard solution was further diluted to create calibration standards with terpenes concentrations of 250 ppm, 100 ppm, 50 ppm, 25 ppm, 5 ppm, and 1 ppm, respectively.

About 500 mg dried sample was extracted with 5.0 mL 1:1 ratio of hexane and ethyl acetate and put on shakers for 20 minutes. After being centrifuged at 10,000 rpm for 5 minutes, a 100 µL of the supernatant

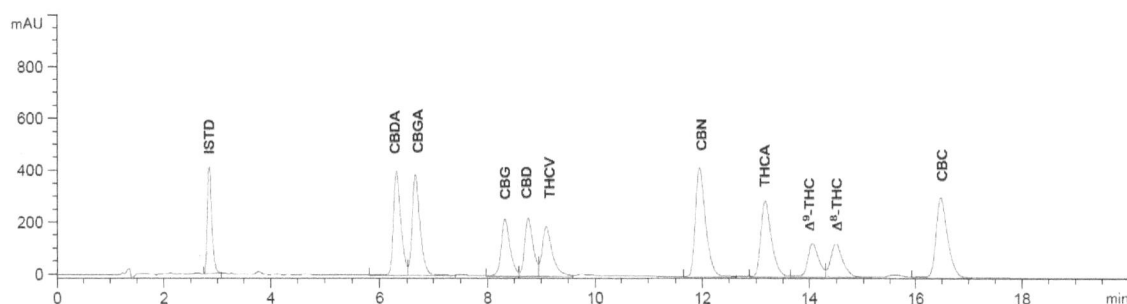

Figure 1: HPLC chromatograms for 10 cannabinoids at 50 ppm.

was transferred into an amber vial with a spring glass insert for GC analysis. Fifteen compounds (including the ISTD tridecane) were divided into 12 groups in the SIM method, with each group assigned with corresponding quantifier and qualifiers. Quantifications of terpenes were achieved by comparing the ratio of sample/ISTD with the ratio of ESTD/ISTD at the target concentration. Terpene analysis results were reported as w/w%.

Method validation

Both HPLC and GC-MS methods were validated for specificity (selectivity), linearity, accuracy, precision (repeatability and intermediate precision), LOD and LOQ as instructed by the ICH Harmonised Tripartite Guideline for Validation of Analytical Procedures Q2 (R1) [28]. Specificity (selectivity) was determined by injecting a solvent blank to confirm that there were no false signal peaks at the targeted retention time. Each cannabinoid and terpene standard was individually injected to determine retention times. A linear regression (calibration) curve for each compound was constructed by plotting the peak-area ratio of STD/ISTD (y) against concentration (x, ppm). The slope, y-intercept and coefficient of determination were calculated from the standard curves.

To test accuracy (recovery), spiked samples were prepared by adding three levels of known concentrations of standards into three replicates of a known sample. Each spiked sample was injected three times (N=3). The spiked levels ranged from low, medium, to high concentrations of each analyte, as the contents of cannabinoids and terpenes vary significantly in the natural samples. Precision includes repeatability and intermediate precision, otherwise referred to as intraday precision and interday precision. Although ICH requires sampling from authentic samples, it was impractical in this case to obtain blank matrices completely free from cannabinoids and terpenes. In addition, contents of these compounds vary significantly in samples and some may below the LOQs. However, it was also impractical to spike every standard into samples to bring each analyte above the LOQ. In this work, we chose to spike mixed cannabinoid standard or mixed terpene standard into a solvent blank as blank matrices. Repeatability was determined by assaying a spiked blank matrices 12 times as intraday precision. Twelve data points (N=12) were used to calculate the relative standard deviation (%RSD) of the set. To obtain inter-day precision, six assays were repeated on two different days. Both intra-day and inter-day data were calculated together to determine intermediate precision (N=12).

Statistical analysis was applied to the linear regression line in order to determine the standard deviation (SD) and slope (S). From these values, LOD and LOQ were calculated by using the following equations:

$$LOQ = 3.3 \times \frac{SD}{S} \qquad (1)$$

$$LOD = 10 \times \frac{SD}{S} \qquad (2)$$

Quantification and classification

Each sample was quantified for both cannabinoid and terpene content, which were then subjected to cluster analysis and PCA in order to enable cannabis cultivar classification. The software used for both analysis was JMP' 13.0.0. Observations (cultivars in this case) were grouped using hierarchical clustering. The distances between clusters were calculated using Ward's minimum variance equation [2]:

$$D_{KL} = \frac{\left\| \overline{X_K} - \overline{X_L} \right\|^2}{\dfrac{1}{N_K} + \dfrac{1}{N_L}} \qquad (3)$$

where

D_{KL} is Ward's distance between clusters K and L;

K and L subscripts are positive integers up to the number of observations;

$\mathbf{x}_K$ is the mean vector for the Kth cluster C_K;

$\mathbf{x}_L$ is the mean vector for the Lth cluster C_L;

$\|\mathbf{x}\|$ is the square root of the sum of the squares of the elements of $\mathbf{x}$ (the Euclidean length of the vector $\mathbf{x}$);

N_K is the number of observations in C_K; and

N_L is the number of observations in C_L.

PCA is a commonly used multivariate technique to detect patterns in high-dimensional data. PCA can also identify the critical compounds for discriminating cannabis cultivars, which is useful in choosing cultivars with specific abundant bioactive components. PCA projects the original chemical data into a new coordinate system, which is produced by calculating eigenvalues and eigenvectors from the covariance matrix of the original matrix. The eigenvectors (principal components, shortened as PCs) are orthogonal to each other and are ordered by significance: the first PC explains the most variance and the last PC explains the least [29]. For better visual interpretation of the data, the first two or three PCs are reserved, resulting in a lossy data compression process.

Results and Discussion

Method validation for cannabinoids (HPLC)

A solvent blank was injected and no false signal peak was observed at the targeted retention time area. Five levels of cannabinoid standard solutions were injected. Specificity was demonstrated by well-separated peaks (Figure 1).

Regression (calibration) curves were visibly linear (Figure 2). Additionally, the correlation coefficients for all 10 cannabinoids ranged from 0.9993 to 1.0000 (Table 2). The %RSD of accuracy (recovery) ranged from 91.3% to 104.4%. The precision (%RSD) for all compounds were less than 1.39%. Inter-day injection precisions were found to be less than 1.45%. The method was precise in terms of repeatability and intermediate precision. Cannabinoids' LODs ranged from 0.07 to 0.99 ppm and LOQs ranged from 0.21 to 3.00 ppm (Table 2).

Method validation for terpenes (GC-MS)

A solvent blank was injected and no junk peak was observed at the targeted retention time area. Six concentrations of terpene standards were injected. Specificity was demonstrated by well-separated peaks (Figure 3).

Regression (calibration) curves were visibly linear (Figure 4). Additionally, the correlation coefficients for all 14 terpenes ranged from 0.9993 to 1.0000 (Table 3). The %RSD of accuracy (recovery) ranged from 93.0 to 104.8%. The method precisions (%RSD) for all compounds were less than 2.1%. Inter-day injection precisions were found to be less than 1.34%. The method was precise in terms of repeatability and intermediate precision. For all terpenes, LODs ranged from 0.82 to 3.69 ppm, and LOQs ranged from 2.47 to 11.2 ppm (Table 3) respectively.

Quantification of cannabinoids and terpenes

In this work, each cultivar was labelled with an identifier for convenience (Table 4). Quantitative data for cannabinoids and terpenes are listed in Tables 4 and 5, respectively. Hierarchical cluster analysis was applied first and PCA was used to confirm the grouping results.

Cluster analysis

Total THC (Δ^9-THC+THCA as T-THC) and total CBD (CBD+CBDA as T-CBD) were calculated and used in cluster analysis because THCA and CBDA become THC and CBD after decarboxylation [24]. The levels of T-THC ranged from 7.08% (LM20) to 0.24% (LM7). The levels of T-CBD ranged from undetectable amounts (LM2, 11, 13, 25, 26) to 5.52% (LM7). Cluster analysis based only on T-THC and T-CBD classified 32 samples into four clusters, which is presented as a constellation plot in Figure 5. Cluster 1 and cluster 2 are THC dominant (chemotype I) with average T-THC more than 3% and average T-CBD less than 1% (Table 6). Cluster 3 has an approximately equal amount

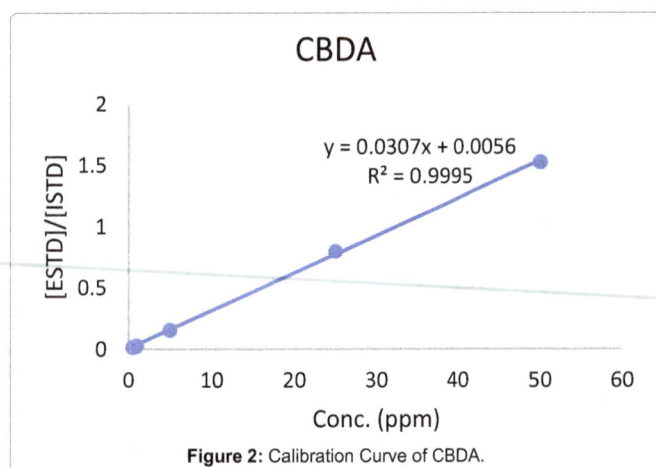

Figure 2: Calibration Curve of CBDA.

| S. No | Cannabinoids | Coefficient of determination (R²) | Repeatability %RSD (N=12) | Intermediate Precision %RSD (N=12) | Recovery% (N=3) | | | LOD (ppm) | LOQ (ppm) |
					Low Conc. 5 ppm	Medium Conc. 10 ppm	High Conc. 25 ppm		
1	CBDA	0.9995	0.31	0.76	99.0	98.4	96.7	0.81	2.46
2	CBGA	0.9993	0.57	0.56	99.2	93.1	92.3	0.99	3.00
3	CBG	0.9997	0.53	1.14	97.4	96.2	94.1	0.60	1.82
4	CBD	0.9997	0.48	1.16	98.0	98.2	95.6	0.59	1.80
5	THCV	1.0000	0.60	0.91	95.1	98.4	94.4	0.09	0.28
6	CBN	0.9998	0.48	1.44	91.3	93.7	93.2	0.48	1.44
7	THCA	0.9995	0.38	0.51	99.0	97.9	98.1	0.77	2.34
8	Δ^9-THC	0.9999	1.22	1.03	104.4	99.4	97.3	0.40	1.21
9	Δ^8-THC	0.9996	0.52	1.10	102.7	100.0	97.5	0.68	2.07
10	CBC	1.0000	1.39	1.45	94.5	93.8	93.9	0.07	0.21

Table 2: Method Validation results for cannabinoids (HPLC).

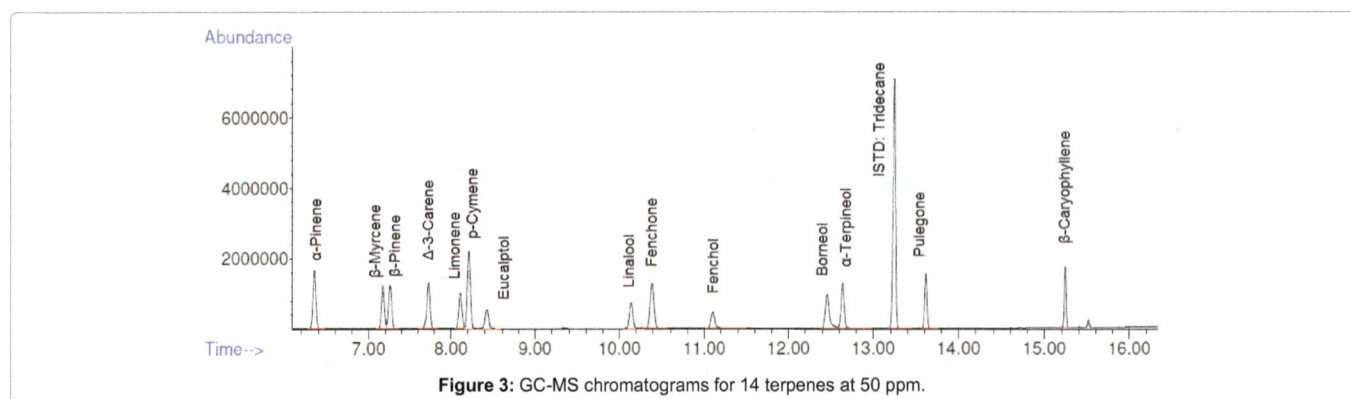

Figure 3: GC-MS chromatograms for 14 terpenes at 50 ppm.

of T-THC and T-CBD (chemotype II) at around 1.5% each (Table 6). Cluster 4 has only one cultivar and is CBD dominant (chemotype III)

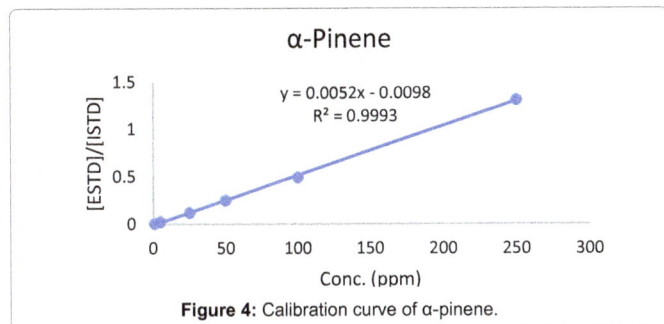

Figure 4: Calibration curve of α-pinene.

with T-CBD more than 5% and T-THC less than 1% (Table 6).

However, if samples are grouped into clusters based on the full chemical profile (10 cannabinoids and 14 terpenes), the classification results changed (Figure 6). After involving more cannabinoids and terpenes, samples were classified into four clusters, with cluster 1, 2, 3 being THC dominant and cluster 4 being CBD dominant (Table 7). Furthermore, cultivars clustered together not only have similar THC and CBD content, but also have similar full profiles. Classification based on the full chemical profile may offer more flexible and reliable choices for clinical researchers and licensed producers in terms of choosing cannabis cultivars. For example, if LM28 is clinically studied and recommended for a particular condition, LM29 in the same cluster likely can be an alternative if LM28 is not available, due to the two cultivars' similarities.

S. No	Terpenes	Coefficient of determination (R²)	Repeatability %RSD (N=12)	Intermediate Precision %RSD (N=12)	Recovery% (N=3)			LOD (ppm)	LOQ (ppm)
					Low Conc. 50 ppm	Medium Conc. 150 ppm	High Conc. 250 ppm		
1	α-Pinene	0.9993	2.1	0.87	93.0	94.4	96.3	3.69	11.2
2	β-Myrcene	0.9994	1.3	0.45	98.8	97.8	95.4	3.52	10.7
3	β-Pinene	0.9993	1.3	0.88	94.0	97.1	93.6	3.70	11.2
4	Δ³-Carene	0.9995	1.2	0.81	94.2	98.6	94.6	3.21	9.72
5	Limonene	0.9998	1.5	0.39	97.2	99.9	96.5	2.20	6.67
6	p-Cymene	0.9996	1.7	1.34	97.3	100.7	95.4	2.82	8.55
7	Eucalptol	0.9997	2.0	0.98	96.6	100.2	94.6	2.36	7.14
8	Linalool	0.9999	1.9	0.37	99.1	102.1	98.0	1.58	2.99
9	Fenchone	1.0000	1.9	0.47	96.9	101.4	97.4	0.99	2.99
10	Fenchol	0.9999	1.7	0.60	98.2	103.4	100.8	1.28	3.86
11	Borneol	0.9997	1.5	1.34	98.5	104.8	100.1	2.62	7.94
12	α-Terpineol	1.0000	1.5	0.50	98.6	103.5	100.4	0.82	2.47
13	Pulegone	0.9999	1.7	0.71	97.0	103.8	97.9	1.73	5.24
14	β-Caryophyllene	0.9999	1.5	0.49	100.0	102.7	102.0	1.30	3.95

Table 3: Method validation results for terpenes (GC-MS).

File ID	Cultivar	CBDA	CBGA	CBG	CBD	THCV	CBN	THCA	Δ⁹-THC	Δ⁸-THC	CBC	T-THC	T-CBD
LM1	AD	0.03	0.11	NQ	NQ	NQ	0.05	4.49	0.18	NQ	NQ	4.67	0.03
LM2	ADEB	*NQ	0.11	NQ	NQ	NQ	0.04	3.03	0.12	NQ	NQ	3.14	NQ
LM3	ATH	0.88	0.06	0.05	0.10	0.06	0.04	1.01	0.16	0.05	NQ	1.17	0.97
LM4	BW	0.03	0.09	0.29	NQ	NQ	0.04	5.72	0.11	0.08	0.03	5.84	0.03
LM5	BD	0.03	0.08	0.02	NQ	NQ	0.03	4.44	0.08	NQ	NQ	4.52	0.03
LM6	BDEB	0.03	0.11	0.04	NQ	NQ	0.04	5.47	0.10	NQ	NQ	5.57	0.03
LM7	CT	5.22	0.17	NQ	0.30	0.34	0.09	0.18	0.05	0.06	0.05	0.24	5.52
LM8	CT+LSD	1.57	0.12	NQ	0.10	0.09	0.06	1.62	0.15	0.09	NQ	1.77	1.67
LM9	CC	1.97	0.07	0.03	0.14	0.25	0.06	1.25	0.19	0.05	NQ	1.44	2.11
LM10	DQ	0.03	0.20	0.05	NQ	NQ	0.04	3.71	0.07	0.08	NQ	3.79	0.03
LM11	EMD	NQ	0.06	NQ	NQ	NQ	0.03	2.57	0.07	NQ	NQ	2.64	NQ
LM12	EMD#1	0.03	0.10	0.04	NQ	NQ	0.05	4.34	0.15	NQ	NQ	4.50	0.03
LM13	EMD#2	NQ	0.05	0.05	NQ	NQ	0.04	2.48	0.13	NQ	NQ	2.61	NQ
LM14	ER	0.03	0.10	NQ	NQ	NQ	0.05	2.54	0.28	NQ	NQ	2.82	0.03
LM15	GLA	0.03	0.19	0.03	NQ	NQ	0.04	6.11	0.25	NQ	NQ	6.36	0.03
LM16	GLA#1	0.03	0.14	NQ	NQ	NQ	0.04	4.22	0.23	0.10	NQ	4.45	0.03
LM17	GLA#2	0.03	0.10	NQ	NQ	NQ	0.05	3.51	0.26	NQ	NQ	3.77	0.03
LM18	KOG	0.04	0.23	0.06	NQ	NQ	0.05	6.60	0.15	0.17	0.05	6.75	0.04
LM19	KOGEB	0.04	0.27	0.06	NQ	NQ	0.05	6.88	0.13	0.21	0.05	7.01	0.04
LM20	KOGEBE	0.04	0.29	0.06	NQ	NQ	0.05	6.95	0.14	0.18	0.05	7.08	0.04
LM21	LAA	0.04	0.05	NQ	NQ	NQ	0.08	5.27	0.38	0.17	0.05	5.65	0.04
LM22	OGK	0.03	0.10	0.08	NQ	NQ	0.05	4.96	0.11	NQ	NQ	5.07	0.03
LM23	PLA	0.03	0.15	0.02	NQ	NQ	0.04	3.63	0.05	0.14	NQ	3.68	0.03
LM24	PLA#1	0.03	0.15	NQ	NQ	NQ	0.05	4.26	0.14	0.13	NQ	4.40	0.03
LM25	PLA#2	NQ	0.04	NQ	NQ	NQ	0.05	2.78	0.13	0.16	NQ	2.91	NQ
LM26	PLA#3	NQ	0.03	NQ	NQ	NQ	0.04	2.42	0.09	0.08	NQ	2.51	NQ

LM27	QK	0.04	0.14	0.09	NQ	NQ	0.04	5.85	0.11	0.10	0.02	5.97	0.04
LM28	SD	0.04	0.47	0.05	0.06	NQ	0.05	6.07	0.59	0.15	0.03	6.66	0.10
LM29	SDEB	0.04	0.47	0.05	NQ	NQ	0.05	5.61	0.59	0.24	NQ	6.20	0.04
LM30	SQ	0.03	0.22	0.04	NQ	NQ	0.04	4.78	0.14	NQ	NQ	4.92	0.03
LM31	STC	0.03	0.20	NQ	NQ	NQ	0.04	4.76	0.16	NQ	NQ	4.92	0.03
LM32	WB	2.78	0.12	0.03	0.16	0.06	0.05	0.63	0.13	0.06	0.04	0.76	2.94

Table 4: Quantitative data for cannabinoids w/w (%).

File ID	Cultivar	*T1	T2	T3	T4	T5	T6	T7	T8	T9	T10	T11	T12	T13	T14
LM1	AD	0.011	0.094	0.018	NQ	0.069	NQ	NQ	0.043	0.003	0.019	0.008	0.016	0.003	0.331
LM2	ADEB	0.010	0.016	0.017	NQ	0.052	NQ	NQ	0.043	0.002	0.018	0.006	0.017	0.002	0.414
LM3	ATH	0.032	0.063	0.014	0.004	NQ	0.007	NQ	0.008	NQ	0.007	0.006	0.010	0.002	0.063
LM4	BW	0.009	0.042	0.016	NQ	NQ	NQ	NQ	0.033	NQ	0.015	0.005	0.014	0.002	0.187
LM5	BD	0.415	0.188	0.190	NQ	0.088	NQ	0.004	0.047	0.003	0.036	0.013	0.031	0.001	0.119
LM6	BDEB	0.458	0.205	0.209	NQ	0.129	NQ	0.005	0.052	0.004	0.044	0.016	0.039	0.002	0.141
LM7	CT	0.047	0.132	0.021	0.007	NQ	0.037	NQ	0.048	0.003	0.020	0.011	0.021	0.002	0.056
LM8	CT+LSD	0.032	0.020	0.019	NQ	0.013	0.005	0.002	0.030	0.002	0.016	0.009	0.022	0.002	0.103
LM9	CC	0.015	0.047	0.005	NQ	NQ	0.002	NQ	0.015	NQ	0.010	0.007	0.010	0.002	0.041
LM10	DQ	0.007	0.018	0.011	NQ	NQ	NQ	NQ	0.039	NQ	0.012	0.005	0.010	0.002	0.066
LM11	EMD	0.028	0.029	0.029	NQ	0.107	NQ	NQ	0.013	0.002	0.033	0.010	0.025	0.002	0.195
LM12	EMD#1	0.052	0.114	0.044	NQ	0.071	NQ	0.006	0.029	0.003	0.038	0.013	0.019	0.003	0.208
LM13	EMD#2	0.023	0.023	0.020	NQ	0.063	NQ	NQ	0.010	0.002	0.026	0.010	0.022	0.003	0.174
LM14	ER	0.039	0.046	0.028	0.006	NQ	0.004	NQ	0.006	NQ	NQ	0.007	0.015	0.002	0.072
LM15	GLA	0.025	0.044	0.044	0.012	0.028	NQ	0.006	0.061	0.001	0.010	0.005	0.033	0.008	0.321
LM16	GLA#1	0.067	0.102	0.048	0.006	0.034	0.002	0.003	0.035	0.002	0.015	0.007	0.022	0.005	0.197
LM17	GLA#2	0.082	0.103	0.046	NQ	0.029	NQ	0.002	0.026	0.002	0.016	0.008	0.016	0.004	0.171
LM18	KOG	0.036	0.283	0.070	NQ	0.239	NQ	NQ	0.094	0.004	0.063	0.016	0.049	0.003	0.363
LM19	KOGEB	0.034	0.155	0.065	NQ	0.187	NQ	NQ	0.104	0.005	0.069	0.019	0.056	0.002	0.640
LM20	KOGEBE	0.038	0.171	0.072	NQ	0.233	NQ	NQ	0.140	0.005	0.074	0.019	0.058	0.003	0.700
LM21	LAA	0.022	0.261	0.041	NQ	0.136	NQ	0.003	0.021	0.004	0.045	0.016	0.018	0.003	0.198
LM22	OGK	0.026	0.045	0.045	NQ	0.156	NQ	0.003	0.071	0.007	0.040	0.010	0.035	0.002	0.202
LM23	PLA	0.029	0.052	0.051	NQ	0.184	NQ	0.002	0.075	0.008	0.041	0.009	0.034	0.002	0.210
LM24	PLA#1	0.024	0.055	0.040	NQ	0.104	NQ	NQ	0.076	0.006	0.043	0.013	0.037	0.003	0.199
LM25	PLA#2	0.015	0.036	0.025	NQ	0.073	NQ	0.003	0.038	0.003	0.027	0.009	0.023	0.003	0.153
LM26	PLA#3	0.013	0.019	0.020	NQ	0.053	NQ	NQ	0.023	0.002	0.018	0.006	0.014	0.003	0.129
LM27	QK	0.012	0.110	0.020	NQ	0.063	NQ	NQ	0.119	0.002	0.020	0.006	0.019	0.003	0.413
LM28	SD	0.337	0.129	0.080	NQ	NQ	0.002	NQ	0.034	NQ	0.011	0.012	0.012	0.002	0.082
LM29	SDEB	0.796	0.301	0.179	NQ	0.056	NQ	0.004	0.058	0.002	0.021	0.022	0.022	0.003	0.156
LM30	SQ	0.021	0.037	0.033	NQ	0.097	NQ	NQ	0.071	0.003	0.033	0.010	0.029	0.003	0.257
LM31	STC	0.123	0.119	0.039	NQ	0.027	NQ	0.002	0.015	0.002	0.013	0.007	0.012	0.003	0.167
LM32	WB	0.035	0.044	0.009	0.003	NQ	0.010	NQ	0.005	NQ	0.004	0.006	0.004	0.002	0.020

*NQ: Not quantifiable – below LOD or LOQ; *T1=α-Pinene, T2=β-Myrcene, T3=β-Pinene, T4=Δ³-Carene, T5=Limonene, T6=p-Cymene, T7=Eucalyptol, T8=Linalool, T9=Fenchone, T10=Fenchol, T11=Borneol, T12=α-Terpineol, T13=Pulegone, T14=β-Caryophyllene

Table 5: Quantitative data for terpenes w/w (%).

Principal component analysis (PCA)

Although Figure 7 gives a clear profile of all cannabinoids and terpenes levels in each cluster, some compounds are more important to the classification. In this case, 10 cannabinoids and 14 terpenes are the original 24 variables (24 dimensions) in PCA. By calculating the covariance matrix between these 24 dimensions, PCA can generate 24 new variables (24 PCs), that are orthogonal to each other and can explain 100% of the total variance of the original data. In this work, the first three PCs explain 65.3% of the total variance. Each PC is correlated with the original 24 variables. The first column in the loading matrix (Table 8) are the correlations of PC1 with each compound. The higher the absolute value, the high the correlation. For example, PC1 is more correlated with THCA, Limonene, Fenchol, Terpineol, Borneol, Linalool, β-Caryophyllene, Fenchone, β-Myrcene, which indicates that PC1 is more of a "THCA+terpenes" item. This conclusion indicates that cultivars within close proximity along PC1 have similar combination of THCA and these terpenes, whereas separated clusters have distinct amounts of these compounds. For instance, Clusters 1, 2 and 3 are separated along PC1 (Figure 8) due to different combinations of THCA and these terpenes – this separation corresponds with cannabinoids and terpenes content in Figure 7. In addition, PC2 is more correlated with CBN, p-Cymene, CBC, CBDA, CBD, and THCV, which makes PC2 a "cannabinoids" item. For example, Clusters 1, 2, 3 (THC dominant) are separated from Cluster 4 (CBD dominant) along PC2 mostly due to the distinct CBDA content in Cluster 4. Additionally, Cluster 4 (LM7) may be related with higher percentages of p-Cymene (0.037%) compared to cultivars in Cluster 1, 2, and 3. However, Cluster 1 only contains one cultivar, which suggests that additional data is required to make a reliable conclusion. Finally, PC3 is more correlated with α-Pinene and Δ⁹-THC, which explains the separation between Cluster 2 and Cluster 3 along PC3 (Figure 9). More specifically, LM20 in Cluster 2 has 0.038% α-Pinene and 0.14% Δ⁹-THC while LM29 in

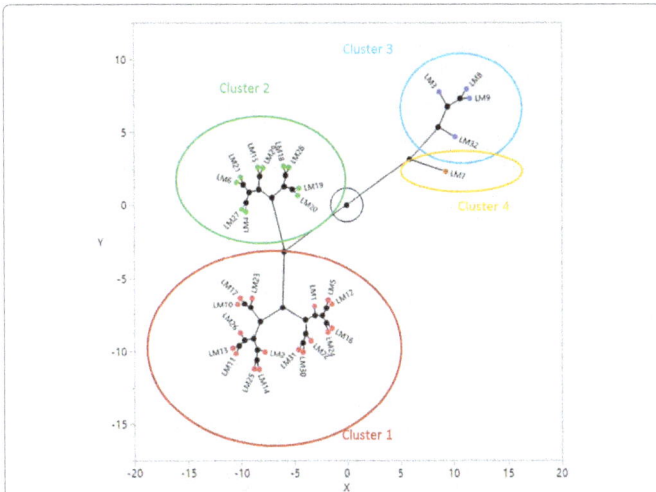

Figure 5: Constellation Plot-cluster analysis based on total THC and total CBD.

Cluster	Cultivar Count	Total THC Ave. (w/w%)	Total CBD Ave. (w/w%)
1	17	3.84	0.02
2	10	6.31	0.04
3	4	1.29	1.92
4	1	0.24	5.52

Table 6: Average levels of total THC and total CBD in each cluster in Figure 5.

Cluster	Cultivar Count	Total THC Ave. (w/w%)	Total CBD Ave. (w/w%)
1	20	3.55	0.40
2	9	5.53	0.04
3	2	6.43	0.07
4	1	0.24	5.52

Table 7: Average levels of total THC and total CBD in each cluster in Figure 6.

S. No		PC1	PC2	PC3
1	THCA	0.8567	-0.1382	0.1237
2	Limonene	0.8432	0.1535	-0.3894
3	Fenchol	0.8335	0.2800	-0.3638
4	α-Terpineol	0.8052	0.2273	-0.2980
5	Borneol	0.7640	0.4396	0.2640
6	Linalool	0.7438	0.2725	-0.2908
7	β-Caryophyllene	0.7070	0.0750	-0.4148
8	Fenchone	0.6332	0.1422	-0.3827
9	β-Myrcene	0.6152	0.4238	0.4825
10	Δ^8-THC	0.5065	0.4541	0.1443
11	CBN	-0.1634	0.8297	-0.0112
12	p-Cymene	-0.5247	0.7599	0.0133
13	CBC	0.2543	0.7580	-0.1432
14	CBDA	-0.5951	0.7568	-0.0153
15	CBD	-0.6232	0.7362	0.0493
16	THCV	-0.5624	0.6991	-0.0224
17	α-Pinene	0.3298	0.0486	0.8362
18	Δ^9-THC	0.1593	0.0418	0.8012
19	β-Pinene	0.5746	0.0329	0.5819
20	CBGA	0.4417	0.3003	0.5266
21	Eucalyptol	0.2127	-0.2522	0.4494
22	Pulegone	0.0372	-0.1933	0.2270
23	Δ^3-Carene	-0.3958	0.1525	0.1674
24	CBG	0.1627	-0.0546	-0.0734

Table 8: Loading matrix for the first three PCs.

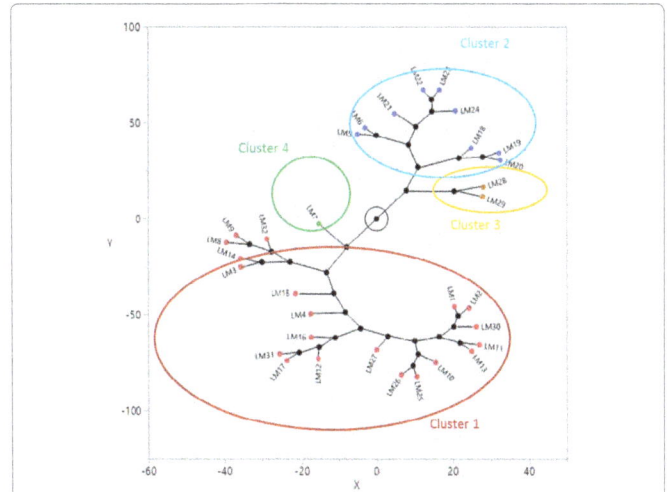

Figure 6: Constellation Plot-cluster analysis based on full chemical profile.

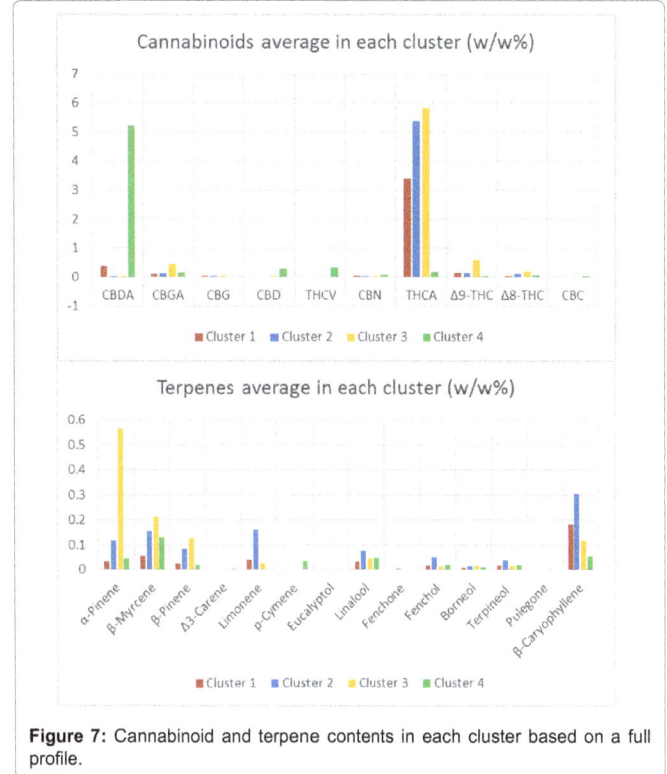

Figure 7: Cannabinoid and terpene contents in each cluster based on a full profile.

Cluster 3 has 0.337% α-Pinene and 0.59% Δ^9-THC, which also matches with average cluster content of α-Pinene and Δ^9-THC in Figure 7. The loading plot for PC1 and PC2 (Figure 8) gives an intuitive explanation whereby the longer the radial separation of the compound from the center, the more important the compound is in distinguishing cultivars in PC1 and PC2. The mathematical explanation is that the radial equals the square sum of the compound's correlations with PC1 and PC2 (Table 8). In conclusion, if cultivars are separated along PC1, they contain a distinct amount of THCA and terpenes (Limonene, Fenchol, Terpineol, Borneol, Linalool, β-Caryophyllene, Fenchone, β-Myrcene). If cultivars are separated along PC2, they contain different amount of cannabinoids (CBN, CBC, CBDA, CBD, and THCV) and p-Cymene. If they are separated along PC3, most likely the α-Pinene and THC contents are differentiable.

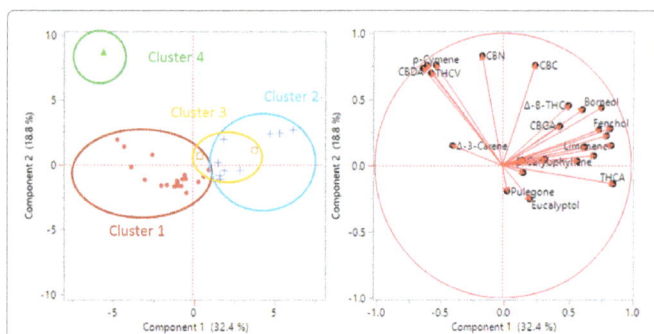

Figure 8: PCA of 32 cannabis cultivars (PC1 and PC2), scatter plot on the left and loading plot on the right. *Dot represents cultivar from Cluster 1 in the cluster analysis. Cross represents cultivar from Cluster 2 in the cluster analysis. Square represents cultivar from Cluster 3 in the cluster analysis. Triangle represents cultivar from Cluster 4 in the cluster analysis.

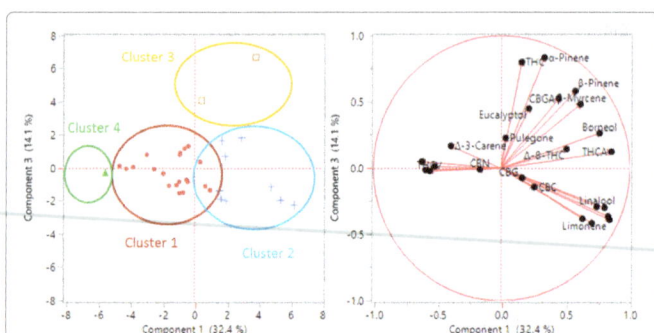

Figure 9: PCA of 32 cannabis cultivars (PC1 and PC3), scatter plot on the left and loading plot on the right. *Dot represents cultivar from Cluster 1 in the cluster analysis. Cross represents cultivar from Cluster 2 in the cluster analysis. Square represents cultivar from Cluster 3 in the cluster analysis. Triangle represents cultivar from Cluster 4 in the cluster analysis.

After grouping these cultivars and visualizing the clusters in fewer dimensions using PCA, the grouping results were compared to the constellation plot from the cluster analysis (Figures 8 and 9). The cultivars circled in each group in the scatter plot were the same cultivars in each cluster as in the constellation plot. Because Cluster 1 and Cluster 2 and Cluster 4 were separated in PC1 and PC2 scatter plot, and Cluster 2 and Cluster 3 were separated in PC1 and PC3 scatter plot, visually the two grouping results match.

Conclusions

Existing approaches for cannabis classification may be inadequate because they analyze cannabis from botanical perspectives or based on only the two primary cannabinoids THC and CBD. In this work, an HPLC method for cannabinoids and a GC-MS method for terpenes were developed and validated. We quantified 10 cannabinoids and 14 terpenes in 32 medical cannabis samples from two licensed producers in Canada. Samples were classified using both cluster analysis and PCA. In cluster analysis, samples were grouped into four clusters, where clusters 1, 2 and 3 are THC dominant, and cluster 4 is CBD dominant. The result was different from cluster analysis using only THC and CBD content, which supports the hypothesis that classification based exclusively on THC and CBD may be insufficient when considering all medically relevant compounds in cannabis. PCA results confirmed the cluster results and also indicated which cannabinoids and terpenes are critical in discriminating cultivars. Currently, a systematic cultivar classification involving all commercially available cultivars in Canada has not been accomplished. However, this is necessary as

these relationships will allow clinicians to identify the right cannabis cultivar with the right components to achieve optimal treatment outcomes. The ultimate goal is to develop a systematic classification and standardization method using chemical and genetic analysis techniques in tandem that can link cultivars with morphological characteristics, chemical composition, and medicinal applications.

Acknowledgements

The authors are highly grateful for full funding and facilities support from Labs-Mart Inc. for conducting sample collection and the experimental process. Labs-Mart Inc. is a third-party testing lab that has no affiliation with any licensed producers.

References

1. Mahmoud El Sohly A (2014) Constituents of Cannabis Sativa. Handbook of Cannabis.

2. http://www.hc-sc.gc.ca/dhp-mps/marihuana/med/infoprof-eng.php

3. Russo EB (2011) Taming THC: potential cannabis synergy and phytocannabinoid-terpenoid entourage effects. British Journal of Pharmacology 163: 1344-1364.

4. Brenneisen R (2007) Chemistry and analysis of phytocannabinoids and other Cannabis constituents. Forensic Science and Medicine: Marijuana and the Cannabinoids. Humana Press Inc., Totowa, New Jersey, USA.

5. http://medicalmarijuana.procon.org/view.resource.php?resourceID=000881

6. http://www.hc-sc.gc.ca/dhp-mps/marihuana/info/list-eng.php

7. http://www.hc-sc.gc.ca/dhp-mps/marihuana/about-apropos-eng.php

8. Anderson LC (1980) Leaf variation among Cannabis species from a controlled garden. Botanical Museum Leaflets 28: 61-69.

9. Small E (1972) Interfertility and chromosomal uniformity in Cannabis. Canadian Journal of Botany 50: 1947-1949.

10. Small E, Cronquist A (1976) A practical and natural taxonomy for Cannabis. Taxon 25: 405-435.

11. Hillig KW (2004) A chemotaxonomic analysis of terpenoid variation in Cannabis. Biochemical systematics and ecology 32: 875-891.

12. Hillig KW (2005) Genetic evidence for speciation in Cannabis (Cannabaceae). Genetic Resources and Crop Evolution 52: 161-180.

13. Hillig KW, Mahlberg PG (2004) A chemotaxonomic analysis of cannabinoid variation in Cannabis (Cannabaceae). American Journal of Botany 91: 966-975.

14. Mandolino G, Carboni A (2004) Potential of marker-assisted selection in hemp genetic improvement. Euphytica 140: 107-120.

15. United Nations Office on Drugs and Crime Vienna (2009) Recommended methods for the identification and analysis of cannabis and cannabis products. United Nations Office on Drugs and Crime: New York, USA.

16. Ross SA, Mehmedic Z, Urphy TP, ElSohly MA (2000) GC-MS analysis of the total δ9-thc content of both drug-and fiber-type cannabis seeds. Journal of Analytical Toxicology 24: 715-717.

17. Turner CE, Elsohly MA, Cheng PC, Lewis G (1979) Constituents of Cannabis sativa L., XIV: Intrinsic problems in classifying Cannabis based on a single cannabinoid analysis. Journal of Natural Products 42: 317-319.

18. Mandolino G, Bagatta M, Carboni A, Ranalli P, de Meijer E (2003) Qualitative and quantitative aspects of the inheritance of chemical phenotype in Cannabis. Journal of Industrial Hemp 8: 51-72.

19. De Meijer E, Hammond K (2005) The inheritance of chemical phenotype in Cannabis sativa L.(II): cannabigerol predominant plants. Euphytica 145: 189-198.

20. De Meijer E, Hammond K, Sutton A (2009) The inheritance of chemical phenotype in Cannabis sativa L. (IV): cannabinoid-free plants. Euphytica 168: 95-112.

21. Zuardi AW, Hallak JEC, Crippa JAS (2012) Interaction between cannabidiol (CBD) and Δ 9-tetrahydrocannabinol (THC): influence of administration interval and dose ratio between the cannabinoids. Psychopharmacology 219: 247-249.

22. Fischedick JT, Hazekamp A, Erkelens T, Choi YH, Verpoorte R (2010) Metabolic fingerprinting of Cannabis sativa L., cannabinoids and terpenoids

for chemotaxonomic and drug standardization purposes. Phytochemistry 71: 2058-2073.

23. Hazekamp A, Fischedickm JT (2012) Cannabis - from cultivar to chemovar, Drug Test. Analysis, Wiley Online Library.

24. American Herbal Pharmacopoeia (2014) Cannabis Inflorescence, Cannabis spp., Standards of identity, analysis and quality control.

25. Brenneisen R (2007) Chemistry and analysis of phytocannabinoids and other Cannabis constituents. From: Forensic Science and Medicine: Marijuana and the Cannabinoids. Humana Press Inc., Totowa, New Jersey, USA.

26. John M, McPartland M, Ethan Russo B (2001) Cannabis and Cannabis Extracts: Greater Than the Sum of Their Parts? Journal of cannabis Therapeutics 1: 103-132.

27. https://www.ich.org/fileadmin/Public_Web_Site/ICH_Products/Guidelines/Quality/Q2_R1/Step4/Q2_R1__Guideline.pdf

28. Milligan GW (1980) An Examination of the Effect of Six Types of Error Perturbation on Fifteen Clustering Algorithms. Psychometrika 45: 325-342.

29. Kimberly Colson L, Jimmy Y, Christian F (2015) Nuclear Magnetic Resonance - A Revolutionary Tool for Nutraceutical Analysis, Botanicals, Methods and Techniques for Quality and Authenticity. CRC Press.

17

Photocatalytic Studies of Tio$_2$/Sio$_2$ Nanocomposite Xerogels

Muhammad Yaseen[1], Zeban Shah[2]*, Renato C.Veses[2], Silvio L. P. Dias[2], Éder C. Lima[2], Glaydson S. dos Reis[2], Julio C.P. Vaghetti[2], Wagner S.D.Alencar[2] and Khalid Mehmood[1]
[1]*Department of Chemistry, Hazara University Mansehra Dhudial 21130, K.P.K Pakistan*
[2]*Federal University of Rio Grande do Sul, Av. Bento Gonçalves, Porto Alegre, RS, Brazil*

Abstract

The use of titania-silica materials in photocatalytic processes has been proposed as an alternative to the conventional TiO2 catalysts, in order to facilitate the separation of products after the reaction. However, despite the large number of research in this field, the mechanism governing the photocatalytic activity of the mixed TiO2/SiO2 oxides is not clear. Titania-Silica nanocomposite xerogels were prepared by sol-gel method. This work has been used to describe the synthesis and the photocatalytic properties of TiO2-SiO2 nanocomposite xerogel. The nanocomposite xerogels were prepared by keeping the molar ratio of TEOS:TTIP:MtOH:DIW at 1: 1:6:14 respectively and the catalysts used were HCl and NH4OH. After the preparation xerogels were characterized by FTIR, XRD, UV and LLS. All these techniques show the amorphous nature of Titania-silica xerogel.

Keywords: Photocatalysis; TiO$_2$ SiO$_2$ mixed oxide; LLS; FTIR; UV; XRD

Introduction

TiO$_2$ is well recognized as a valuable material with application as a white pigment in paints, as filler in paper, textile and in rubber/plastics [1]. Due to low cost, non-toxicity, stability and other best characteristics TiO$_2$ attracts a great attention. TiO$_2$ has wide applications in various fields like antireflection optics, coatings, waste water purifications, catalyst supporting, ceramics senser element, as a photocatalyst, in electric devices like (in lithium based battery), as a base in high quality paints, paper, plastics [2]. Titania has excellent biocompatibility with respect to bones implants and applications in electrochromic devices [3]. Titania shows good photocatalytic applications due to which it gained tremendous demands and green energy and environmental protection. Many other oxides like iron oxides, zinc oxides etc. also shows the similar behavior due to photocatalytic activity of titanium dioxide it play a wide role in different fields like air, waste water purification, good UV blocking properties weakening of the organic fibers [4]. Silica doped in to the titania matrix increase the photocatalytic activity because the silica doping decrease particle size and also increase the specific surface area and thermal stability of titania particle towards anatase to rutile phase conversion [5]. SiO$_2$-TiO$_2$ materials are used in different fields like as catalyst supporting materials, acidic catalyst for many reactions, selective reduction, as an anti-reflective materials for coatings or sensing nanoimprints photonic crystals [6-9]. Dielectric mirrors and low loss waveguides solids of low thermal expansion coefficient, bioactive solids self-cleaning coatings solids of controlled acidity and photocatalysts [10-14]. The TiO$_2$-SiO$_2$ mixed oxides catalytic activity was studied and observed that the TiO$_2$-SiO$_2$ have better photocatalytic activity as compared to TiO$_2$ and SiO$_2$ which was confirmed through LLS and UV results.

Experimental

Sample preparation

The xerogels can be synthesized by Sol-Gel process in which metal alkoxide is used as a precursor source that undergoes catalyzed hydrolysis and condensation to get nano scale materials of that metal [15]. TTIP was used as a precursor and TEOS was added as an organic solvent. In this synthesis HCl and NaOH were used as catalysts.

TiO$_2$ SiO$_2$ mixed oxide

TiO$_2$-SiO$_2$ xerogels was synthesized by Sol-Gel process. TEOS (Tetraethylorthosilicate) and TTIP (Titanium tetra isopropoxide) were used as precursor. It was observed that the TTIP hydrolysis rate is much faster than TEOS [16]. The synthesis consists of two steps. The first step is the synthesis of SiO$_2$ sol, in which 1:6 TEOS and Methanol were mixed together i.e. 7 ml of TEOS and 43 ml of methanol. TEOS were added drop wise slowly to the methanol with continuous stirring. Then 0.05 mol L^{-1} HCl was added drop wise to the sol in order to adjust the acidic pH at 2. Then the sol was allowed to stirrer for about 2 an hour in order to get the homogeneous sol. In the second step TiO$_2$ Sol synthesis, 1:14 solution of TTIP and DIW were mixed with continuous stirring until the homogeneous sol of TiO$_2$ were obtained. Then in 1:1 of TiO$_2$ and SiO$_2$ sols were mixed with continuous stirring in order to get homogeneous sol and then 0.05 mol L^{-1} NH$_4$OH solution was added to this mixture drop wise for the adjustment of pH. The pH of the homogeneous mixtures (sol) was observed by the pH-meter continuously. At last the mixtures (sol) were allowed to stirrer for some time to get the homogeneous mixture (sol).

Gel and xerogel preparation

In this method 3ml of TEOS, the silica precursor was taken in the reaction beaker and 6ml of ethanol was added dropwise with continue stirring. Further 3 ml of water, 4 ml of acetic acid and 3 ml of TTIP (Titania precursor) were added with continuous stirring for 15 min at room temperature. The prepared sol changed into a gel which was placed in the oven at 65 for 1 h which resulted in the conversion of gel in to xerogel.

---

*Corresponding author: Dr. Zeban Shah, Federal University of Rio Grande do Sul, Av. Bento Gonçalves, Porto Alegre, RS, Brazil
E-mail: zeban.shah@ufrgs.br or zs_zaib77@yahoo.com

Characterization

Structure, quality, photocatalytic studies and morphological characteristics of Titania-silica gel were studied by different characterization techniques like, LLS, FTIR, UV and XRD etc.

Results and Discussions

FTIR result

The FTIR spectra of the synthesized nanocomposite (TiO_2/SiO_2) were recorded by Perkin Elmer series 100 FTIR spectrometer with a 5 cm^{-1} resolution. FTIR spectrum was recorded at 4000-450 cm^{-1}. The absorption at 1074 cm^{-1} (Figure 1) is the characteristics peak for Si-O-Si. The peak observed at 801 cm^{-1} is due to Si-O-Si symmetric stretching. The broad absorption at 1633 cm^{-1} match with OH bending vibrations and is attributed to chemisorbed water. The well defined peak at 3441 cm^{-1} shows OH stretching vibrations. The peak observed at 923 cm^{-1} corresponded Si-O-Ti vibrations. The band observed at 450-610 cm^{-1} is due to Ti-O Stretching.

Many characteristic FTIR peaks were observed in Figure 2. The bands observed at 3391 cm^{-1} and 1557cm^{-1} were due to the OH bending and stretching vibrations respectively. The peak observed at 1165 cm^{-1} correspond to Si-O-Si antisymmetric stretching vibration. The peak obtained at 923 cm^{-1} correspond to Si-O-Ti vibration. The peak observed at 801 cm^{-1} is due to the Si-O-Si symmetric stretching vibration. The absorption at 1410 cm^{-1} is due to C-H interaction of Si-R structure unit (Figure 3).

XRD measurement

The XRD pattern of synthesized nanocomposites (TiO_2/SiO_2) was collected in the range of 10-60 2θ (degree) shown in Figure 4. The XRD patterns of the synthesized materials indicate that the TiO_2-SiO_2 nanocomposites are essentially non-crystalline and have amorphous structure which can be accomplished from the broad characteristic diffraction peak between 2θ ~ 20θ° and 30 θ°.

The Figures 5 and 6 also show the same XRD results of the synthesized TiO_2/SiO_2 nanocomposites. These samples were also synthesized by sol-gel method with different concentrations of precursors and solvents used. The hump in XRD pattern indicate that the TiO_2-SiO_2 nanocomposites are essentially non-crystalline and have amorphous structure which can be accomplished from the broad characteristic diffraction peak between 2θ ~ 20θ° and 30 θ° while the Figure 2 shows very low crystallinity at 2θ and 25 θ°.

Catalytic activity

To check the catalytic activity of TiO_2-SiO_2 the p-nitrophenol was reduced by $NaBH_4$ to p-aminophenol in aqueous medium. The procedure adopted was used as a given amount of mixed oxides was added to 1 ml of p-nitrophenol (0.08 mmol L^{-1}) for initiation of reduction. For mixture preparation 1 ml of aqueous solution of $NaBH_4$ (1.5 mmol L^{-1}) was added to reaction chamber. The p-nitrophenol reduction was determined by studying the absorbance at 440 nm with respect to time. This reaction was selected due to simplicity and formation of single product. By using time dependence uv-vis spectra the reduction process was checked. The appearance of a new peak at 300 nm was observed which confirm the p-nitrophenol reduction to p-aminophenol which is shown in Figure 6.

LLS result (hydrodynamic radii)

The LLS results (hydrodynamic radii) of the three samples are given

in Table 1 below. All the three samples have different hydrodynamic radii. The two samples TSX_2 and TSX_3 have same precursors and solvent but used at different ratio. The sample TSX_{17} have same precursors but different solvents and precursor ratio were used. The LLS results (hydrodynamic radii) are different because the LLS result

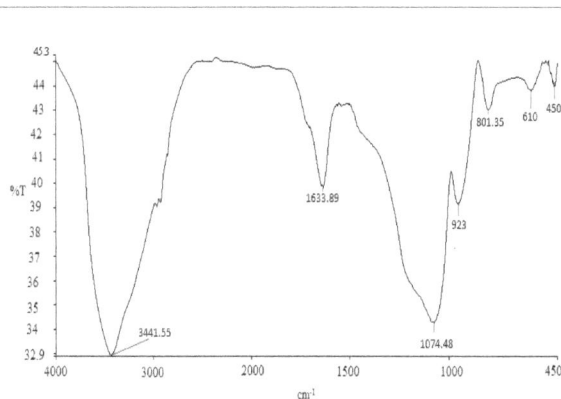

Figure 1: FTIR spectrum of synthesized TiO_2/SiO_2 nanocomposite for procedure 1.

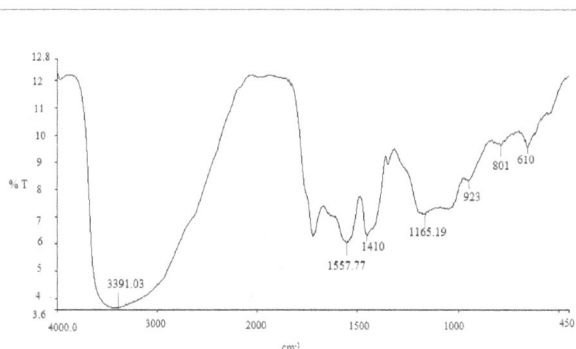

Figure 2: FTIR spectrum of synthesized TiO_2/SiO_2 nanocomposite for procedure 2.

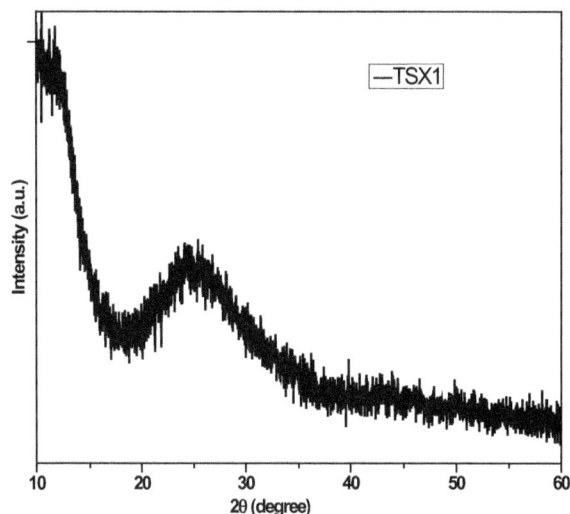

Figure 3: XRD results of TiO_2-SiO_2 nanocomposite xerogels of the sample TX_1.

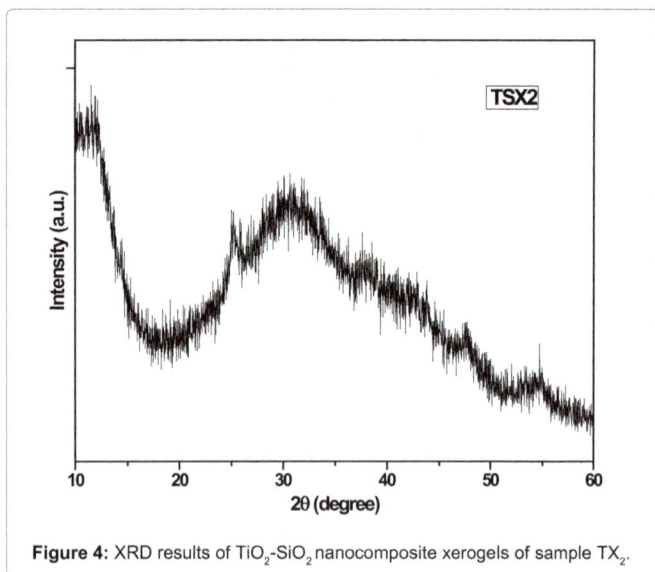

Figure 4: XRD results of TiO$_2$-SiO$_2$ nanocomposite xerogels of sample TX$_2$.

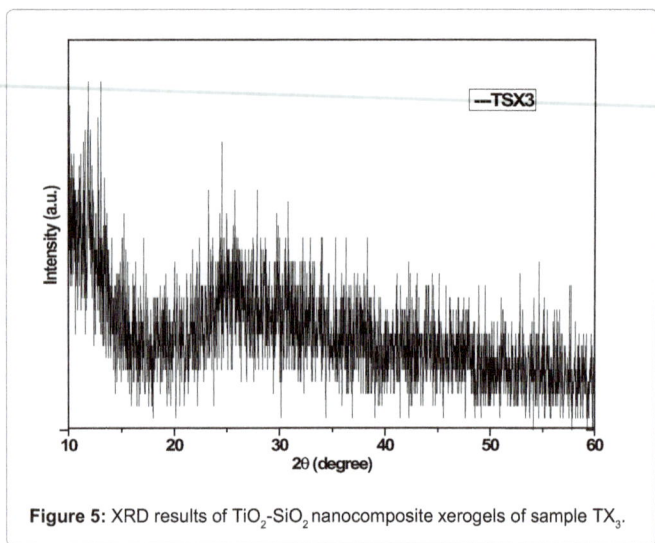

Figure 5: XRD results of TiO$_2$-SiO$_2$ nanocomposite xerogels of sample TX$_3$.

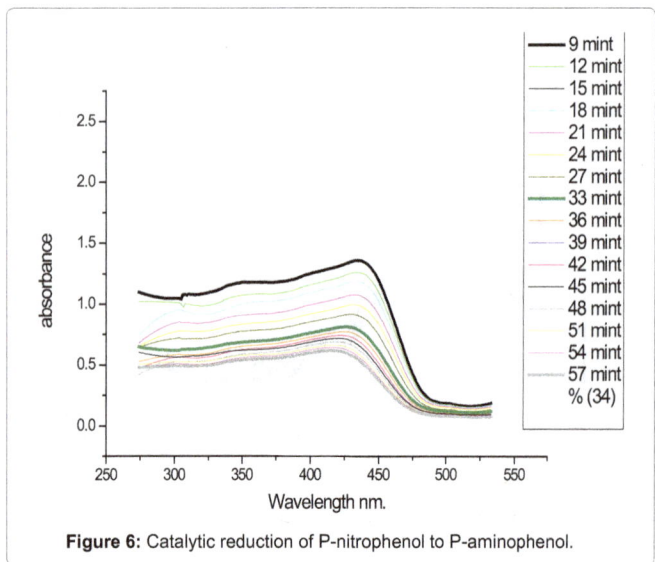

Figure 6: Catalytic reduction of P-nitrophenol to P-aminophenol.

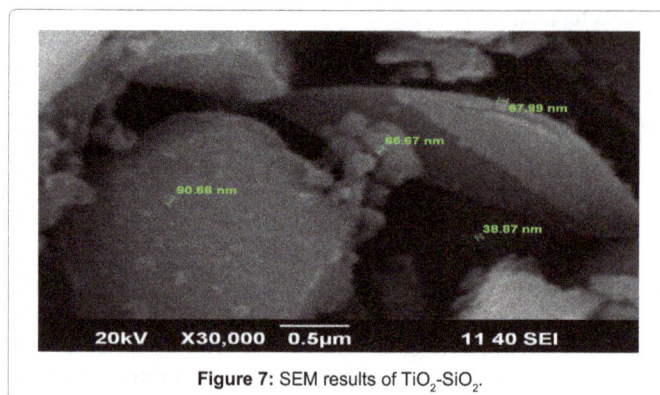

Figure 7: SEM results of TiO$_2$-SiO$_2$.

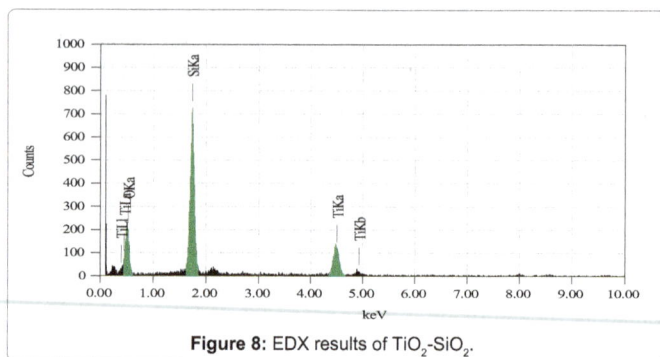

Figure 8: EDX results of TiO$_2$-SiO$_2$.

(hydrodynamic radius) of the sample TSX$_3$ were measured during the conversion of sol in to gel and the LLS results (hydrodynamic radii) of the TSX$_2$ and TSX$_{17}$ were calculated after the gel was formed.

SEM results

The Figure 7 shows the SEM image of synthesized containing TiO$_2$-SiO$_2$. From SEM image, it is clear that in most part of the sample large particles is obtained, which gives the evidence of the agglorimation of small particles. Along large agglomarited particle small size individual particles can also be absorved.

EDX results

The EDX spectrum is showing the elemental composition of the three elements in different percentage which is silicon (Si) and titanium (Ti) and oxygen (Figure 8). The EDX spectra has confirmed all the constituents that were added in the synthesized sample.

Conclusion

(TiO$_2$/SiO$_2$) nanocomposites were synthesized by sol-gel process where Tetraethylorthosilicate (TEOS) and Titanium tetraisopropoxide (TTIP) were used as precursors in the presence of organic solvent. The XRD results confirmed that the synthesized TiO$_2$-SiO$_2$ mixed oxide xerogel is non-crystalline and having amorphous nature. In the TiO$_2$/SiO$_2$ nanocomposites, materials, in the amorphous SiO$_2$ matrix TiO$_2$ nanocrystals are present in highly dispersed form. The amorphous SiO$_2$ and Ti–O–Si bond formation and TiO$_2$-SiO$_2$ mixed oxides give rise effectively stability of TiO$_2$ anatase form. It also bound crystallites growth, significantly increase surface area. So, the increase in surface area caused the improvement in photocatalytic activity of TiO$_2$-SiO$_2$ mixed oxides nanocomposi[te xerogels. FTIR spectra show the presence of Ti-O-Si crosslinks and revealed interactions between TiO$_2$ and SiO$_2$ at a molecular scale. Ti-O-Si bonds and interactions may

S No	Sample	Hydrodynamic radius (nm)
1	TSX$_{17}$	170.3
2	TSX$_2$	138.49
3	TSX$_3$	52.6

Table 1: LLS Results (hydrodynamic radii) of the three samples prepared by sol gel process.

enhance surface properties, catalytic and photoactivity. The TiO$_2$-SiO$_2$ mixed oxides catalytic activity was also studied and observed that the TiO$_2$-SiO$_2$ have better photocatalytic activity as compared to TiO$_2$ and SiO$_2$ which was confirmed through LLS and UV results.

Acknowledgements

The authors acknowledge CNPq for financial support.

References

1. Jalava JP (2006) Size and Shape Dependence of the Electronic and Spectral Properties in TiO2 Nanoparticles. Part Syst Charact 23: 159-164.

2. Hashimoto K, Irie H, Fujishima A, Jpn J (2005) TiO2 photocatalysis: a historical overview and future prospects. Appl Phys 44: 8269.

3. Guo YG, Hu YS, Maier J (2006) Synthesis of hierarchically mesoporous anatase spheres and their application in lithium batteries. Chem commun 26: 2783-2785.

4. Tanaka K, Capule MF, Hisanaga T (1991) Effect of crystallinity of TiO2 on its photocatalytic action. Chem Phys Lett 187: 73-76.

5. Meng X, Qian Z, Wang H, Gao X, Zhang S, et al. (2008) Sol–gel immobilization of SiO2/TiO2 on hydrophobic clay and its removal of methyl orange from water. J Sol-Gel Sci Tech 46: 195-200.

6. Wang X, Shen J (2010) Sol–gel derived durable antireflective coating for solar glass. J sol-gel sci tech 53: 322-327.

7. Courtney L, Sermon PA, Towers J, Halepoto D, de Namor AFD, et al. (2004) Sol-gel chemistry for Ba 2+ sensors to allow oil production engineered nanometrically. J sol-gel sci tech 32: 229-236.

8. Li M, Tan H, Chen L, Wang J, Chou SY (2003) Large area direct nanoimprinting of SiO 2–TiO 2 gel gratings for optical applications. J Vacuum Sci Tech B: Microelectronics and Nanometer Structures Processing, Measurement, and Phenomena 21: 660-663.

9. Kanehira S, Kirihara S, Miyamoto Y (2005) Fabrication of TiO2–SiO2 photonic crystals with diamond structure. J Am Ceramic Soc 88: 1461-1464.

10. Lin W, Wang GP, Zhang S (2005) Design and fabrication of omnidirectional reflectors in the visible range. J Modern Optics 52: 1155-1160.

11. Sermon PA, Leadley JG, MacGibbon RM, Ruzimuradov O (2012) Tuning X/(TiO2) x–(SiO2) 100–x (0<x<40) xerogel photocatalysts. Ionics 18: 455-459.

12. Miyata N, Kamitakahara M, Kawashita M, Kokubo T, Nakamura T (2003) Mechanical properties of bioactive PDMS-CaO-SiO2-TiO2 and PTMO-CaO-TiO2 hybrids soaked in a simulated body fluid. In Key Engineering Materials, Trans Tech Publications 240: 943-946.

13. Miller JB, Ko EI (1997) Control of mixed oxide textural and acidic properties by the sol-gel method. Catalysis today 35: 269-292.

14. Yoon KH, Noh JS, Kwon CH, Muhammed M (2006) Photocatalytic behavior of TiO 2 thin films prepared by sol–gel process. Mater Chem Phy 95: 79-83.

15. Privman V, Goia DV, Park J, Matijević E (1999) Mechanism of formation of monodispersed colloids by aggregation of nanosize precursors. J Colloid Interface Sci 213: 36-45.

16. Cheng P, Zheng M, Jin Y, Huang Q, Gu M (2003) Preparation and characterization of silica-doped titania photocatalyst through sol–gel method. Mater Lett 57: 2989-2994.

18

Unusual Post-Spray Proton Transfer to Protein Using Acetone Spray in Desorption Electrospray Ionization

Anna Warnet[1]*, Nicolas Auzeil[2] and Jean-Claude Tabet[1]

[1]CSOB-Institut Parisien de Chimie Moléculaire (UMR 8232-UFR 926), CNRS, Université Pierre et Marie Curie, Paris, France

[2]EA4463: Laboratoire de chimie et toxicologie analytique et cellulaire, Université Paris Descartes, Faculté des Sciences Pharmaceutiques 4 avenue de l'Observatoire, Paris, France

Abstract

Although acetone, in DESI ionisation generally leads to protein aggregation, in this study we report unexpected multi-proton transfers to lysozyme using this aprotic solvent as a charged spray. The DESI/acetone mass spectrum of lysozyme displays (i) a significant increase in the average charge state (Z_{av}) and (ii) an incomplete H^+/Ca^{2+} exchange, even though the overall contribution of cationised species is high, relative to those from spraying with a methanol/water solvent. This behavior is contrary to that expected from gas phase basicity, because $GB_{acetone} > GB_{methanol}$. Decreasing the amount of sample deposited on the target (from 50 to 0.050 pmole) leads to a charge state increase, as seen in ESI, but not in the extent of cationisation. Moreover, the DESI signal duration is extended with sprayed acetone even though the total ionic current is significantly lowered. With a d6-acetone spray, no incorporation of a deuteron occurs, and the ionization yield is strongly decreased for multi-protonated lyso[i+] lysozyme. This is in contrast to that observed with a d4-methanol spray, which displays a distribution of 48 deuterons in the lyso[9+] ion as shown in high resolution with a LTQ/Orbitrap instrument. This unexpected behavior of the $(CD_3)_2CO$ spray suggests that protons do not originate from acetone. Furthermore, dry argon post-flow on the target surface results in the lysozyme signal suppression, whereas with a humid argon flow, the signal is regenerated. On the other hand, an argon stream bubbling in heavy water, yields incorporation of several deuterons. The interpretation of this behavior is explained by considering the acetone radical ions at the surface of the primary droplets (and/or offspring droplets and/or at the wet sample surface), being able to react with ambient moisture (or with traces of water adsorbed at liquid phase). Under these conditions, enough protons are produced to generate multi-charged solvated lysozyme aggregates which then become desolvated in the reduced pressure in the skimmer area.

Keywords: MS; DESI; Aprotic solvent; Lysozyme; Cationization; Labelling

Introduction

Desorption electrospray ionization mass spectrometry (DESI-MS), an ambient ionization method first reported by Takats et al. in 2004 [1], permits fast analysis of bulk samples without pre-treatment. Many applications appeared rapidly [1-6] and their number still increases [7]. Since, over forty ambient ionization methods have now been developed, based on sprays with [1] or without high voltage [8], acoustic mode [9-12], chemical ionization [10], laser [13], plasmas [14], heating [15] and combined techniques [6]. Several of these ambient ionization modes can lead to imaging analysis [16]. Based on the extractive approaches [7], they permit monitoring reaction [17] and intermediate studies in fast reactive processes [18-20]. Furthermore, it is possible to combine them with an electrochemistry device [21,22]. Among the various ambient ionization/desorption modes, DESI-MS [1] evolved as the most popular mode and is suitable for analysis of a large panel of organic compounds. It has been reported in a wide area of applied analyses including forensic [3-6,8,23,24], homeland security [4,5,25-28], food contaminants [29] and agrochemicals [30,31], metabolites [4,32-34], drugs, and drugs of abuse [34-39]. DESI mode also contributes to biomolecule detection [40-46], as well as direct analysis of heterogeneous biological material [3,4,47-50] and in imaging [40,48,51]. For larger analytes, DESI produces efficiently multicharged proteins [44,52-56] as well as non-covalent complexes [54,57-62].

The proposed DESI mechanism based on the sample extraction from a thin target layer and transmission solvated ionic species to the transfer capillary, known as "droplet pick-up" [5,7,63], may be regarded as a process of four distinct steps occurring from the micro-electrospray source assisted by auxiliary gas flow and yielding charged solvent micro-sized droplets (d ≤ 10 µm). As described in literature [1-5], these steps involve: (i) the impacting of the charged droplets to the target surface with velocities of the order of 100 m/s at an angle ranging from 20 to 90 degrees, (ii) the wetting and subsequent partial dissolution of the sample deposited on the target, (iii) the *momentum* transfer from the multiple impacting droplets, originating from a shock wave-like phenomenon [4], which leads to production of analyte offspring droplets (characterized by a distribution in sizes and in velocities) from target surface layer to move toward the transfer capillary. Those droplets are significantly smaller than those generated in ESI before the droplet Coulomb explosions [64] (except for a large for which mainly the "charge residue model" mechanism must be considered [65]), (iv) Finally, this is followed by the release of solvated analyte ions, similar in charge but probably smaller than those generated in the ESI process [32]. The DESI mechanism briefly presented above specially emphasizes the importance of: (i) properties of the surface used for the sample deposition, and (ii) properties of the selected electrospray solvent and additives to enhance ionization efficiency [66,67]. The surface must have a weak affinity for analyte [68] to facilitate its extraction/dissolution in the formed thin solvent layer. In contrast, generally a high affinity between the surface and the sprayed solvent increases the signal intensity and its stability. Moreover, due to the ESI electrical field, the insulated surface undergoes electrostatic charge closely related to its conductivity, and is responsible for the microdroplet ejection from the thin solvent layer [68-70].

*Corresponding author: Anna Warnet, CSOB-Institut Parisien de Chimie Moléculaire (UMR 8232-UFR 926), CNRS, Université Pierre et Marie Curie (UPMC), 75005 Paris, France, E-mail: anna.warnet@upmc.fr

Depending on the analyte, one can use various surfaces for sample deposition. The most popular materials in analysis of protein, carbohydrates, and synthetic organic compounds are: polymethyl-methacrylate (PMMA) [42,43,71], polytetra-fluoro-ethylene (PTFE) [1,42,71-73], glass [42,71,74], and paper [71,74] (different to the Paper Spray mode, another ambient ionization mode) [75], with a limit of detection (LOD) [26,42-44,68] from 0.1 to 2000 pg.mm^{-2}. More recently, nanoporous silicon and ultra-thin layer chromatography UTLC led to improve LOD values compared with PMMA and PTFE surfaces [68]. Proteomic analysis uses commonly nanoporous alumina surface [76].

The sprayed solvent is also a crucial parameter in the DESI process [7,59]. In fact, polarity, boiling point and viscosity are the major macroscopic properties that affect DESI performances [66,67]. When it is required, other properties as volatility and capacity to dissolve the sample, guide towards particular solvents. Stability and thickness of the solvent layer, correlated with intensity and stability of the signal depend on the analyte interactions with the surface and the evaporation rate. Moreover, the solvent polarity influences dissolution efficiency in solvent layer, and hence analyte concentration in the offspring droplet evaporation [26,69,77]. It is known that protic solvents favor the stabilization of the analyte charge during the offspring droplet evaporation, although using non-aqueous solvents is possible if, at first, the analyte is dissolved. Indeed that method generates ions containing weaker internal energies [66,67]. In practice, the most commonly employed solvent in DESI-MS consists of a methanol-water mixture [1]. However, depending on the analyte, an aprotic solvent such as acetonitrile may advantageously replace methanol [30,52,66,67,77].

Furthermore, addition of a small amount (0.1% v/v) of a protonating agent, e.g., formic, acetic, or trifluoroacetic acids, facilitates analyte protonation [72,76]. In some cases, pure solvents were proposed, for instance in the targeting of plant alkaloids separated on a thin-layer chromatography plate. Van Berkel et al. reported higher ionization efficiency with acetonitrile than with methanol [78,79]. Recently, Badu-Tawiah et al. demonstrated that the acetonitrile/chloroform (1/1) or tetrahydrofuran/chloroform (1/1) mixtures exhibit an improved efficiency compared to methanol/ water (1/1) for hydrophobic analytes detected under DESI-MS conditions [66,67]. Note that the addition of particular reagents (e.g., m-nitrobenzyl alcohol (m-NBA) or sulfolane) permits the supercharging of protein analytes [79].

In the present work, we chose oxidized lysozyme as model in order to explore the potentiality of unusual aprotic solvent for desorption of small proteins within an enough efficiency to detect their characteristic ionized species. Its medium size and its multiple disulfide bridges prevent protein denaturation but the large amount of conformational freedom for the protein motivates this choice. Moreover, lysozyme includes 19 basic residues, which promote protonation in DESI. In this work, among the less used aprotic solvents, we selected anhydrous acetone, although it is considered as an inefficient solvent for protein solubilization. We opted for a PTFE target surface because of its hydrophobic character, which manifests strong interactions with acetone and results in stabilization of the thin solvent layer. Therefore, the recorded mass spectra of lysozyme were carefully examined, especially in terms of charge state distribution (CSD) [80] and average charge states Z_{av} [81-83] of cationized and multi-protonated lysozyme species. Systematically, we compared them to those acquired with commonly sprayed protic solvents, e.g., methanol/water. In addition, in order to contribute to understand the details of the ionization mechanism i.e., the origin of ionizing protons when an aprotic solvent

such as acetone is used, we modified the DESI ambient conditions by using post-flow gas. We performed DESI experiments by using sprayed anhydrous acetone (labelled or not) either in conventional ambient conditions or combined to an additional post-flow gas: (i) dry argon, and (ii) humidified argon/water (labeled or not).

Experimental

Chemicals and reagents

HPLC grade methanol, d4-methanol (CD$_3$OD) and d6-acetone [(CD$_3$)$_2$CO] were acquired from Sigma (St Quentin Fallavier, France). The d6-acetone was used extemporaneously. HPLC grade acetone and formic acid, were supplied from WWR International (Fontenay-sous-Bois, France), and argon, nitrogen were purchased from Air-Liquide (Nanterre, France). Anhydrous acetone was prepared in accordance to the protocol of Yves Baratoux prior DESI/MS analysis [84]. Water was purified to 18.2 MΩ.cm with a milli-Q water system, from Millipore (El Paso, TX, USA). Hen Egg White lysozyme, supplied by Sigma (Saint-Quentin Fallavier, France), was purified according to the procedure of Thomas et al. [84] and was partially desalted to its isoionic state [85]. Finally, the pH of the isoionic protein solution was adjusted to pH 4.5 with acetic acid, and Lysozyme concentration (approximately 100 µM) was determined using UV-spectrophotometry. This solution was then diluted to appropriate concentrations (0.001 to 100 µM) with pure water, before DESI-MS analysis. Polytetra-fluoro-ethylene surfaces (PTFE plates 1/16 inch, 2.95 inch × 0.98 inch) were purchased from Isoflon SAS (Diemoz, France).

DESI mass spectrometry

DESI-MS experiments were performed using a LTQ Orbitrap™ from Thermo Fisher Scientific (Courtaboeuf, France). The analyser was operated in the FTMS mode (high resolving power fixed at 10^5). Two micro scans were used to record one scan and the maximum injection time was 0.200 s. DESI mass spectrum acquisition was done from m/z 200 to m/z 4000 (i.e., high m/z ratio range selection mode). Xcalibur™ software (Thermo Fisher Scientific) was used for data acquisition and analysis. DESI experiments, in positive ion polarity mode, were carried out using an Omni Spray™ Ion source from Prosolia, Inc. (Indianapolis, IN, USA) equipped with a manual X-Y-Z positioner. A double charge-coupled device (CCD) camera was used for positioning and retaining in place as accurately as possible the deposited sample. The following optimized values for experimental parameters were: spray voltage, 3.8 kV; capillary temperature, 300°C; capillary voltage, 49 V; and tube lens, 250 V. For DESI source, preferred parameters hereafter were: solvent flow rate, 5 µL.min^{-1}; spray angle, 37°; distance from sprayer to PTFE surface, approximately 0.5 mm; distance from sample deposit to mass spectrometer inlet, 1-2 mm; dry nitrogen gas pressure, 72 psi.

In a typical DESI-MS experiment, seven aliquots of aqueous Lysozyme solution (0.5 µL-10 µM) deposited on PTFE plate, were left for approximately 10 min, at room temperature until total dry. Successively, each deposit (approximately 5 pmoles) was analyzed by a manual sweep. Before use, the PTFE plate was sonicated for 5 min in 0.1% aqueous formic acid, rinsed with water and dried under a nitrogen flow. DESI mass spectra are the scan average obtained from the spots.

Notation and thermochemistry

The large ionic species (produced at the end of droplet lifetime), with a large and excess of charge number on the studied protein, stabilized by a lot of solvent molecules, are herein called charged aggregates. However, it does not arise with protein aggregates which are very minor species. The charged aggregates are macromolecular non covalent systems which are desolvated under reduced pressure in the skimmer area [86].

To simplify the ion notation, the multi-protonated lysozyme species with n protons, $[Lyso + nH]^{n+}$ (or $LysoH_n^{n+}$), were denoted as $Lyso^{n+}$ [87]. The multi-protonated/cationized forms, e.g., $[Lyso + (n-kN)H + kC]^{n+}$ (with C=cation, N=cation valence number, k=number of cations), were denoted as $LysoC_k^{n+}$. The main adduct ion corresponding to an m/z shift relative to that of $Lyso^{n+}$ by 38 k/n m/z (with k=1) is observed. This means that cationization occurs mainly with one (or more) K^+ and/or Ca^{2+} cations. It results the formation either $[Lyso + (n-p)H + pK]^{n+}$ and/or $[Lyso + (n-2q)H + qCa]^{n+}$ (p and q as number of alkali and alkaline earth cations, respectively) noted as $LysoK_p^{n+}$ and $LysoCa_q^{n+}$. The average charge state [81-83] corresponding to the following ratio was noted as Z_{av}, with I_{i+}, related to the charge i, as the sum of the peak intensities $I_{(h)i+}$ and $\Sigma I_{(h/c)i+}$ (h and c, in subscript letters, characterize peak intensities corresponding to ions constituted by protons and metallic cation(s)) of the $Lyso^{i+}$ and $\Sigma LysoC_k^{i+}$ ions, respectively :

$$Z_{av} = \Sigma\left(i.I_{i+}\right) \; / \; \Sigma I_{i+}$$

These values are related in particular to the protein conformations (seen supplementary S2 material with included references [88-93] for Lysozyme). Thus, the Z_{av} expression is composed by the sum of two terms ($Z_{(h)av} + Z_{(h/c)av}$), which were explored mainly when sprayed acetone was used:

$$Z_{(h)av} = \Sigma\left(i.I_{(h)i+}\right) \; / \; \Sigma I_{i+} \, and \, Z_{(h/c)av} = \Sigma\left(i.I_{(h/c)i+}\right) \; / \; \Sigma I_{i+}$$

To compare on one hand, the stability of the multi-charged species related to its environment, and on the other hand, the possible proton exchanges with solvent, the apparent gas phase [94,95], $GB_{app}(Lyso^{n+})$, was used (Equation 1):

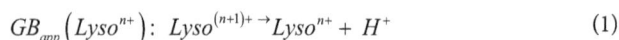

$$GB_{app}\left(Lyso^{n+}\right): \; Lyso^{(n+1)+} \rightarrow Lyso^{n+} + H^+ \qquad (1)$$

This definition is based on the gas phase basicity GB(M) definition, a thermochemical state $\Delta G°$ basicity(M) function, the Gibbs energy change related to the $[MH^+ \rightarrow M + H^+]$ fictive proton desolvation reaction. Its $\Delta H°$ term is called proton affinity PA(M) [95,96] and for multiplied charged Lysozyne, this term is $PA_{app}(Lyso^{(n-1)+})$ [91,92] (with some of their GB_{app} values from literature are provided) [83,97-102].

By analogy to $PA_{app}(Lyso^{n+})$, apparent alkaline earth cation affinity (e.g., for Ca^{2+}, cation chosen for our data, seen supplementary material S3 with included references [104-112]) of multi-protonated $Lyso^{(n-1)+}$ species is noted as $CaCA_{app}(Lyso^{(n-1)+})$ and $CaCB_{app}(Lyso^{(n-1)+})$ for the Gibbs energy change of Equation 2 [103], whereas the gas phase basicity of cationized multi-protonated $LysoCa^{n+}$ Lysozyme noted as $GB_{app}(LysoCa^{n+})$ (Equation 3), are considered to be:

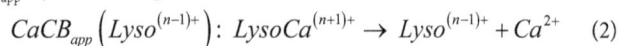

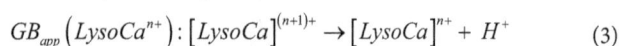

$$CaCB_{app}\left(Lyso^{(n-1)+}\right): \; LysoCa^{(n+1)+} \rightarrow Lyso^{(n-1)+} + Ca^{2+} \quad (2)$$

$$GB_{app}\left(LysoCa^{n+}\right): [LysoCa]^{(n+1)+} \rightarrow [LysoCa]^{n+} + H^+ \quad (3)$$

The GB_{app} state function is introduced from the CB cation affinity particularly known for amino-acid neutrals (AA) [104-108].

Results and Discussion

As previously stated, the solvent is one of important factors [7,66,67] among several experimental parameters concerned about CSD (seen experimental part and supplementary material S2) of ions in gas phase, with a maximum charge state of solvent-free proteins, resulting from

desolvation of multi-charged aggregates in ESI mode. Indeed, it implies at the same time, the droplet charge evolution by macroscopic effects, and intra-aggregate protons/alkali (or alkaline earth) cations exchange reactions by macromolecular effects. The macroscopic effects influence the surface tension, and thus, the droplet size/shape, their analyte concentration, and the formation of large charged aggregates either by "ion evaporation" [113-115] desorbed or produced from "charged residue" process [65]. The macromolecular effects and gas phase thermochemistry, act on the CSD, as on the maximum of the charge state of the multiply-charged solvent-free proteins. The situation somewhat differs in DESI mode, because the charged offspring droplets are significantly smaller and distorted in a shell-shape due to their high velocity [2,4,63,65,67]. This results in higher speed favoring their fast fission through the produced dragging force.

Despite these minor differences, Myung et al. [43] demonstrated similar charge distributions in both the ESI and DESI modes in ion mobility experiments. This may involve a compensation of the macroscopic and macromolecular effects, which relative importance varies with the mode of desorption, resulting in similar ESI (2.1 μM concentration of the used lysozyme solution) and DESI mass spectra (recorded from 5 pmoles of deposited lysozyme). Although, the discussion of the macroscopic effect influence on the dynamic of plume should deserve a fundamental interest, this aspect will not be discussed in this study any more. Actually, the processes involved in the aggregate ion formation from the offspring droplets have not been particularly scrutinized in contrast to the role of the agent providing available protons for charging the droplets (as with the charged aggregates).

In order to investigate the solvent role in providing charges, we studied the influence of the lysozyme amount in the offspring droplets on CSD and Z_{av} of $Lyso^{n+}$ (and $LysoC_k^{n+}$), using first sprayed protic solvent (methanol/water mixture). Besides, distribution of cations (e.g., Na^+, and/or K^+ ...) was notably scrutinized, according to lysozyme CSD, just like the deposited lysozyme amount. Secondarily, the effects, while changing sprayed protic solvent by an aprotic as acetone (an inappropriate solvent for lysozyme) were explored on the previous characteristics (i.e., CSD, Z_{av} and cation distribution).

CSD and Z_{av} values for $Lyso^{n+}$ produced in DESI-MS from sprayed aqueous protic solvent

Under DESI conditions with acidified methanol/water mixture (8/2) (Figure 1a) with 5 pmole of deposited lysozyme, three peaks of the lysozyme mass spectrum dominate at m/z 1431.49, m/z 1590.45, and m/z 1789.12, corresponding to ions with 10^+, 9^+ and 8^+ charge states, respectively. Consequently, a narrow CSD value (i.e., from 7^+ to 11^+) characterizes this profile, with a calculated average charge state as $Z_{av}=(9.0 \pm 0.1)$, which is slightly lower than that obtained in ESI i.e., $Z_{av}=(9.8 \pm 0.1)$ for the lysozyme sample consumption considered approximately similar (Supplementary material S1 and Figure S1) in both the experiments (i.e., instrument, in skimmer desolvation and ion transmission conditions). Takas et al. reported already this behavior [4]. The weak differences observed between these ESI and DESI experiments may signify that the yield of ionization/desorption are not very different, although the latter is somewhat gentler than the former. This result is consistent with those provided from the study of Myung et al. [43], which shown similar conclusion from IMS experiments performed using sprayed protic solvents.

For the production of solvent-free multi-charged lysozyme $Lyso^{n+}$ species, it can be assumed that formation of the multi-protonated aggregates occurs from the small offspring droplets followed by "in skimmer" desolvation steps, as it is described by an ESI-like mechanism [4]. However, to explain the origin of the small variation of Z_{av} between the profiles of ESI (Supplementary material S1 and Figure S1) and DESI (Figure 1), the net charge carried by lysozyme/solvent aggregates has been qualitatively scrutinized by considering the $Lyso^{n+}$ ions.

In DESI mode, the Z_{av} decrease was previously ascribed to the small sizes of the offspring droplets [64,68]. Very likely, these latter carry a charge number lower than those produced in ESI. This led to a positive net charge decrease on the solvated and folded protein surface, and thus, after its solvent release from the charged lysozyme/solvent aggregates, the solvent-free ionized proteins are characterized by a slightly reduced average charge state. On the other hand, this slight Z_{av} discrepancy between the ESI and DESI mass spectra may be due to the nature (protons vs. cations) of the charge transfer mechanism of DESI overall process, in which protic solvent properties do not strongly affect both the ionization steps and production of solvent-free protein ions [43,67,80]. This interpretation may rationalize the observed effect of the lysozyme amount deposited at the target, on the proton/cation distribution, for a given lysozyme charge state in DESI (Figure 1a) compared that obtained in ESI (Figure S1). Despite the nascent offspring droplet heterogeneity, their initial lysozyme concentration contributes essentially to the CSD as well as to the distribution of the proton/metallic cation ratio. The signal "zooms" for the 8^+ lysozyme species, around m/z 1780-m/z 1800 (Insets of Figures S1 and 1a) display the cationized $[Lyso + 6H + Ca]^{8+}$ forms (i.e., $LysoCa^{8+}$) rather than its isobaric $[Lyso + 7H + K]^{8+}$ form (i.e., $LysoK^{8+}$) (supplementary material S2), in addition to $Lyso^{8+}$.

Interestingly, for each charge state, the normalized relative abundances of the $Lyso^{n+}$ and $LysoC_k^{n+}$ ions [i.e., $(ILyso^{n+} + \Sigma ILysoCa_q^{n+} = 100\%]$ depend on the charge (n) (Figure S2). So, for the charge state i^+ which increases from 7^+ to 13^+ in ESI (Figure S2a), the $Lyso^{n+}$ and $LysoC_k^{n+}$ relative abundances are respectively equal to $(96 \pm 3)\%$ and $(4 \pm 3)\%$, with a $\left\{ \left[I_{(h)i+} \right] / \left[\Sigma I_{(h/c)i+} \right] \right\}$ ratio (noted as R_{i+}) almost constant within the experimental errors.

Consequently, in ESI, this ratio can be considered almost constant within the experimental errors. This trend is not that of the DESI experiments with methanol/water (Figure S2b). Indeed, when the charge state increases from 7^+ to 11^+, a significant enlargement of the $Lyso^{n+}$ relative abundance from 62% to 90%, and vice versa from that of $LysoC_k^{n+}$ which present an abundance decrease from 38% to 10%. It results in a R_{i+} ratio multiplied by a factor of # 5.5. Furthermore, the larger contribution of cationized multi-protonated Lysozyme is clearly illustrated by deconvolution of the DESI mass spectrum of Figure 1a, which indicates that: (i) approximately 8% of ions are cationized, and (ii) the cationization is reinforced for the lower charged species. From these features, it appears a significant variation of the $Z_{(h)av}$ and $Z_{(h/c)av}$ values (calculated from $Z_{av} = 9.0$) which are 9.1 and 7.9, respectively. This behavior is consistent with different studies which enlightened this reinforced cationization of the proteins in DESI [4,36].

In the ESI experiments, independently of the maximum number of present protons on the micro-droplet surface, the fast exchanges of metallic ion/proton taking place in solution (or into the charged aggregates) can explain the almost constant cationization, maintained at low level. This yields formation of strongly charged lysozyme/solvent aggregates, which are then desolvated under reduced pressure conditions. On the other hand, this means that, in ESI, the residual alkaline/alkaline earth ions (naturally present in the native proteins) are almost completely exchanged by protons in droplet/aggregate systems,

leading to reduction of the metallic ion contribution into the naked multi-charged proteins. This behavior cannot occur in the DESI mode since the offspring droplets (emitted from the thin layer on the bulk sample onto the target) [43] are unlikely saturated by a lot of protons. Consequently, in such secondary droplets, the number of protons is not enough large [113,114] to allow an almost complete displacement of all alkaline/alkaline earth ions and to provide cation-free- multi-protonated lysozyme. Thus, after desolvation of the desorbed multi-charged aggregates, residual metallic cations can be carried by the multi-protonated lysozyme. This explanation is consistent with the droplet pickup DESI mechanism model [4], which considers that secondary microdroplets are significantly smaller than those formed in ESI [43,64,116], and thus, should very likely carry less protons (vide supra). Consequently, to achieve the final ionization, cation/proton exchanges are significantly larger in ESI than in DESI from the charged aggregates (Supplementary material S2, S3).

Effects of the sample dilution in protic solvent on the charge state distribution, and on the extent of proton/cation exchange

For this purpose, we studied the evolution of the CSD and its corresponding Z_{av} relative values depending on the amount of lysozyme deposited on the PTFE target of DESI experiments (Figures S3 and 2a). When the lysozyme amount decreased from 50 pmoles to 0.05 pmol, it appears: (i) a charge state shifting towards higher values up to 12^+ (Figure S3) and (ii) a monotonic increase of Z_{av} from 8.5 to 10.2 (Figure 2a). Such behavior does not differ strongly from that observed in the ESI mode [81-83,88-91,117]. However, the ESI signal is maintained constant in time, due to continuous droplet renewal, contrasting to that occurred from the DESI experiments which involve three periods. Indeed, due to the lysozyme solubility in water/methanol mixture, we can consider that the lysozyme concentration (i) reaches its maximum as soon as the first acetone layers are deposited on the target, (ii) remains constant during the period of the continuous consumption, and (iii) decreases rapidly when the sample bulk disappears. Furthermore, the average size and net charge distribution of offspring droplets should be approximately constant in time because of the momentum transfer from the multiple impacting primary droplets of solvent mixture with a constant composition (the layer surface being constantly renewed). Consequently, under these conditions, the charge number on the emerging droplets remains almost independent of analyte concentration; the decrease of the deposited lysozyme amount should result in a Z_{av} increasing. This is consistent with the monotone and slight lowering Z_{av} evolution observed by increasing the deposited lysozyme amount on the PTFE target (Figure 2a).

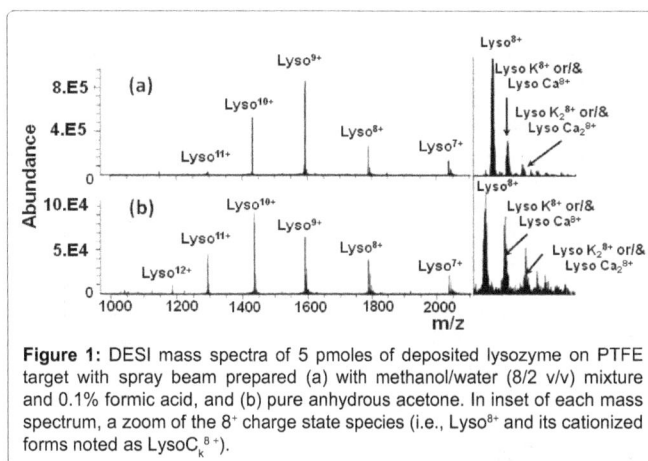

Figure 1: DESI mass spectra of 5 pmoles of deposited lysozyme on PTFE target with spray beam prepared (a) with methanol/water (8/2 v/v) mixture and 0.1% formic acid, and (b) pure anhydrous acetone. In inset of each mass spectrum, a zoom of the 8^+ charge state species (i.e., $Lyso^{8+}$ and its cationized forms noted as $LysoC_k^{8+}$).

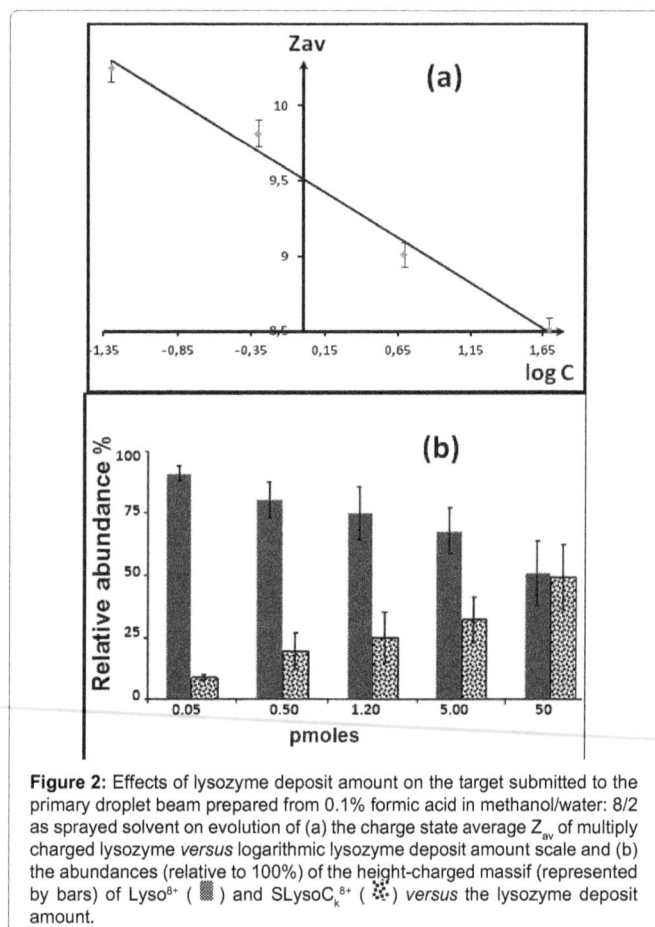

Figure 2: Effects of lysozyme deposit amount on the target submitted to the primary droplet beam prepared from 0.1% formic acid in methanol/water: 8/2 as sprayed solvent on evolution of (a) the charge state average Z_{av} of multiply charged lysozyme *versus* logarithmic lysozyme deposit amount scale and (b) the abundances (relative to 100%) of the height-charged massif (represented by bars) of $Lyso^{8+}$ (■) and $SLysoC_k^{8+}$ (▒) *versus* the lysozyme deposit amount.

In addition, this explanation is also consistent with the increased contribution of cationized molecules with higher deposited lysozyme amount (Figure 2b), which results in a higher proton consumption. Indeed, e.g., the relative $Lyso^{8+}$ abundance decreases while that of the cationized $LysoCa_q^{8+}$ forms (q equal to 1 or 2) increases as the lysozyme deposit enhances from 0.050 pmol to 50 pmoles as shown in Figure 2b. This effect is consistent with the

interpretation which considers that an increase in native lysozyme deposited (in the form of salts) leads to a less proton/metal cation exchanged in the layer and offspring droplets, and therefore, reinforces the relative contribution of cationized multi-protonated $LysoCa_q^{n+}$ forms of lysozyme (Figure 2b). This interpretation supports the observed effects of the deposited lysozyme amount on the CSD, and the proton/cation distribution for a given charge state carried out by lysozyme in the DESI mass spectrum (Figure 2b). Conversely, it is possible to apply the latter particular effect to explore qualitatively the evolution of lysozyme concentration into the offspring droplets when an aprotic solvent such as acetone is sprayed in DESI experiments.

Unusual lysozyme ionization from primary droplet beam of aprotic solvent as anhydrous acetone

Replacement of the protic spray solvent (i.e., methanol) by an aprotic one, as acetonitrile, led to a dramatic suppression of the lysozyme signal by more than two orders of magnitude in DESI mode. This behavior was somewhat unexpected because in ESI experiments, acetonitrile [42,52] is currently used as a very effective solvent for lysozyme ionization (no more discussion herein) in DESI. This

degradation of the signal is probably due to a slower solubilization than that achieved with a mixture of methanol-water. However, this possibility is not confirmed with the anhydrous acetone use as shown in Figure 1b, since an important signal appears, even if solubilization of lysozyme is less effective than with acetonitrile. Indeed, acetone presents very weak solubilization efficiency for proteins and rather favors their aggregation and, their precipitation [118]. The multiply-charged lysozyme production (Figure 1b) led us to explore deeper the influence of properties of fine droplets prepared from sprayed solvent on the ionization of proteins in DESI, especially with an aprotic one such as anhydrous acetone. This should enlighten certain features about the ionization mechanism under sprayed acetone conditions. Note that acetone is not commonly used as an ESI solvent. Its mixture with water, in 50/50 or 99/1 ratios, shows desorption/ionization of ferrocene derivatives [119] and oligomeric compounds [120], respectively. As far as we know, under ambient ionization conditions, anhydrous acetone has never been successfully applied to protein analysis. From the characteristic properties of acetone, the nascent offspring droplets should contain lysozyme within a very lower concentration than that with offspring droplets provided from the sprayed methanol/water. However, due to the fast evaporation of the acetone, the size of the survivor droplets in front of the transfer capillary will be significantly reduced. That can lead either to an increase of the yield of protein ionization/desorption, or to a faster evaporation of the offspring droplets reaching the formation of their charged residue with lysozyme as aggregates characterized by both the charge and size distributions.

With a primary sprayed acetone droplet beam, the DESI mass spectrum of 5 pmoles deposited lysozyme (Figure 1b) displays an unexpected broad CSD from $Lyso^{7+}$ to $Lyso^{12+}$ with $Lyso^{10+}$ as main charged species and a Z_{av} charge state average of (9.4 ± 0.3), whereas with acidified methanol/water mixture, with the same deposited amount, the Z_{av} value slightly decreases to (9.0 ± 0.1). *A priori*, this moderate value seems to be inconsistent with that expected by considering both the methanol [95,96] and water [95,121] gas phase basicities (i.e., $GB(CH_3OH)=724.5$ kJ.mol^{-1} and $GB(H_2O)=660.4$ kJ.mol^{-1}), which are significantly lower than that of acetone [$GB(acetone)=789.6$ kJ.mol^{-1})] [95,122]. Consequently, a reverse trend should be observed if charge state depends mainly on the relative GB (and PA) values as evidenced in ESI various studies [50,123-127].

The Z_{av} variation from 9.0 to 9.4 suggests that the "equivalent lysozyme deposit" in methanol/water, into the thin target layer (or in the offspring droplets), could roughly be estimated, according to Figure 2a, to 1.41 pmole   (or much less). In addition, from the spraying anhydrous acetone DESI experiments, the [$LysoCa_q^{8+}$/$Lyso^{8+}$] ion abundance ratio (Inset of Figure 1b) is significantly higher than that provided from experiments performed with acidified methanol/water DESI spray (Inset of Figure 1a). Indeed, the cationized species contribution is 30% i.e., $Z_{(h)av}=9.6$, and $Z_{(h/c)av}=8.8$, calculated from deconvoluted acetone spray mass spectrum, significantly higher than the 8% values (i.e., $Z_{(h)av}=9.1$, and $Z_{(h/c)av}=7.9$) characterizing DESI mass spectrum performed with CH_3OH/H_2O (Figure 1a).

Consequently, because of the limitation in the charge number carried by acetone droplets, the alkaline/alkaline earth cation/proton exchanges are also limited. In this way, when using anhydrous acetone instead of methanol/water mixture, the DESI mass spectrum displays a reduction of the absolute intensity of the base peak from 8.10^5 a.u. (a.u. is arbitrary unit) to 9.10^4 a.u. (Figure 1a and 1b). This is due to the reduced available proton number relatively less numerous with anhydrous acetone, a particular aprotic solvent, than with protic solvents. This conclusion was expected because of the high volatility of solvent, which in DESI mode plays a significant effect, and thus, must be considered in the formation of multi-charged aggregates with lysozyme.

These considerations are consistent with the highest Z_{av} values (i.e., 9.4), if it is considered that the previous deposited lysozyme equivalent of 1.41 pmole (Figure 2a) was strongly overestimated on nascent offspring droplets (or in thin surface layer). Furthermore, this analysis explains the shift of the CSD values towards higher values, despite the lower number of available protons in the nascent offspring droplets, as regards numerous present protons released by the sprayed protic solvent, as well as the incomplete exchange of the metallic cations by the available protons.

Comparison of the lysozyme ion signal duration according to the sprayed solvent in DESI

Knowing the aprotic character of sprayed anhydrous acetone and its relative low boiling point (56°C), the resulting offspring droplets will undergo faster evaporation than those generated from the 8/2 methanol/water mixture (boiling point higher than that of acetone). As it was demonstrated from the ESI process [127], one can expect a similar trend in DESI, e.g., that an elevated rate of acetone evaporation is associated with more offspring droplets bearing a high surface charge density and then, production of higher charged lysozyme ions since the lower lysozyme concentration (weak solubility in acetone). Finally, the average equivalent concentration of lysozyme (dissolved and/or adhered to the droplet surface) in the nascent offspring droplets of acetone can be roughly estimated from: (i) the amount of deposited lysozyme (~5 pmoles), (ii) the solvent flow rate (5 μL.min^{-1}), and (iii) the average duration for a total spot desorption close to (5 ± 1) min (i.e., until the entire signal intensity extinction). Thus, the average concentration of lysozyme in the nascent acetone offspring droplets was roughly estimated to be 0.2 μM i.e., ten times less than that in methanol/water (considered as 2.1 μM, supplementary material S1) for a shorter signal duration (0.48 min for methanol/water vs. 5 min for acetone anhydrous).

The corresponding total ion current (TIC) of 1.54×10^6 a.u. (with sprayed anhydrous acetone) was calculated by the signal integration during 5 min. Comparison with DESI experiments based on the acidified spray methanol/water mixture [i.e., duration of (0.48 ± 0.2) min for a TIC value of 10.3×10^6 a.u.] leads to a significant ion abundance diminution, corroborating the key role of lysozyme dissolution (and droplet surface adhesion) in the DESI process as previously evoked [63,67,92]. This justifies the larger duration required, for a sufficient volume of primary sprayed acetone anhydrous to impact the target, and to completely dissolve the analyte, leading to offspring droplets released from the thin solvent layer [66,67].

By using the TIC values and signal durations, the relative averages of ion production rates were estimated at $(21.5 ± 2).10^6$ a.u./min and $(3.1 ± 0.6).10^5$ a.u./min, for the sprayed acidified methanol/water and dry acetone, respectively. Consequently, with the latter, the lysozyme signal is reduced by more than 69 times. This emphasizes the importance of the analyte dissolution/solvation as well as the available proton number in offspring. It is very low in dry acetone compared to acidified methanol/water mixture. Despite its inefficient solubilizing capacity, acetone anhydrous produced significant ionic signal duration in DESI process, due to its low solubility and weak sample consumption.

The acetone effectiveness on the lysozyme desorption/ionization in the DESI experiments can undoubtedly be related to the surface

PTFE properties, with its strong hydrophobic character [128]. Thus, a limited sprayed solvent dispersion improves the film homogeneity on PTFE (sample/solvent amount per unit of surface). Moreover, the fast acetone evaporation supports the formation of small ionic aggregates with enough charges, although the available charge number is lower than that obtained with methanol/water mixture, leading to a more efficient desolvation at the skimmer.

Origin of protons required for multi-protonation of lysozyme in DESI acetone anhydrous

In these experiments, a relevant question appears about the origin of the protons carried by multi-protonated lysozyme. A first assumption may be based on formation of the $CH_3COCH_3^{+\bullet}$ molecular ions into the dry primary charged droplets. The formation of the odd-electron molecular ions is considered to take place through an electrochemical process in the sprayer. In addition to the detection of the protonated acetone, presence of its odd-electron molecular ions in the background of DESI, is mainly observed, but in very low abundance results not reported). This behavior is similar to that observed in the acetone/ESI [129-132]. If in sprayed dry acetone, this ion can, in solution, indirectly tautomerize into the less stable ionized enol (i.e., $H_2C=C(OH)CH_3^{+\bullet}$) thanks to the protic solvent. Such tautomerization can occur in the gaseous phase by the formation of an adduct-ion resulting from ion-molecule reaction with neutral acetone via "self-catalysis" [133] and through an homo-dimer linked to the radical ion by hydrogen bonds [133-136]. This dimeric species, as the drawing force, orients dissociation towards the $CH_2=C(O^\bullet)CH_3$ radical release [137,138]. Consequently the $(CH_3)_2C=OH^+$ formation could be the origin of the lysozyme multi-protonation process.

To examine this hypothetical pathway, d6-acetone (i.e., CD_3COCD_3) was used as solvent spray. Surprisingly, three interesting features can be underlined from the mass spectrum (Supplementary material, Figure S4a) of lysozyme and its deconvolution (Supplementary material, Figure S4b): (i) detection of multi-protonated lysozyme without incorporation of several deuterons. Indeed, the lysozyme average molecular mass provided from the sprayed light anhydrous acetone (i.e., $Mw_{ave,exp}$=14305.20 u, vide infra) compared to that obtained from deconvoluted DESI/d6-acetone mass spectrum (i.e., $Mw_{ave,exp}$=14305.9183 u values) displays a shift of less than one u on the average molecular mass; (ii) a spectacular shifting of the average of the charge state Z_{av}, decreasing from 9.4 (Figure 1b) to 6.7 (i.e., $Z_{(h,d)av}$=7.0, and $Z_{(h,d/c)av}$=6.6 (Figure S4a), although the presence of only one deuteron at maximum, (vide infra); and finally, (iii) an increase of the cation contribution (calculated from deconvoluted mass spectra) (Figure S4b), and enlarged from 30% to 70% for light and heavy sprayed acetone, respectively.

The lowering of the Z_{av} value in (ii) (i.e., a shift of the charge state distribution of multi-protonated lysozyme from $Z_{(h)av}$=9.6 to $Z_{(h,d)av}$=7.0) must reflect a decrease of the available proton number when d6-acetone spray is used. This interpretation is consistent with the increase of the cationized form contribution from 30% to 70% in (iii), since the available proton number is significantly weakened. In addition, as for $Z_{(h,d)av}$ lower than $Z_{(h)av}$, $Z_{(h/c)av}$ (i.e., 8.8) is decreased to $Z_{(h,d/c)av}$=6.6. More importantly is the incorporation of almost 9 protons for Lyso^{9+} in (i) rather than the expected nine deuterons. This restriction could be explained by fast H/D exchanges from labeled aggregates occurring in the gaseous phase and/or in the surrounding wet environment of target.

In order to explore this possible H/D back stepwise exchange pathway in gaseous phase experiments with sprayed anhydrous CD_3OD solvent (without D_2O) were performed instead of the sprayed

CD_3COCD_3 use. Indeed, the presence of heavy water may prevent the detection of back D/H exchange that could be *a priori* possible with the ambient humidity. In fact, a broad H/D exchange distribution is observed with, e.g., an average of 48 deuterons introduced in the $Lyso^{9+}$ ion, $9D^+$ for ionization and 39 for H/D exchange (non-reported data). This shows that if the direct H/D exchanges take place, back reactions with the atmospheric ambient humidity do not occur. Thus, such D/H back exchanges must be ruled out to explain the quasi-absence of the labeled multi-charged lysozyme with the (d6) labeled acetone spray. On the other hand, this confirms that the above mechanism, involving possible formation of protonated acetone (or deuterated d6-acetone) does not take place in these conditions, and is not directly responsible for multiple-proton transfers to lysozyme.

Interestingly, the spraying of labeled acetone implicates a significant decrease of available protons, as shown by both the Z_{av} value and H^+/Ca^{2+} exchange reduction (*vide infra*) i.e., (ii) and (iii)). The decrease of the deuteron/proton number can be attributed to primary isotopic effects which slow down the formation of the deuterated species of lysozyme via multi-deuteron/proton transfers. Consequently, one may ask: what is (are) the entity(ies), responsible for the formation of multi-protonated lysozyme ($Lyso^{n+}$) and its cationized counter-part, observed in DESI with sprayed dry acetone?

Let us to recall that in APPI, acetone is known as a doping agent to assist APPI, because otherwise photon-energy (~10 eV) of Kr VUV lamp [139,140] is insufficient to ionize usual protic solvents (e.g., water and methanol having high ionization energy). Doping participate to solvent protonation (e.g., IE_{water}:12.62 eV), presumably by exothermic stepwise consecutive collisions on acetone molecular ion with at least $2H_2O$ (-63 kJ.mol^{-1}) as reported in Equation 4 [95,96,141].

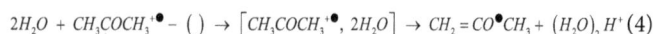

$$2H_2O + CH_3COCH_3^{+\bullet} - (\ ) \rightarrow \left[CH_3COCH_3^{+\bullet}, 2H_2O \right] \rightarrow CH_2 = CO^{\bullet}CH_3 + (H_2O)_2 H^+ \ (4)$$

However, only the production of mono-protonated molecules in gas phase would take place (due to same charge polarity repulsion in the ion-ion interaction) in contrast to that observed in DESI, because of multi-protonated forms with sprayed dry acetone. Thus, APPI-like process cannot be considered. Despite the process taking place in APPI, the ambient water trace role could be finely observed in DESI process even with a spray of anhydrous acetone. Such atmospheric pressure conditions do not prevent from the surrounding moisture, which could be the source of protons responsible for the multi-protonation processes in DESI when anhydrous acetone is sprayed. Note that in the APCI, protonation of acetone does not occur when a dry compressed gas (e.g., D-nitrogen) is used, whereas with the T-air gas, protonation of acetone appears [142], likely similar adsorption of the ambient water takes place with singly charged small aggregates.

In order to check this assumption, three different experiments based on sprayed anhydrous acetone, were performed under rarefied air conditions on the target of deposited lysozyme, by introducing post-flow argon constituted by: (i) dry argon flushing the closely surrounded target. It causes a total extinction of the lysozyme signal (Figure 3a). Taking into account the very low GB of argon, (GB_{Ar}=345.8 kJ.mol^{-1}, estimated from proton affinity) [143] the proton transfer from ambient multi-protonated lysozyme to Ar is thus excluded. Then, a question arises, why dry argon led to such an ionic signal removal? (ii) preliminary humidified argon stream (Ar/H_2O, Figure 3b). The signal is restored with reduction of the charge state average from (9.4 ± 0.3) (Figure 1b) to (8.5 ± 0.3), with $Z_{(h)av}$=8.7 and $Z_{(h/c)av}$=8.1, although the argon flow was moistened prior to. This experiment strongly supports the ambient humidity role in DESI source environment in lysozyme multi-

protonation. Indeed, the dilution of the water vapor into argon (larger dilution than that in the ambient atmosphere) results in the decrease of the previous Z_{av} values. Thus, the protons would originate mainly from ambient humidity rather than directly from the sprayed anhydrous acetone as shown from the d6-acetone experiments (*vide supra*); (iii) labeled Ar/D_2O post-flow (Figure 3c) to confirm the ambient water role. The mass spectrum exhibited a narrow CSD from 11^+ to 5^+ related to a Z_{av} decrease from (8.5 ± 0.3) with Ar/H_2O stream to (7.7 ± 0.2) with Ar/D_2O. Furthermore, it appears a large deuteron incorporation confirming the role of ambient water (or heavy water) for lysozyme multi-protonation (or multi-deuteration). A deconvoluted DESI/dry acetone mass spectrum, with Ar/D_2O post-flow, displays an increase of the metallic ion contribution (alkali/alkaline earth cations) as 43% compared to 35% observed in the unlabelling post-flow experiments. On the other hand, the $Z_{(h)av}$ and $Z_{(h/c)av}$ terms equal to 8.7 and 8.1, with the Ar/H_2O experiment, respectively decrease to $Z_{(h,d)av}$=7.9 and $Z_{(h,d/c)av}$=7.4, under the labeling post flow conditions. This trend is consistent with a lowering of the available proton/deuteron number with the surrounding D_2O, very likely, due to the previous considered isotopic effect during proton/deuteron transfers (and/or exchanges).

For each charge state, the natural isotopic clusters are shifted to higher m/z ratios and the isotopic pattern distribution is broader than that corresponding to the natural one (Figure 4). For instance, the estimated centroid of the non-cationized $Lyso^{9+}$ ion shifted from m/z 1590.5044 to m/z 1591.3182 by 0,8138 (i.e., 7.3241 u, corresponding on average to introduction of more than 7 deuterons, Figure 4). Furthermore, the isotopic signal distribution width, measured at 10% of isotopic pattern height, presents a peak number increasing from 12 (natural isotopic distribution with Ar/H_2O, Figure 3b) to 24 (distribution enlargement with Ar/D_2O, Figure 3c) i.e., a maximum of 12 deuterons, far from completion, although several mobile protons were exchanged in addition to the $9D^+$ charging $Lyso^{9+}$. Since deuterons are not directly supplied from d6-acetone (see above), the shift of the isotopic clusters observed with post-Ar/D_2O flow (Figure 4) allows to consider the main part of the labeled surrounding water for adding $9D^+$ (plus the H/D exchanges). These clearly indicate that the multi-protonated lysozyme, arising from the multi-step post-spray, is promoted by the air moisture.

Under Ar/D_2O post-flow conditions, the overall duration of both the target layer and offspring droplet production, can be estimated about 10 to 100 µsec. It is enough to introduce on average D^+ and 12 at the maximum (i.e., $9D^+$ and 3 H/D exchanges) and to form labeled $Lyso^{9+}$ ions. Several studies [91,144] on the gas-phase H/D exchanges in ESI with multi-protonated lysozyme were performed by ion storage. After one-second storage of the $Lyso^{9+}$ ion under labeling gas phase conditions, approximately 60 H/D exchanges were introduced [144]. This result is comparable to that obtained after 10 s of the $Lyso^{9+}$ ion storage, since 63 D are introduced from gas phase H/D exchanges [145]. Without ion storage, the $Lyso^{9+}$ ions yield only 26 H/D exchanges, during ion accumulation and analysis cycles, corresponding to a few hundreds of milliseconds in the labeled gas phase environment [145]. As noted from the DESI/CD_3OD experiments (*vide supra*), 48 D were introduced into the $Lyso^{9+}$ ion. Those corresponded to a similar gas phase exchange of $Lyso^{9+}$ provided with same labeled reagent, after ion storage of 3-4 sec into an ion trap [145]. This means a larger ion reactivity takes place in DESI as enlightened using nucleophilic reagent in DESI [8,27,69, 144-146].

In DESI, it can be considered from the previous results, that prompt adsorption of ambient water traces (or heavy water) on

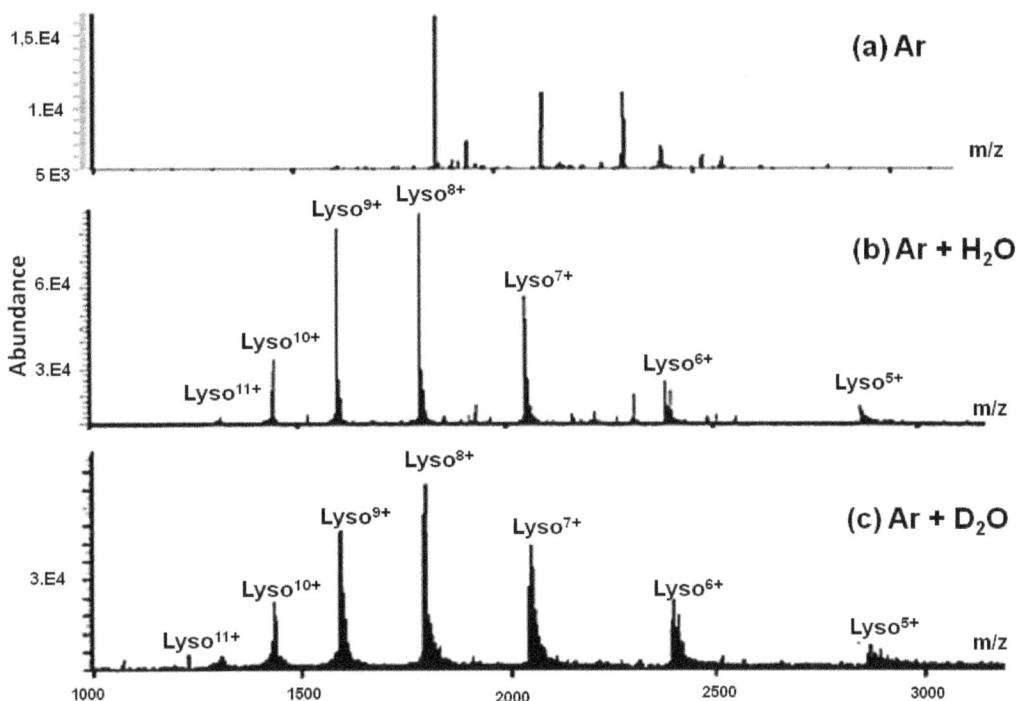

Figure 3: DESI mass spectra of 5 pmole lysozyme deposited on target and desorbed using acetone spray under post flow conditions: (a) signal extinction with Ar as ambient gas, (b) with argon preliminary bubbled in light water, and (c) with Ar preliminary bubbled in D_2O.

charged droplets or/and the impacted surface layer may lead to primary charged species hydration (i.e., the odd-electron acetone ions). Similar reactions, described by Momoh [136], can also occur on the charged surface (or on droplets) subjected to the atmospheric moisture. It results in the provided aqueous layer (or micro-droplets) of acetone on sample surface enhancing lysozyme extraction and thus, the ion abundance increase. Consequently, in the Ar/D_2O post-flow experiments, the $(H_3C)_2CO^+$ molecular ions present in droplets, after reaction with adsorbed water, provide solvated protons, which are carried out, e.g., by labeled water into charged aggregates (Equation 4). Furthermore, the solvated proton/deuteron in the possible $(D_2O)_nH^+$ clusters are randomized and/or exchanged by multiple collisions with D_2O to give rise to formation of a formal $(D_2O)_nD^+$ and $(D_2O)_nH^+$ mixture. The result is an accumulation of protons/deuterons either at the surface of droplets, or at the target thin layer to produce charged aggregates which give rise, after stepwise desolvation, to formation of multi-deuterated/protonated lysozyme species.

Finally, unlike reactions of charged aggregates through ion-molecule reactions in gaseous phase, macroscopic processes could be considered, especially those implicating multi-solvated systems constituted by charged aggregates of acetone/protein with water provided by the atmospheric moisture. It may result in charged aggregates of water-acetone-lysozyme, odd-electron acetone ion enolisation assisted by water, which yields in the protein protonation. Similar mechanism was described for small size ions, in higher vacuum experiment of a FTICR instrument. Indeed, assisted radical-ion isomerization yielding distonic ions [134-137] from short life-time adduct ions were achieved with trace of water.

However, if such a mechanism explains the proton origin, it cannot rationalize multiple-proton transfers to lysozyme from the ion-molecule processes (*vide supra*). Only by considering that, after the adsorption

of water molecules on the primary droplets, and more likely in the liquid layer wetting the target, where the charges are accumulated, the required solvated proton formation (formed as described above) occurs. The role of water is even more enhanced than the evaporation of the acetone is faster than that of water, so that the water enrichment occurs in offspring droplets when these ones approach the step of the charged aggregate emergence. This leads directly (or *via* offspring droplets) to fast desorption of the multi-protonated lysozyme aggregates or fast production of solvated and charged residues which are desolvated at the reduced pressure skimmer zone. Thus, this means that the multi-protonated aggregates can be first formed from the offspring droplets, and then, released from the wet surface layer on the target. This implies that a significant number of protons are already available in the wet layer on the target. On the other hand, this means that water molecules are rapidly adsorbed in primary droplets (or at the target layer) to react with ionized acetone and give protonated reagent. It results a large number of protons yielding a sufficient number of protonated water molecules to produce the multi-protonated lysozyme from desorbed charged aggregates.

Conclusions

The DESI potentiality to produce large size ions in gas phase is important, since the use of aprotic solvents is possible, even if they do not directly provided protons for ionization. Thus, proteins can be multi-protonated by inappropriate solvents such as aprotic solvent, e.g., anhydrous acetone. An unusual origin of protonated agent seems arise for generating the multi-protonated molecules in the case of deposited lysozyme on target within a long duration. Indeed, when using dry d6-acetone spray, $Lyso^{9+}$ does not incorporate more than one deuteron over the nine expectable, whereas with sprayed CD_3OD, 48 deuterons (nine ionizing deuterons and H/D exchanges) are incorporated thanks to the direct *in situ* $CD_3OD_2^+$ formation (from the initial $CD_3OD^{+\boxtimes}$

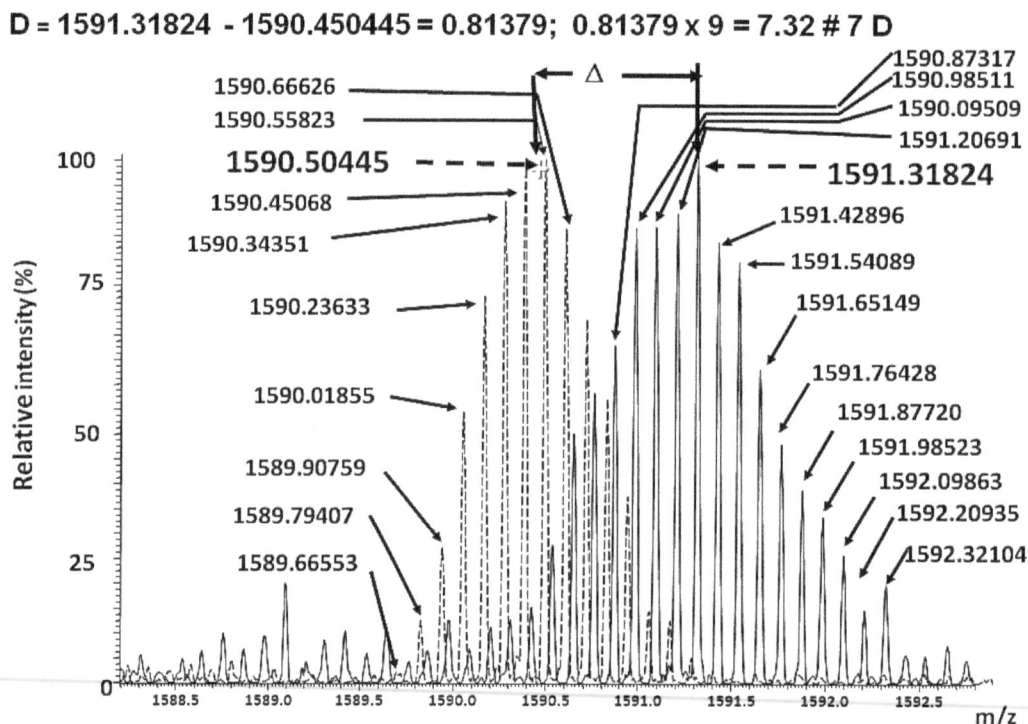

D = 1591.31824 - 1590.450445 = 0.81379; 0.81379 x 9 = 7.32 # 7 D

Figure 4: Superimposition of lysozyme 9$^+$ charge state (Lyso^{9+}) isotopic cluster distribution from the DESI experiments using sprayed light acetone using post-flow argon preliminary bubbled in light water (- - -) and heavier D$_2$O in (——). Comparison of the isotopic cluster distributions for estimating contribution of deuteron isotopes to lysozyme multi-protonation showing a shift by D=0.81 m/z of the broad isotopic distribution centroid of the Lyso^{9+} ion.

ion reacting with the CD$_3$OD neutrals). Such as sequence is hindered with odd-electron acetone because its reaction with neutral acetone does not provide protonation. This indicates that from sprayed dry acetone, ionizing solvated protons are not directly generated. This abnormal behavior was interpreted by considering the role of the ambient moisture as the cause of the lysozyme multi-protonation. This takes place through: (i) condensed phase by adsorption of the water molecules on the micro-droplets (or/and at the thin bulk surface layer), or/and (ii) the macromolecular systems as multi-charged aggregates submitted to multiple solvations by thermal collision cascades with the ambient water molecules, where the solvated D$^+$ agent is accumulated.

Such indirect reactions, *via* the odd-electron acetone ion into the charged complex aggregates, give rise to formation of a lot of available solvated protons (or deuterons) yielding intra-aggregate proton (or deuteron) transfers for the multi-protonation (or deuteration) of lysozyme. Otherwise, convincing experiments give evidence such a process.

They are based on the dry argon post-spray flow introduction which leads to the lysozyme ion suppression by rarefying the moisture around the target. Reversely, the wet argon experiment results in the recovery of the multi-protonated lysozyme signals which, by using post-flow of Ar/D$_2$O, is shifted due to the multi-deuteron incorporation (i.e., an average of 7D with a maximum of 12D for Lyso^{9+}). Note, that no large gas phase D/H back exchange takes place as evidenced by introducing irreversibly 48D, using protic CD$_3$OD solvent spray. On the other hand, preservation of cationization is shown as depending upon the available protons/deuterons on the thin target layer or/and on the offspring droplets. The average charge state of cationized multi-protonated lysozyme is useful as probe for available proton comparison in function of sprayed solvent and experimental conditions. All these results evidence that solvated H$^+$/D$^+$ found their origin from light (heavy) water from ambient humidity rather than directly from sprayed anhydrous acetone in DESI. From the sprayed protic solvent (here, methanol or water), the ambient water traces is not needed since solvated protons are directly produced in the primary spray (or in the target layer). Most likely, the offspring droplets carried out enough protons to promote production of the multi-protonated aggregates. Finally, if acetone is a wrong solvent for proteins, it allows indirectly multi-protonation steps with a large charge distribution and a long duration of the signal.

Acknowledgements

Authors Anna Warnet and Nicolas Auzeil contributed equally to this work. Thanks to UPMC and CNRS and the SM3P platform where analyzes were conducted. Thanks to Antony Mallet, for his advices, his precious help and his fruitful discussions.

References

1. Takáts Z, Wiseman JM, Gologan B, Cooks RG (2004) Mass spectrometry sampling under ambient conditions with desorption electrospray ionization. Science 306: 471-473.

2. Venter A, Nefliu M, Cooks RG (2008) Ambient desorption ionization mass spectrometry. Trends Anal Chem 27: 284-290.

3. Cotte-Rodríguez I, Mulligan CC, Cooks RG (2007) Non-proximate detection of small and large molecules by desorption electrospray ionization and desorption atmospheric pressure chemical ionization mass spectrometry: instrumentation and applications in forensics, chemistry, and biology. Anal Chem 79: 7069-7077.

4. Takáts Z, Wiseman JM, Cooks RG (2005) Ambient mass spectrometry using desorption electrospray ionization (DESI): instrumentation, mechanisms and applications in forensics, chemistry, and biology. J Mass Spectrom 40: 1261-1275.

5. Cooks RG, Ouyang Z, Takats Z, Wiseman JM (2006) Ambient Mass Spectrometry. Science 311: 1566-1570.

6. Cotte-Rodríguez I, Takáts Z, Talaty N, Chen H, Cooks RG (2005) Desorption electrospray ionization of explosives on surfaces: sensitivity and selectivity enhancement by reactive desorption electrospray ionization. Anal Chem 77: 6755-6764.

7. Badu-Tawiah AK, Eberlin LS, Ouyang Z, Cooks RG (2013) Chemical aspects of the extractive methods of ambient ionization mass spectrometry. Annu Rev Phys Chem 64: 481-505.

8. Haddad R, Sparrapan R, Eberlin MN (2006) Desorption sonic spray ionization for (high) voltage-free ambient mass spectrometry. Rapid Commun Mass Spectrom 20: 2901-2905.

9. Dixon RB, Sampson JS, Muddiman DC (2009) Generation of multiply charged peptides and proteins by radio frequency acoustic desorption and ionization for mass spectrometric detection. J Am Soc Mass Spectrom 20: 597-600.

10. Haapala M, Pol J, Saarela V, Arvola V, Kotiaho T, et al. (2007) Desorption atmospheric pressure photoionization. Anal Chem 79: 7867-7872.

11. Heron SR, Wilson R, Shaffer SA, GoodlettDR, Cooper JM (2010) Surface Acoustic Wave Nebulization of Peptides as a MicrofluidicInterface for Mass Spectrometry. Anal Chem 82: 3985-3989.

12. Chen TY, Lin JY, Chen JY, Chen YC (2010) Ultrasonication-assisted spray ionization mass spectrometry for theanalysis of biomolecules in solution. J Am Soc Mass Spectrom 21: 1547-1553.

13. Nemes P, Vertes A (2007) Laser ablation electrospray ionization for atmospheric pressure, in vivo, and imaging mass spectrometry. Anal Chem 79: 8098-8106.

14. Cody RB, Laramée JA, Durst HD (2005) Versatile new ion source for the analysis of materials in open air under ambient conditions. Anal Chem 77: 2297-2302.

15. McEwen CN, McKay RG, Larsen BS (2005) Analysis of solids, liquids, and biological tissues using solids probe introduction at atmospheric pressure on commercial LC/MS instruments. Anal Chem 77: 7826-7831.

16. Wu C, Dill AL, Eberlin LS, Cooks RG, Ifa DR (2013) Mass spectrometry imaging under ambient conditions. Mass Spectrom Rev 32: 218-243.

17. Ma X, Zhang S, Zhang X (2012) An instrumentation perspective on reaction monitoring by ambient mass spectrometry. Trends Anal Chem 35: 50-66.

18. Chipuk JE, Brodbelt JS (2008) Transmission mode desorption electrospray ionization. J Am Soc Mass Spectrom 19: 1612-1620.

19. Espy RD, Badu-Tawiah A, Cooks RG (2011) Analysis and modification on surfaces using molecular ions in the ambient environment. Curr Opin Chem Biol 15: 741-747.

20. Peters KC, Comi TJ, Perry RH (2015) Multistage Reactive Transmission-Mode Desorption Electrospray Ionization Mass Spectrometry. J Am Soc Mass Spectrom 26: 1494-1501.

21. Liu P, Zheng Q, Dewald HD, Zhou R, Chen H (2015) The study of electrochemistry with ambient mass spectrometry. Trends Anal Chem 70: 20-30.

22. Lu M, Liu Y, Helmy R, Martin GE, Dewald HD et al. (2015) On line Investigation of Aqueous-Phase Electrochemical Reactions by Desorption Electrospray Ionization Mass Spectrometry. J Am Soc Mass Spectrom 26: 1676-1685.

23. Justes DR, Talaty N, Cotte-Rodriguez I, Cooks RG (2007) Detection of explosives on skin using ambient ionization mass spectrometry. Chem Commun (Camb): 2142-2144.

24. Morelato M, Beavis A, Kirkbride P, Roux C (2013) Forensic applications of desorption electrospray ionisation mass spectrometry (DESI-MS). Forensic Sci Int 226: 10-21.

25. D'Agostino PA, Chenier CL, Hancock JR, Lepage CRJ (2007) Desorption electrospray ionisation mass spectrometric analysis of chemical warfare agents from solid phase microextraction fibers. Rapid Commun Mass Spectrom 21: 543-549.

26. Hagan NA, Cornish TJ, Pilato RS, Van Houten KA, Antoine MD et al. (2008) Detection and identification of immobilized low-volatility organophosphates by desorption ionization mass spectrometry. Int J Mass Spectrom 278: 158-165.

27. Song Y, Cooks RG (2007) Reactive desorption electrospray ionization for selective detection of the hydrolysis products of phosphonate esters. J Mass Spectrom 42: 1086-1092.

28. Wells JM, Roth MJ, Keil AD, Grossenbacher JW, Justes DR, et al. (2008) Implementation of DART and DESI ionization on a fieldable mass spectrometer. J Am Soc Mass Spectrom 19: 1419-1424.

29. Nielen MWF, Hooijerink H, Zomer P, Mol JGJ (2011) Desorption electrospray ionization mass spectrometry in the analysis of chemical food contaminants. Trends Anal Chem 30: 165-180.

30. García-Reyes JF, Jackson AU, Molina-Díaz A, Cooks RG (2009) Desorption electrospray ionization mass spectrometry for trace analysis of agrochemicals in food. Anal Chem 81: 820-829.

31. Schurek J, Vaclavik L, Hooijerink H, Lacina O, Poustka J (2008) Control of Strobilurin Fungicides in Wheat Using Direct Analysis in Real Time Accurate Time-of-Flight and Desorption Electrospray Ionization Linear Ion Trap Mass Spectrometry. Anal Chem 80: 9567 -9575.

32. Mulligan CC, MacMillan DK, Noll RJ, Cooks RG (2007) Fast analysis of high-energy compounds and agricultural chemicals in water with desorption electrospray ionization mass spectrometry. Rapid Commun Mass Spectrom 21: 3729-3736.

33. Jackson AU, Werner SR, Talaty N, Song Y, Campbell K, et al. (2008) Targeted metabolomic analysis of Escherichia coli by desorption electrospray ionization and extractive electrospray ionization mass spectrometry. Anal Biochem 375: 272-281.

34. Williams JP, Scrivens JH (2008) Coupling desorption electrospray ionisation and neutral desorption/extractive electrospray ionisation with a travelling-wave based ion mobility mass spectrometer for the analysis of drugs. Rapid Commun Mass Spectrom 22: 187-196.

35. Kauppila TJ, Talaty N, Kuuranne T, Kotiaho T, Kostiainen R, et al. (2007) Rapid analysis of metabolites and drugs of abuse from urine samples by desorption electrospray ionization-mass spectrometry. Analyst 132: 868-875.

36. Siebenhaar M, Kuellmer K, Fernandes NM, Huellen V, Hopf C (2015) Personalized monitoring of therapeutic salicylic acid in dried blood spots using a three-layer setup and desorption electrospray ionization mass spectrometry. Anal Bioanal Chem 407: 7229-7238.

37. Bailey MJ, Bradshaw R, Francese S, Salter TL, Costa C, et al. (2015) Rapid detection of cocaine, benzoylecgonine and methylecgonine in fingerprints using surface mass spectrometry. Analyst 140: 6254-6259.

38. Liu J, Kennedy JH, Ronk M, Marghitoiu L, Lee H, et al. (2014) Ambient analysis of leachable compounds from single-use bioreactors with desorption electrospray ionization time-of-flight mass spectrometry. Rapid Commun Mass Spectrom 28: 2285-2291.

39. Stojanovska N, Tahtouh M, Kelly T, Beavis A, Fu S (2015) Qualitative analysis of seized cocaine samples using desorption electrospray ionization- mass spectrometry (DESI-MS). Drug Test Anal 7: 393-400.

40. Manicke NE, Nefliu M, Wu C, Woods JW, Reiser V, et al. (2009) Imaging of lipids in atheroma by desorption electrospray ionization mass spectrometry. Anal Chem 81: 8702-8707.

41. Bereman MS, Williams TI, Muddiman DC (2007) Carbohydrate analysis by desorption electrospray ionization fourier transform ion cyclotron resonance mass spectrometry. Anal Chem 79: 8812-8815.

42. Shin YS, Drolet B, Mayer R, Dolence K, Basile F (2007) Desorption electrospray ionization-mass spectrometry of proteins. Anal Chem 79: 3514-3518.

43. Myung S, Wiseman J, Valentine S, Takats Z, Cooks RG, et al. (2006) Coupling Desorption Electrospray Ionization with Ion Mobility/Mass Spectrometry for Analysis of Protein Structure: Evidence for Desorption of Folded and Denatured States. J Phys Chem B 110: 5045-5051.

44. Miao Z, Chen H (2009) Direct analysis of liquid samples by desorption electrospray ionization-mass spectrometry (DESI-MS). J Am Soc Mass Spectrom 20: 10-19.

45. Qiu B, Luo H (2009) Desorption electrospray ionization mass spectrometry of DNA nucleobases: implications for a liquid film model. J Mass Spectrom 44: 772-779.

46. Hollenhorst MI, Lips KS, Wolff M, Wess J, Gerbig S, et al. (2012) Luminal cholinergic signalling in airway lining fluid: a novel mechanism for activating chloride secretion via Ca$^+$-dependent Cl$^-$ and K$^+$ channels. Br J Pharmacol 166: 1388-1402.

47. Wiseman JM, Evans CA, Bowen CL, Kennedy JH (2010) Direct analysis of dried blood spots utilizing desorption electrospray ionization (DESI) mass spectrometry. Analyst 135: 720-725.

48. Wiseman JM, Ifa DR, Zhu Y, Kissinger CB, Manicke NE, et al. (2008) Desorption electrospray ionization mass spectrometry: Imaging drugs and metabolites in tissues. Proc Natl Acad Sci USA 105: 18120-18125.

49. Song Y, Talaty N, Tao WA, Pan Z, Cooks RG (2007) Rapid ambient mass spectrometric profiling of intact, untreated bacteria using desorption electrospray ionization. Chem Commun (Camb): 61-63.

50. Montowska M, Rao W, Alexander MR, Tucker GA, Barrett DA (2014) Tryptic digestion coupled with ambient desorption electrospray ionization and liquid extraction surface analysis mass spectrometry enabling identification of skeletal muscle proteins in mixtures and distinguishing between beef, pork, horse, chicken, and turkey meat. Anal Chem 86: 4479-4487.

51. Dill AL, Eberlin LS, Zheng C, Costa AB, Ifa DR, et al. (2010) Multivariate statistical differentiation of renal cell carcinomas based on lipidomic analysis by ambient ionization imaging mass spectrometry. Anal Bioanal Chem 398: 2969-2978.

52. Bereman MS, Nyadong L, Fernandez FM, Muddiman DC (2006) Direct high-resolution peptide and protein analysis by desorption electrospray ionization Fourier transform ion cyclotron resonance mass spectrometry. Rapid Commun Mass Spectrom 20: 3409-3411.

53. Yang SH, Wijeratne AB, Li L, Edwards BL, Schug KA (2011) Manipulation of Protein Charge States through Continuous Flow-Extractive Desorption Electrospray Ionization: A New Ambient Ionization Technique. Anal Chem 83: 643-647.

54. Ferguson CN, Benchaar SA, Miao Z, Loo JA, Chen H (2011) Direct ionization of large proteins and protein complexes by desorption electrospray ionization-mass spectrometry. Anal Chem 83: 6468-6473.

55. Douglass KA, Venter AR (2013) Protein analysis by desorption electrospray ionization mass spectrometry and related methods. J Mass Spectrom 48: 553-560.

56. Rao W, Celiz AD, Scurr DJ, Alexander MR, Barrett DA (2013) Ambient DESI and LESA-MS analysis of proteins adsorbed to a biomaterial surface using in-situ surface tryptic digestion. J Am Soc Mass Spectrom 24: 1927-1936.

57. Przybylski C, Gonnet F, Hersant Y, Bonnaffé D, Lortat-Jacob H, et al. (2010) Desorption electrospray ionization mass spectrometry of glycosaminoglycans and their protein noncovalent complex. Anal Chem 82: 9225-9233.

58. Moore BN, Hamdy O, Julian RR (2012) Protein structure evolution in liquid DESI as revealed by selective noncovalent adduct protein probing. Int J Mass Spectrom 330-332: 220-225.

59. Yao C, Na N, Huang L, He D, Ouyang J (2013) High-throughput detection of drugs binding to proteins using desorption electrospray ionization mass spectrometry. Anal Chim Acta 794: 60-66.

60. Liu P, Zhang J, Ferguson CN, Chen H, Loo JA (2013) Measuring Protein-Ligand Interactions Using Liquid Sample Desorption Electrospray Ionization Mass Spectrometry. Anal Chem 85: 11966-11972.

61. Yao C, Wang T, Zhang B, He D, Na N et al. (2015) Screening of the Binding of Small Molecules to Proteins by Desorption Electrospray Ionization Mass Spectrometry Combined with Protein Microarray. J Am Soc Mass Spectrom 26: 1950-1958.

62. Yao Y, Shams-Ud-Doha K, Daneshfar R, Kitova DR, Klassen JS (2015) Quantifying Protein-Carbohydrate Interactions Using Liquid Sample Desorption Electrospray Ionization Mass Spectrometry. J Am Soc Mass Spectrom 26: 98-106.

63. Costa AB, Cooks RG (2008) Simulated splashes: Elucidating the mechanism of desorption electrospray ionization mass spectrometry. Chem Phys Lett 464: 1-8.

64. Volný M, Venter A, Smith SA, Pazzi M, Cooks RG (2008) Surface effects and electrochemical cell capacitance in desorption electrospray ionization. Analyst 133: 525-531.

65. De la Mora JF (2000) Electrospray ionization of large multiply charge species proceeds via Dole's charge residue mechanism. Anal Chim Acta 406: 93-104.

66. Badu-Tawiah A, Bland C, Campbell DI, Cooks RG (2010) Non-aqueous spray solvents and solubility effects in desorption electrospray ionization. J Am Soc Mass Spectrom 21: 572-579.

67. Badu-Tawiah AK, Cooks RG (2010) Enhanced ion signals in desorption electrospray ionization using surfactant spray solutions. J Am Soc Mass Spectrom 21: 1423-1431.

68. Kauppila TJ, Talaty N, Salo PK, Kotiaho T, Kostiainen R, et al. (2006) New surfaces for desorption electrospray ionization mass spectrometry: porous silicon and ultra-thin layer chromatography plates. Rapid Commun Mass Spectrom 20: 2143-2150.

69. Ifa DR, Wu C, Ouyang Z, Cooks RG (2010) Desorption electrospray ionization and other ambient ionization methods: current progress and preview. Analyst 135: 669-681.

70. Gao L, Li G, Cyria J, Nie Z, Cooks RG (2010) Imaging of Surface Charge and the Mechanism of Desorption Electrospray Ionization Mass Spectrometry. J Phys Chem C114: 5331-5337.

71. Manicke NE, Wiseman JM, Ifa DR, Cooks RG (2008) Desorption electrospray ionization (DESI) mass spectrometry and tandem mass spectrometry (MS/MS) of phospholipids and sphingolipids: ionization, adduct formation, and fragmentation. J Am Soc Mass Spectrom 19: 531-543.

72. Bereman MS, Muddiman DC (2007) Detection of attomole amounts of analyte by desorption electrospray ionization mass spectrometry (DESI-MS) determined using fluorescence spectroscopy. J Am Soc Mass Spectrom 18: 1093-1096.

73. Ifa DR, Manicke NE, Rusine AL, Cooks RG (2008) Quantitative analysis of small molecules by desorption electrospray ionization mass spectrometry from polytetrafluoroethylene surfaces. Rapid Commun Mass Spectrom 22: 503-510.

74. Takáts Z, Cotte-Rodriguez I, Talaty N, Chen H, Cooks RG (2005) Direct, trace level detection of explosives on ambient surfaces by desorption electrospray ionization mass spectrometry. Chem Commun (Camb): 1950-1952.

75. Wang H, Manicke NE, Yang Q, Zheng L, Shi R, et al. (2011) Direct analysis of biological tissue by paper spray mass spectrometry. Anal Chem 83: 1197-1201.

76. Sen AK, Nayak R, Darabi J, Knapp DR (2008) Use of nanoporous alumina surface for desorption electrospray ionization mass spectrometry in proteomic analysis. Biomed Microdevices 10: 531-538.

77. Loriau M, Alves S, Churlaud F, Tabet JC (2009) Solvent Effects on the DESI mass spectra of Industrial polymers. Proceedings of the 57th ASMS conference, Philadelphia, USA.

78. Van Berkel GJ, Tomkins BA, Kertesz V (2007) Thin-layer chromatography/desorption electrospray ionization mass spectrometry: investigation of goldenseal alkaloids. Anal Chem 79: 2778-2789.

79. Liu Y, Miao Z, Lakshmanan R, Ogorzalek Loo RR, Loo JA, et al. (2012) Signal and Charge Enhancement for Protein Analysis by Liquid Chromatography-Mass Spectrometry with Desorption Electrospray Ionization. Int J Mass Spectrom 325-327: 161-166.

80. Samalikova M, Grandori R (2003) Role of opposite charges in protein electrospray ionization mass spectrometry. J Mass Spectrom 38: 941-947.

81. Wang G, Cole RG (1994) Effect of Solution Ionic Strength on Analyte Charge State Distributions in Positive and Negative Ion Electrospray Mass Spectrometry. Anal Chem 66: 3702-3708.

82. Kuprowski MC, Konermann L (2007) Signal response of coexisting protein conformers in electrospray mass spectrometry. Anal Chem 79: 2499-2506.

83. Samalikova M, Grandori R (2005) Testing the role of solvent surface tension in protein ionization by electrospray. J Mass Spectrom 40: 503-510.

84. Thomas BR, Vekilov PG, Rosenberger F (1996) Heterogeneity determination and purification of commercial hen egg-white lysozyme. Acta Crystallogr D Biol Crystallogr 52: 776-784.

85. Ries-Kautt M, Ducruix A (1997) Interferences Drawn from Physicochemical Studies of Crystallogenesis and PrecrystaUine State. Methods in Enzymology 276: 23-59.

86. Kebarle P, Verkerk UH (2009) Electrospray: from ions in solution to ions in the gas phase, what we know now. Mass Spectrom Rev 28: 898-917.

87. Deng L, Sun N, Kitova EN, Klassen JS (2010) Direct quantification of protein-metal ion affinities by electrospray ionization mass spectrometry. Anal Chem 82: 2170-2174.

88. Gross DS, Schnier PD, Rodriguez-Cruz SE, Fagerquist CK, Williams ER (1996) Conformations and folding of lysozyme ions in vacuo. Proc Natl Acad Sci U S A 93: 3143-3148.

89. Konermann L, Douglas DJ (1998) Unfolding of proteins monitored by electrospray ionization mass spectrometry: a comparison of positive and negative ion modes. J Am Soc Mass Spectrom 9: 1248-1254.

90. Loo RR, Loo JA, Udseth HR, Fulton JL, Smith RD (1992) Protein structural effects in gas phase ion/molecule reactions with diethylamine. Rapid Commun Mass Spectrom 6: 159-165.

91. Takamizawa A, Fujimaki S, Sunner J, Hiraoka K (2005) Denaturation of lysozyme and myoglobin in laser spray. J Am Soc Mass Spectrom 16: 860-868.

92. Schnier PD, Gross DS, Williams ER (1995) On the maximum charge state and proton transfer reactivity of peptide and protein ions formed by electrospray ionization. J Am Soc Mass Spectrom 6: 1086-1097.

93. Reimann CT, Sullivan PA, Axelsson J, Quist AP, Altmann S, et al. (1998) Conformation of Highly-Charged Gas-Phase Lysozyme Revealed by Energetic Surface Imprinting. J Am Chem Soc 120: 7608-7616.

94. Verkerk UH, Peschhke M, Kebarle P (2003) Effect of buffer cations and of H3O+ on the charge states of native proteins. Significance to determinations of stability constants of protein complexes. J Mass Spectrom 38: 618-631.

95. Hunter EP, Lias SG (1998) Evaluated Gas Phase Basicities and Proton Affinities of Molecules: An Update. J Phys Chem Ref data 27: 413-656.

96. Lias SG, Liebman JF, Levin RD (1984) Evaluated gas phase basicities and proton affinities of molecules; heats of formation of protonated molecules. J Phys Chem 13: 695-808.

97. Petrie S, Javahery G, Wincel H, Bohme DK (1993) Attaching Handles to C60 2+: The Double-Derivatization of C60 2+. J Am Chem Soc 115: 629-634.

98. Kaltashov IA, Fabris D, Fenselau CC (1995) Assessment of Gas Phase Basicities of Protonated Peptides by the Kinetic Method. J Phys Chem 99: 10046-10051.

99. Williams ER (1996) Proton transfer reactivity of large multiply charged ions. J Mass Spectrom 31: 831-842.

100. Miteva M, Demirev P, Karshikoff D (1997) Multiply-Protonated Protein Ions in the Gas Phase: Calculation of the Electrostatic Interactions between Charged Sites. J Phys Chem B 101: 9645-9650.

101. Touboul D, Jecklin MC, Zenobi R (2008) Investigation of deprotonation reactions on globular and denatured proteins at atmospheric pressure by ESSI-MS. J Am Soc Mass Spectrom 19: 455-466.

102. Touboul D, Jecklin MC, Zenobit R (2008) Exploring deprotonation reactions on peptides and proteins at atmospheric pressure by electro-sonic spray ionization-mass spectrometry (ESSI-MS). Chimia 62: 282-286.

103. Burk P, Tammiku-Taul J, Tamp S, Sikk L, Sillar K, et al. (2009) Computational study of cesium cation interactions with neutral and anionic compounds related to soil organic matter. J Phys Chem A 113: 10734-10744.

104. Ai H, Bu Y, Li P, Zhang C (2005) The regulatory roles of metal ions (M = Li, Na, K, Be, Mg, and Ca) and water molecules in stabilizing the zwitterionic form of glycine derivatives. New J Chem 29: 1540-1548.

105. Dunbar RC (2000) Complexation of Na+ and K+ to Aromatic Amino Acids: A Density Functional Computational Study of Cation-p Interactions. J Phys Chem A 104: 8067-8074.

106. Ryzhov V, Dunbar RC, Cerda B, Wesdemiotis C (2000) Cation-pi effects in the complexation of Na+ and K+ with Phe, Tyr, and Trp in the gas phase. J Am Soc Mass Spectrom 11: 1037-1046.

107. Tsang Y, Wong CL, Wong CHS, Cheng JMK, Ma Nlet al. (2012) Proton and potassium affinities of aliphatic and N-methylated aliphatic-amino acids: Effect of alkyl chain length on relative stabilities of K+ bound zwitterionic complexes. Int J Mass Spectrom 316-318: 273-283.

108. Jover J, Bosque R, Sales J (2008) A comparison of the binding affinity of the common amino acids with different metal cations. Dalton Trans : 6441-6453.

109. Nemirovskiy O, Giblin DE, Gross ML (1999) Electrospray ionization mass spectrometry and hydrogen/deuterium exchange for probing the interaction of calmodulin with calcium. J Am Soc Mass Spectrom 10: 711-718.

110. Malisauskas M, Zamotin V, Jass J, Noppe W, Dobson CM, et al. (2003) Amyloid protofilaments from the calcium-binding protein equine lysozyme: formation of ring and linear structures depends on pH and metal ion concentration. J Mol Biol 330: 879-890.

111. Permyakov SE, Khokhlova TI, Nazipova AA, Zhadan AP, Morozova-Roche LA (2006) Calcium-Binding and Temperature Induced Transitions in Equine Lysozyme: New Insights From the pCa-Temperature Phase Diagrams. Proteins 65: 984-998.

112. Morozova-Roche LA (2007) Equine lysozyme: the molecular basis of folding, self-assembly and innate amyloid toxicity. FEBS Lett 581: 2587-2592.

113. Thomson BA, Iribarne JV (1979) Field induced ion evaporation from liquid surfaces at atmospheric pressure. J Chem Phys 71: 4451.

114. Smith JN, Flagan RC, Beauchamp JL (2002) Droplet Evaporation and Discharge Dynamics in Electrospray Ionization. J Phys Chem A106: 9957-9967.

115. Kaltashov I, Mohimen A (2005) Estimates of Protein Surface Areas in Solution by Electrospray Ionization Mass Spectrometry. Anal Chem 77: 5370-5379.

116. Green FM, Salter TL, Gilmore IS, Stokes P, O'Connor G (2010) The effect of electrospray solvent composition on desorption electrospray ionisation (DESI) efficiency and spatial resolution. Analyst 135: 731-737.

117. Przybylski M, Glocker MO (1996) Electrospray mass spectrometry of biomacromolecular complexes with noncovalent interactions-new analytical perspectives for supramolecular chemistry and molecular recognition processes. Angew Chem Int Ed Engl 35: 806-826.

118. Green-Church KB, Nichols JJ (2008) Mass spectrometry-based proteomic analyses of contact lens deposition. Mol Vis 14: 291-297.

119. Hartinger CG, Nazarov A, Chevchenko V, Arion VB, Galanski M et al. (2003) Synthesis, crystal structures, and electrospray ionisation mass spectrometry investigations of ether-and thioether-substituted ferrocenes. Dalton Trans 15: 3098-3102.

120. Saf R, Schitter R, Mirtl C, Stelzer F, Hummel K (1996) Electrospray Ionization Mass Spectrometry Investigation of Oligomers Prepared by Ring-Opening Metathesis Polymerization of Methyl N-(1-Phenylethyl)-2-azabicyclo[2.2.1]hept-5-ene-3-carboxylate. Macromolecules 29: 7651-7656.

121. Leito I, Koppel IA, Burk P, Tamp S, Kutsar M, et al. (2010) Gas-phase basicities around and below water revisited. J Phys Chem A 114: 10694-10699.

122. Szulejko JE, Luo Z, Solouki T (2006) Simultaneous determination of analyte concentrations, gas-phase basicities, and proton transfer kinetics using gas chromatography/Fourier transform ion cyclotron resonance mass spectrometry (GC/FT-ICRMS). Int J Mass Spectrom 257: 16-26.

123. Ogorzalek Loo RR, Winger BE, Smith RD (1994) Proton transfer reaction studies of multiply charged proteins in a high mass-to-charge ratio quadrupole mass spectrometer. J Am Soc Mass Spectrom 5: 1064-1071.

124. Ogorzalek Loo RR, Smith RD (1994) Investigation of the gas-phase structure of electrosprayed proteins using ion-molecule reactions. J Am Soc Mass Spectrom 5: 207-220.

125. Hunter AP, Severs JC, Harris FM, Games DE (1994) Proton-transfer reactions of mass-selected multiply charged ions. Rapid Commun Mass Spectrom 8: 417-422.

126. Schnier PD, Price WD, Williams ER (1996) Modeling the maximum charge state of arginine-containing Peptide ions formed by electrospray ionization. J Am Soc Mass Spectrom 7: 972-976.

127. Nguyen S, Fenn JB (2007) Gas-phase ions of solute species from charged droplets of solutions. Proc Natl Acad Sci U S A 104: 1111-1117.

128. Pasilis SP, Kertesz V, Van Berkel GJ (2007) Surface scanning analysis of planar arrays of analytes with desorption electrospray ionization-mass spectrometry. Anal Chem 79: 5956-5962.

129. Heaton J, Jones MD, Legido-Quigley C, Plumb RS, Smith NW (2011) Systematic evaluation of acetone and acetonitrile for use in hydrophilic interaction liquid chromatography coupled with electrospray ionization mass spectrometry of basic small molecules. Rapid Commun Mass Spectrom 25: 3666-3674.

130. Keppel TR, Jacques ME, Weis DD (2010) The use of acetone as a substitute for acetonitrile in analysis of peptides by liquid chromatography/electrospray ionization mass spectrometry. Rapid Com. Mass Spectrom 24: 6-10.

131. Fountain KJ, Xu J, Diehl DM, Morrison D (2010) Influence of stationary phase chemistry and mobile-phase composition on retention, selectivity, and MS response in hydrophilic interaction chromatography. J Sep Sci 33: 740-751.

132. Trikoupis MA, Burgers PC, Ruttink PJA, Terlow JK (2002) Self-catalysis in the gasphase: enolization of the acetone radical cation. Int J Mass Spectrom 217: 97-108.

133. Matsuda Y, Yamada A, Hanaue K, Mikami N, Fujii A (2010) Catalytic action of a single water molecule in a proton-migration reaction. Angew Chem Int Ed Engl 49: 4898-4901.

134. Van der Rest G, Mourgues P, Fossey J, Audier HE (1997) The [+CH2OH, H2O] and [+CH2OH, 2H2O] solvated ions. Int. Mass Spectrom. Ion Process 160: 107-115.

135. Trikoupis MA, Terlouw JK, Burgers PC (1998) Enolization of Gaseous Acetone Radical Cations: Catalysis by a Single Base Molecule. J Am Chem Soc 120: 12131-12132.

136. Momoh PO, El-Shall MS (2008) Gas phase hydration of organic ions. Phys Chem Chem Phys 10: 4827-4834.

137. Norrman K, Sølling TI, McMahon TB (2005) Isomerization of the protonated acetone dimer in the gas phase. J Mass Spectrom 40: 1076-1087.

138. Abdel Azeim S, van der Rest G (2005) Thermochemical properties of the ammonia-water ionized dimer probed by ion-molecule reactions. J Phys Chem A 109: 2505-2513.

139. Robb DB, Covey TR, Bruins AP (2000) Atmospheric pressure photoionization: an ionization method for liquid chromatography-mass spectrometry Anal Chem 72: 3653-3659.

140. Marchi I, Rudaz S, Veuthey JL (2009) Atmospheric pressure photoionization for coupling liquid-chromatography to mass spectrometry: a review. Talanta 78: 1-18.

141. Lias SG, Bartness JE, Liebman JF, Homes JL, Levin R.D et al. (1988) Gas-Phase Ion and Neutral Thermochemistry. J Phys Chem Ref data Suppl 17: 861.

142. Kolakowski BM, Grossert JS, Ramaley L (2004) Studies on the positive-ion mass spectra from atmospheric pressure chemical ionization of gases and solvents used in liquid chromatography and direct liquid injection. J Am Soc Mass Spectrom 15: 311-324.

143. Schul RJ, Passarella R, Upschulte BL, Keesee RG, Castleman AW (1987) Thermal energy reactions involving Ar+ monomer and dimer with N2, H2, Xe, and Kr. J Chem Phys 86: 4446-4451.

144. Sparrapan R, Eberlin LS, Haddad R, Cooks RG, Eberlin MN, et al. (2006) Ambient Eberlin reactions via desorption electrospray ionization mass spectrometry. J Mass Spectrom 41: 1242-1246.

145. Wu C, Ifa DR, Manicke NE, Cooks RG (2009) Rapid, direct analysis of cholesterol by charge labeling in reactive desorption electrospray ionization. Anal Chem 81: 7618-7624.

146. Xu G, Chen B, Guo B, He D, Yao S (2011) Detection of intermediates for the Eschweiler-Clarke reaction by liquid-phase reactive desorption electrospray ionization mass spectrometry. Analyst 136: 2385-2390.

19

Quantitative Determination of Topiramate in Human Breast Milk

Cristina Cifuentes, Sigrid Mennickent* and Marta De Diego
University of Concepción, Concepción, Chile

Abstract

A thin-layer chromatographic (HPTLC) method for quantification of topiramate in human breast milk was developed using liquid –liquid extraction with n-hexane and methanol as extraction solvents, fluorescence activation with ninhidrine (1% ethanolic solution) and chlorpromazine as internal standard.

Thin-layer chromatographic separation was performed on precoated silica gel F 254 HPTLC plates using a mixture of toluene: ethanol (25:10, v/v), as mobile phase. Densitometric detection was done at 326 nm. The method was validated for linearity, precision, selectivity, LOD and LOQ, and accuracy. Linear calibration curves in the range of 0.30 to 50.00 µg/mL showed correlation coefficient of 0.991. The intra-assay and inter-assay precision, expressed as the relative standard deviation (RSD), were in the range of 3.04% - 3.14% (n=3) and 1.81%-4.10% (n=9), respectively. The limit of detection was 0.24 µg/mL, and the limit of quantification was 0.30 µg/mL Accuracy, calculated as percentage recovery, was between 101.65% and 109.51%, with a RSD not higher than 0.41%. Topiramate is well resolved from others antiepileptic drugs and from the internal standard (Rs=5.20). In conclusion, the method is precise, accurate, reproducible and selective for the analysis of topiramate in human breast milk.

Keywords: Topiramate; Breast milk; Antiepileptics; Thin-layer chromatography

Introduction

Topiramate (TPM) is a monosaccharide drug with a molecule of fructopyranose in its chemical structure [1] (Figure 1). TPM is an anticonvulsant drug indicated for the treatment and control of partial seizures and severe tonic-clonic (grand mal) seizures in adults [2], including Lennox-Gastant Syndrome in children [3,4]. It is also used for prophylaxis of migraine [5,6], mood instability disorder [7], post-traumatic stress disorder, bulimia nervosa and binge-eating disorders [8]. It is also investigated as promising agent for obesity treatment [9], release of neuropathic pain [10], treatment of sleep related eating disorder [11], alcohol and drug addiction therapy [12,13] and smoking cessation [14-16].

These effects of TPM results from its multiple pharmacological actions, as the enhancement of the activity of γ-aminobutyric acid type A, inhibition of some voltage-gated sodium and calcium channels, inhibition of some glutamate receptors, as well a little inhibition of carbonic anhydrase [17-19].

Food and Drug Administration (FDA) suggest avoid the use of TPM in pregnant woman and in her babies, because the reports of risk oral birth defects (cleft lip and clef palate) in children born to mothers taking TPM (FDA New Release, March 2011). However, sometimes is more difficult to change the antiepileptic drug to some patients, including pregnant women, increasing the risks for their babies.

Some studies shown when the mother was taking 200 mg of topiramate daily had average milk levels of 0.6 µg/mL to 1.2 µg/mL, and these works estimated that the infants received doses between 0.1 and 0.7 mg/kg/day which was between 3 and 23% of the mother's weight-adjusted dose. Serum levels in a 24-day-old infant whose mother was taking 150 mg daily were about 0.5 µg/mL. Another infant whose mother was taking 200 mg daily had an average serum level of 0.6 µg/mL at 20 days of age and 0.7 µg/mL at 97 days of age. Overall, their plasma levels were about 10 to 20% of maternal plasma levels. Baby blood topiramate concentration is higher as baby age is increased, because the breast –feeding increased too, therefore, the blood concentration of topiramate. Moreover, this drug is a lipophilic compound, increasing its body accumulation [19,20].

The review of literature revealed that TPM has no ultraviolet, visible or fluorescence absorption [21,22]. Analysis of topiramate in pharmaceutical formulation and in biological fluids has been reported by HPLC with pre-column or post column derivatization, and by capillary electrophoresis [23-30], and by LC-MS [31-34]. None method by HPTLC was founded by quantitative determination of TPM in breast milk.

High performance thin-layer chromatography (HPTLC) is a technique carried out within a short period of time, requires few

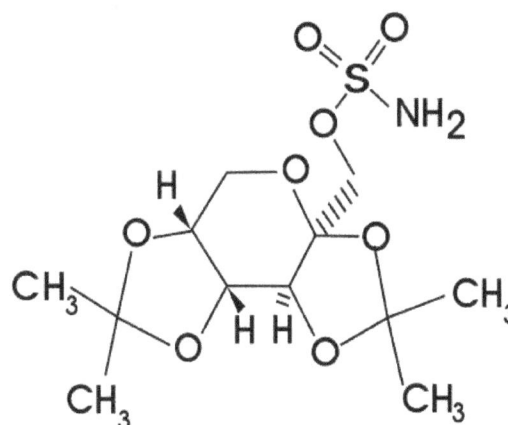

Figure 1: Chemical structure of TPM.

*Corresponding author: Sigrid Mennickent, Department of Pharmacy, Faculty of Pharmacy, University of Concepción, PO Box 237, Concepción, Chile
E-mail: smennick@udec.cl

mobile phase and allows for the analysis of a large number of samples simultaneously. A plate of 10 × 20 cm. allows applying up to 33 spots (27 samples plus standards). HPTLC allows to detect quantities in the order of micrograms and of nanograms (in UV absorbance mode) and smaller than picograms (in fluorescence mode). Therefore, HPTLC allows a fast analysis, with minor cost than other techniques, and with a high selectivity, accuracy and reproducibility.

The developed method can be used to quantitative determination of TPM in human breast milk, suitable for therapeutic drug monitoring of this drug in this matrix, considering as the woman in lactation period as her baby. From these values, could be estimate the baby blood topiramate concentration without the risks to take blood sample from his (her).

Experimental

Instrumentation and reagents

USP standards of topiramate and chlorpromazine were purchased from Sigma- Aldrich, St. Louis, MO. Methanol, ethanol, toluene and ninhidrine, were obtained from Merck, Darmstadt, Germany. All of the reagents were pro-analysis quality.

Preparation of standard solutions

Stock solutions containing 200 µg/mL of TPM were prepared in methanol. Separate solutions were prepared for the calibration standards and quality control samples. These solutions were diluted immediately before use with methanol, to obtain working solutions of 0.3 – 1.2 – 1.4 – 2.0 – 5.0 – 10.0 – 20.0 – 30.0 – 40.0 and 50.0 µg/mL All of solutions were stored at 4°C for about two days.

Spiking procedure for calibration and quality control (QC) samples

The calibration samples were prepared immediately before use by spiking 1 mL of pooled human breast milk with 0.1 mL of a convenient working solution in methanol. Quality control samples were used to determine the intra and inter-assay precision and accuracy of the method. Human breast milk used for the validation of the method was obtained from healthy volunteers.

Sample preparation

Human breast milk samples were stored at -20°C until required for analysis. Calibration and quality control samples were thawed at room temperature. Immediately after thawing, 2 mL of sample was processed by adding initially 150 µL of chlorpromazine (internal standard) (50 µg/mL) and 100 µL de NaOH 0.1 M. to the solution, which was subsequently vortexed and centrifuged for 5 minutes at 3000 rpm. Then, the supernatant was transferred to other recipient and 100 µL of methanol was added. Aqueous phase was evaporated under a gentle steam of dry nitrogen at 37°C, and the residue was dissolved in 300 µL of methanol. An aliquot of 1 µL of this solution was spotted for analysis, to obtain the required quantity/spot. All the procedure was accomplished under safety conditions.

Instrumentation and chromatographic conditions

The HPTLC system consisted of a TLC Scanner 3 (CAMAG, Muttenz, Switzerland), equipped with software winCATS 1.4.2 (CAMAG); band application device Linomat V (CAMAG); twin trough chromatographic chamber (CAMAG) 10 × 10 cm. and 20 × 10 cm; and HPTLC glass backed plates 10 × 10 cm. and 20 × 10 cm. Precoated with silica gel F 254, layer thickness 0.2 mm (Merck, Darmstadt, Germany), previously washed with methanol and activated at 120°C during 20 minutes.

Serum samples, calibration and quality control samples application volumes were 1 µL of each of them. Sample application was done on 4 mm bands. Number of tracks depended of each assay. Mobile phase consisted of a mixture of toluene: ethanol (25:10, v/v). The chamber was previously saturated. Migration distance was 8 cm. Derivatization of topiramate was done with ninhidrine (1% ethanolic solution), by immersion of the plates into Deeping Device (Camag). This procedure makes topiramate fluorescent and the background dark. Densitometry scanning was performed at 326 nm.

Stability study

To establishment the stability of TPM samples, in normal storage conditions, the study was performed as follows: six extractions solutions with derivative agent were used at four different concentrations: 0.3, 1.2 µg/mL, 10 µg/mL and 50 µg/mL. These solutions were storage at three different conditions: freezer temperature, room temperature with light protection, and room temperature without light protection. Concentration determination was evaluated at 0, 1, 4, 7, 11 and 15 days of storage. Each sample was determined by duplicated.

Method validation

Validation of developed LC method was carried out as per the International Conference on Harmonization (ICH) guidelines, and as per FDA [33,34].

Results and Discussion

During method development different conditions for sample extraction and chromatographic conditions was tried to achieve optimal results.

The selection of the mobile phase was carried out on the basis of polarity i.e., choice a solvent system that would give dense and compact spots with appropriate Rf value, as well as a satisfactory separation of TPM and internal standard and good peak symmetry. Several trials for optimization of mobile phase were taken and finalized as toluene: ethanol (25:10, v/v). This solvent system gave compact spots for topiramate.

Some wavelengths were tried, chosen 326 as working wavelength. Complete resolution of the peaks with clear baseline separation was obtained of this way.

Sample extraction was optimized to eliminate the laborious extraction steps, with minimal losses of carbamazepine and very good recoveries from spiked serum samples.

Calibration curves

Calibration curves were constructed over the concentration range of 0.3 to 50.0 µg/mL. Each solution was spotted three times. This range of solution concentrations include the TPM concentrations expected in human breast milk [19,20].

The mean equation (curve coefficients ± standard deviation) for the calibration curve (n=5), obtained from five points, was y=0.3577x(± 0.1) x+0.7893(± 0.2) with a correlation coefficient, r=0.99. ANOVA assay showed a relationship between the ratio of peak area of the analyte to the internal standard as a function of the concentration added, with p<0.005.

Precision and accuracy

The intra assay precision and accuracy was calculated at low (L), medium (M), and high (H) quality control levels for three replicates each of the same analytical run (each replicate was spotted three times), and inter assay precision and accuracy was calculated after repeated analysis in three different analytical runs.

Each experiment included the sample extraction step. The precision and the accuracy of the assay were measured by the relative standard deviation (RSD) over the concentration range. TPM solutions were of 0.3, 10.0 and 50.0 µg/mL. RSD for intra–assay study was between 3.04% and 3.14%, and for inter-assay was between 1.81% and 4.10%.

Accuracy was calculated from the test results as the percentage of analyte recovered by the assay, determined by linear regression equation of peak area vs. drug concentration. Accuracy was between 101.65% and 109.51%, expressed as analyte recovery percentage. The results are presented in Tables 1 and 2 (precision) and (accuracy) respectively.

Detection and quantification limits

The limit of detection (LOD) and limit of quantification (LOQ) were calculated preparing solutions at three concentrations (0.1-0.2-0.3 µg/mL) in the lower range of linear regression curve for both biological matrices.

LOD was 0.24 µg/mL and LOQ was 0.30 µg/mL, determined using the equations [35,36]: LOD=3.3 σ/b; LOQ=10 σ/b, where σ is the standard deviation of the values, and "b" corresponds to the slope obtained from the curve peak area versus concentration of the analyte. These values were experimentally verified.

Selectivity

The selectivity of the assay was checked by analyzing three independent blank human breast milk samples. The chromatograms of these blanks samples were compared with chromatograms obtained by analyzing the biological fluid samples spiked with TPM and chlorpromazine, the internal standard, and with other antiepileptic drugs: carbamazepine and phenytoin.

The solutions were prepared at a concentration of 20 µg/mL of each compound. TPM and chlorpromazine were well separated, with a resolution (R) value between both peaks of 5.20 (Figure 2). Carbamazepine and phenytoin run with the solvent line. Moreover, no interference was observed in drug free samples, indicating the high selectivity of the developed method.

Concentration µg/mL	RSD Intra-assay	RSD Inter-assay
0.3	3.04	3.46
10.0	3.10	1.81
50.0	3.14	4.10

Table 1: Precision of Topiramate in human breast milk.

Intra-assay			
Initial concentration (µg/mL)	Founded concentration (µg/mL)	Recovery (%)	RSD
0.3	1.34 ± 0.07	109.51	0.41
10.0	10.40 ± 0.01	101.65	0.01
50.0	49.96 ± 0.03	101.97	0.04
Inter-assay			
Initial concentration (µg/mL)	Founded concentration (µg/m)	Recovery (%)	RSD
0.3	1.35 ± 0.50	105.14	0.30
10.0	11.21 ± 0.09	108.47	0.08
50.0	50.93 ± 0.05	101.90	0.04

Table 2: Accuracy of Topiramate in human breast milk.

Extraction recovery

The extraction recovery of TPM from human breast milk and that of the internal standard were quantified using the concentrations of 0.3, 10.0 and 50 µg/mL for the drug, and 10 µg/mL of the internal standard. The extraction recoveries were calculated by comparing the observed concentrations obtained from the processed standard samples to the concentrations obtained from the standards solutions added to the human breast milk after the extraction, which represented 100% recovery. The extraction recovery of TPM from biological fluid ranged from 92.3% to 95.7%. The internal standard extraction recoveries were found between 91.6% and 94.8%.

Stability

Stability of TPM human breast milk was assessed with concentration solutions of 0.3, 1.2, 10.0 and 50.0 µg/mL, stored at -20°C for up to 12 weeks. Reference solutions for corresponding calibration curves were prepared freshly on each day of measurement (days 0, 10, 20, 40, 60 and 84). Observed concentrations of TPM during this time ranged between 96.5%-97.7%.

Freeze-thaw stability of TPM in biological fluid was studied using solutions exposed to three cycles of freezing-thawing versus regularly treated quality control samples. The measured concentrations of TPM ranged between 95.4% and 97.8%.

Application of the method

The developed method was linear between the concentrations range expected, precise, accurate, sensible, and selective for the quantitative determination of TPM in studied matrix: human breast milk. It is very important because TPM can produce severe damage to the fetus, and not always is possible that the epileptic woman change this drug when she is pregnant, because sometimes no other antiepileptic drug is effective in some people.

Using human breast milk we can estimate the drug quantity in baby blood, without the ethic aspects to withdrawal blood of them. It is possible to establish a relationship between the drug quantity in breast milk of epileptic woman with topiramate as her medication for the disease, and dug quantity in baby blood. These steps will be a continuation of this work, including another biological fluid studied before: human serum and umbilical cord blood. Of this way, we can compare the TPM quantity in these three biological fluids, establishing a correlation between drug quantity in serum of the mother, the drug quantity in umbilical cord blood, in breast milk, and the drug quantity in baby blood, using the values founded in serum´s mother, umbilical cord and in breast milk.

Conclusion

The most significant advantage of the present HPTLC method is this allows the quantitation of TPM in human breast milk with the aim of predict the drug concentration in the baby blood using a relationship between these levels. Of this way, it is not necessary to obtain blood from the baby to quantify the drug levels.

The chromatographic conditions are simple, the analysis requires a short period of time (HPTLC separation was obtained within twelve minutes), and the method allows for the analysis of many samples simultaneously with a very good accuracy, sensitivity and precision.

None method was founded by quantitative determination of TPM in breast milk.

Track 3, ID: Standard1

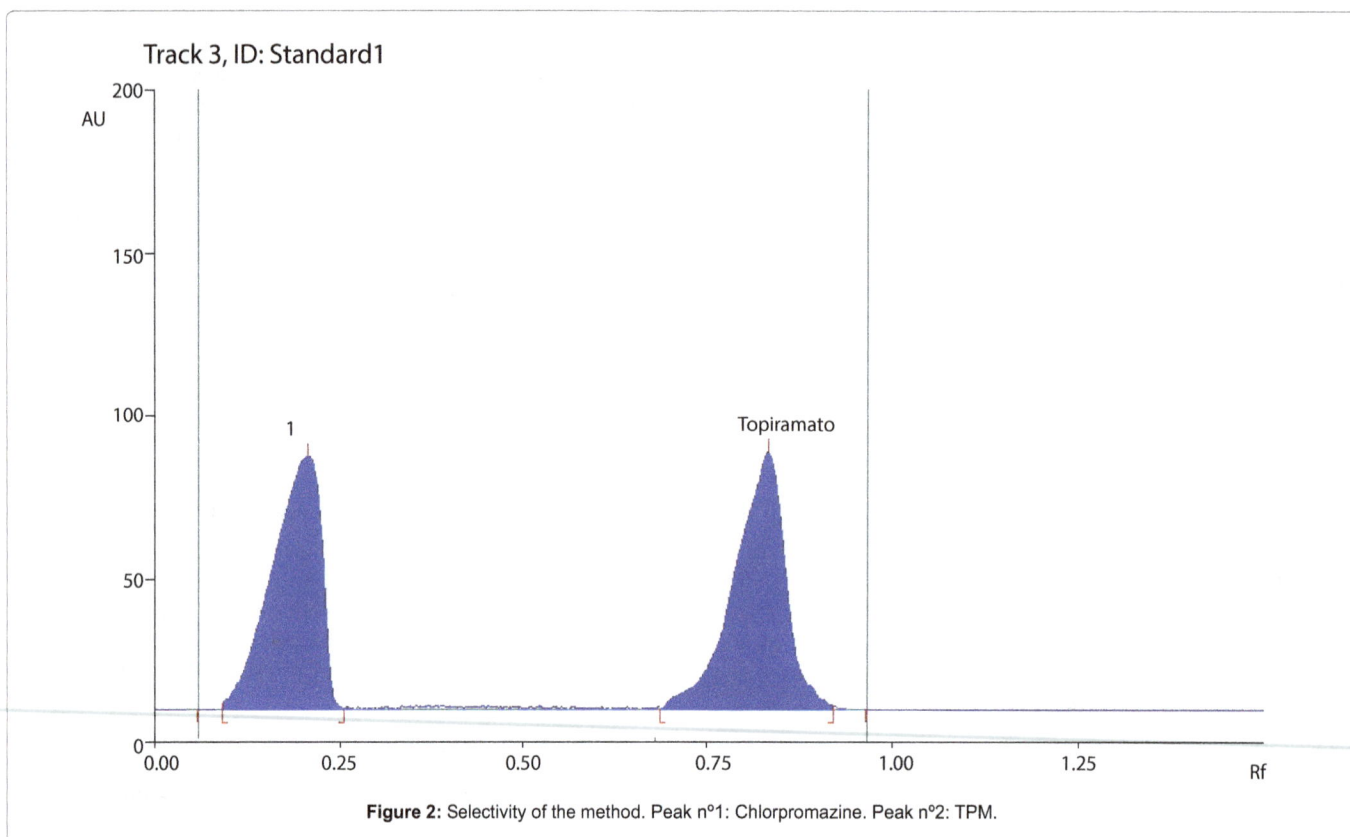

Figure 2: Selectivity of the method. Peak n°1: Chlorpromazine. Peak n°2: TPM.

Conflict of Interest Statement

The authors declare that they have no financial/commercial conflicts of interest.

Acknowledgements

The authors would like to thank the Research Council at the University of Concepción (Project UCO 12.01). This work is part of the Thesis of Master in Pharmaceutical Sciences (University of Concepción, Chile) of Miss Cristina Cifuentes.

References

1. Summary of Product Characteristics-Topamax® (2009) 15 mg 25 mg 50 mg tablets (topiramate). Ortho-McNeil-Janssen Pharmaceuticals Inc.

2. Sachdeo RC (1998) Topiramate Clinical profile in epilepsy. Clinical Pharmacokinetics 34: 335-346.

3. Glauser TA, Levisohn PM, Ritter F, Sachdeo RC (2000) Topiramate in Lennox-Gastaut syndrome: open-label treatment of patients completing a randomized controlled trial. Epilepsia 41: S86-S90.

4. Kugler SL, Sachdeo RC (1998) Topiramate efficacy in infancy. Pediatric Neurology 19: 320-322.

5. Huang WY, Lo MC, Wang SJ, Tsai JJ, Wu HM (2010) Topiramate in prevention of cluster headache in the Taiwanese. Neurology India 58: 284-287.

6. Kanemura H, Sano F, Tando T, Sugita K, Aihara M (2015) Effects of topiramate on headache in children with epilepsy. Brain & development 47: 18-22.

7. Sahraian A, Bigdeli M, Ghanizadeh A, Akhondzadeh S (2014) Topiramate as an adjuvant treatment for obsessive compulsive symptoms in patients with bipolar disorder: a randomized double blind placebo controlled clinical trial. Journal of affective disorders 166: 201-205.

8. Correll CU, Sheridan EM, Del Bello MP (2010) Antipsychotic and mood stabilizer efficacy and tolerability in pediatric and adult patients with bipolar I mania: a comparative analysis of acute, randomized, placebo-controlled trials. Bipolar disorders 12: 116-141.

9. Alfaris N, Minnick AM, Hopkins CM, Berkowitz RI, Wadden TA (2015) Combination phentermine and topiramate extended release in the management of obesity. Expert opinion on pharmacotherapy 16; 1263-1274.

10. Wiffen PJ, Derry S, Lunn MP, Moore RA (2013) Topiramate for neuropathic pain and fibromyalgia in adults. CDSR 8: CD008314.

11. Chiaro G, Caletti MT, Provini F (2015) Treatment of sleep-related eating disorder. Current Treatment Options in Neurology 17: 361.

12. Guglielmo R, Martinotti G, Quatrale M, Ioime L, Kadilli I, et al. (2015) Topiramate in Alcohol Use Disorders: Review and Update. CNS drugs 29: 383-395.

13. Hammond CJ, Niciu MJ, Drew S, Arias AJ (2015) Anticonvulsants for the treatment of alcohol withdrawal syndrome and alcohol use disorders. CNS drugs 29: 293-311.

14. Khazaal Y, Zullino DF (2009) Topiramate for smoking cessation and the importance to distinguish withdrawal-motivated consumption and cue-triggered automatisms. Journal of Clinical Psychopharmacology 29: 192-193.

15. Oncken C, Arias AJ, Feinn R, Litt M, Covault J, et al. (2014) Topiramate for smoking cessation: a randomized, placebo-controlled pilot study. Nicotine & tobacco research 16: 288-296.

16. Reid MS, Palamar J, Raghavan S, Flammino F (2007) Effects of topiramate on cue-induced cigarette craving and the response to a smoked cigarette in briefly abstinent smokers. Psychopharmacology 192: 147-158.

17. Hardman J, Limbird L (2006) Las Bases Farmacológicas de la Terapéutica. Mc Graw-Hill, Mexico, pp. 550.

18. McEvoy G (2012) AHFS Drug Information American Society of Health - System Pharmacists. Bethesda, pp: 2219-2223.

19. Ohman I, Vitols S, Luef G (2002) Topiramate kinetics during delivery, lactation, and in the neonate: preliminary observations. Epilepsia 43: 1157-1160.

20. Froscher W, Jurges U (2006) Topiramate used during breast feeding. Aktuel Neurol 33: 215-217.

21. Sweetman S (2006) Martindale, Guía Completa de Consulta Farmacoterapéutica. Pharma Editores SL, Barcelona, pp: 500-501.

22. Guerrini R, Parmeggiani L (2006) Topiramate and its clinical applications in epilepsy. Expert Opinion on Pharmacotherapy 7: 811-823.

23. Bahrami G, Mirzaeei S, Mohammadi B, Kiani A (2005) High performance liquid chromatographic determination of topiramate in human serum using UV detection. Journal of Chromatography B: Biomedical Sciences and Applications 822: 322-325.

24. Walker RB, Mohammadi A, Rezanour N, Ansari M (2010) Development of a Stability-Indicating High Performance Liquid Chromatographic Method for the Analysis of Topiramate and Dissolution Rate Testing in Topiramate Tablets. Asian Journal of Chemistry 22: 3856-3866.

25. Styslo-Zalasik M, Li W (2005) Determination of topiramate and its degradation product in liquid oral solutions by high performance liquid chromatography with a chemiluminescent nitrogen detector. Journal of Pharmaceutical and Biomedical Analysis 37: 529-534.

26. Jalili R, Majnooni MB, Mohammadi B, Fakhri S, Mirzaei S, et al. (2014) Development and validation of a new method for determination of topiramate in bulk and pharmaceutical formulation using high performance liquid chromatography UV detection after precolumn derivatization. Journal of Reports in Pharmaceutical Sciences 3: 179-183.

27. Mandrioli R, Musenga A, Kenndler E, De Donno M, Amore M, et al. (2010) Determination of topiramate in human plasma by capillary electrophoresis with indirect UV detection. Journal of Pharmaceutical and Biomedical Analysis 53: 1319-1323.

28. Roskar R, Milosheska D, Vovk T, Grabnar I (2015) Simple and Sensitive High Performance Liquid Chromatography Method with Fluorescence Detection for Therapeutic Drug Monitoring of Topiramate. Acta Chimica. Slovenica 62: 411-419.

29. Bahrami G, Mohammadi B (2007) A novel high sensitivity HPLC assay for topiramate, using 4-chloro-7-nitrobenzofurazan as pre-column fluorescence derivatizing agent. Journal of Chromatography B: Biomedical Sciences and Applications 850: 400-404.

30. Bahrami G, Mirzaeei S, Kiani A (2004) Sensitive analytical method for Topiramate in human serum by HPLC with pre-column fluorescent derivatization and its application in human pharmacokinetic studies. Journal of Chromatography B: Biomedical Sciences and Applications 813: 175-180.

31. Contin M, Riva R, Albani F, Baruzzi A (2001) Simple and rapid liquid chromatographic-turbo ion spray mass spectrometric determination of topiramate in human plasma. Journal of Chromatography B: Biomedical Sciences and Applications 761: 133-137.

32. Mercolini L, Mandrioli R, Amore M, Raggi MA (2010) Simultaneous HPLC-F analysis of three recent antiepileptic drugs in human plasma. Journal of Pharmaceutical and Biomedical Analysis 53: 62-67.

33. Popov TV, Maricic LC, Prosen H, Voncina DB (2013) Determination of topiramate in human plasma using liquid chromatography tandem mass spectrometry. Acta Chimica Slovenica 60: 144-150.

34. Kuchekar SR, Zaware BH, Kundlik ML (2010) Rapid and Specific Approach for Direct Measurement of Topiramate in Human Plasma by LC-MS/MS: Application for Bioequivalence Study. Journal of Bioanalysis & Biomedecine 2: 107-112.

35. The Sixth ICH International Conference on Armonization of Technical Requirements for Registration of Pharmaceuticals for Human Use (2003) Osaka, Japan.

36. The United States Pharmacopeia (USP30) (2007) United States Pharmacopeial Convection, Inc., Rocville, USA.

20

Selected Reaction Monitoring: A Valid Tool for Targeted Quantitation of Protein Biomarker Discovery

Subodh Kumar* and Priyanka Mittal

Central Research Laboratory, Multi-disciplinary Research Unit, University College of Medical Sciences (University of Delhi) and GTB Hospital, Delhi, India

Abstract

Biomarker discovery is relying on the sensitivity and specificity of the detected biomolecules in clinical samples. Several potential biomarkers against the particular disease cannot be detected because of the unavailability of either specific procedure or sensitive instrument or both, which hamper the diagnosis. In spite of that some of the potential biomarkers are available in trace amount in biological fluids like serum, urine, buccal swab, sputum etc. and need a sensitive method to detect precisely those biomolecules. Now days Selected Reaction Monitoring (SRM) is Mass Spectrometry based approached who can overcome the problems associated with biomarker discovery. India have adequate and variety of clinical samples but some time due to lack of infrastructure and knowledge we are unable to utilise those samples for human welfare. In this review, we are discussing about the experimental setup and procedure of SRM experiment.

Keywords: LC-MS/MS; Triple quad; Selected reaction monitoring; Biomarker; Diagnosis

Abbreviations: LC-MS: Liquid Chromatography-Mass Spectrometry; SRM: Selected Reaction Monitoring; ELISA: Enzyme Linked Immuno-Sorbent Assay; ICT: Immunochromatography Test; SDS-PAGE: Sodium do-decyl sulphate-Poly acrylamide gel electrophoresis; DTT: Dithiothreitol; IAA: Indole Acetic Acid; MS: Mass Spectrometer; ESI: Electro Spray Ionization.

Introduction

Mass spectrometry (MS) is an analytical technique which is used to measure sample's molecular mass by measuring mass to charge ratio. In this method, a soft ionization technique like ESI (Electro Spray Ionization) is used to generate charge ions. Now days, this is the most popular method for protein identification and quantitation. In shotgun proteomics proteins are digested into smaller peptides followed by analysis of complex mixture of peptides on high performance liquid chromatography (HPLC) coupled with mass spectrometry. MS based proteomics approach can be designed to use either non-targeted (shotgun) or targeted (SRM) proteomics. Bottom up shotgun MS approach is commonly used in discovery proteomics. Shotgun proteomics is a powerful tool for high-throughput peptide/protein identification and relative quantification but unsuitable for absolute quantification. In contrast to this, MS based targeted proteomics is mostly used for absolute quantification of small set of protein or peptide.

Targeted Proteomics

In contrast to shotgun proteomics [1], targeted proteomics is now days play a very important role in the accurate measurement of protein of interest in complex biological sample which can be helpful in dissecting the protein network and pathway. Selected Reaction Monitoring (SRM) is one such approach.

SRM-MS (Selected Reaction Monitoring-Mass Spectrometry)

Selected Reaction Monitoring (SRM) is a non-scanning technique, generally performed on triple quadrupole (QQQ) instruments in which fragmentation is used as a means to increase selectivity. In SRM, the first and third quadrupoles act as filters to specifically select predefined m/z values corresponding to the peptide ion and a specific fragment ion of the peptide, whereas the second quadrupole serves as collision cell (Figure 1).

SRM is a revolutionary tool for clinical application, for biomarker validation of blood plasma [2]. The method proposed in this review is a targeted proteomics approach which combines multiple approaches investigating a certain set of proteins in more detail. One such targeted proteomics approach is the combination of liquid chromatography and selected reaction monitoring or multiple reactions monitoring (SRM/MRM). The robustness of this methodology not only allows for consistent measurement of different sample conditions, but also obtains reliable results across various laboratories with coefficient of variation less than 20% [3]. SRM-MS requires the prior knowledge of fragmentation pattern of the targeted peptide because the analytes in a sample is determined by measuring m/z values of predefined precursor and fragment ions.

The two level of mass selection with narrow mass windows results in a high selectivity as co-eluting background ions are filtered out very effectively. These dual filters provide a higher sensitivity for monitoring specific analytes as supposed to conventional shotgun experiment. The set of precursors and their corresponding fragment ion is usually referred as **"TRANSITIONS"**. Unlike in other MS-based proteomics techniques, no full mass spectra are recorded in QqQ-based SRM analysis resulting in an increased sensitivity by one or two orders of magnitude compared with conventional full scan techniques. Typically, positive detection of specific precursor/peptide of interest is usually based on the positive detection of atleast three to five coeluting transition ions of particular precursor/peptide of interest, hence, SRM-MS does not depend on single spectrum for positive identification, but on coeluting transitions [4]. In an unscheduled MRM, the number of transitions is defined by the *dwell time* and *cycle time*. The dwell time is the time to take each transition and cycle time is the times to complete

---

*Corresponding author: Subodh Kumar, Research Scientist, Multi-Disciplinary Research Unit, UCMS and GTB Hospital, University of Delhi, Delhi-110 095, India E-mail: subodh_bt2003@yahoo.co.in

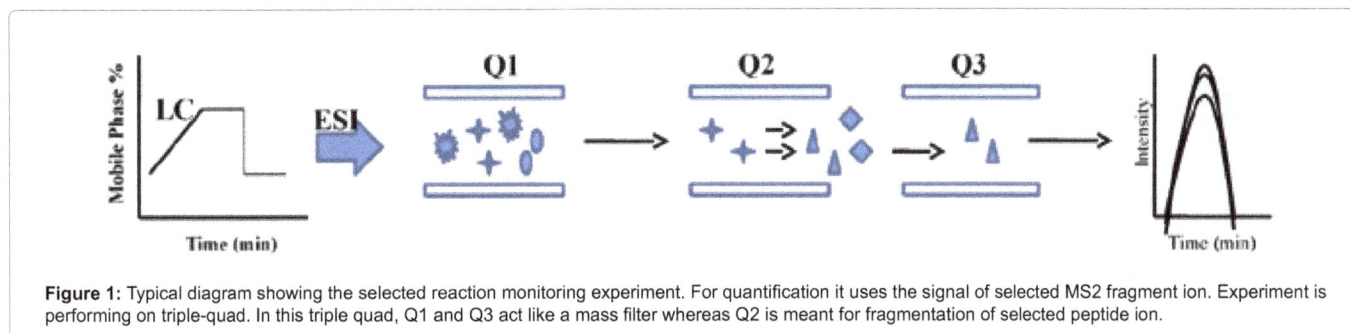

Figure 1: Typical diagram showing the selected reaction monitoring experiment. For quantification it uses the signal of selected MS2 fragment ion. Experiment is performing on triple-quad. In this triple quad, Q1 and Q3 act like a mass filter whereas Q2 is meant for fragmentation of selected peptide ion.

all the transitions listed in the methods. The higher the dwell-time, the higher the signal-to-noise ratio and thus lower the limit of detection [5,6]. Therefore, it's very important to set the dwell time 2-4 minutes time window for specific and reproducible detection for peptide/protein of interest. In spite of the dwell and cycle time, some instrument specific set up time is also considered that switch the *m/z* of the filter. For example if each transition is measured for 30 ms dwell time and instrument set up time is 3 ms, in 2000 ms approximately 60 transitions can be measured. In an unscheduled SRM-MS, only 12 peptides can be measured with 5 transitions per peptide. Since the numbers of peptide detected is low therefore scheduled SRM-MS could be a choice but the limitation with this method is that we should know the retention time of the peptide. If we wish to effectively monitor hundreds of assays in a single injection during LC-SRM-MS run, the window should be set between 2-4 minutes.

In the proposed review we discussed in brief about the instrumentation, parameters etc. of targeted proteomics approach for biomarker discovery.

General procedure to establish a proteomic SRM experiment

In contrast to the conventional shotgun proteomic studies, in LC-SRM-MS, three main selection processes should be kept in mind before going to start validation of the experiment and these are; first the information about the protein that is to be targeted; second, the selection of those peptide that present good MS response and uniquely identify the targeted protein, such peptide are known as proteotypic peptide (PTPs) [7] and third, identified those PTP (Proteotypic peptide) who have optimal signal intensity. The overall workflow of the experiment is mentioned in Figure 2. The major steps that are involved in the experiment are explained in the following sections.

Selection of target protein

In LC-SRM-MS, the researcher should have prior knowledge of protein and their peptide behaviours which includes the retention time and fragmentation pattern. The major issue is with the selection of those target proteins which are present in low abundance in a complex sample. This information is either present in the repositories data base available online or have to be determined before SRM-MS experiment through bottom up shotgun LC-MS/MS experiment. Several, sites are available freely to select the target proteins (Table 1a) [8-14]. The success of shotgun proteomics relies on the sample preparation. In this section, we are discussing about the proteomics sample preparation.

Sample preparation for shotgun proteomics: Add four volume of chilled acetone to one volume of plasma/serum sample followed by vortex for complete breakdown. Store the sample at -20°C for overnight. After incubation, spin and the resulting pellet should wash away with chilled acetone. Add one volume of buffer containing 8 M

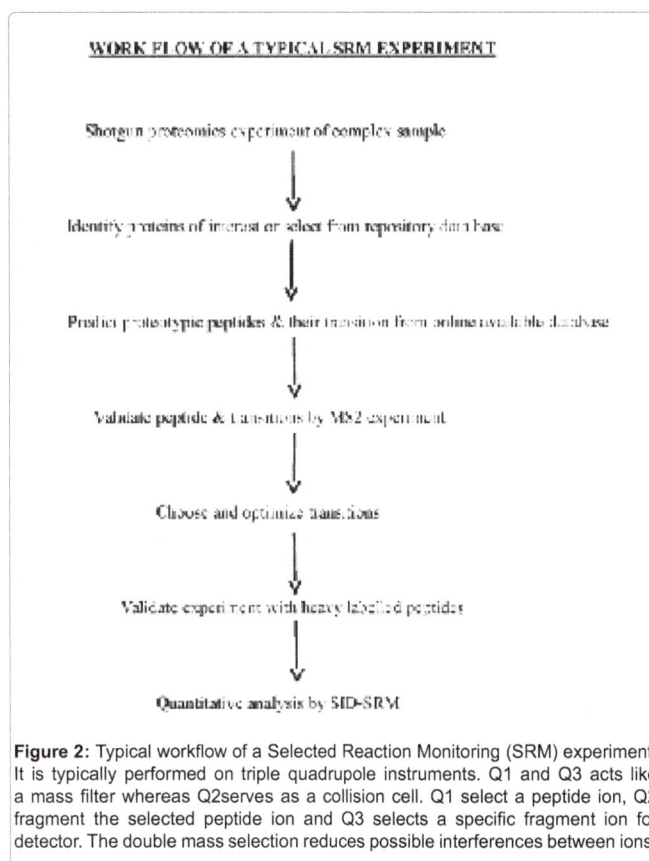

Figure 2: Typical workflow of a Selected Reaction Monitoring (SRM) experiment. It is typically performed on triple quadrupole instruments. Q1 and Q3 acts like a mass filter whereas Q2serves as a collision cell. Q1 select a peptide ion, Q2 fragment the selected peptide ion and Q3 selects a specific fragment ion for detector. The double mass selection reduces possible interferences between ions.

urea in 100 mM ammonium bicarbonate (pH 8.0) to initial volume of the sample. Subsequently add protease and phosphatase inhibitor in the sample. Vortex thoroughly and centrifuge the dissolve protein. Collect the supernatant in a fresh tube and do the protein estimation either through Lowry or BCA method. Dilute sample to 1 M urea concentration with 100 mM ABB buffer followed by reduction (with 20 mM DTT, at 60°C for 20-30 min) and alkylation (with 60 mM IAA, in dark for 30 minutes). The resulting sample will digest two times with trypsin at 3-4 hrs interval followed by overnight incubation at 37°C incubator. The reaction will be stop by adding 10% acetic acid to the digested sample. The digested sample desalt with C18 cartridge. The desalted samples now do the fractionation with either 1D or 2D separation on HPLC and collect the fractionated samples. Speedvac the samples to dry in vacuum and reconstitute with the appropriate solvent. The reconstituted sample will be vortex vigorously followed by centrifugation to collect the upper layer of the sample. Now the sample is ready to inject in mass spectrometer for proteomic study.

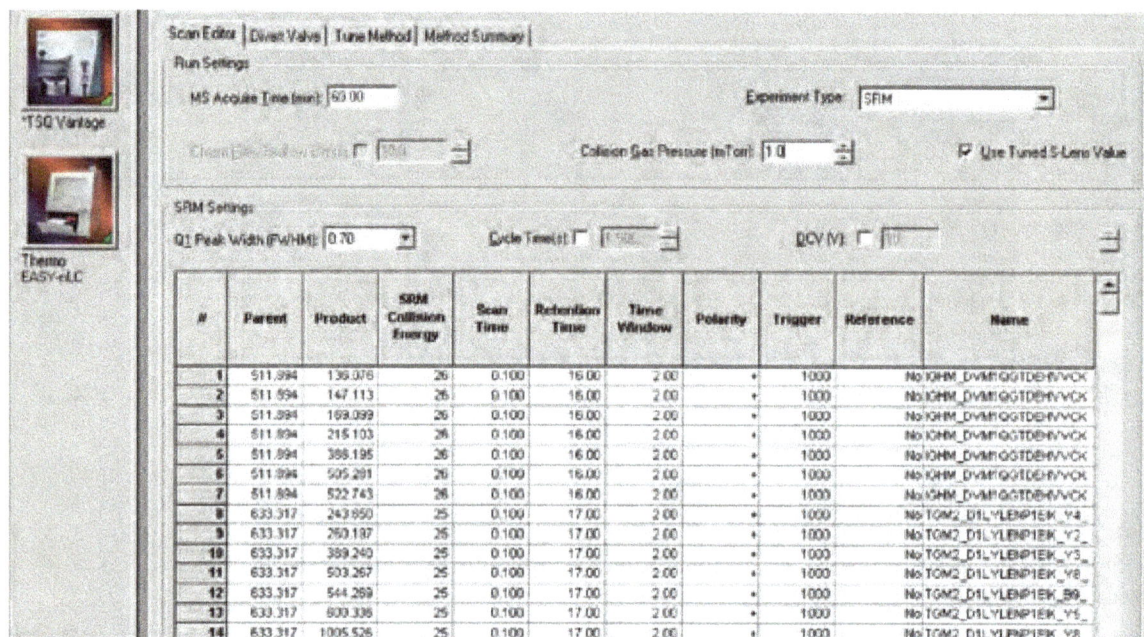

Figure 3: A typical example of setting the SRM's experimental parameters in Instrument (TSQ Vantage, Thermofischer Scientific Inc.) method development program. It's just an example, that how to set the method for SRM-MS experiment, which will be helpful to the users for their experiments.

***In gel* digestion:** The SDS-PAGE gel containing desired protein band will be place in petridish and fill it with MilliQ water. Remove the gel from pertidish and place it on glass slide. Cut the gel into their respective bands and dice the gel slice into uniform pieces and place them in eppendorf tube. The tube containing dice piece of gel is soaked in 0.5 ml of 25 mM ABB (Ammonium Bicarbonate) buffer and vortex for 10-15 min. To destain the gel, add 1 ml of 25 mM ABB buffer containing 50% ACN. Vortex for 20-30 minutes at RT (Room Temperature) followed by several washing. Dehydrate the gel with 100% ACN (Acetonitrile) solution by vortexing. For reduction and alkylation incubate the tube with 10 mM DTT (for 1 h at 56°C in water bath) and 55 mM IAA (for 45 minutes at RT in dark), respectively. Digest the gel with freshly prepare trypsin solution. Put on ice for 15 minutes until the gel particles absorb the buffer completely. Transfer gel to 37°C and incubate overnight. Finally the peptides will extract by adding 2 volume of 50% ACN containing 0.1% TFA. Vortex and collect the supernatant in a fresh tube and speed vac to dry the supernatant in vaccum. Desalt the extracted peptide on C18 cartridge. After desalting, dissolve the peptide in appropriate solvent for LC-MS/MS analysis.

Peptide selection

Upon trypsin digestion, each protein generates ten to hundreds of peptides [15]. The sensitivity of the MRM assay relies on the selection of peptides with favourable mass spectrometry properties and these are;

MS properties: A number of peptide generates after digestion with trypsin but only a small subset is routinely observed because they satisfy the required MS properties [16]. For assay development, the time can be reduced significantly if previous information generated from multiple shotgun experiment, deposited in online repositories is used to select those peptide that are most likely observed in the experiment and provide the strongest specific signals (Table 1a).

Uniqueness: For targeted MS analysis, it is essential to ensure that the peptide selected uniquely identify the targeted protein or one isoform thereof.

Post Translational Modification (PTM): Modified peptide can't be detected by SRM unless specifically targeted. For reliable quantification, at least two should be monitored for each targeted protein.

Chemically induced modification: The potential source of error in quantitative MS experiments, during sample processing is the introduction of artifactual chemical modification in the sequence. Therefore, avoid those peptides with a high propensity for artifactual modifications. Generally, avoid peptides containing Methionine or Tryptophan and Glutamine or Asparagine in the sequence, as they are prone to oxidation and deamidation, respectively.

Cleavage site: Data analysis in shotgun proteomics experiment revealed a number of missed cleavage due to incomplete digestion of trypsin. These peptide sequences should be avoided for absolute quantification experiment. In general, peptide with two neighbouring basic amino acids at either cleavage site of the protein sequence should be avoid as those sites are predestined for a high rate of missed cleavage.

Selection of SRM Transitions

In general, for SRM experiment per peptide atleast 2-3 transition will be selected whose intensity is comparatively more so that it can be quantitate effectively. There is commercially available software which generates peptide transition *in-silico*. The commercially available software that helps in the selection of peptide transitions is listed in Table 1b [17,18].

Validation of transitions

Peptides with precursor/fragment ion pairs of similar masses shows unspecific signal. This is because of the closely related sequences so that parts of the transitions are identical. The first step of validation is the parallel acquisition of several transitions for a targeted peptide (Figure 3). At the time of peptide elution, such transition yield a perfect set of co-eluting intensity peaks if they are derived from same peptide.

S No	Software	Link	References
1	Gene expression GEO	http://www.ncbi.nlm.nih.gov/geo/	[8]
2	Protein expression Protein Atlas	http://www.proteinatlas.org/	[9]
3	Gene ontology groups GO	http://www.geneontology.org/	[10]
4	Functional groups KEGG	http://www.genome.jp/kegg/	[11]
5	Protein–protein interactions IntAct	http://www.ebi.ac.uk/intact/	[12]
6	Protein–protein interactions MINT	http://mint.bio.uniroma2.it/	[13]
7	Network expansion PhosphoPep	http://www.phosphopep.org/	[14]

Table 1a: Online information resources relevant to selection of a set of proteins of interest [4].

S No	Software	Link	Vendor/Lab	Reference
1	MRMaid	http://www.ncbi.nlm.nih.gov/pubmed/19011259	Bessant/Cranfield University	[17]
2	MRMPilot	http://mrmpilot.software.informer.com/	AB Sciex	Commercial
3	Pinpoint	https://www.thermofisher.com/order/catalog/product/IQLAAEGABSFALDMAXF	Thermo Scientific	Commercial
4	Skyline	https://skyline.gs.washington.edu/labkey/project/home/begin.view?	MacCoss Lab, U Washington	[18]

Table 1b: List of software's uses for selected Reaction Monitoring (SRM) Experiment's development.

Quantification of SRM transitions

Two emerging core strategies for targeted biomarker quantitation are

1) LC-SID-MRM-MS and

2) The combination of the so called SISCAPA technology.

The gold standard for MS-based quantitative method is to use stable-isotope-labelled peptide standard spiked into the sample of interest which then allow quantitation from a simple ratio of the labelled peak height or peak area to the unlabelled. However, some analogue IS can also work well when stable isotope labelled IS is not available. This stable isotope dilution (SID) method is the basis for SRM or MRM method. When some of the peptides are spiked with isotopic reference peptides, the corresponding endogenous peptides with the same sequences can be accurately quantified.

Discussion

An ideal biomarker has high sensitivity and specificity against a particular state of the disease and it should be validated on atleast 200 control group samples. On the other hand antigen based diagnosis is more advantageous over the antibody based detection. The major drawback of antibody based detection is their cross reactivity with other similar diseases, showing false positivity with endemic healthy controls and unable to differentiate between past and active infection etc. especially in infectious diseases. In infectious diseases, during active infection, several host or parasite specific proteins either up regulate or down regulate and these proteins can be a biomarker for the particular disease, who can overcome the above mentioned problem but the limitation is their detection and quantitation in clinical samples as they are present in trace amount. On the other hand, in non-communicable diseases like cancer, atherosclerosis, diabetes etc. several diagnostic biomarkers had been identified in the past by using mass spectrometry based proteomics approach [19-21] but their quantitative estimation in real samples had not been followed. Now days, targeted proteomics approach of triple quadrupole based Selected Reaction Monitoring (SRM) is experiment of choice which is able to quantitate effectively and sensitively in clinical sample. In this review, we have discussed about the importance and experimental procedures which will be helpful in designing the biomarker discovery experiment and will be helpful in achieving the goal of elimination program of communicable and non-communicable diseases.

Conclusion

SRM is a revolutionary MS based approach for a confident quantitation of small set of proteins/peptides in complex biological samples. This method is highly sensitive and specific as double mass selection reduces the possible Ion interference. Previously, due to lack of such a sensitive technique we were unable to analyse and validate the known biomarker against a specific disease across different set of samples. This analytical based method will surely be helpful in near future in improving the human health by identifying those biomarkers which is either present is trace amount or differentially express. After the successful identification, in long term goal, from the final outcome, we can develop simple and easy detection format, either in the form of Immunochromatographic strip (ICT) or ELISA (Enzyme Linked Immunosorbent Assay) format for commercial use. Therefore, we can utilize this technique for biomarker discovery against several deadly diseases to support the elimination program.

Conflict of Interest

The authors declare that there is no conflict of interest.

Acknowledgements

Author would like thanks to Department of Health Research-Indian Council of Medical Research, New Delhi, India and Central Research Laboratory, UCMS and GTB Hospital for their support to complete this review. Author would also like thanks to Dr. SK Sze, Newman, Nanyang Technological University, Singapore, for their guidance in understanding the selected reaction monitoring (SRM) experiment.

References

1. Hayes McDonald W, John Yates R (2002) Shotgun proteomics and biomarker discovery. Disease Markers 18: 99-105.

2. Anderson L, Hunter CL (2006) Quantitative mass spectrometric multiple reaction monitoring assays for major plasma proteins. Mol Cell Proteomics 5: 573-588.

3. Prakash A, Tomazela DM, Frewen B (2009) Expediting the development of targeted SRM assays: using data from shotgun proteomics to automate method development. J Proteome Res 8: 2733-2739.

4. Barnidge DR, Dratz EA, Martin T (2003) Absolute quantification of the G protein-coupled receptor rhodopsin by LC/MS/MS using proteolysis product peptides and synthetic peptide standards. Anal Chem 75: 445-451.

5. Lange V, Picotti P, Domon B (2008) Selected reaction monitoring for quantitative proteomics: a tutorial. Mol Syst Biol 4: 222.

6. Picotti P, Aebersold R (2012) Selected reaction monitoring-based proteomics: workflows, potential, pitfalls and future directions. Nat Methods 9: 555-566.

7. Mallick P, Schirle M, Chen SS (2007) Computational prediction of proteotypic peptides for quantitative proteomics. Nat Biotechnol 25: 125-131.

8. Barrett T, Troup DB, Wilhite SE (2007) NCBI GEO: mining tens of millions of expression profiles database and tools update. Nucl Acids Res 35: D760-D765.

9. Uhlen M, Ponten F (2005) Antibody-based proteomics for human tissue profiling. Mol Cell Proteomics 4: 384-393.

10. Ashburner M, Ball CA, Blake JA (2000) Gene ontology: tool for the unification of biology. The Gene Ontology Consortium. Nat Genet 25: 25-29.

11. Kanehisa M, Goto S (2000) KEGG: Kyoto Encyclopedia of Genes and Genomes. Nucl Acids Res 28: 27-30.

12. Kerrien S, Alam-Faruque Y, Aranda B (2007) IntAct-open source resource for molecular interaction data. Nucl Acids Res 35: D561-D565.

13. Chatraryamontri A, Ceol A, Palazzi LM (2007) MINT: The Molecular INTeraction database. Nucl Acids Res 35: D572-D574.

14. Bodenmiller B, Malmstrom J, Gerrits B (2007) PhosphoPep–a phosphoproteome resource for systems biology research in Drosophila Kc167 cells. Mol Syst Biol 3: 139.

15. Picotti P, Aebersold R, Domon B (2007) The implications of proteolytic background for shotgun proteomics. Mol Cell Proteomics 6: 1589-1598.

16. Kuster B, Schirle M, Mallick P (2005) Scoring proteomes with proteotypic peptide probes. Nat Rev Mol Cell Biol 6: 577-583.

17. Mead JA, Bianco L, Ottone V (2009) MR Maid, the web-based tool for designing multiple reaction monitoring (MRM) transitions. Mol Cell Proteomics 8: 696-705.

18. MacLean B, Tomazela DM, Shulman N (2010) Skyline: an open source document editor for creating and analyzing targeted proteomics experiments. Gene expression 26: 966-968.

19. Prieto DA, Johann DJ, Bih-Rong W (2014) Mass spectrometry in cancer biomarker research: a case for immuno depletion of abundant blood-derived proteins from clinical tissue specimens. Biomark Med 8: 269-286.

20. Tessitore A, Gaggiano A, Cicciarelli G (2013) Serum Biomarkers Identification by Mass Spectrometry in High-Mortality Tumors. International Journal of Proteomics.

21. Dang VT, Werstuck GH (2016) Metabolomics - Based Biomarkers of the Pathogenesis of Atherosclerosis. Biomark J 2: 10.

21

Sodium Dodecyl Sulphate/Poly(Brilliant Blue)/Multi Walled Carbon Nanotube Modified Carbon Paste Electrode for the Voltammetric Resolution of Dopamine in the Presence of Ascorbic Acid and Uric Acid

Ganesh PS and Kumara Swamy BE*

Department of PG Studies and Research in Industrial Chemistry, Kuvempu University, Jnana Sahyadri, Shankaraghatta, Shimoga, Karnataka, India

Abstract

Sodium dodecyl sulphate/poly(brilliant blue)/multi walled carbon nanotube modified carbon paste electrode was fabricated for the electroanalysis of dopamine in the presence of ascorbic acid and uric acid in phosphate buffer solution of pH 7.4. The key parameters such as sensitivity, selectivity, antifouling property and stability were achievedby the modified electrode. The redox peaks obtained at modified electrode shows good electrocatalytic activity towards the oxidation of dopamine. From the effect of scan rateand concentration the electrode phenomenon was confirmed to be adsorption-controlled process. The lower limit of detection of dopamine was 2.69×10^{-7}M, and the simultaneous analysis shows a good result with peak to peak separation between dopamine and other two analytes ascorbic acid and uric acid by both cyclic voltammetry and differential pulse voltammetric techniques.

Keywords: Electroanalysis; Dopamine; Poly(brilliant blue); Multi walled carbon nanotube; Sodium dodecyl sulphate; Voltammetry

Introduction

In the last few decades the electrochemical methods are most widely studied and accepted for the determination of electroactive compounds in pharmaceutical samples and physiological fluids due to its simple, sensitive, rapid and economical properties [1,2]. Particularly the development of voltammetric sensors for both the analysis and determination of biologically important molecules such as dopamine (DA), ascorbic acid (AA) anduric acid (UA) has been received much more importance for the scientific growth of electroanalytical research [3-6]. DA is one of the naturally occurring neurotransmitter in the human brain belongs to the family of catecholamine and plays a very important role in the normal activity of the central and peripheral nervous systems [7,8]. Extreme abnormalities of DA levels are the major drawbacks of several diseases, such as Parkinsonism and Schizophrenia [9,10]. A patient suffering from this disease shows a low level of DA. UA is the primary product of purine metabolism in the human body and major nitrogenous compound inthe urine [11]. Its abnormality in human body leads to many diseases, such as gout, hyperuricaemia and Lesch-Nyan disease [12,13]. Increased urate level also leads to pneumonia and leukaemia [14]. AA is also known as vitamin-c and is water soluble compound that take part in many important life processes, it has been used as a medicament for the treatment of common cold, mental illness and cancer [15]. It can be chemically or electrochemically oxidized to dehydroascorbic acid [16]. Hence monitoring the concentration of these biological compounds is veryimportant in clinical diagnosis. Normally in the electrochemical detection of DA the major problem was the coexistence of the common interference AA and UA. Because of the similar oxidation potential of AA there is an always interference in analysing the DA electrochemically and gives overlapped and broad voltammetric response at bare carbon paste electrode (BCPE). To improve the performance of the BCPE the properties such as sensitivity, selectivity, antifouling property, reproducibility and stability are the key parameters [17-21]. Carbon nanotubes (CNTs) have been one of the most actively studied electrode material due to their unique electronic and mechanical properties and also the availability of high accessible surface area and low resistivity [22-25]. Recently carbon nanotubes modified electrodes were utilised

to investigate the direct electrochemistry of several biomolecules [26-29]. A surfactant is a linear molecule with amphiphilic or amphipathic behaviour and they bear an ionicor non-ionic polar head group and a hydrophobic portion. Due to their unique molecular structure, surfactants have been employed extensively in the field of electroanalytical chemistry for various purposes [30-33]. In the present work the modification of the bare carbon paste electrode was achieved by using different quantity of multi walled carbon nanotube (MWCNT) along with carbon powder and a silicon oil binder by mechanical grinding method. In order to enhance both the sensitivity and selectivity it was further modified by immobilising the sodium dodecyl sulphate (SDS) to the surface of the electrode followed by electropolymerisation brilliant blue [34]. The fabricated electrode was employed for the electroanalysis of dopamine in presence of ascorbic acid and uric acid at physiological pH.

Experimental Section

Reagents

Multi walled carbon nanotube (110-170 nm) are received from Sigma Aldrich. Dopamine hydrochloride (DA), Uric acid (UA), Ascorbic acid (AA) were purchased from Himedia and the stock solutions of 25×10^{-4}M, 25×10^{-3}M and 25×10^{-3}M were prepared in 0.1M perchloric acid, 0.1M NaOH and double distilled water respectively. Buffer used was 0.2M phosphate buffer solution (PBS) of pH 7.4. Graphite powder of 50 μm particle size was purchased from

*Corresponding author: Kumara Swamy BE, Department of PG Studies and Research in Industrial Chemistry, Kuvempu University, Jnana Sahyadri, Shankaraghatta-577 451, Shimoga, Karnataka, India
E-mail: kumaraswamy21@yahoo.com

Merck and silicone oil from Himedia was used to prepare carbon paste electrode (CPE). All the chemicals mentioned were all of analytical grade used as received without any further purification.

Apparatus

The electrochemical experiments were performed using a model CHI-660c (CH Instrument-660 electrochemical workstation) at ambient temperature. A traditional three electrode compartment was used with a saturated calomel electrode (SCE) as a reference, a platinum counter electrode, and bare or Sodium dodecyl sulphate/poly(brilliant blue)/multi walled carbon nanotube modified carbon paste electrode (SDS/Poly(brilliant blue)/MWCNT/MCPE) as working electrode.

Preparation of the BCPE and MWCNT/MCPE

The bare carbon paste electrode (BCPE) was prepared by hand mixing of 70% graphite powder and 30% silicone oil in an agate mortar until a homogeneous paste was formed. The paste was then packed into a cavity of PVC tube of 3 mm internal diameter and smoothened on a tissue paper. The electrical contact was provided by a copper wire connected to the end of the tube. MWCNT/MCPE was prepared by homogeneously grinding different amounts of MWCNT in milligrams along with 70% graphite powder and 30% silicone oil.

Preparation of Poly(brilliant blue)/MWCNT/MCPE

The reported procedure was used for the preparation of poly(brilliant blue) film onto the MWCNT/MCPE surface [34]. Electrochemical polymerization of brilliant blue on the surface of MWCNT/MCPE was carried out using cyclic voltammetric method in aqueous solution containing 0.5mM brilliant blue in 0.1M NaOH solution. The electropolymerisation was achieved by the formation of film that grew between -0.5 V to +1.5 V at the scan rate of 0.1 Vs⁻¹ for 10 cycles. After this the electrode was rinsed thoroughly with double distilled water.

Preparation of SDS/poly(brilliant blue)/MWCNT/MCPE

10 µL of SDS solution (0.1mM) was added by using micropipette on the surface of the poly(brilliant blue)/MWCNT/MCPE and allowed it for about 12 min at room temperature. The electrode was later thoroughly rinsed with double distilled water to remove unadsorbed SDS to get the SDS/poly(brilliant blue)/MWCNT/MCPE.

Results and Discussion

Effect of quantity MWCNT on the peak current of DA

The MWCNT MCPE was prepared by adding different amount of MWCNT to the carbon paste electrode and was employed for the oxidation of 0.1mM DA in 0.2M PBS of pH 7.4 using cyclic voltammetric (CV) technique. By increasing the quantity of MWCNT in the modification, the electrochemical cathodic peak current (Ipc) and anodic peak current (Ipa) goes on increasing at particular ratio. The modification procedure was calibrated from 2 mg to 14 mg. The redox peak currents were increased up to 8 mg MWCNT in carbon paste electrode. After this, the redox peak current was decreased as shown in Figure 1. Further increase in the quantity of MWCNTboth Ipa and Ipcwere decreased. Therefore, 8 mg MWCNT was chosen as optimum for the modification procedure. In order to enhance the sensitivity of detection, the electropolymerisation of brilliant blue on the surface of MWCNT/MCPE was carried out using CV methodin an aqueous solution containing 0.5mM brilliant blue in 0.1M NaOH solution. The electropolymerisation was achieved by the formation of

film that grew between -0.5 V to +1.5 V at the scan rate of 0.1 Vs⁻¹ for 10 cycles. After this the poly(brilliant blue)/MWCNT/MCPE electrode was rinsed thoroughly with double distilled water.

Micellar effect on Poly(brilliant blue)/MWCNT/MCPE for the oxidation of DA

Surfactants are proven to enhance the sensitivity of the electrode [35-37]. In the present study 0.1mM sodium dodecyl sulphate (SDS) an anionic, 0.1mM Triton X-100 (TX-100) a non-ionic, 0.1mM cetyl trimethyl ammonium bromide (CTAB) a cationic surfactant solutions of different concentration (5 µL to 25 µL) are immobilized on the surface of poly(brilliant blue)/MWCNT/MCPE for about 5 min and it is employed for the oxidation of 0.5mM DA in 0.2M PBSof pH 7.4 at the scan rate 0.05 Vs⁻¹ using CV technique. All three surfactants shows noticeable enhancement in the peak currents of DA as shown in Figure 2(A), 2(B) and 2(C). However, SDS shows remarkable enhancement as compared with other two surfactantsnamely TX-100 and CTAB. At the concentration of 10 µL of SDS both anodic and cathodic peak currents was maximum as already illustrated in Figure 2(A). In order to calibrate the sensitivity of the electrode, again the influence of immobilization time was checked in the interval of 2 min each upto 16 min. The Ipa and Ipc go on increasing upto 12 min and later remains almost constant (Figure 2D). Hence the concentration of 10 µL SDS and immobilization time of 12 min was fixed as optimum to fabricate a stable working electrode to investigate all other remaining parameters.

Electrocatalytic oxidation of DA at SDS/poly(brilliant blue)/MWCNT/MCPE

DA being an easily oxidizable electroactive catecholamine, its voltammogram was recorded in the potential range from -0.2 to 0.6 V. Figure 3A shows the cyclic voltammograms recorded for 0.1mM DA at BCPE (curve a) MWCNT/MCPE (curve b) poly(brilliant blue)/MWCNT/MCPE (curve c) SDS/poly(brilliant blue)/MWCNT/MCPE (curve d) in 0.2M PBS of pH 7.4 with the scan rate 0.05 Vs⁻¹. At BCPE (curve a) the oxidation potential was occurred at 0.15 V with poor voltammetric response and for the MWCNT/MCPE the oxidation occurred at 0.14 V vs. SCE as shown in inserted Figure 3B. However, at SDS/poly(brilliant blue)/MWCNT/MCPE (curve d) the oxidation potential was observed at 0.180 V with a slight shift in the oxidation potential towards positive side with the significant enhancement in the redox peak current signals. This enhancement of current signal reflects the electrocatalytic activity of SDS/poly(brilliant blue)/MWCNT/MCPE towards the detection of DA.

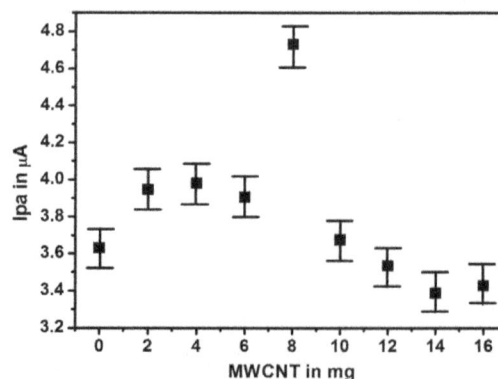

Figure 1: Graph of anodic peak current versus quantity of MWCNT in carbon paste electrode.

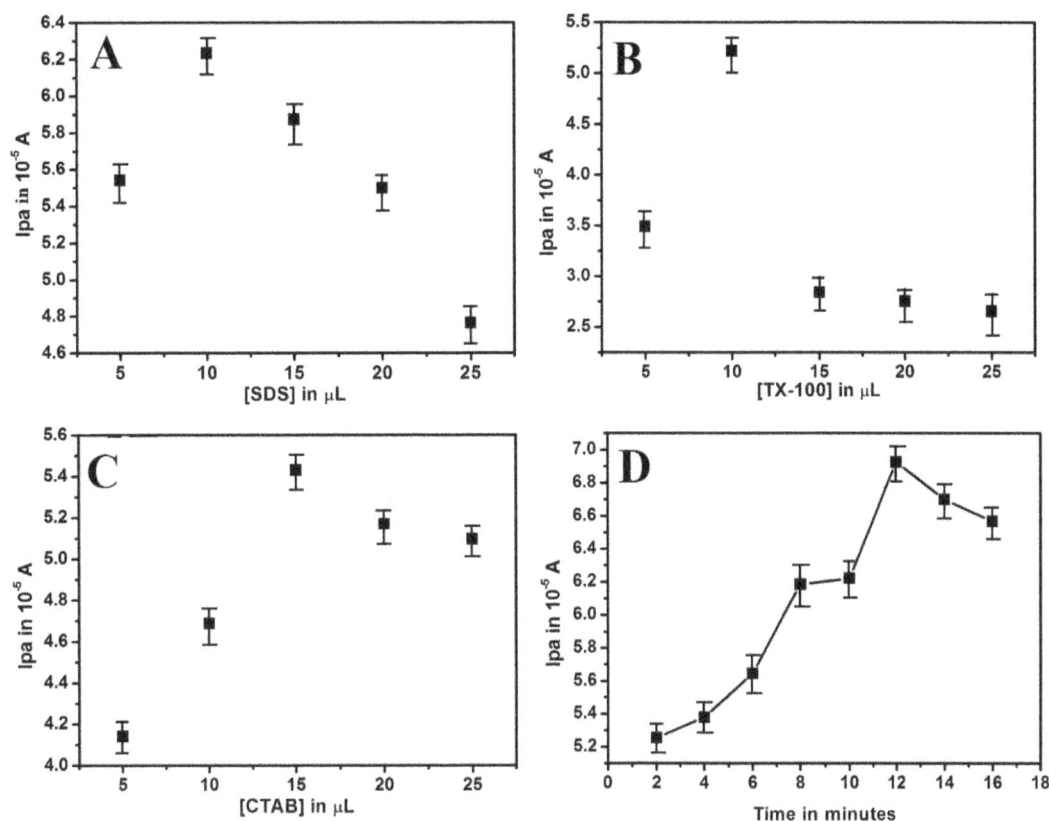

Figure 2: The micellar effect on poly(brilliant blue)/MWCNT MCPE for the oxidation of 0.5mM DA in 0.2M PBS of pH 7.4 **(A)** SDS **(B)** TX-100 **(C)** CTAB **(D)** Anodic peak current of oxidation of 0.5mM DA in 0.2M PBS of pH 7.4 versus immobilisation time.

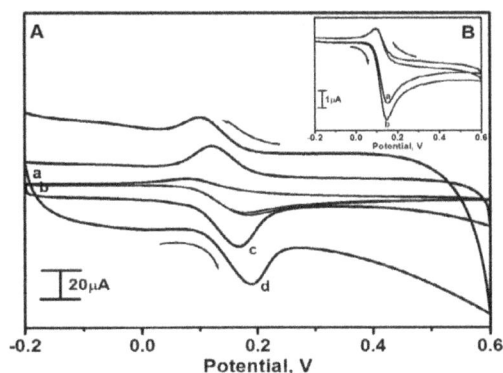

Figure 3: Cyclic voltammograms of 0.1mM DA in 0.2M PBS solution of pH 7.4 at the scan rate 0.05 Vs⁻¹. **(A)** Shows the results obtained for (a) BCPE (curve a) (b) MWCNT/MCPE (curve b) and (c) poly(brilliant blue)/MWCNT/MCPE (curve c) (d) SDS/Poly(brilliant blue)/MWCNT/MCPE (curve d) **(B)** Shows the results of (a) BCPE (curve a) (b) MWCNT/MCPE (curve b).

Effect of scan rate

The effect of scan rate for 0.5mM DA in 0.2M PBS of pH 7.4 was studied by CV technique at SDS/poly(brilliant blue)/MWCNT/MCPE as shown in Figure 4A. The modified electrodeobeys Randles-Sevcik equation and showed increase in the redox peak currents with increase in the scan ratewith the small shifting of the redox peak potential. In order to confirm the electrode process, the graph of peak current (Ip) vs. scan rate (υ) was plotted and the obtained graph is a straight line with

good linearity in the range from 0.05-0.5 Vs⁻¹ as shown in Figure 4B the correlation coefficient (r^2) was 0.9988 and 0.9984. The Ip vs. square root scan rate ($\upsilon^{1/2}$) were plotted as shown in Figure 4C with the correlation coefficient (r^2) 0.9946 and 0.9946 this suggests the electrode process was adsorption controlled and in support to this logarithm of anodic peak current vs. logarithm of scan rate (Figure 5) was plotted and the determined slope was 0.8686 which confirms the electrode process was adsorption controlled process [38]. This was again supported by previously reported literatures [39].

According to an equation previously reported [40] for determining the value of heterogeneous rate constant (k^0) from experimental peak potential difference (ΔEp) values, equation (1) was used.

$$\Delta Ep = 201.39 \log (\upsilon/k^0) - 301.78 \qquad (1)$$

From the experimental ΔEp values as shown in Table 1 and equation (1) the values of the k^0 for the DA oxidation was determined. The value of k^0 obtained at a scan rate of 0.05 Vs⁻¹ for the SDS/poly(brilliant blue)/MWCNT/MCPE exhibits larger heterogeneous rate constant compared with those determined in other scan rate variation studies. All the parameters are tabulated in Table 1.

Effect of DA concentration

The electrocatalytic oxidation of DA was carried out by varying its concentration at SDS/poly(brilliant blue)/MWCNT/MCPE. Figure 6A shows by increasing the concentration of DA from 0.60×10^{-4} M to 1.19×10^{-4} M, the Ipa and Ipc goes on increasing with shifting Epa towards less positive and Epc towards least negative side. The graph of Ipa vs. concentration of DA was plotted as shown in Figure 6B it shows almost

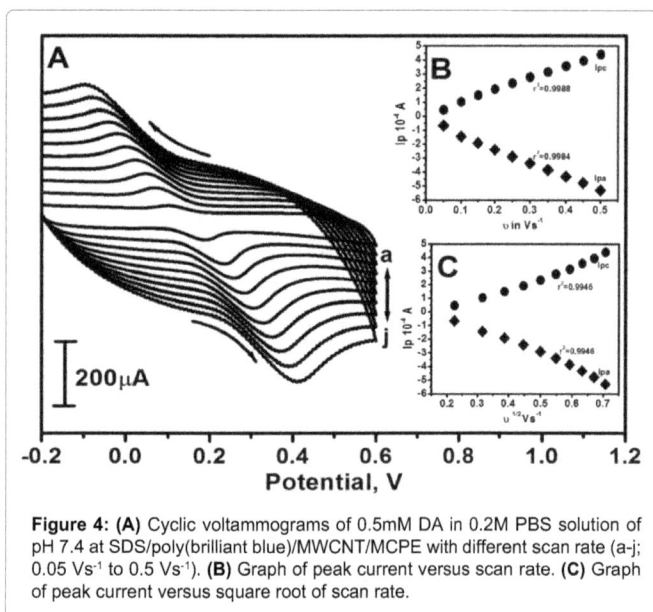

Figure 4: (A) Cyclic voltammograms of 0.5mM DA in 0.2M PBS solution of pH 7.4 at SDS/poly(brilliant blue)/MWCNT/MCPE with different scan rate (a-j; 0.05 Vs^{-1} to 0.5 Vs^{-1}). **(B)** Graph of peak current versus scan rate. **(C)** Graph of peak current versus square root of scan rate.

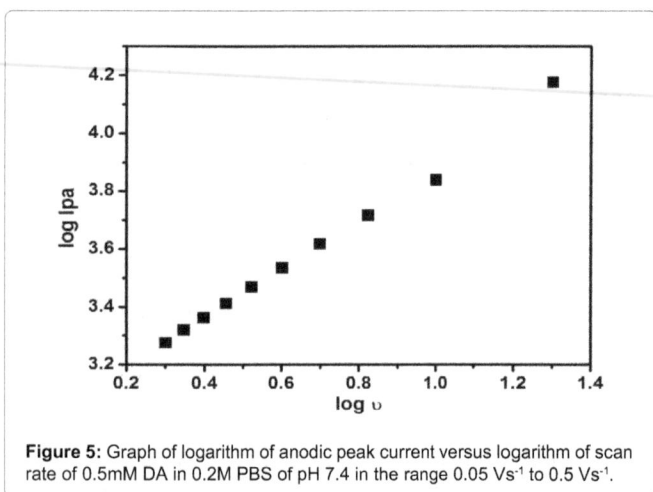

Figure 5: Graph of logarithm of anodic peak current versus logarithm of scan rate of 0.5mM DA in 0.2M PBS of pH 7.4 in the range 0.05 Vs^{-1} to 0.5 Vs^{-1}.

u/mVs^{-1}	ΔE_p/mV	k^0/ s^{-1}
0.05	87.6	0.5827
0.10	165.4	0.4789
0.15	210.6	0.4286
0.20	256.9	0.3365
0.25	302.0	0.2511
0.30	309.2	0.2780
0.35	314.0	0.3069
0.40	311.0	0.3625
0.45	320.8	0.3650
0.50	314.0	0.4385

Table 1: Variation of the voltammetric parameters gathered from the plots shown in Figure 4 as a function of the potential scan rate.

straight line with linear regression equation Ipa(10^{-5}A)=6.2192(C$_0$10^{-4}M/L)+1.6157, (r^2=0.9975). The limit of detection was calculated [41,42] and the detection limit on the lower concentration range for DA was 2.69 × 10^{-7} M for the SDS/poly(brilliant blue)/MWCNT/MCPEand limit of quantification was 8.97 × 10^{-7}M. The proposed electrode exhibited a relatively lower detection limit than those reported [43-46] as shown in Table 2.

Electrochemical oxidation of AA at SDS/poly(brilliant blue)/MWCNT/MCPE

Figure 7 showed the cyclic voltammograms of 1mM AA at the BCPE (curve a), SDS/poly(brilliant blue)/MWCNT/MCPE(curve b)in 0.2M PBS solution of pH 7.4 with the scan rate 0.05 Vs^{-1}. At the BCPE the oxidation peak occurred at around 0.242 V and generally oxidation of AA at bare electrode was irreversible, seldom broad and required high over potential due to fouling of the electrode surface by the adsorption of oxidized product of AA. However, at the SDS/poly(brilliant blue)/MWCNT/MCPEthe oxidation peak potential of AA was obtained at around 0.020 V which shifted to least positive potential and showed faster electron transfer kinetics of AA when compared to that of BCPE, which indicated that the SDS/poly(brilliant blue)/MWCNT/MCPE lowers the over potential and favoured the oxidation process of AA.

Figure 8A shows the cyclic voltammograms of AA at SDS/poly(brilliant blue)/MWCNT/MCPEfor 1mM AA in 0.2M PBS of pH 7.4 in the scan raterange of 0.05 Vs^{-1} to 0.5 Vs^{-1} by increase the scan rate there was an increase in the anodic peak current (Ipa) and the oxidation peak potential was observed to shift positively with the increase in scan rate [47], in addition to this, the graph of Ipa versus υ and Ipa versus υ$^{1/2}$ were plotted the graph obtained was linearly straight line shown in Figure 8B and Figure 8C respectively. A good linearity with correlation coefficients 0.9974 and 0.9970 indicated that the electrode transfer reaction was adsorption-controlled process on the SDS/poly(brilliant blue)/MWCNT/MCPEsurface.

Effect of AA concentration

The electrochemical oxidation of AA was carried out by varying its concentration at SDS/poly(brilliant blue)/MWCNT/MCPE by using CV technique at scan rate 0.05 Vs^{-1}. Figure 9A shows the voltammograms obtained for AA at different concentrations. By increasing the concentration of AA from 2.77 × 10^{-3}M to 4.31 × 10^{-3}M the Ipa was also increased. The graph of Ipa vs. different concentration of AA was plotted in Figure 9B the result showed linear increase in peak current with increase in the AA concentration with thelinear regression equation Ipa(10^{-5}A)=1.280(C$_0$10^{-3}M/L)+1.938, (r^2=0.9926). The LOD and LOQ were 1.31 × 10^{-6} and 4.36 × 10^{-6} respectively.

Electrochemical response of UA at SDS/poly(brilliant blue)/MWCNT/MCPE

Figure 10 shows the cyclic voltammograms of 1mM UA for BCPE (curve a) and SDS/poly(brilliant blue)/MWCNT/MCPE(curve b)in 0.2M PBS of pH 7.4 with the scan rate 0.05 Vs^{-1}. It is could be noticed that voltammetric peak appeared at about 0.304 V for BCPE, the peak was less sensitive rather broad suggesting slow electron transfer kinetics. However, at SDS/poly(brilliant blue)/MWCNT/MCPE the UA showed a significant increment in oxidation peak current and located at 0.307 V. By this it can be confirmed that there wasanoccurrence of electrocatalytic reaction between the SDS/poly(brilliant blue)/MWCNT/MCPE and UA.

Figure 11A shows the cyclic voltammograms of UA at SDS/poly(brilliant blue)/MWCNT/MCPE for 1mM UA in 0.2M PBS of pH 7.4 and scan rate from 0.05 to 0.5 Vs^{-1}. The graph of Ipa versus υ was plotted in the range from 0.05 to 0.5 Vs^{-1} The graph obtained was linearly straight line shown in Figure 11B with correlation coefficient 0.9976. And a plot of Ipa versus υ$^{1/2}$ in the same scan rate range showed correlation coefficient of 0.9930 as in Figure 11C. Therefore, it was confirmed that there was an adsorption complications of analytes on the surface of the SDS/poly(brilliant blue)/MWCNT/MCPE.

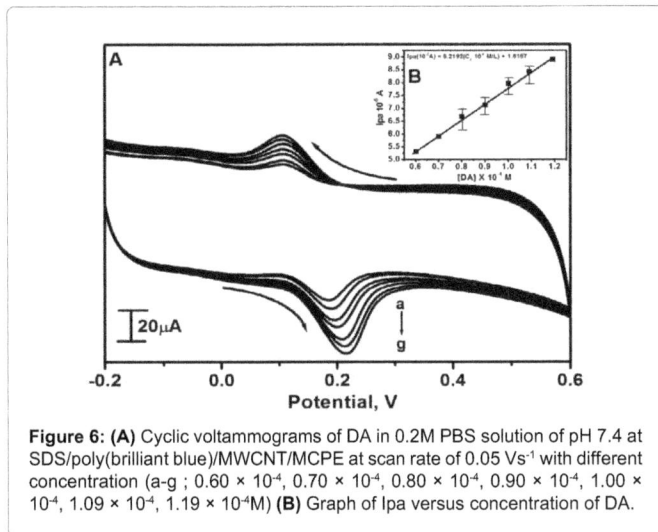

Figure 6: (A) Cyclic voltammograms of DA in 0.2M PBS solution of pH 7.4 at SDS/poly(brilliant blue)/MWCNT/MCPE at scan rate of 0.05 Vs^{-1} with different concentration (a-g ; 0.60 × 10^{-4}, 0.70 × 10^{-4}, 0.80 × 10^{-4}, 0.90 × 10^{-4}, 1.00 × 10^{-4}, 1.09 × 10^{-4}, 1.19 × 10^{-4}M) **(B)** Graph of Ipa versus concentration of DA.

Working Electrode	Limit of Detection (mol/L)	Method	References
Bicopper complex modified GCE	1.4 × 10^{-6}	DPV	[43]
CPE modified with SDS micelles at pH 7	3.70 × 10^{-6}	DPV	[44]
Ionic liquid modified Carbon paste electrode	7.0 × 10^{-7}	CV	[45]
CTAB/CPE	11 × 10^{-7}	DPV	[46]
SDS/poly(brilliant blue)/MWCNT/MCPE	2.69 × 10^{-7}	CV	This work

Table 2: Comparison of detection limits of different modified electrodes.

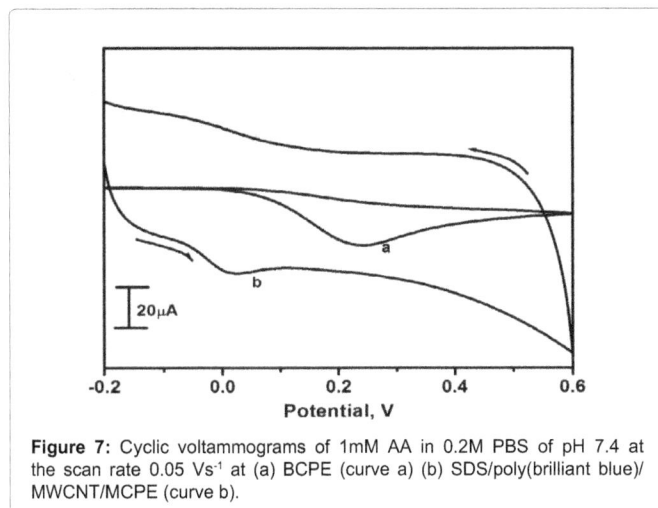

Figure 7: Cyclic voltammograms of 1mM AA in 0.2M PBS of pH 7.4 at the scan rate 0.05 Vs^{-1} at (a) BCPE (curve a) (b) SDS/poly(brilliant blue)/MWCNT/MCPE (curve b).

Effect of UA concentration

The cyclic voltammograms were recorded for the oxidation of UA with varying concentration in 0.2M PBS of pH 7.4 at scan rate 0.05 Vs^{-1}. The cyclic voltammogram of different concentration of UA (1.65 × 10^{-3} M to 2.52 × 10^{-3} M) as shown in the Figure 12A which shows the increase in anodic peak current due to increase in the concentration of UA. The plot shown in the Figure 12B shows the linear relationship between Ipa and the concentration of UA with the linear regression equation Ipa(0.1 mA)=0.5045(C$_0$10^{-3}M/L)+0.4297, r^2=0.9945. The detection limit on the lower concentration range for UA was 4.36 × 10^{-6}M for the SDS/poly(brilliant blue)/MWCNT/MCPE and limit of quantification was 1.10 × 10^{-5}M.

Simultaneous electroanalysis of DA, AA and UA

In mammalian brain AA and UA were present along with DA. Since the oxidation potential of both AA and UA were nearly same as that of DA results in a broad and overlapped voltammetric response at BCPE. Figure 13 shows CV recorded for 0.5 × 10^{-4}M DA, 1.0 × 10^{-3}M AA and 0.5 × 10^{-3} UA in 0.2M PBS of pH 7.4 at scan rate 0.05 Vs^{-1}. At BCPE (curve a) the oxidative separation of all the three analytes was impossible due to fouling of the surface and gives poor voltammetric response. However, in the same condition the SDS/poly(brilliant blue)/MWCNT/MCPE(curve b) taken this task and separated all three analytes into well distinguished voltammetric signals, the oxidation potential of AA, DA and UA were located at 0.012 V, 0.173 V and 0.299 V respectively. The peak to peak separation of DA-AA was 0.161 V and that of DA-UA was 0.126 V. This result was sufficient to identify DA in presence of probable interferenceUA and AA atSDS/poly(brilliant blue)/MWCNT/MCPE.

Differential pulse voltammetry (DPV) was used for the determination of DA, AA and UA at SDS/poly(brilliant blue)/MWCNT/MCPEdue to its higher current sensitivity and absence of background current. The simultaneous study was carried out in the potential range from -0.2 to 0.6 V versus SCEthe Figure 14(A) shows the simultaneous determination of 0.61 × 10^{-4}M DA, 0.51 × 10^{-3}M AA, 0.21 × 10^{-3}M UA in 0.2M PBS of pH 7.4 with well separated voltammetric signals corresponding to their oxidation at SDS/poly(brilliant blue)/MWCNT/MCPE. The oxidation potential of DA, AA and UA was located at 0.126 V, -0.048 V and 0.234 V respectively. The peak to peak separation between DA-AA was 0.174 V and that of DA-UA was 0.108 V.

Interference study

The interference study was carried out in themixture of samplesin an electrochemical cell containing DA, AA and UA. In their mixtures DPV was performed at the SDS/poly(brilliant blue)/MWCNT/MCPEwhen the concentration of one species is changed, whereas the concentration of the other two species was maintained constant. From the Figure 14(B) it can be noticed that the peak current of DA was increased from 0.421 × 10^{-4}M to 1.02 × 10^{-4}M by constant keeping of the AA and UA concentration to 0.5 × 10^{-}

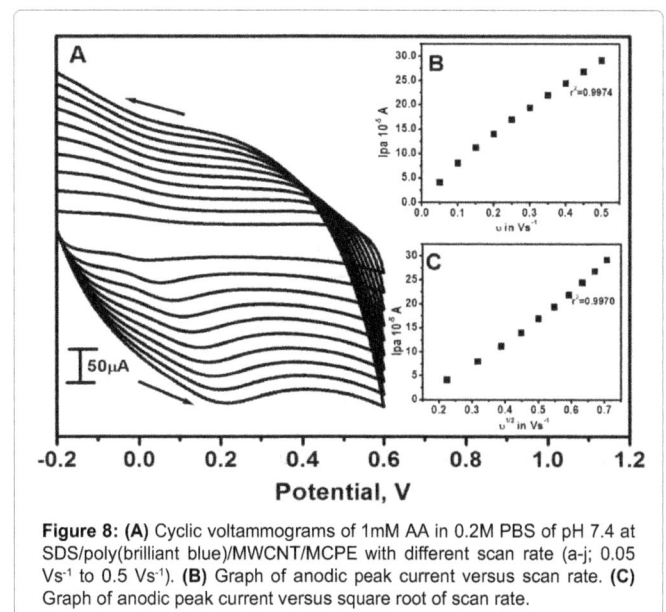

Figure 8: (A) Cyclic voltammograms of 1mM AA in 0.2M PBS of pH 7.4 at SDS/poly(brilliant blue)/MWCNT/MCPE with different scan rate (a-j; 0.05 Vs^{-1} to 0.5 Vs^{-1}). **(B)** Graph of anodic peak current versus scan rate. **(C)** Graph of anodic peak current versus square root of scan rate.

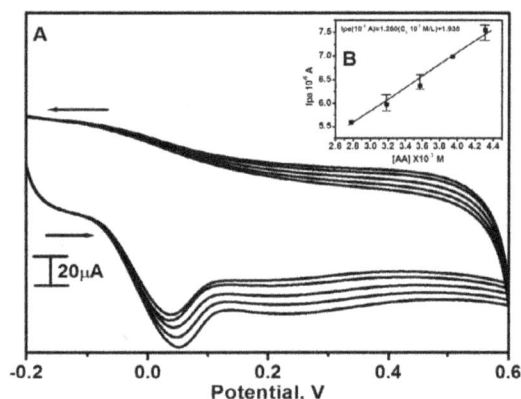

Figure 9: (A) Cyclic voltammograms of AA in 0.2M PBS of pH 7.4 at SDS/ poly(brilliant blue)/MWCNT/MCPE at scan rate of 0.05Vs^{-1} with different concentration (a-e; 2.77 × 10^{-3}, 3.18 × 10^{-3}, 3.57 × 10^{-3}, 3.94 × 10^{-3}, 4.31 × 10^{-3} M). **(B)** Graph of anodic peak current versus concentration of AA.

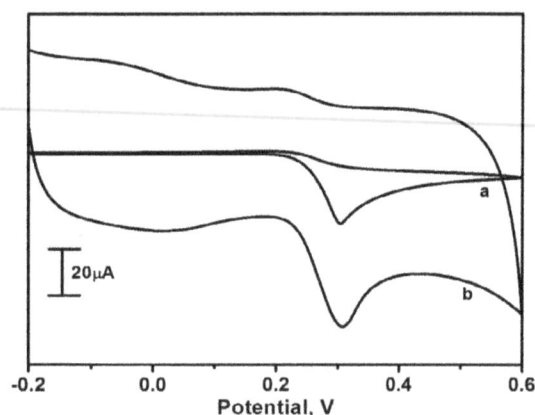

Figure 10: Cyclic voltammograms of 1mM UA in 0.2M PBS of pH 7.4 at the scan rate 0.05 Vs^{-1} at (a) BCPE (curve a) (b) SDS/poly(brilliant blue)/ MWCNT/MCPE (curve b).

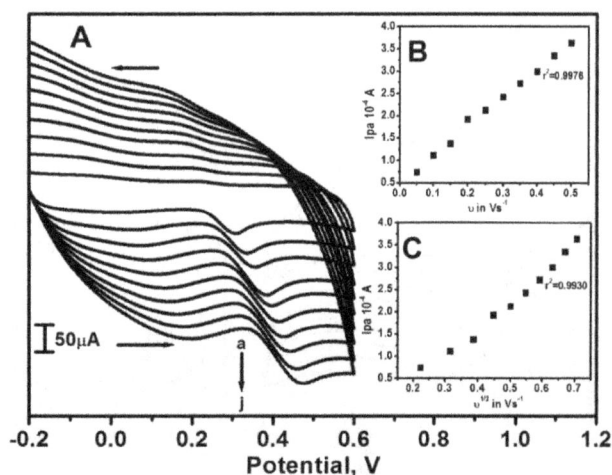

Figure 11: (A) Cyclic voltammograms of 1mM UA in 0.2M PBS solution of pH 7.4 at SDS/poly(brilliant blue)/MWCNT/MCPE with different scan rate (a-j; 0.05 Vs^{-1} to 0.5 Vs^{-1}). **(B)** Graph of anodic peak current versus scan rate. **(C)** Graph of anodic peak current versus square root of scan rate.

^{3}M and 0.2 × 10^{-3}M respectively. From the Figure 14(C) and Figure 14(D) it is seen that by keeping the concentration of other two analytes constant the anodic peak current of AA or UA increased upto a certain concentration range.

Conclusion

A simple and convenient method for the modification of the BCPE was proposed. The preparedSDS/poly(brilliant blue)/ MWCNT/MCPE shows excellent sensitivity, reproducibility, antifouling property and electrocatalytic activity towards the electrochemical oxidation of DA in the mixture of solutions contains large excess of AA and UA at physiological pH of 7.4 by using both CV and DPV techniques. Because of the distinguished voltammetric response obtained at SDS/poly(brilliant blue)/ MWCNT/MCPE the peak to peak separation of DA-AA was 0.161 V and that of DA-UA was 0.126 V by CV technique. This result was more enough for the electroanalysis of DA in presence of common interferences AA and UA. The proposed method can be used for other neurotransmitters. The modified electrode acts as very good sensor for the detection of dopamine.

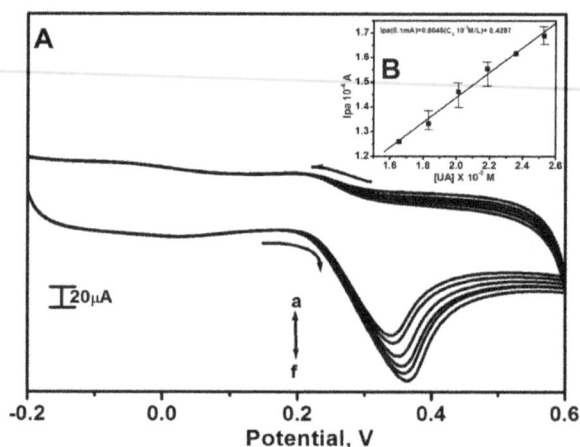

Figure 12: (A) Cyclic voltammograms of UA in 0.2M PBS solution of pH 7.4 at SDS/poly(brilliant blue)/MWCNT/MCPE at scan rate of 0.05 Vs^{-1} with different concentration (a-f ; 1.65 × 10^{-3}, 1.83 × 10^{-3}, 2.01 × 10^{-3}, 2.18 × 10^{-3}, 2.35 × 10^{-3}, 2.52 × 10^{-3} M). **(B)** Graph of anodic peak current versus concentration of UA.

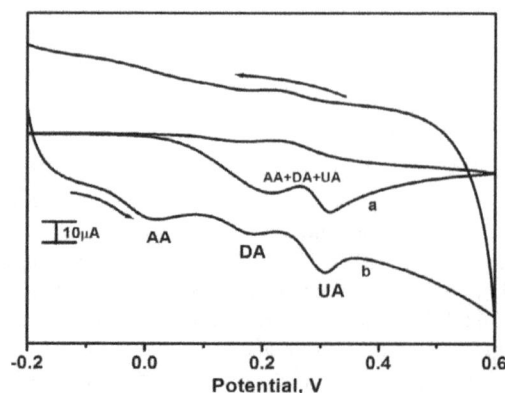

Figure 13: Cyclic voltammograms for mixture of 0.5 × 10^{-4} M DA, 1.0 × 10^{-3}M AA and 0.5 × 10^{-3} M UA in 0.2M PBS of pH 7.4 at the scan rate 0.05 Vs^{-1} at (a) BCPE (curve a) (b) SDS/poly(brilliant blue)/MWCNT MCPE (curve b).

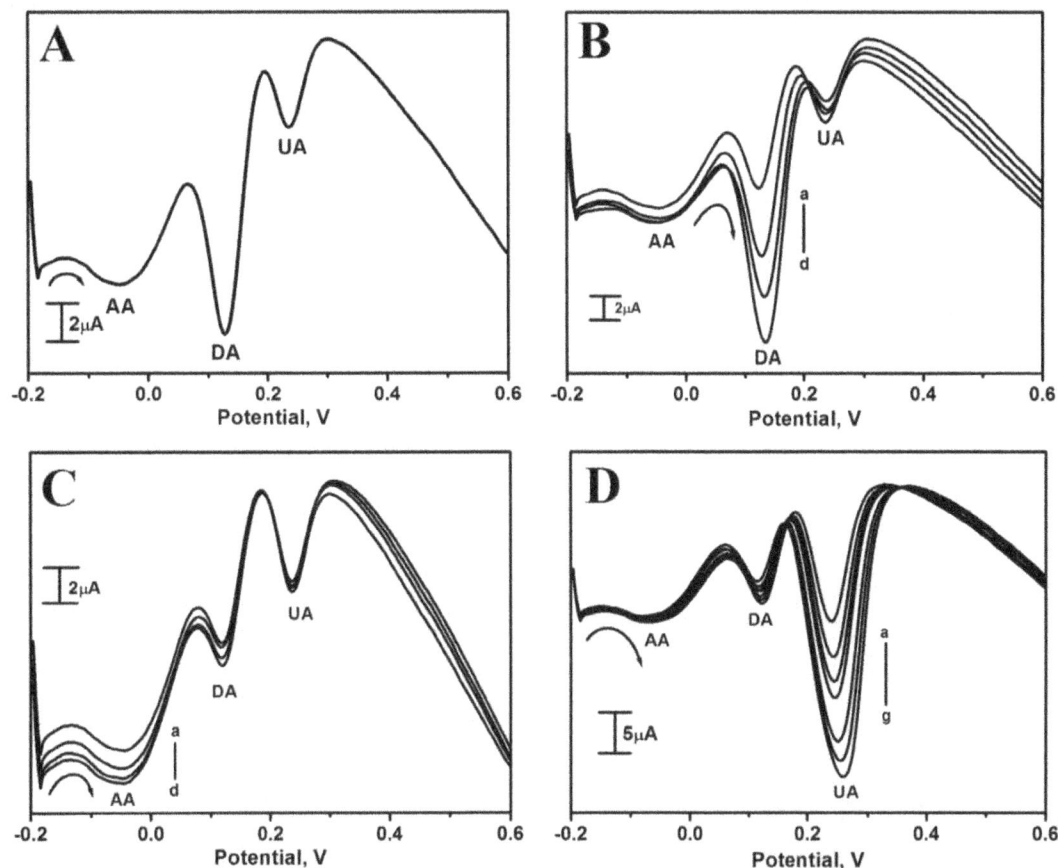

Figure 14: (A) Differential pulse voltammogram obtained for 0.61×10^{-4} M DA, 0.51×10^{-3}M AA and 0.21×10^{-3} M UA in 0.2M PBS solution of pH 7.4 at SDS/poly(brilliant blue)/MWCNT/MCPE **(B)** Differential pulse voltammograms of (a) 0.421×10^{-4} M (b) 0.627×10^{-4} M (c) 0.823×10^{-4} M (d) 1.02×10^{-4} M DA in 0.2M PBS of pH 7.4 in presence of 0.5×10^{-3}M AA and 0.2×10^{-3}M UA. **(C)** Differential pulse voltammograms of (a) 0.529×10^{-3}M (b) 0.735×10^{-3}M (c) 0.937×10^{-3}M (d) 1.136×10^{-3}M AA in 0.2M PBS of pH 7.4 in presence of 0.3×10^{-4} M DA and 0.2×10^{-3} M UA **(D)** Differential pulse voltammograms of (a) 0.630×10^{-3}M (b) 0.833×10^{-3}M (c) 1.033×10^{-3}M (d) 1.229×10^{-3}M (e) 1.422×10^{-3}M (f) 1.612×10^{-3}M (g) 1.800×10^{-3}M UA in 0.2M PBS of pH 7.4 in presence of 0.3×10^{-4}M DA and 0.5×10^{-3}M AA at SDS/poly(brilliant blue)/MWCNT/MCPE.

References

1. Yao H, Sun Y, Lin X, Tang Y, Liu A, et al. (2007) Selective determination of epinephrine in the presence of ascorbic acid and uric acid by electrocatalytic oxidation at poly(eriochrome black T) film-modified glassy carbon electrode. Anal Sci 23: 677-682.

2. Jin GP, Lin XQ (2004) The electrochemical behavior and amperometric determination of tyrosine and tryptophan at a glassy carbon electrode modified with butyrylcholine. Electrochem Comm 6: 454-460.

3. Razmi H, Agazadeh M, Habibi-A B (2003) Electrocatalytic oxidation of dopamine at aluminum electrode modified with nickel pentacyanonitrosylferrate films, synthesized by electroless procedure. J Electroanal Chem 547: 25-33.

4. Zhang Y, Cai Y, Su S (2006) Determination of dopamine in the presence of ascorbic acid by poly(styrene sulfonic acid) sodium salt/single-wall carbon nanotube film modified glassy carbon electrode. Anal Biochem 350: 285-291.

5. Xu F, Gao M, Wang L, Shi G, Zhang W, et al. (2001) Sensitive determination of dopamine on poly(aminobenzoic acid) modified electrode and the application toward an experimental Parkinsonian animal model. Talanta 55: 329-336.

6. Ganesh PS, Swamy BEK (2015) Simultaneous electroanalysis of norepinephrine, ascorbic acid and uric acid using poly(glutamic acid) modified carbon paste electrode. J Electroanal Chem 752: 17-24.

7. Wightman RM, May LJ, Michael AC (1988) Detection of dopamine dynamics in the brain. Anal Chem 60: 769A-779A.

8. Raoof JB, Kiani A, Ojani R, Valiollahi R, Nadimi SR (2010) Simultaneous voltammetric determination of ascorbic acid and dopamine at the surface of electrodes modified with self-assembled gold nanoparticle films. J Solid State Electrochem 14: 1171-1176.

9. Zhao GH, Li MF, Li ML (2007) Differential Pulse Voltammetric Determination of Dopamine with Coexistence of Ascorbic Acid on Boron-Doped Diamond Surface. Central European Journal of Chemistry 5: 1114-1123.

10. Damier P, Hirsch EC, Agid Y, Graybiel AM (1999) The substantia nigra of the human brain. II. Patterns of loss of dopamine-containing neurons in Parkinson's disease. Brain 122: 1437-1448.

11. Premkumar J, Khoo SB (2005) Electrocatalytic oxidations of biological molecules (ascorbic acid and uric acids) at highly oxidized electrodes. J Electroanal Chem 576: 105-112.

12. Fox IH (1981) Metabolic basis for disorders of purine nucleotide degradation. Metabolism 30: 616-634.

13. Raj CR, Kitamura F, Ohsaka T (2002) Square wave voltammetric sensing of uric acid using the self-assembly of mercaptobenzimidazole. Analyst 127: 1155-1158.

14. Miland E, Miranda Ordieres AJ, Tuñón Blanco P, Smyth MR, Fágáin CO (1996) Poly(o-aminophenol)-modified bienzyme carbon paste electrode for the detection of uric acid. Talanta 43: 785-796.

15. Dursun Z, Pelit L, Taniguchi I (2009) Voltammetric Determination of Ascorbic Acid and Dopamine Simultaneously at a Single Crystal Au(111) Electrode. Turk J Chem 33: 223-231.

16. Wang Z, Liu J, Liang Q, Wang Y, Luo G (2002) Carbon nanotube-modified electrodes for the simultaneous determination of dopamine and ascorbic acid. Analyst 127: 653-658.

17. Pacios M, Valle MD, Bartroli J, Esplandiu MJ (2008) Electrochemical behavior of rigid carbon nanotube composite electrodes. J Electroanal Chem 619: 117-124.

18. Torres DS, Huerta F, Montillaa F, Morallon E (2011) Study on electroactive and electrocatalytic surfaces of single walled carbonnanotube-modified electrodes. Electrochim Acta 56: 2464-2470.

19. Zare HR, Nasirizadeh N, Ardakani MM (2005) Electrochemical properties of a tetrabromo-p-benzoquinone modified carbon paste electrode. Application to the simultaneous determination of ascorbic acid, dopamine and uric acid. J Electroanal Chem 577: 25-33.

20. Ganesh PS, Swamy BEK (2014) Electrochemical Determination of Dopamine in Presenceof Ascorbic Acid at Brilliant Blue Modified Carbon Paste Electrode: A Voltammetric Study. Journal of Chemical Engineering and Research 2: 113-120.

21. Vasantha VS, Chen SM (2006) Electrocatalysis and simultaneous detection of dopamineand ascorbic acid using poly(3,4-ethylenedioxy)thiophenefilm modified electrodes. J Electroanal Chem 592: 77-87.

22. Nugent JM, Santhanam KSV, Rubio A, Ajayan PM (2001) Fast Electron Transfer Kinetics on Multiwalled Carbon Nanotube Microbundle Electrodes. Nano Lett 1: 87-91.

23. Gong K, Yan Y, Zhang M, Su L, Xiong S, et al. (2005) Electrochemistry and electroanalytical applications of carbon nanotubes: a review. Anal Sci 21: 1383-1393.

24. Wang J (2005) Carbon-Nanotube Based Electrochemical Biosensors: A Review. Electroanal 17: 7-14.

25. Patil RH, Hegde RN, Nandibewoor ST (2011) Electro-oxidation and determination of antihistamine drug, cetirizine dihydrochloride at glassy carbon electrode modified with multi-walled carbon nanotubes. Colloids Surf B Biointerfaces 83: 133-138.

26. Umasankar Y, Thiagarajan S, Chen SM (2007) Nanocomposite of functionalized multiwall carbon nanotubes with nafion, nano platinum, and nano gold biosensing film for simultaneous determination of ascorbic acid, epinephrine, and uric acid. Anal Biochem 365: 122-131.

27. Tsai YC, Chiu CC (2007) Amperometric biosensors based on multiwalled carbon nanotube-Nafion-tyrosinasenano biocomposites for the determination of phenolic compounds. Sens Actuat B Chem 125: 10-16.

28. Zeng J, Gao X, Wei WZ, Zhai X, Yin J, et al. (2007) Fabrication of carbon nanotubes/poly(,2-diaminobenzene) nanoporous composite via multipulse chronoamperometric electropolymerization process and its electrocatalytic property toward oxidation of NADH. Sens Actuat B Chem 120: 595-602.

29. Xu JM, Wang YP, Xian YZ, Jin LT, Tanaka K (2003) Preparation of multiwall carbon nanotubes film modified electrode and its application to simultaneous determination of oxidizable amino acids in ion chromatography. Talanta 60: 1123-1130.

30. Rusling JF (1991) Controlling electrochemical catalysis with surfactant microstructures. Accounts of Chemical Research 24: 75-81.

31. Gao JX, Rusling JF (1998) Electron transfer and electrochemical catalysis using cobalt-reconstituted myoglobin in a surfactant film. J Electroanal Chem 449: 1-4.

32. Chowdappa N, Swamy BEK, Niranjana E, Sherigara BS (2009) Cyclic Voltammetric Studies of Serotonin at Sodium Dodecyl Sulfate Modified Carbon Paste Electrode. Int J Electrochem Sci 4: 425-434.

33. Zhang SH, Wu KB (2004) Square Wave Voltammetric Determination of Indole-3-acetic Acid Based on the Enhancement Effect of Anionic Surfactant at the Carbon Paste Electrode. Bull Korean Chem Soc 25: 1321-1325.

34. Ganesh PS, Swamy BEK (2015) Voltammetric Resolution of Dopamine in Presence of Ascorbic Acid and Uric Acid at Poly (Brilliant Blue) Modified Carbon Paste Electrode. J Anal Bioanal Tech 5: 229.

35. Sathisha TV, Swamy BEK, Chandrashekar BN, Thomas N, Eswarappa B (2012) Selective determination of dopamine in presence of ascorbic acid and uric acid at hydroxy double salt/surfactant film modified carbon paste electrode. J Electroanal Chem 674: 57-64.

36. Sathisha TV, Swamy BEK, Reddy S, Chandrashekar BN, Eswarappa B (2012) Clay modified carbon paste electrode for the voltammetric detection of dopamine in presence of ascorbic acid. J Mol Liq 172: 53-58.

37. Mahanthesha KR, Swamy BEK, Chandra U, Reddy S, Pai KV (2014) Sodium dodecyl sulphate/polyglycine/phthalamide/carbon paste electrode based voltammetric sensors for detection of dopamine in the presence of ascorbic acid and uric acid. Chemical Sensors 4: 10.

38. Gosser Jr DK (1993) Cyclic Voltammetry Simulation and Analysis of Reaction Mechanisms (VCH, Weinheim).

39. Gilbert O, Swamy BEK, Chandra U, Sherigara BS (2009) Electrocatalytic Oxidation of Dopamine and Ascorbic Acid at Poly (Eriochrome Black-T) Modified Carbon Paste. Electrode Int J Electrochem Sci 4: 582-591.

40. Ganesh PS, Swamy BEK (2015) Simultaneous electroanalysis of hydroquinone and catechol at poly (brilliant blue) modified carbon paste electrode: A voltammetric study. J Electroanal Chem 756: 193-200.

41. Mahanthesha KR, Swamy BEK (2013) Pretreated/Carbon paste electrode based voltammetric sensors for the detection of dopamine in presence of ascorbic acid and uric acid. J Electroanal Chem 703: 1-8.

42. Zhu Z, Qu L, Guo Y, Zeng Y, Sun W, et al. (2010) Electrochemical detection of dopamine on a Ni/Al layered double hydroxide modified carbon ionic liquid electrode. Sens Actu B 151: 146-152.

43. Wang M, Xu X, Gao J (2007) Voltammetric studies of a novel bicopper complex modifiedglassy carbon electrode for the simultaneous determination of dopamine and ascorbic acid. J Appl Electrochem 37: 705-710.

44. Orozco EC, Silva MTR, Avendano SC, Romo MR, Pardave MP (2012) Electrochemical quantification of dopamine in the presence of ascorbic acid and uric acid using a simple carbon paste electrode modified with SDS micelles at pH 7. Electrochim Acta 85: 307-313.

45. Sun W, Yang M, Jiao K (2007) Electrocatalytic oxidation of dopamine at an ionic liquid modified carbon paste electrode and its analytical application. Anal Bioanal Chem 389: 1283-1291.

46. Avendano SC, Silva MTR, Pardave MP, Martinez LH, Romo MR, et al. (2010) Influence of CTAB on the electrochemical behavior of dopamine and on its analytic determination in the presence of ascorbic acid. J Appl Electrochem 40: 463-474.

47. Zhao Y, Bai J, Wang L, Xu E, Hong P, et al. (2006) Simultaneous Electrochemical Determination of Uric Acid and Ascorbic Acid Using L-Cysteine Self-Assembled Gold Electrode. Int J Electrochem Sci 1: 363-371.

22

Purification, Determination Molecular Weight and Study Kinetic Properties of G6PD from Diabetes Patient

Abdulkader Rasheed W[1], Firas Maher T[1] and Akeel Al-Aisse H[2]*

[1]Department of Chemistry, College of Science, University of Tikrit, Tikrit, Iraq
[2]Department of Biology, College of Science, University of Tikrit, Tikrit, Iraq

Abstract

This study was conducted to purification G6PD enzyme from diabetic patients by using simple and cheap method the technique gel filtration on Sephadex G100 and determine molecular weight of enzyme and compare it with true molecular weight of enzyme and determine kinetic constant (Km, Vmax) and study the effect of temperature and substrate and pH and known the best condition to give optimum work of enzyme. study contain (60) patients with diabetes and (60) control Glucose and activity of G6PD were measured and the enzyme precipitated by Ammonium Sulfate with concentration (75%) and purification enzyme gel filtration on Sephadex G-100 with dimensions (1.5 × 30) cm and using the buffer solution from (Tris-HCl) at pH 8.2 to isolate the enzyme and determine molecular weight with same method. Specific activity was calculated (21.5 UI/mg), total activity (706.8 UI), number of purification (3.45) enzyme yield (23.188%) and enzyme activity (17.67 UI/ml). and the molecular weight was calculated with using same technique (57.82) kD. Effect of increased concentration of substrate on enzyme activity and found the activity increase with increase substrate and amount constant level not change however increase of concentration of substrate when drawing relation between activity and concentration of substrate format appear exchange excess and after study effect of pH found the optimum value (8.4) and study effect of temperature on activity found (38 C) the optimum temperature. Study of kinetic constant was done and the and the Michaelis-Menten (Km) value was (3.8 mM) and Vmax value (8 IU/ml).

Keywords: G6PD; Diabetes mellitus; Purification; Molecular weight

Introduction

Diabetes mellitus is a metabolic disease clinically and genetically heterogeneous group characterized by hyperglycemia due to defects in insulin metabolism. If the hyperglycemia of diabetes is not managed properly, it causes long-term damage, dysfunction, and failure of different organism [1]. Notably the eyes, kidneys, nerves, heart, and blood diabetes mellitus is a multifactorial disease resulting from interaction of both genetic and environmental It has been stated that oxidative stress and impaired release of nitric oxide may be the contributory factors in the pathogenesis of diabetes [2-4].

One of the main causes of diabetes is functional causes such as pancreatic disorders, this occurs when the pancreas is infected with tumors (benign and malignant), internal bleeding or when the pancreas is removed leading to an absolute inability to secrete insulin to lead to diabetes induced by excessive use of thiazide diuretics, anti-inflammatory drugs and antiviral [2]. Genetic causes Individuals with parents who have diabetes or individuals from a family with a family history of the disease are more likely to develop diabetes than others obesity is a cause of diabetes, Obese people who store high amounts of fat in the abdomen are more likely to develop diabetes [5]. Those who accumulate fat in the limbs and increase the amount of fats affect the blood sugar level because it is one of the main factors that cause insulin resistance of cells by reducing the sensitivity of insulin receptors on the surface of target cells [4]. One of the other reasons is also lack of physical activity as it affects the increased incidence of diabetes in some cases, emotional emotions have an effect on diabetes, such as anxiety, fear or sudden shocks, and susceptibility to disease [6]. Viral infections have a major role in the development of type 1 diabetes, Self-infection due to viral infection or by the destruction of beta cells in the pancreas [7] (Table 1).

Diagnosis of diabetes

Clinical diagnosis

Patients with diabetes of all types have the following symptoms: Polyuria, polydipsia, lethargy, boils, slow healing wounds, frequent infections persist for a long time. Patients with Type I diabetes suffer from weight loss dehydration, ketonuria, and hyperventilation [6]. Symptoms of type 1 diabetes tend to be short-term, whereas patients with type 2 diabetes tend to have chronic symptoms with longer duration of symptoms [5]. This is a significant difference between the two. Lack of secretion of insulin also causes excessive metabolism of free fatty acids, and this leads to confusion in fat metabolism [7].

Laboratory diagnosis: High blood sugar hyperglycemia and the emergence of glucose in the blood glucose is a distinctive phenomenon of diabetes, so some tests are used to diagnose diabetes, including: Fasting Plasma Glucose (FPG), Random plasma Glucose (RPG), Oral glucose Tolerance test (OGTT), Glycosylated Haemoglobin (HbA$_1$c), Glucose in Urine, Ketone body in Blood or Urine [8,9].

Glucose-6-Phosphate Dehydrogenase (G6PD)

Enzyme: Are vital catalysts that accelerate the rate of chemical reactions. They have a high molecular weight protein structure. Like other proteins, the enzyme is composed of a combination of a large

---

*Corresponding author: Akeel Al-Aisse H, Department of Chemistry, College of Science, University of Tikrit, Tikrit, Iraq, E-mail: akeel@yahoo.com

number of amino acids that have one or more polypeptides. Is a three-dimensional form of the protein? Amino acids are found in these sequences according to a particular sequence of each enzyme, leading to a specific vacuum structure that enables the enzyme to accelerate its own reaction [10-12].

Glucose-6-Phosphate Dehydrogenase (G6PD ((Oxidoreductase, EC1.1.1.1-49) (G6PD) is one of the cytoplasmic enzymes is spread throughout the body, especially in the red blood cells, which is one of the most important enzymes of the egg, as it is the main enzyme and the key to the Pentose phosphate pathway [13]. It stimulates the oxidation process of the glucose-6-phosphate (G-6-P) (NADP) and to convert it to an effective reduced form (NADP) to preserve the life-producing pathways of several important substances, particularly in red blood cells because they have no other source of production (NADP) to preserve the life-producing pathways of several important substances, particularly in red blood cells because they do not have another source of NADPH production. NADPH product produced by G6PD is complementary to the reduced triglyceride enzyme (GRG), which converts and converts oxidative glutathione (GSSG) (GSH) [7], which protects human red blood cells from partial and planned (1), shows the pathway of pentose phosphate sugar and rule of G6PD in reactions of pathway.

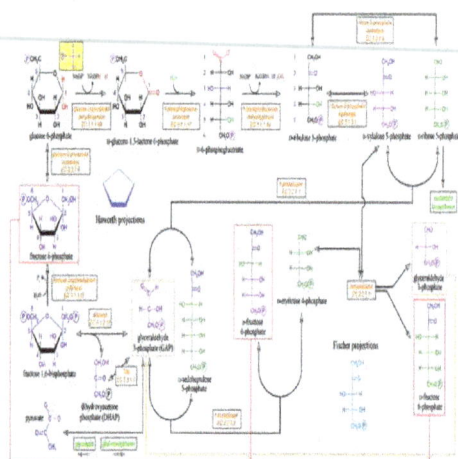

Planning (1) Pentose phosphate pathway [8].

The enzyme was first discovered by the scientists Warburg (Christian) in 1931 in the red blood cells of the horse and since then studies and research have been conducted to extract and purify the enzyme from various sources [9,14-16].

There are other important reasons to study as the change in the effectiveness of this enzyme G6PD enzyme in the body's various tissues is linked to many diseases in humans and these diseases is the disease of jaundice in children and hemolytic anemia [17]. G6PD deficiency is a disease that is prevalent in different parts of the world. The number of infected people is 400 million males, females, neonates and other ages, according to the World Health Organization (WHO) report [1]. More than 442 types of enzyme (G6PD variants) have been identified using a large number of biological techniques, including molecular analytical methods, which identify genetic mutations that occur in the genes responsible for the biological processing of different types of enzyme [10]. Gene found that the enzyme gene was found to be carried on the sex chromosome (X) [12]. The shortening of enzyme efficacy is associated with genetic and hereditary disorders [11]. This shortage is widespread in the world, especially in the Mediterranean region Patients with this

type of hemorrhagic deficiency are generally affected by certain drugs and foods in the case of a paroxysmal or neonatal jaundice found in natural erythrocytes, enzymatic activity decreases with age [11]. Many mutations of this enzymatic deficiency are widespread in the world and by geographic location. Moreover, genetic defects and age can lead to Enzymatic Deficiency An enzyme deficiency leads to the production of some red blood cell anemia due to exposure to certain chemical agents or certain infections and wounds [12].

Classification

Numerous G6PD variants have been described These have been classified by the World Health Organization according to the magnitude of the enzyme deficiency and the severity of hemolysis. This classification gives some approximation of the magnitude of hemolysis an individual might incur in the setting of an oxidative stress. Only class I, II, and III are of clinical significance.

Class I – Class I variants have severe enzyme deficiency (<10 percent of normal) relation with chronic hemolytic anemia.

Class II – Class II variants also have severe enzyme deficiency (<10 percent of normal), but there is usually only intermittent hemolysis, typically on exposure to oxidant stress such as fava bean exposure or ingestion of certain drugs and the classic example is (G6PD Mediterranean).

Class III – Class III variants have moderate enzyme deficiency (10 to 60 percent of normal) with intermittent hemolysis, typically associated with significant oxidant stress the classic example is (G6PD A⁻).

Class IV – Class IV variants have no enzyme deficiency or hemolysis the wild-type (normal) enzyme is considered a class IV variant, as are numerous other genetic changes that do not alter levels of the enzyme and these variants are of no clinical significance.

Class V – Class V variants have increased enzyme activity (more than twice normal). These are typically uncovered during testing for G6PD deficiency and they are of no clinical significance [13].

Clinical Significance of the Enzyme

Acute hemolytic anemia

Some individuals with G6PD deficiency have acute hemolytic anemia at the site of the wound when some medications, acute diseases, and certain foods are taken [15].

Neonatal jaundice

Anemia and jaundice are most often observed in newborns in individuals with severe enzyme deficiency [18-21].

Neutrophil dysfunction

The enzyme is used in addition to red blood cells in white cells to reduce the oxidizing factors. Some people with severe enzyme deficiency suffer from a defect in the function of white blood cells, which causes weak respiratory resistance to diseases and also weakens the presence of beneficial bacteria in the body [14].

Diabetic mellitus-induced hemolysis

In people with an enzyme deficiency, hemolyticysis starts with an increase in ketone content in diabetics and has the lowest levels when blood glucose levels are abnormal in diabetics [22]. Studies have indicated that high sugar leads to the deposition of decomposed blood in patients with deficiency [15].

Relationship Between (G6PD) Enzyme and Diabetes

An epidemiological study from suggested a positive correlation between diabetes and deficiency [16]. (G6PD) has conducted a study on Indian society being the most potential to give this relationship Serum samples were collected for healthy people and of both sexes. A higher incidence among Indians provided an excellent opportunity to study the possible association of G6PD deficiency in diabetes mellitus [5]. G6 PD deficiency is one of the common enzymopathy in human being affecting about 400 million people worldwide. It is suggested that there may be a positive association of G6PD deficiency with diabetes mellitus. Although G6PD deficiency is not uncommon in our country but there is scarcity of data on this regard especially on diabetes. Therefore, this study was undertaken to observe the G6PD status in patients with type 2 diabetes mellitus in order to explore the role of this enzyme deficiency as one of the risk factor for diabetes mellitus [7]. A positive association of G6PD deficiency with diabetes mellitus Although G6PD deficiency is not uncommon in our country but there is scarcity of data on this regard especially on diabetes. Therefore, this study was undertaken to observe the G6PD status in patients with type 2 diabetes mellitus in order to explore the role of this enzyme deficiency as one of the risk factor for diabetes mellitus [7].

Materials and Methods

Collection of sample

The total number of these samples was (60) samples, serum samples were collected for people with diabetes and both sexes. They reached (60) satisfactory samples of both types of diabetes (type I and type II) Diagnosis of the disease using a blood glucose test.

Blood was drawn from the vein using a 5 ml plastic syringe with one use. The blood was placed in clean, sterile plastic tubes free of anticoagulant EDTA. And left to coagulate at room temperature. The blood serum was then separated from the centrifuged portion of the centrifuge and at a velocity of 5000 G for 15 minutes to ensure adequate serum red blood cell extraction, Micro pipette. The effectiveness of the enzyme was measured directly and the study was done outside the body (in vitro) [23-28].

Diagnosis kits

The kits Clarifiers in Table 2 were used in procedures of this study.

Estimation of Biochemical Parameters in Blood Serum

Estimation of glucose concentration in serum

Principle: glucose level in serum was measured by using (kit Aflu Italia) depending on enzyme method that stated on Trinder reaction [29].

Determination of total protein in serum

Total protein level in serum was measured by using (kit Aflu Italia) depending on enzyme method [30].

Determination of G6PD activity in serum

G6PD activity in serum was measured by using Biolabo kit according this equation [31]:

Where; A=Absorbance; V=enzyme volume in ml; 6.22=absorbance confection of (NADPH) in length 340 nm.

Separation and Purification of the Enzyme G6PDH from Serum Diabetes Patients

G6PDH was purified from the serum of diabetic patients using the following steps:

Addition ammonium sulphate

Serum proteins were precipitated using gradual concentrations of ammonium sulphate until 75%. 3.75 gm of ammonium sulphate was added to 5 ml of serum during 60-45 minutes by placing the serum in a snow bath with constant stirring, and then dissolved Deposition using 4 ml of 1M Tris-HCl regulated solution (pH=7.8) [32].

Dialysis

Which is one of the most important methods used in the purification of enzymes and the oldest, and the goal is to remove the remaining ammonium sulphate added to the deposition of proteins by placing the dissolved protein in the above step in the membrane bag dialysis bag after measuring the effectiveness of the enzyme G6PD and protein concentration, and immerses the bag in the solution [33-35]. 1 M Tris-HCl pH 7.8 The regulator solution was changed from time to time for 16 hours. This step was performed at (4°C ± 1°C) to maintain the efficacy of G6PD. After the membrane separation process After the membrane separation process, G6PD and protein concentration were measured.

Gel filtration chromatography

Gel Filtration technology is one of the most important techniques in the field of biochemistry, one of the methods used in the separation of compounds depending on the size of their molecules and their molecular weights. The proteins with large molecular weights are not carried out through the gel but move outside the gel layer with the solvent that is removed in sequence. This solvent is often distilled water or a dilute regulator. Therefore, the large particles first filter during the separation. Small molecules can penetrate the gel granules finally. The fraction of the gel filtration is collected using the Fraction Collector. The volume of distilled water or solution is calculated to displace each

Year 2030	Year 2010	Country
2.009.000	968.000	Iraq
6.725.000	2.873.000	Egypt
680.000	315.000	Jordan
2.523.000	895.000	Saudi Arabia
99.000	38.000	Bahrain
378.000	149.000	Lebanon
1.138.000	437.000	Morocco

Table 1: Statistical represents the number of patients with diabetes in the Arab States in 2010, and it is expected in 2030 [4].

1	Glucose 6-Phosphate dehydrogenase (G6PDH) kit	Biolabo- France
2	Total protein kit	Aflu-Italia
3	Glucose kit	Aflu-Italia

Table 2: The kits and its source.

protein from the separation column, and the protein is isolated by reading the absorption at 280 nm [17,36-38].

The gel filtration technology is also used as a method for estimating the approximate molecular weight of protein by drawing a graph showing the relationship between the size of the elution and the known molecular weight. The protein with the molecular weight is then passed through the separation column and calculated. The size of the Rogan accurately and in comparison with the known molecular weights can estimate the approximate molecular weight of the unknown protein [18,19].

Used solutions

buffer solution 0. 1M Tris-HCl pH 7.8. Prepare to dissolve 15.76 gm of Tris-HCl per liter of distilled water and adjust pH at 7.8.

Sephadex G100 Prepare to dissolve 2.5 gm of the Sephadex G100 column filler in 200 ml of the 0.1 M Tris-HCl pH solution 7.2 and leave the solution for 28-24 hours at 4°C. During this time, the solution was changed several times to remove the soft minutes from the solution Because its presence reduces the velocity of the flow of the liquid solution through the column [39].

Sodium chloride solution at 500 mM concentration Prepare 29.25 g of NaCl per liter of 0.1 M Tris-HCl pH 7.8 solution.

Procedure

Use a glass column with a diameter of 1.5 cm and a length of 30 cm. A small amount of glass wool is placed at the end of the column to prevent the gel particles from leaking out of the column. The gel solution is slowly and homogeneously poured into the column to prevent air bubbles from forming. (11 cm), wash the column with sufficient amounts of 0.1M Tris-HCl pH solution until a flow velocity of 2.5 ml/min was obtained [20].

Add 5 ml of enzyme after membrane separation slowly over the surface of the G100 Sephadex gel and leave for 5 minutes to soak in the gel column.

The process of separation was started using 150 ml of the structured solution containing 500 ml of NaCl, collecting 5 ml per part.

After collecting the extracting parts of the separation column, the efficacy of the G6PD enzyme was evaluated by paragraph and protein concentration by the method (kit Aflu).

Kinetics of G6PD

The kinetics of G6PD were studied after its separation and partially purified from the serum of diabetic patients by gel filtration. These included:

Effect of Glucose 6-phosphate concentration (G6P)

The effect of different concentrations of G6P on the activity of G6PD was studied by using different concentrations (0.6, 0.0512, 0.0256, 0.0128, 0.064.0.048, 0.024, 0.012) M to determine the effect of the concentration of substrate on the work of the enzyme G6PD, G6PD reaction (G6PD kit Biolabo), and plotting the relationship between the reaction rate and the concentration of substrate to determine that the enzyme is subject to the Michaelis-Menten equation. The Km values were obtained using the Linover-Burke graphical method, which links the inverse values of both velocity and G6P concentration (1/(S)vs. 1/V) [40].

Optimized pH mapping

The pH effect of the regulated solution (0.1M Tris-HCl pH 7.8) was studied at the velocity of the G6PD reaction. Different pH solutions (11, 10, 9, 8, 7.6) were used with G6P at 0.6 mM and 37°C (G6PD kit Biolabo), and by plotting the relationship between reaction velocity and pH, the optimal pH was identified.

Effect of temperature

G6PD kit Biolabo was used to measure the effectiveness of G6PD. The reaction was conducted at different temperatures (57, 47, 37, 27, 17 and 7) with the regulated solution (0.1M Tris - HCl pH 7.8) Basically (G6P) 0.6 mM, and then painted the relationship between the reaction velocity and the temperature to find out the optimal temperature of the reaction [41].

Results and Discussion

Purification of enzyme

Precipitation by ammonium sulphate: The basic principle of the method is to equalize the charges on the surface of the protein and the degradation of the water layer surrounding the protein and reduce the degree of watering and reduce the solubility of the protein and sedimentation [23]. Add the stages to get rid of some of the protein content with the enzymatic extract [24]. The result in Table 3 showing the efficiency of quality (10.77 units/mg) with the number of times of purification (1.73) and the yield (35.02) during the satisfaction rate of ammonium sulfate sulfate estimated (75%). The results differed with the studies of enzyme extraction. In a study involving red blood cell extraction, the specific activity was 1.251 (units/mg) with 121.5 purification number and 53.8% enzymatic yield when using ammonium sulphate salt at 35-65% [25]. Another study showed that the specific efficacy was 0.37 mg/ml and the number of purification times was 39.79 times and the yield of 79.18 units/mg after adding the salt with saturation concentration (40-60%) of the red blood cells of the geese [26]. In the Penicillium duPonti fungi, the salt was used by saturation (45-60%), giving the result a quality efficacy of (1.04 units/mg) and the number of times the purification of 8.67 times and proceeds 63.4% [33].

Gel filtration: The different methods used to purify the enzyme from bacterial, fungal, plant, or animal sources were obtained in obtaining high purity of the enzyme, during this research the use of gel filtration technique with the Sephadex G-100 was the result of the specific efficacy (21.5 units/mg), total activity (706.8 units), purification number 3.45, enzymatic yield (23.188%) and enzymatic efficacy (17.67 units/ml). Results were obtained with other studies of enzyme purification. In one study, using the Sephacryl-S200 column to purify the enzyme from rat liver it was found that the specific efficacy was 24.75 units/mg and the total efficiency was 198 units with 6.17 times purification and an enzyme yield of 60.57% [28]. In a study that included several steps, first transfer the concentrated enzymatic extract with ammonium sulphate salt on the calcium phosphate column and then transfer the extract to the ion exchanger DEAE-Cellulose and then the gel filtration column Bio-Gel A-150 these steps gave a quality effect of 470 units/mg with a frequency of 2.42 times and an enzyme yield of 10% [29]. A study of the purified enzyme from calf tissue using the gel filter column Sephadex G-25 an enzyme yield of 91% with 450 times purification. Another study included enzyme purification from Coriander Leaves The first two-step purification was performed with ammonium sulphate deposition and the use of the Sephadex G-200 gel. The enzymatic activity was 1.82 units/mg and the enzymatic yield was

26.4% and the number of purification times was 74 [30]. Also using gelatin filtration of the Sephadex G-200 column and concentration of sulphate salts prior to filtration, the results included (326 unit/mg) the enzymatic yield was 19.9% and the frequency of purification was 2.5 when purifying the enzyme from the pituitary gland of caw [31].

In another study to Saccharomyces cerevisiae include using saturated concentration of ammonium sulphate salt (40-80) and gel filtration column on Sephacryl S-200 gave specific activity 65.68 unit/mg and the enzymatic yield was 20.62% and the number of purification times was 2.94 [29].

Determination of Molecular Weight in Gelatin Filtration Technology

The researcher's method [26] was based on the gel filter method in estimating the approximate molecular weight of the G6PD enzyme from the protein package (18) which showed the highest concentration of the protein. and passed a number of known molecular weight indicated in Table 4 and the molecular weight ranges between (20000000-13700) Dalton for the purpose of specifying the characteristics of the column. In terms of internal volume (Elution Volume Ve) for each material as well as the free or empty size of the granules (Void Volume Vo), which was estimated from the standard curve of blue dextran at 3 mL per part. The Vo value was equal to 33 mL and the recovery volume of the standard Ve proteins to the volume of the recovery of dextran blue Vo, represented by the Ve/Vo relationship [28]. As shown in Table 4 (Sephadex G-100), molecular weight and recovery volume The Elution Volume of each material versus its molecular weight logarithm shows the appearance of a straight line in which the approximate molecular weight of the protein packet separated by the gel filtration technique is shown in Figure 1. The recovery volume of the package (18 ml) To approximate molecular weight (Figure 2), the approximate molecular weight of the G6PD is (57.82) kDa for the enzymatic extract using gel filtration technique.

We see these results contrasted with a range of studies and studies conducted to estimate the molecular weight of the enzyme from its various sources [34]. The molecular weight was 133 kDa for the purified enzyme from pig liver using ammonium sulphate and Sephadex G-200 for the bilateral body the single molecule had a molecular weight of 67.50 and other studies showed that the molecular weight of the enzyme was purified from the liver of the mice, and the pituitary gland for cows

had a molecular weight of 64 kDa [32]. The molecular weight of the purified enzyme from human red blood cells was estimated to be 43 kDa [34]. and in another study using gelatin filtration with the Sephadex G150 the molecular weight was (40 KDa) from Bean plant [35]. The molecular weight of the purified enzyme from diabetics was within the range mentioned [41], ranging from 22-58 KDa. Due to the difference in the scientific basis in the different methods used in the Dalton of the enzyme purified from yeast Saccharomyces cerevisiae Purification of enzymes The molecular weights vary from one study to another [40]. The difference is also due to the length of the purification steps of some of the different studies which may lead to the enzyme breaking through these long stages and therefore the resulting molecular weights are less than the real G6PD Study of Enzyme kinetics [42-44].

G6PD Study of Enzyme Kinetics

Effect of the pH

pH has a significant effect on the enzyme's effectiveness for controlling ionization Ionic aggregates at the active site of the enzyme The optimal pH of enzyme stability is an important characteristic of enzymes [38]. The results of the kinetic study of the enzyme showed that the optimal pH of the enzymatic extract was in the range (8-8.4) as shown in Figure 3. The optimal basis for the stability of the enzyme was (8-9) Extreme pH values affect substrate and the ionic state of the enzyme and lead to enzyme protein mutagenesis by altering the enzyme body [23]. In a study of [43] people with enzymatic deficiency and a group of healthy patients, pH was (8.5-7) as in Figure 4.

Filter column the gel filtration column using the Sephadex gel and the solution of regulating the hydrochloride gear with different hydrogen numbers ranging from 5-11, respectively.

Effect of temperature on enzyme activity

The optimum temperature of the activity, the highest temperature, where the rate of enzymatic reaction rate is maximal, while the enzyme is highly effective, and is affected by pH and other factors [39]. The optimum temperature of the enzyme activity when the pH was confirmed and the concentration of Substrate was (37-38°C). The results varied with the studies carried out, including a study on the enzymatic extract extracted from coriander leaves (30C) [31]. and another study indicated that the optimum degree of the enzyme purified from bacteria Azotobacter and human placenta is (50C) [36].

Step of purification	Elute (ml)	Activity (IU/ml)	Protein conc. mg/ml	Specific Activity (IU/mg)	Total Activity (IU)	Purification (fold)	Yield (%)	Total Protein conc. (mg)
Crud serum	5	30.48	24.5	6.22	3048	1	100	4.9
Ammonium sulphate (75%)	4	21.54	8	10.77	1077	1.73	35	2
Dialysis	3.5	19.1	6.37	10.49	1049	1.68	34.4	1.82
Gel filtration (Sephadex G-100)	3	17.67	2.46	21.5	706.8	3.45	23.188	0.82

Table 3: Steps of enzyme purification.

Standard protein	Molecular weight (Dalton)	Log Molecular weight	Number of fraction (ml)	Elution volume (Ve) in (ml)	Ve/Vo
Blue Dextran	2000000	6.3	11	33	1
BSA	67000	4.82	15	45	1.36
alpha-Amylase Enzyme	58000	4.7	20	60	1.3
ovalbumin	43000	4.63	24	72	2.18
Chymotrypsinogen	23000	4.39	27	81	2.45
Ribonuclease	13700	4.13	31	93	2.81

Table 4: Relation between molecular weight and elution volume to standard protein.

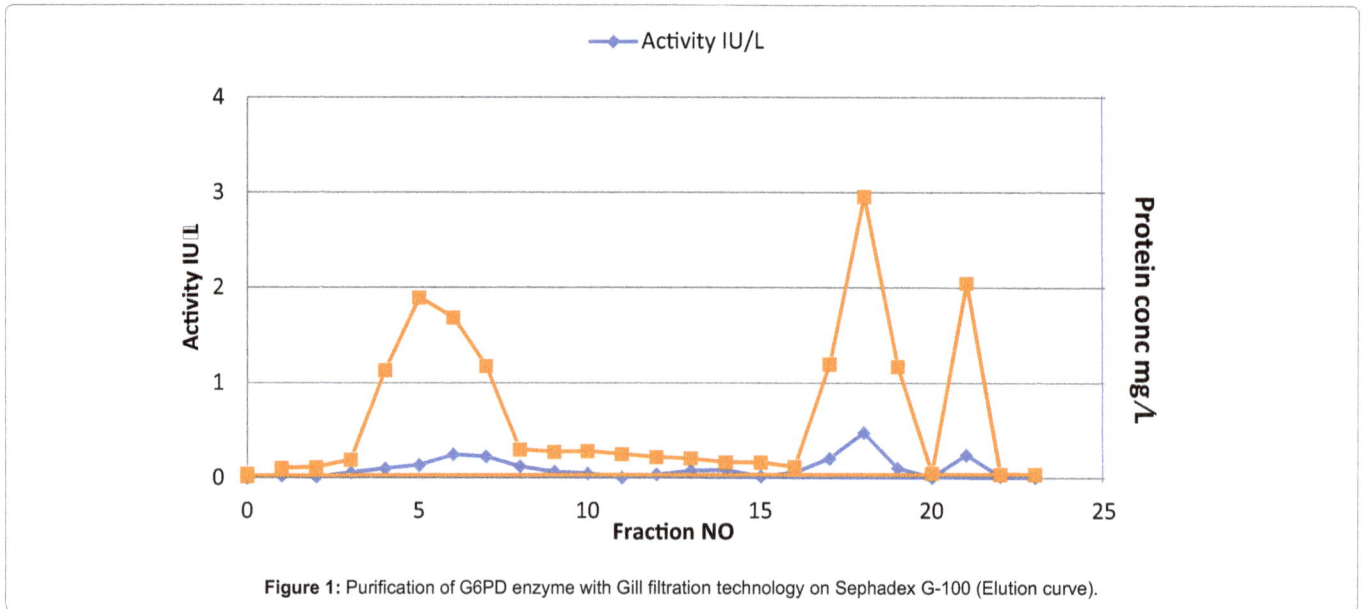

Figure 1: Purification of G6PD enzyme with Gill filtration technology on Sephadex G-100 (Elution curve).

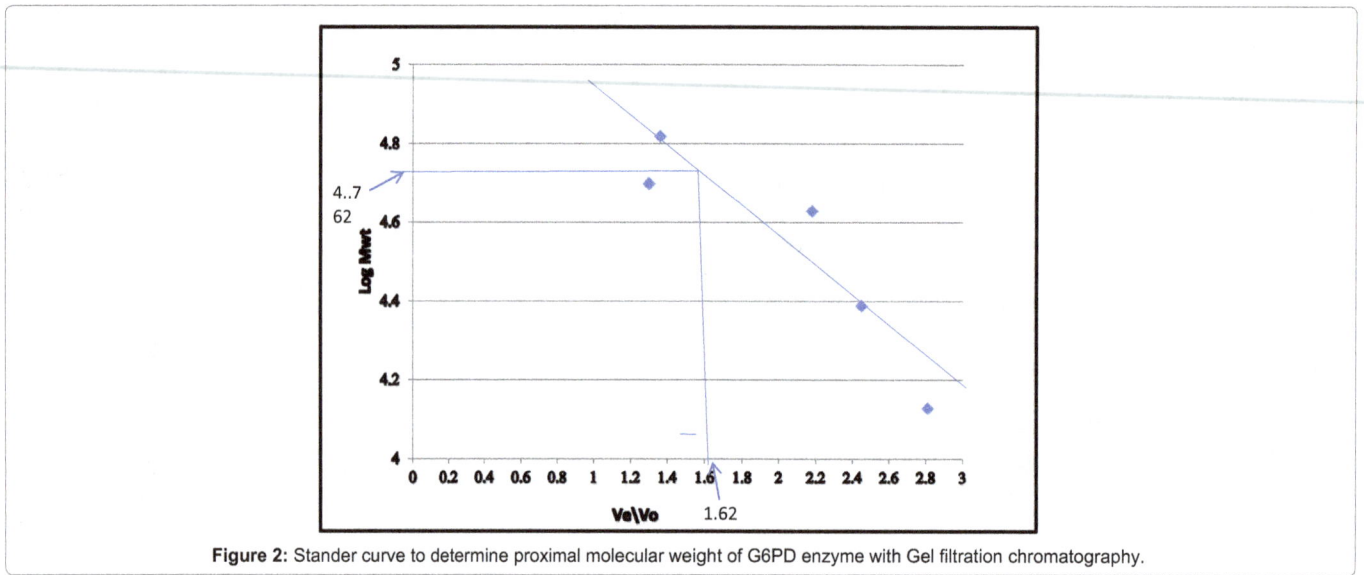

Figure 2: Stander curve to determine proximal molecular weight of G6PD enzyme with Gel filtration chromatography.

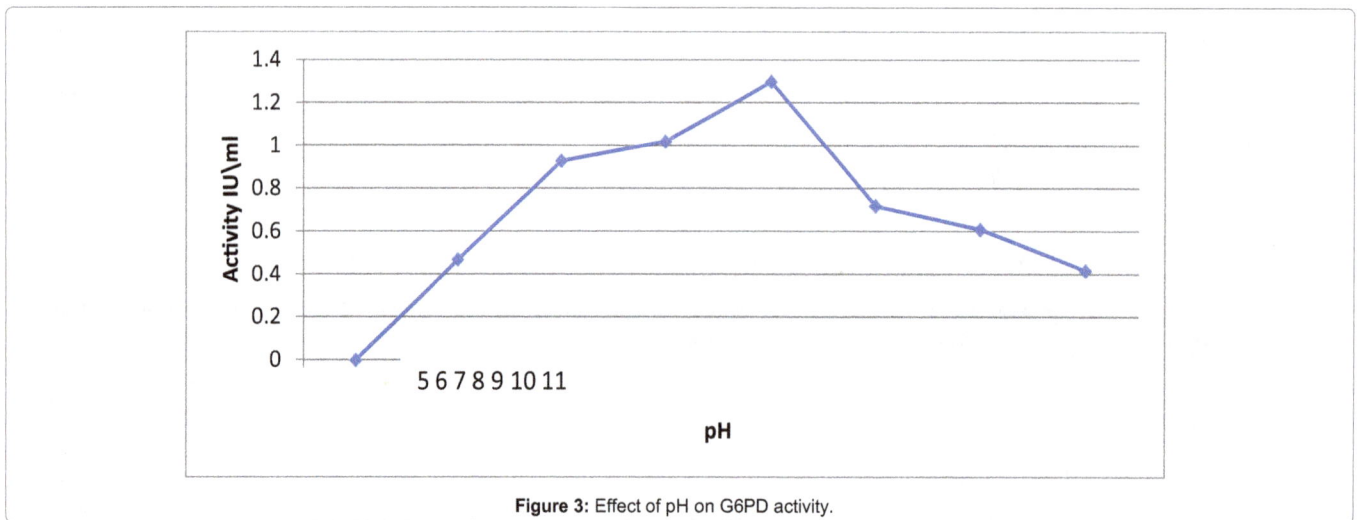

Figure 3: Effect of pH on G6PD activity.

In a study of the purified extract from the lamb marrow, the optimum grade was (38C) [37].

Effect of substrate concentration

The Michaelis-Menten constant (Km) was defined as the affinity between the enzyme and substrate. The higher its value, the less the value of the substance is reduced. In order to stimulate the biological reactions and to determine the stability of enzymes and the effect of inhibitory and activating substances on enzymatic efficacy [40]. The results of the kinetic constants estimated for the enzymatic extract were shown in Figures 5 and 6. The constant value of the Michaelis-Menten of substrate (G6P) was (3.8mM) and the maximum velocity value Vmax (8 UI/ml). The differences between all these studies were clear and almost natural as a result of the different sources of the enzyme which were cleared and the different methods In the Iraqi study to purify the enzyme from human blood, the value of Michaelis-Menten was found

in the substrate after the use of two purification methods (103 and 114 micro mole) and the maximum velocity (362 and 403 micromole/min/mg) respectively [41]. In another study on coriander leaves, the value of the Michaelis-Menten constant was 0.11 mmol and the maximum velocity was 0.038 units/ml. In a study to purify the enzyme from the pituitary gland for cows, the values of the Michaelis-Menten constant were 0.042 mmol and the maximum velocity was 9 units/ml [32]. In a WHO report for 2015, Michaelis Menten - purified from malaria patients is an average (30-50 micro mole) [1]. In a study on fungus Penicillium duPonti the Michaelis constant and the maximum velocity were respectively (0.43 mmol, 9 unit/mg) for the purified enzyme of the fungus [33]. The value of the Michaelis constant (Km) from the human placenta was (0.4 mmol) and the maximum velocity was (8 unit/mg) [36]. study of the enzyme purification of the local isolation of yeast Saccharomyces cerevisiae showed that the value of the Michaelis-Menten (Km) constant was (0.343 mmol) and the maximum velocity of substrate G6P (4.08 mmol/min) [45].

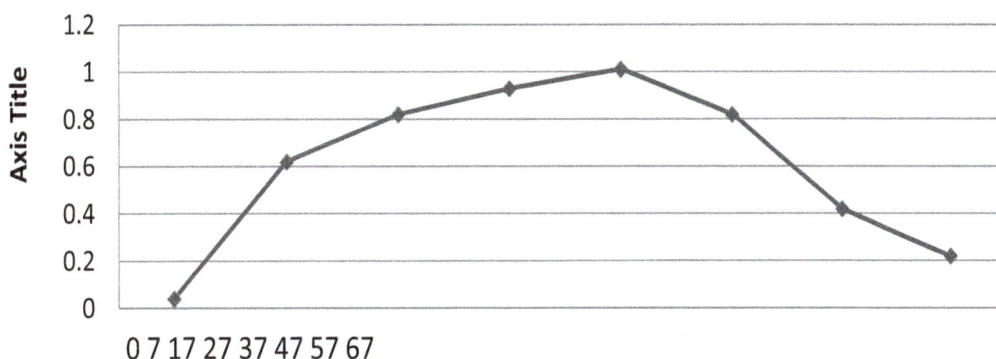

Figure 4: Effect of temperature on G6PD activity.

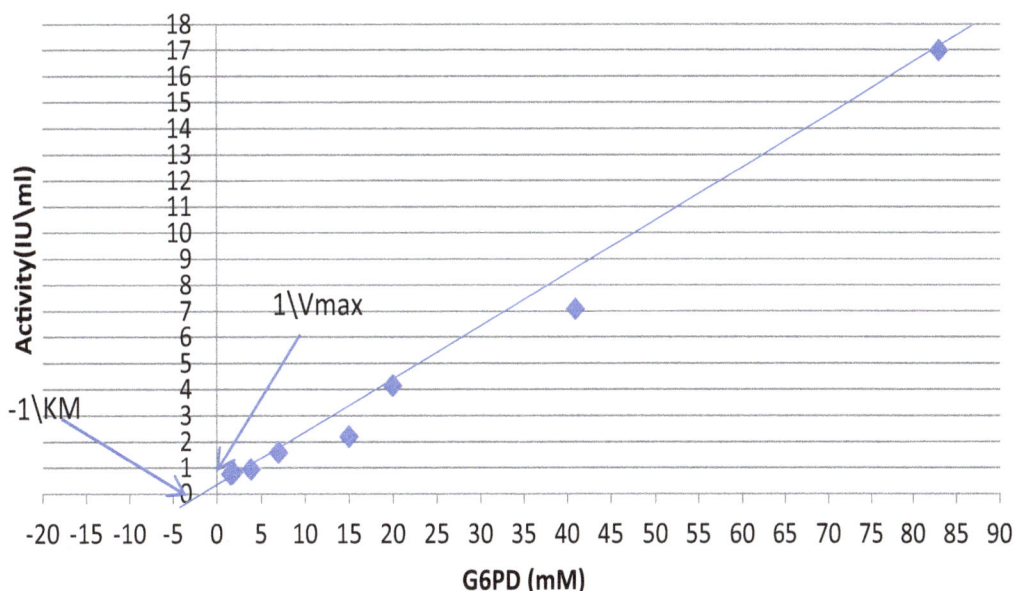

Figure 5: Lineweaver-Burk to calculate (Km) and (Vmax) to G6PD enzyme that purification from diabetes patient with Gill filtration on Sephadex G-100 towards substrate G6P with concentration between (0.6-0.0512 mM) with Coenzyme (NADP⁺=0.2 Mm).

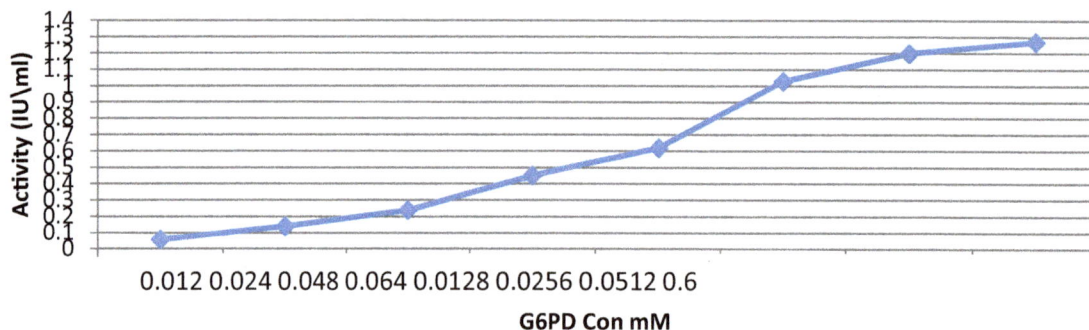

Figure 6: Effect of substrate concentration on enzyme activity.

References

1. Deo SS, Gore SD, Deobagkar DN, Deobagkar DD (2016) Study of inheritance of diabetes mellitus in western Indian population by pedigree analysis. JAPI 441: 43-64.

2. American Diabetes Association (2016) Diagnosis and classification of diabetes mellitus. Diabetes Care 33: S62-S98.

3. Engelgan MM (2014) Diabetes diagnostic criteria and impaired glycemic states. Evol Evid base Clin Diab 22: 69-77.

4. Jun L, Yuying G, Meixiang G, Chen P, Wang H (2016) Serum Magnesium Concentration Is Inversely Associated with Albuminuria and Retinopathy among Patients with Diabetes, p: 4-9.

5. Nadir Akter A, Noorzahan B, Sultana F (2016) Glucose-6-Phosphate Dehydrogenase (G6PD) status in Female Type 2 Diabetes Mellitus and Its Relationship with HbA1C. University of Tehran, Iran, pp: 73-79.

6. Kirkman H, Hendrickson E (2012) G6PD from Human erythrocytes II. Subactive states of the enzyme from normal persons. Journal of Biological Chemistry 237: 2371-2376.

7. Fiorelli G, Cappellini D (2008) Glucose-6-phosphate dehydrogenase deficiency. Lancet 371: 64-74.

8. Yadollah Z, Mousa Ahmadpour K, Haleh Akhavan N, Roya F (2013) Comparison of Molecular Mutations of G6PD Deficiency Gene Between Icteric and Non icteric Neonates. Int J of mol cel med pp: 20-27.

9. Mandel JL, Monaco AP, Nilson DL, Schlessinger D, Willerd H (1992) Genome Analysis and the Human X chromosome. Science 258: 103-109.

10. Trask BJ, Massa H, Kenwrick S, Gitschier J (1991) A 1.6-Mb contig of yeast artificial chromosomes around the human factor VIII gene reveals three regions homologous to probes for the DXS115 locus and two for the DXYS64 locus. Am J Hum Genet 48: 1-15.

11. Pamba A, Richardson ND, Carter N (2012) Clinical spectrum and severity of hemolytic anemia in glucose 6-phosphate dehydrogenase-deficient children receiving dapsone. Blood 120: 4123.

12. Gellady AM, Greenwood RD (1972) G-6-PD hemolytic anemia complicating diabetic ketoacidosis. J Pediatr 80: 1037.

13. Saha N (2013) Association of glucose-6-phosphate dehydrogenase deficiency with diabetes mellitus in ethnic groups. of Singapore. Journal of Medical Genetics 16: 431-434.

14. Plummer TD (1978) An introduction of practical Biochemistry. McGraw-Hill Book Co., UK 54.

15. Andrews P (1965) The gel filtration behavior of proteins related to their molecular weight over a wide range. J Biol Chem 96: 595.

16. Robyt FJ, White JB (1987) Biochemical Techniques. Theory and practice. Books/cole publishing Co., USA 141: 235-236.

17. Morris CJ, Morris P (1998) Separation method in Biochemistry. Pitman publishing, pp: 443-444.

18. Liu H, Liu W, Tang X, Wang T (2015) Association between G6PD deficiency and hyperbilirubinemia in neonates: a meta-analysis. Pediatr Hematol Oncol 32: 92.

19. Shalev O, Wollner A, Menczel J (1984) Diabetic ketoacidosis does not precipitate haemolysis in patients with the Mediterranean variant of glucose-6-phosphate dehydrogenase deficiency. Br Med J (Clin Res Ed) 288: 177.

20. Roth G, Nunes JES, Rosado LA, Bizarrol CV, Volpato G, et al. (2005) Purified and used by Erwinia carotovora MM-3 from L-asparaginase ani bacteria, Production of Enzyme-231. Inhibition of cancer cells (outside the body of the organism). PhD thesis - Faculty of Science - Anbar University.

21. White W, Ahandler P, Smith E (1973) Principle of Biochemistry. McGrew-hill book company. New York.

22. Beydemir S, Gulcin I, Kufrevioglu OI, Ceftice M (2003) G6PD for normal and G6PD in Vitro and in vivo effect of Dantrolene Sodium. Pol J Pharmacol 55: 787-792.

23. Beydemir S, Ylmaz H, Ceftice M, Bakan E, Kufrevioglu OL (2003) Purification and kinetics of glucose-6-phosphate From Erythrocyte Goose. Turk J Vet Anim Sci 27: 1179-1185.

24. Lee WT, Levy HR (1992) Lysine -21-of leuconostoc mesentoides G6PD participates in substrate binding through charge – charge interaction. Protein Sci 1: 327-336.

25. Karlsoon E, Ryden L, Brewer J (1998) Ion exchange chromatography: Introduction to protein purification. A John Whiley and Sons Inc., New York, USA, pp: 40-67.

26. Scott WA, Tatum EL (1971) Purification and partial characterization G6PD from Neurospora crassa. Journal of Biological Chemistry 246: 6347-6352.

27. Haghighi B, Atabi F (2003) Reassociation and reactivation of glucose 6-phosphate dehydrogenase from streptomyces aureofaciens after denaturation by 6 m urea. Journal of Sciences 14: 103-111.

28. Özdemir H, Türkoglu V, Çiftçi M (2007) Purification and Characterization of Glucose-6- phosphate Dehydrogenase from Lake Van Fish (Chalcalburnus tarichii Pallas, 1811) Erythrocytes. Asian Journal of Chemistry 19: 5695-5702.

29. Criss WE, Mckerns KW (1968) Purification characterization of G6PD from cow Adrenal cortex. Biochemistry I, pp: 125-134.

30. Malcolm A, Shepherd M (1972) Purification and properties of Penicillium G6PD. Biochemical Journal 128: 817-831.

31. Kanji M, Toews M, Carper W (1976) A kinetic study of G6PD. Journal of Biological Chemistry 251: 2258-2262.

32. Semenikhina AV, Popova TN, Matasova LV (1999) Catalytic properties of glucose-6-phosphate dehydrogenase from pea leaves. Biochemistry (Moscow) 64: 863-866.

33. Ozer N, Aksoy Y, Ogus I (2001) Kinetic properties of human placental G6PD. International Journal of Biochemistry and Cell Biology 33: 221-226.

34. Kaplan M, Hammerman C (2002) G6PD deficiency: a potential source of severe neonatal hyperbilirubinaemia and kernicterus. Seminars in Neonatology 2: 121-128.

35. Dulaimi D (2002) Microbial enzymes and biotechnologies. Philadelphia University Press, Jordan.

36. Kamel DB (1983) Understanding Enzymes. University of Mosul Press, Mosul University.

37. Segal IH (1976) Biochemistry calculation. John Wiley and Sons. New York, USA.

38. Rafat GI (1982) Study of human erythrocyte G6PD characters in Basrah Area. MSc Thesis, Basrah, Iraq, pp: 88-97.

39. Al-Soufi MAA (2005) Purification and characterization and utilization of G6PD from locally isolated yeast Saccharomyces cerevisiae. Thesis, College of Agriculture, University of Baghdad, Iraq.

40. Roos D, Van Zwieten R, Juul T (1999) Molecular Basis and Enzymatic properties of G6PD volendam, Leading to chronic Nonspherocytic Anemia, Granulocyte Dysfunction, and increased susceptibility infection. Blood 94: 2954-2971.

41. Livingstone FB (1971) Malaria and human polymorphisms. Ann Rev Genet 5: 33-64.

42. Nguyen K, David A, Lee P, Anderson J, Epstein DL (1986) G6PD of Calf trabecular meshwork. Invest Opthalmol Vis Sci 27: 991-998.

43. Bregmeyer HU (1984) Method of enzymatic analysis. Verlag Chemie. Weinheim.

44. Grandall GD (1983) Biochemistry Laboratory. Oxford University Press, New York 29: 83.

45. Fiengezer C, Ulusu NN (2007) Three Different Purification Protocols in Purification of G6PD from Sheep Brain Cortex. J Pharm Sci 32: 65-72.

23

Validation and Development of HPTLC Method for Simultaneous Estimation of Apigenin and Luteolin in Selected Marketed Ayurvedic Formulations of 'Dashmula' and in Ethyl Acetate Extract of *Premna integrifolia* L.

Attarde DL[1]*, Pal SC[2] and Bhambar RS[1]

[1]*Department of Pharmacognosy, Mahatma Gandhi Vidyamandir's Pharmacy College, Panchavati, Nashik, Maharashtra, India*
[2]*Department of Pharmacognosy, RG Sapakal College of Pharmacy, Kalyani Hills, Trimbakeshwar, Nashik, Maharashtra, India*

Abstract

Dashmula are specific ayurvedic combination of ten roots used for various disorders of liver, kidney, uterus. Standardisation of herbals are necessity for efficacy and quality parameter as per WHO guidelines. Therefore, through this original research attempt was made to standardise one of 'Dashmula', P. integrifolia and selected batches of marketed formulation using Apigenin and Luteolin as active biological marker for simultaneous quantification and fingerprinting through developed and validated HPTLC techniques. Developed mobile phase Toluene: Ethyl acetate: Formic acid (6:4:0.15) gave Rf (Retention factor) 0.39 and 0.29 for Standard Apigenin and Luteolin respectively at 347 nm iso absorptive wavelength. The ethyl acetate extract of P. integrifolia (PI-ET), 'Dashmularishtha': Manufactuer 1; 3 coded batches as DF1, DF2, DF3, Manufacturer 2; 3 coded batches -as BF4, BF5, BF6, 'Dashmulkadha': Manufacturer 3; 3 coded batches as KF, KF8, KF9 were found to contain : 12.8% w/w, 0.294 mg/ml%, 0.429 mg/ml%, 0.314 mg/ml%, 0.077 mg/ml%, 0.071 mg/ml%, 0.145 mg/ml%, 0.176 mg/ml%, 0.242 mg/ml%, 0.098 mg/ml% of Apigenin and 4.7% w/w, 0.542 mg/ml%, 0.365 mg/ml%, 0.569 mg/ml%, 0.343 mg/ml%, 0.311 mg/ml%, 0.607 mg/ml%, 0.812 mg/ml%, 0.828 mg/ml%, 0.439 mg/ml% of Luteolin respectively. In stem powder of P. integrifolia 19.84 mg/gm% Apigenin and 7.433 mg/gm% Luteolin was calculated. The estimation shows variance in manufacturers and even within batches, but will be quality control parameter. The method was validated for specificity, linearity, accuracy, precision, and robustness. It was found to be linear in range of 40-120 ng/band with regression coefficient 0.9983, 0.9997 for Apigenin and luteolin. Percentage recovery study carried out for extract PI-ET and DF1 with spike of Apigenin and Luteolin at 80, 100 and 120% level, carried out for inter and intraday precision, subjected for one way ANOVA and found F value is below tabulated $F_{(2,6)}$ value 5.14, therefore there is no significance variance of obtained values.

Keywords: Apigenin; Luteolin; HPTLC validation; Dashmoola; *Premna integrifolia*

Introduction

Dashmool (dashmul, dashamula) are specific combination of ten roots, famous Ayurvedic remedy mainly for strengthening body, promotes healthy elimination of toxins, tonic for kidney, liver, tonic in infertility, uterine tonic, used for anorexia, edema, anaemia, potent antioxidant and in disorders of nerves, bones, joints and muscles. It contains ten roots in equal propotion under category as '**Brihat Panchmul** includes; *Aegle marmelos, Premna integrifolia, Oroxylum indicum, Stereospermum suveolens, Gmelia arborea* While **Laghu panchmul** includes; *Solanum indicum, Solanum xanthocarpum, Uraria picta, Desmodium gangeticum* and *Tribulus terrestris* [1].

More number of people worldwide switching once again towards the alternative system of medicines to avoid harmful side effects of synthetic drugs. As per WHO it is essential to establish quality standards for herbals. Thus, standardization of crude drugs and their formulations are at urgent need [2].

P. integrifolia synonym(s)

P. serratifolia L., Large, thorny, deciduous shrub a member of Brihatpanchamula groups, known as Agni-manth, in Sanskrit; Agetha in Hindi, distributed along the coasts and islands of tropical and subtropical country. The leaves and roots have cardiotonic, astringent, anti-inflammatory, antibacterial, anti- hypoglycemic, anticoagulant, anti-arthritics and cardio protective properties [3-7].

As per literature, diterpenoids reported from the root bark, a verbascoside iridoid glucoside from leaves [8], two alkaloids, Premnine and Ganiarine, have been isolated from it and their physical and chemical properties described [9]. Volatile oil from flower bud, distillate and its fractions were analyzed by GC and GC/MS and the major components are 1-octen-3-ol (16.9%), (Z)-3-hexenol (10.2%) [10-12]. Isolation and evaluation of flavonoid i.e., Luteolin 7-0-methyl ether and apigenin 5,7-0-dimethyl ether from *P. serratifolia's* leaves [13-18].

Cardiac stimulant activity of bark and wood of *P. serratifolia* [14], root extract for its anti-inflammatory, antioxidant activities, antimicrobial, Anti-Arthritic Activity [16], Cardioprotective effect of ethanol extract of stem-bark and stem-wood of *P. serratifolia* Lin., (Verbenaceae) were evaluated Tables 1 and 2.

Apigenin, 4', 5, 7-tetrahydroxy flavone, is recognized as a bioactive flavonoid shown to possess exerts anxiolytic effects by acting as a benzodiazepine ligand, anti-inflammatory, antioxidant, anticancer properties particularly cancers of the breast, digestive tract, skin,

*Corresponding author: Attarde DL, Department of Pharmacognosy, Mahatma Gandhi Vidyamandir's Pharmacy College, Panchavati, Nashik-422 003, Maharashtra, India, E-mail: daksha511@rediffmail.com

Parameter	Apigenin	Luteolin
Wavelength, nm	340 nm	353 nm
Isoabsorptive Wavelength	347 nm	347 nm
RF Value	0.39	0.29
Linearity Range ng/band	40-120	40-120
Regression equation	Y=37.395x+636.27	Y=30.455x-588.62
Correlation Coefficient	0.9983	0.9997
Limit of Detection ng/band LOD	2.01 ng/Band	1.82 ng/Band
Limit of Quantification ng/band LOQ	5.98 ng/Band	5.52 ng/Band
Specificity	Specific- Overlay Spectra, Retention factor, Derivatized Colour Parrot green at 366 nm	Specific- Overlay Spectra, Retention Factor, Derivatized Colour bright yellow at 366 nm

Table 1: Validation parameter for estimation of apigenin and luteolin in extract and formulation.

Analyte	Apigenin Concentration Obtained	Luteolin concentration Obtained
PI-ET	12.8 % w/w	4.7 % w/w
Powder of *P. integrifolia*	19.84 mg/gm %	7.433 mg/gm%
Dashmul Arishtha		
Manufacturer 1 Bottle size: 225 ml, Label Claim: Each 100ml contains 0.52 gm of *'Dashmula'* each.		
DF1*	0.294 mg/ml %	0.542 mg/ml %
DF2*	0.429 mg/ml %	0.368 mg/ml %
DF3*	0.314 mg/ml %	0.509 mg/ml %
Dashmul Arishtha		
Manufacturer 2 Bottle size : 220 ml, Label Claim: Each 10 ml contains 0.50 gm of *'Dashmula kwath'*		
BF4**	0.077 mg/ml %	0.343 mg/ml%
BF5**	0.071 mg/ml %	0.311 mg/ml%
BF6**	0.145 mg/ml %	0.607 mg/ml%
Dashmul Kadha		
Manufacturer 3 Bottle size: 227 ml, Label Claim: Each 100ml contains 418.91 mg of *'Dashmula'* each		
KF7***	0.176 mg/ml%	0.812 mg/ml%
KF8***	0.242 mg/ml%	0.828 mg/ml%
KF9***	0.098 mg/ml%	0.439 mg/ml

PI-ET- *Premna integrifolia* ethyl acetate extract; *Manufacturer -1 -Three Batches Coded as DF1, DF2, DF3; **Manufacturer -2- Three Batches Coded as BF4, BF5, BF6; ***Manufacturer-3 -Three Batches Coded as KF7, KF8, KF9

Table 2: Estimation of apigenin and luteolin in PI-ET extract and selected marketed *Dashmula* formulations.

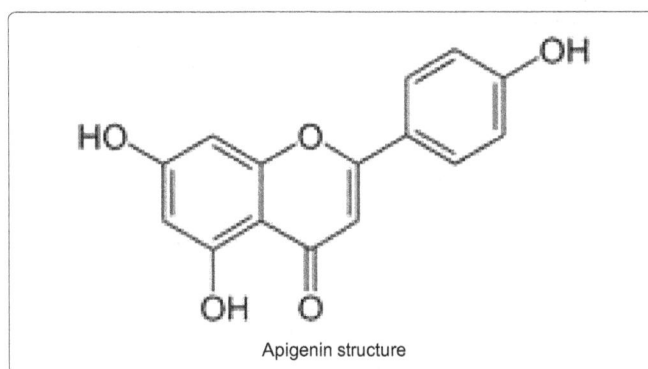

Apigenin structure

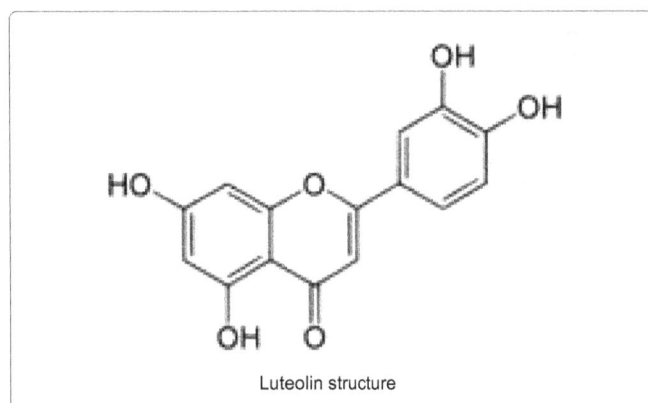

Luteolin structure

prostate and certain hematological malignancies and even protective for cardiovascular and neurological disorders [4].

Luteolin 3',4',5,7-tetrahydroxyflavone, bioflavonoid used in Chinese traditional medicine for treating various diseases such as hypertension, inflammatory disorders, and cancer [5].

P. integrifolia, therefore selected here for biomarking using sophisticated analytical techniques HPTLC using bioactive markers like Apigenin and Luteolin.

This research is first attempt for standardisation of 'dashmul' using two flavonoids simultaneously, along with raw plant extract.

Materials and Methods

Equipment

HPTLC Instrument: CAMAG Linomet Syringe V, CAMAG TLC scanner V, CAMAG Digistore- Reprostar 3, CAMAG Twin Trough Chamber, Win CATS version 1.4.2. Software (CAMAG, Switzerland), UV chamber (CAMAG, Switzerland), Pre-coated silica gel 60 F_{254} alumiFtablenium plates (0.2 mm thick, Merck, Germany) [19].

Analytical reference compound and chemicals

Apigenin, (A3145-5 mg, Sigma Aldrich), Luteolin (L9283-10 mg, Sigma Aldrich), Analytical grade solvents: Pet. Ether (60-80°C), chloroform, ethanol, ethyl acetate, Toluene, formic acid (Merck), Natural product reagent (2-aminoethyl diphenylborinate) (D9754- 5 gm, Sigma Aldrich).

Plant material and marketed formulation

Stem powder of *P. integrifolia*: Agnimanth are distributed through flora of Trimbakeshwar forest area of Nashik District. Few of them identified, spotted stem branches were collected and herbarium was prepared. These was deposited to Botanical Survey of India, Pune for identification purpose. Certificates were issued as i.e., Ref.: BSI / WRC/Tech./2012/DVR-1 dt. 02/1/2012 for *Premna integrifolia* L. (Verbenaceae). Stem branches were cut into small pieces, dried under shade for about 15 days, milled, shifted through sieve 25/30 no.-600 μ. Dried powder packed in air tight container and stored to cool, dry condition.

Preparation ethyl acetate extract of P. integrifolia: Accurately weighed 200 gm of stem powder of P. integrifolia charged in Soxhlet assembly and extracted exhaustively with alcohol. Alcoholic extract conc. over rotary evaporator, dried, made hydro alcoholic and fractionated with pet. ether (60-80°C), chloroform and ethyl acetate consecutively. Ethyl acetate extract concentrated over rotary evaporator, dried, yield noted and designated as PI-ET.

Marketed formulation: 'Dashmula arishtha': Manufacturer 1;3 batches, Manufacturer 2;3 batches, Dashmula Kadha: Manufacturer 3;3 batches, selected, batch no., volume, label claim noted and designated with codes.

Treatment to selected marketed formulation: Above Selected and coded batches were concentrated, made hydro alcoholic and fractionated with ethyl acetate extraction. All ethyl acetate extracts were concentrated, dried, yield noted and designated as for Manufacturer 1, 2, 3; 3 batches as DF1, DF2, DF3, BF4, BF5, BF6, KF7, KF8 and KF9 respectively.

Standard apigenin solution: 5 mg of standard apigenin was dissolved in 5 ml of alcohol in volumetric flask (1000 ppm), sonicated, and marked as Stock solution. Working stock prepared to obtain 20 ppm. (20 ng/μl) from it, designated as- AS.

Standard luteolin solution: 5 mg of standard Luteolin was dissolved in 5 ml of alcohol in volumetric flask (1000 ppm), sonicated and marked as Stock solution- Working stock prepared to obtain 20 ppm. (20 ng/μl) from it, designated as - LS.

Ethyl acetate extract of P. integrofolia solution: 10 mg of ethyl acetate extract was dissolved in 10 ml of volumetric flask (1000 ppm), sonicated, filtered through whatman filter paper No.1 and marked as stock solution. Working stock prepared as 0.5 mg/ml, designated as PI-ET.

Chromatographic condition: It was as follow-

Stationary phase: Pre-coated Silica Gel G 60 F_{254} HPTLC Plates (0.2 mm Thick, Size-20 × 10 cm),

Mobile Phase: Toluene: Ethyl acetate: Formic Acid (6:4:0.15)

Saturation Time: 10 min

Wavelength: 347 nm (Iso-absorptive Point between Apigenin and Luteolin)

Lamp: Deuterium, Scanning slit width 6 × 0.45 mm, Scanning speed 20 mm/s.

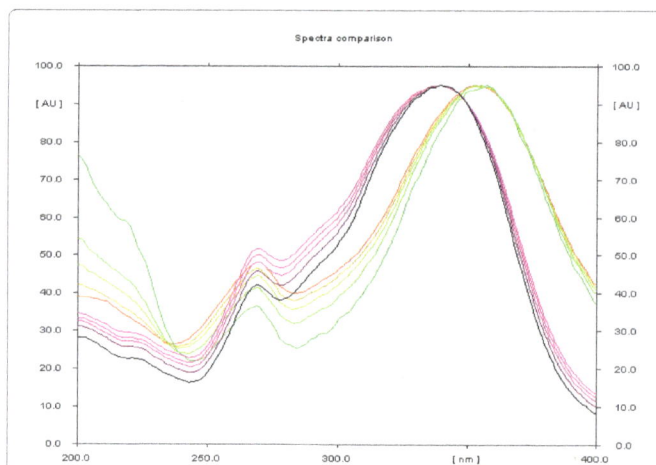

Figure 1a: Overlay spectra of apigenin and luteolin with PI-ET, DF1, BF4 and KF7.

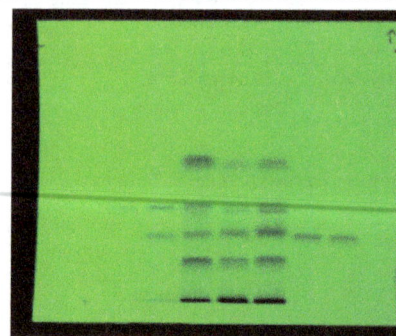

Figure 1b: Finger printing HPTLC chromatogram of standard apigenin, PI-ET, DF1, BF4, KF7 and standard luteolin at 254 nm.

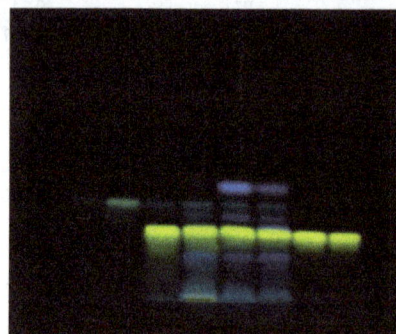

Figure 1c: After derivatization with NP-PEG reagent at 366 nm.

Sample Band length-8 mm, Solvent Run -90 mm, Area Temperature-22 ± 2°C.

Applicator Syringe-100 μl, X-8 mm, Y-15 mm, Spray gas- Nitrogen Inert gas,

Sample Application speed 0.2 μl/s.

Scanner- TLC scanner 5 (1.14.26), Photo documentation- Digistore- Reprostar 3

The HPTLC analysis was performed using above chromatographic condition, using Linomat 5 syringe 5 μl, 10 μl of AS and 5 μl, 10 μl of LS applied on pre coated TLC Silica gel 60 F_{254} plates as band length of 8 mm, sample were air dried, mobile phase Toluene: Ethyl acetate: Formic acid (6:4:0.3) poured to CAMAG twin trough chamber, saturation allowed for 10 min., spotted plates were developed till 90 mm, dried with dryer, scanned over TLC scanner 5 (1.14.26) with absorption/remission mode at scan speed 20 mm/s, spectral scan done in between 200-400 nm with 100 nm/s speed, spectra of apigenin and luteolin over layed and iso-absorptive wavelength noted, again plates were rescanned at 347 nm in detection scanner mode. Retention Factor (Rf), AUC for both standard AS and LS noted, used the data for further detection.

Fingerprinting and specificity of apigenin and luteolin standard in PI-ET and in DF1, BF4, KF7

The HPTLC analysis for fingerprinting and specificity of Apigenin and Luteolin standard done using developed parameter as above, applying on track- AS, PI-ET, DF1, BF4, and KF7, LS, plates developed as above and scanned in detection (347 nm) and spectral mode (200-400 nm). Rf obtained for AS and LS, identified and marked in PI-ET, DF1, BF4, KF7. The identified bands spectral scan over laid for confirmation

of specificity. The purity of the bands was confirmed at start, middle and end position of chromatogram. The plate was derivatized with NP-PEG reagent and chromatogram visualized at 366 nm in florescence mode as shown in Figures 1a-1c.

Linearity calibration curve for standard apigenin and standard luteolin and analysis of PI-ET

For simultaneous estimation, over same plate samples applied as AS-2, 2, 3, 4, 5, 6 μl (40 to 120 ng), PI-ET- 1, 1.5, 3 μl (0.5 mg/ml), LS-2, 3, 4, 5, 6, 6 μl (40 to 120 ng) over track 1 to 15 with band length 8 mm. Chromatogram developed till 90 mm in above mobile phase, scanned at 347 nm and recorded as per above developed condition at Rf- 0.39 for Apigenin and Rf-0.29 for Luteolin. Such procedure repeated thrice and average Area Under Curve(AUC) noted. Standard linearity calibration curve for both standard prepared separately by plotting AUC vs conc. in ng, using linearity calibration curve of apigenin and luteolin concentration in PI-ET calculated as per Figures 2a and 2b.

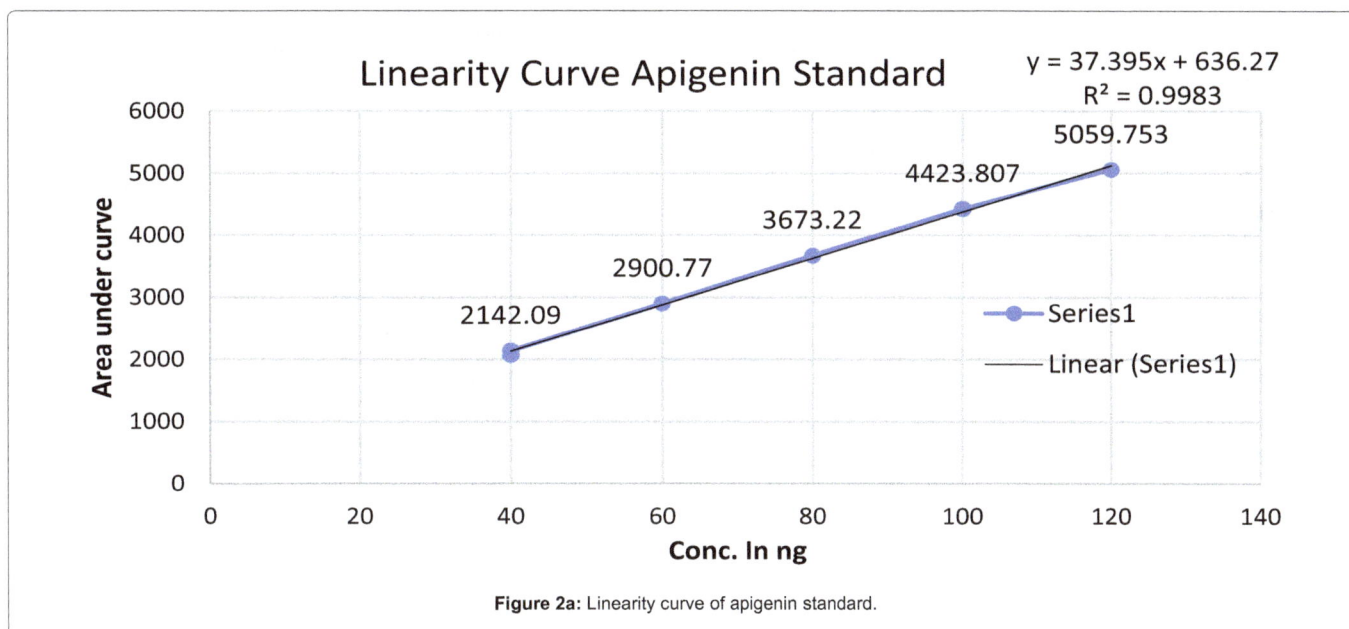

Figure 2a: Linearity curve of apigenin standard.

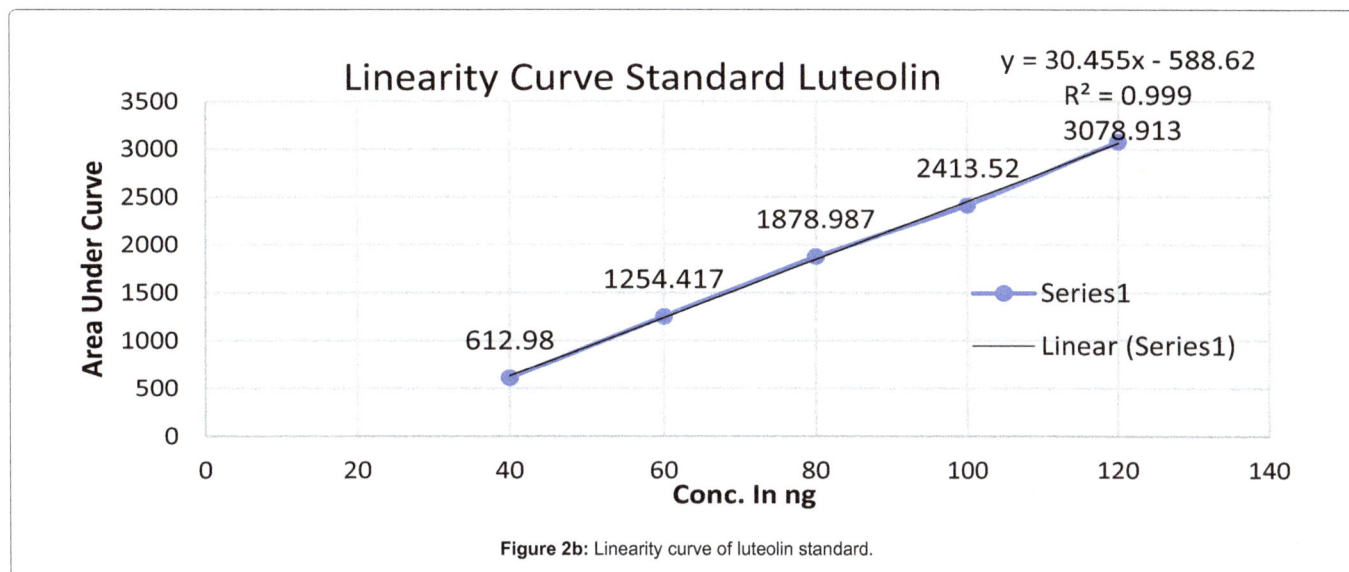

Figure 2b: Linearity curve of luteolin standard.

Concentration^T (ng/band)	Average Concentration^a (n=3) Intraday (ng/band)	Intra day % RSD	Average Concentration^a (n=3) Inter day (ng/band)	Inter day % RSD
40	37.85	1.3403	37.38	1.8197
80	77.37	0.7785	77.12	0.4437
120	115.51	0.5200	114.85	0.7487

T- Therotical, a- Obtained, %RSD- Relative Standard Deviation

Table 3a: Intermediate precision study for standard apigenin.

Concentration ^{T} (ng/band)	Average Concentration^a (n=3) Intraday (ng/band)	Intra day % RSD	Average Concentration^a (n=3) Inter day (ng/band)	Inter day % RSD
40	39.11	0.5920	38.96	0.3632
80	77.68	0.5661	77.04	1.0246
120	119.95	1.3306	119.24	0.5218

T- Therotical, a- Obtained, % RSD- Relative Standard Deviation

Table 3b: Intermediate precision study for standard luteolin.

	Level of % Recovery	Amount In PI-ET (ng)	Amount of Standard Added (ng)	Total amount of Standard taken (ng)	Total amount of Standard obtained (ng)	% recovery ± SD (n=3)
In PI-ET						
AS	80%	44.83	35.86	80.69	88.18	109.29 ± 0.345
	100%	44.83	44.83	89.66	90.64	101.10 ± 0.567
	120%	44.83	53.79	98.62	102.81	107.25 ± 0.234
LS	80%	43.16	34.52	77.68	82.42	106.11 ± 0.385
	100%	43.16	43.16	86.12	81.03	94.09 ± 0.583
	120%	43.16	51.79	94.95	92.65	97.58 ± 0.241
In DF1						
AS	80%	39.27	31.42	70.69	68.87	97.43 ± 0.365
	100%	39.27	39.27	78.54	81.52	103.80 ± 0.765
	120%	39.27	47.12	86.39	88.12	102.01 ± 0.376
LS	80%	36.21	28.96	65.17	64.98	99.72 ± 0.287
	100%	36.21	36.21	72.42	72.65	100.33 ± 0.456
	120%	36.21	43.45	79.66	76.76	96.37 ± 0.345

PI-ET- *Premna integrifolia* ethyl acetate extract, DF1- Manufacture 1 -coded Batch, AS- Apigenin Standard, LS- Luteolin Standard.

Table 4a: Accuracy (% Recovery) for PI-ET extract and DF1 extract with spike of standard apigenin and luteolin at 80%, 100% and 120% level [18].

(ng) Amount	(n=3)	Amount of Standard Obtained in ng			WMS	BMS	F value
PI-ET+AS		Day 1	Day 2	Day 3			
80 %	Mean	91.54	90.34	91.02	0.994011	1.086944	**1.093493**
(80.69)	%RSD	0.493	1.428	1.159			
100%	Mean	95.63	96.69	97.89	1.800944	0.732044	**0.406478**
(89.66)	%RSD	1.176	1.613	1.279			
120%	Mean	104.31	103.39	105.16	2.706011	2.341944	**0.86546**
(98.62)	%RSD	1.465	1.649	1.168			
PI-ET+LS							
80%	Mean	81.153	81.08	81.33	0.341867	0.0499	**0.145963**
(77.68)	%RSD	0.878	1.034	0.429			
100%	Mean	83.23	82.85	82.88	0.36	0.132844	**0.369012**
(86.12)	%RSD	1.102	0.864	0.244			
120%	Mean (n=3)	90.16	90.02	89.26	1.081278	0.707078	**0.653928**
(94.95)	%RSD	0.477	1.190	1.548			

PI-ET- *Premna integrifolia* ethyl acetate extract, AS- Apigenin Standard, LS- Luteolin standard, WMS- Within Mean Square, BMS- Between Mean Square, F Value=BMS/WMS

Table 4b: Precision study for PI-ET with spike of standard apigenin and luteolin at 80%, 100% and 120% level (One Way ANOVA) [18].

(ng) Amount	(n=3)	Amount of Standard Obtained in ng			WMS	BMS	F value
DF1+AS		Day 1	Day 2	Day 3			
80%	Mean	63.21	63.11	64.65	0.641533	2.217678	**3.45684**
(70.69)	%RSD	1.2026	0.7697	1.6299			
100%	Mean	78.94	80.75	79.34	0.665444	2.697411	**4.053548**
(78.54)	%RSD	0.8183	1.0619	1.1576			
120%	Mean	85.16	85.56	87.31	3.287878	3.935511	**1.196976**
(86.39)	%RSD	1.86	1.27	1.84			
DF1+LS							
80%	Mean	65.59	65.99	65.44	0.265833	0.2425	**0.912226**
(65.17)	%RSD	0.5755	1.003	0.7116			
100%	Mean	70.08	70.00	69.80	0.120867	0.064011	**0.529601**
(72.42)	%RSD	0.6060	0.5103	0.3345			
120%	Mean	75.27	75.38	75.20	0.513667	0.024844	**0.048367**
(79.66)	%RSD	0.9726	0.8613	1.015			

WMS- Within Mean Square, BMS- Between Mean Square, F Value=BMS/WMS, DF1- Manufacture-1-coded Batch, AS- Apigenin Standard, LS- Luteolin standard

Table 4c: Precision study for DF1 with spike of standard apigenin and luteolin at 80% ,100% and 120% level (One Way ANOVA) [19].

HPTLC analysis of marketed formulation: Samples were applied over TLC plates as AS-2, 4, 6 µl, DF1-4.2 µl, DF2-4.2 µl, DF3-4.2 µl, BF4-4.0 µl, BF5-4.0 µl, KF7-3.5 µl, KF8- 3.5 µl, KF9- 3.5 µl, LS-2,4,6 µl on track 1 to 15 tracks respectively. Chromatogram developed till 90 mm in above mobile phase, scanned at 347 nm and recorded as per above developed condition at Rf 0.39 for Apigenin and Rf-0.29 for Luteolin. Such procedure repeated thrice and average AUC noted. As per standard linearity calibration curve for both standard apigenin and luteolin concentration in DF1, DF2, DF3, BF4, BF5, BF6, KF7, KF8 and KF9 calculated as Figures 2a and 2b.

Validation of HPTLC method: The method was validated as per International Conference on Harmonization Guidelines- Q2 (R1), 2005. Linearity for Apigenin and luteolin standard were studied in range of 40-120 ng /band for both. The calibration curve developed by plotting Peak area vs Conc. in ng, regression equation with slope, intercept and coefficient of correlation was calculated Table 3.

The Limit of detection (LOD) and limit of quantification (LOQ) were calculated using equation:

$$LOD = \frac{3.3 \times \text{Standard deviation of the Y - intercept}}{\text{Slope of the caliberation curve}}$$

$$LOQ = \frac{10 \times \text{Standard deviation of the Y - intercept}}{\text{Slope of the calibration curve}}$$

Accuracy (% recovery): The accuracy of the method was established by performing recovery experiments over known concentration on Extract -PI-ET and formulation DF1 at 80%, 100% and 120% spike of AS and LS separately in triplicate for each experiment and analyzing it with %RSD. As per Table 4a.

Intermediate precision: Precision study for Standard AS and LS done for three concentration as 40, 80, 120 ng/band, done in triplicate for intraday and inter day precision.

Precision study over recovery method for PI-ET and DF1 done separately in triplicate for 80%, 100% and 120% spike of AS and LS triplicate for intraday and inter day precision with 3 × 3 model for each experiment and analyzing it with %RSD.

Further result obtained were subjected for one way analysis of variance and with-day mean square compared to between -day mean square by F test (Tables 4b and 4c).

Repeatability precision: The repeatability of the method was assessed by three concentration 40,80,120 ng/band in triplicate for each AS and LS done. The percentage relative standard deviation was expressed as Tables 3a and 3b.

Robustness: Mid concentration of 80 ng/band in triplicate of both standards subjected for robustness study using variability like wavelength 347-5 nm, slit width change as 6 × 0.3 *mm*, scan speed change as 40 mm/s.

Results and Discussion

In fingerprinting and specificity study with developed chromatographic condition in mobile phase Toluene: Ethyl Acetate: formic acid (6:4:0.3), Rf- obtained 0.39 and 0.29 for standard apigenin and standard luteolin respectively. There is single spot over track for standard. Track of PI-ET, DF1, BF4, KF7 shows spots with similar Rf for Apigenin and luteolin. Spectral scan of this selected spots gave specific overlay for Apigenin and Luteolin. Upon derivatization with NP-PEG reagent Spot of apigenin and luteolin appears bright yellow visually in day light, parrot green and bright yellow in florescence mode at 366 nm respectively.

That confirms presences and helped for fingerprinting of Apigenin and luteolin bioactive flavonoids in PI-ET and in selected formulations.

The calibration curve of Standard Apigenin and Standard Luteolin was found to be linear in range of 40-120 ng/band for both, with good regression coefficient 0.9983, 0.9997 for Apigenin and luteolin respectively. Table 1 summaries the validation parameter.

Estimation analysis for Extract PI-ET and formulation as shown in Table 2 indicates high %value for PI-ET extract due to highly purified flavonoid fraction from extract. Variability observed between manufactures and even within batches.

Resolution of extract and formulations are better while estimating apigenin and luteolin simultaneously as shown in Figures 3a-3l.

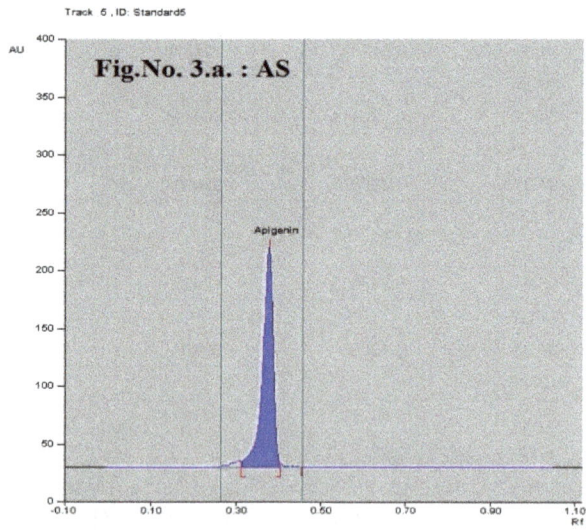

Fig.No. 3.a. : AS

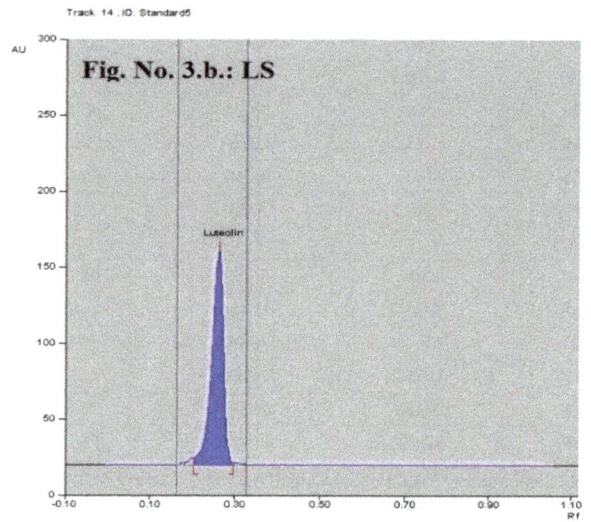

Fig. No. 3.b.: LS

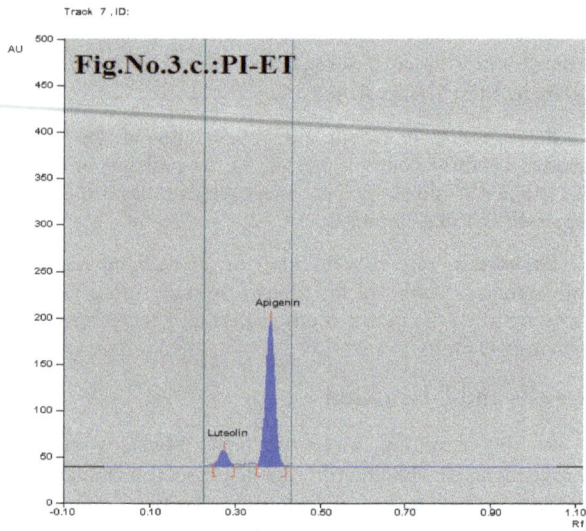

Fig.No.3.c.:PI-ET

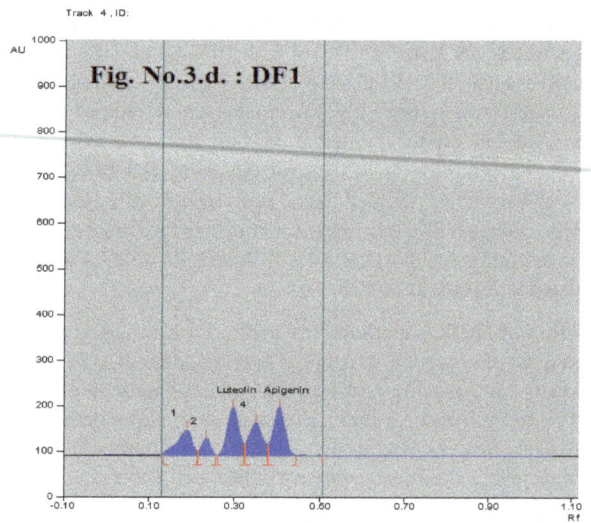

Fig. No.3.d. : DF1

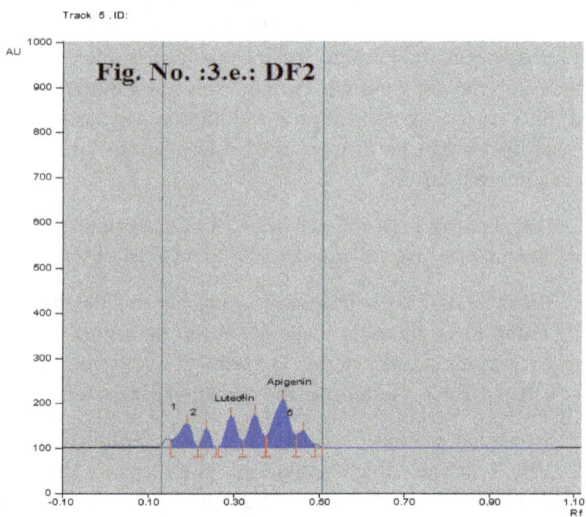

Fig. No. :3.e.: DF2

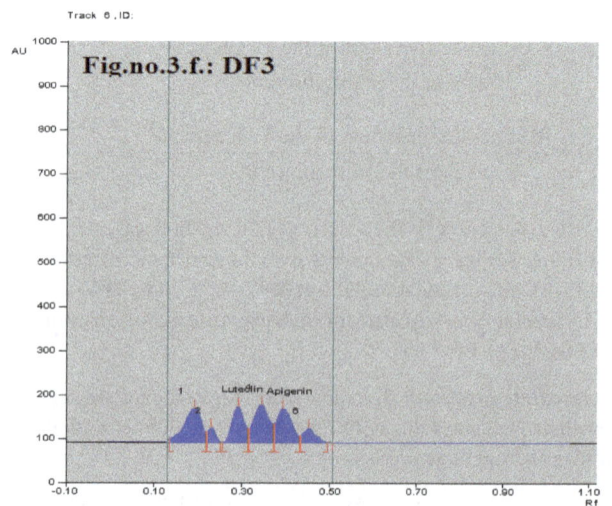

Fig.no.3.f.: DF3

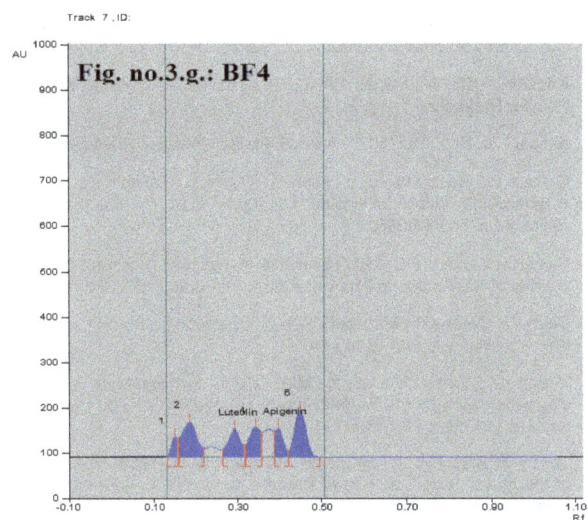

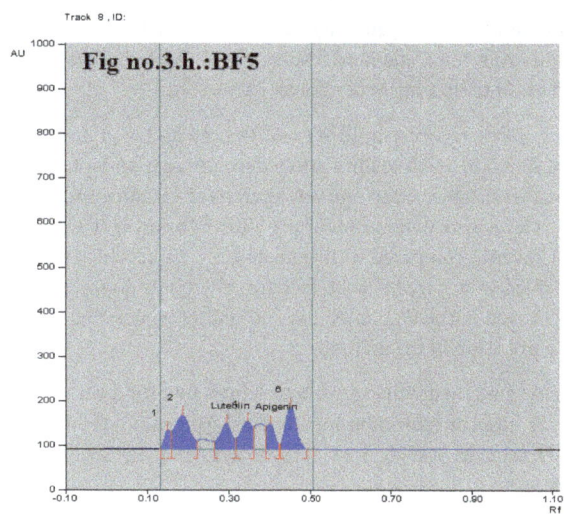

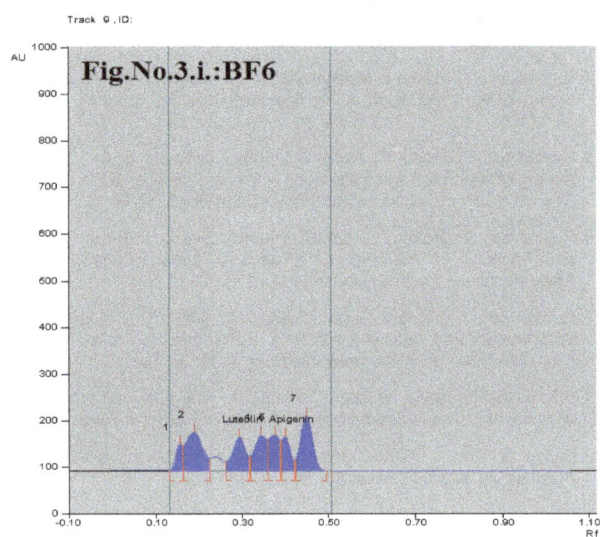

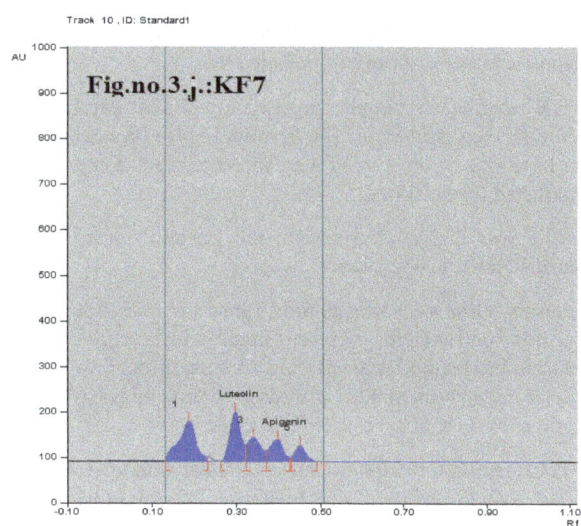

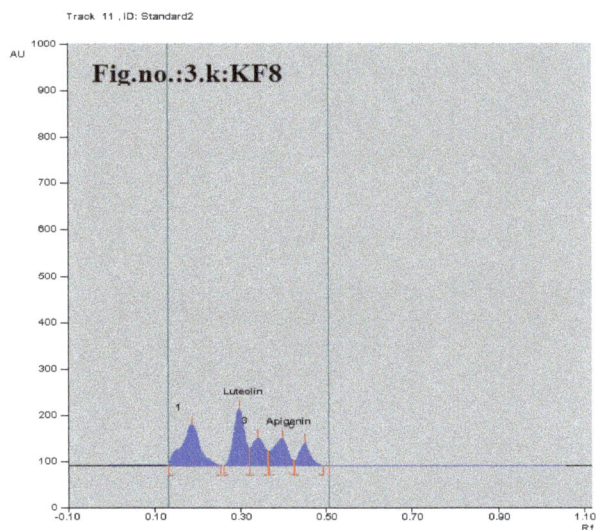

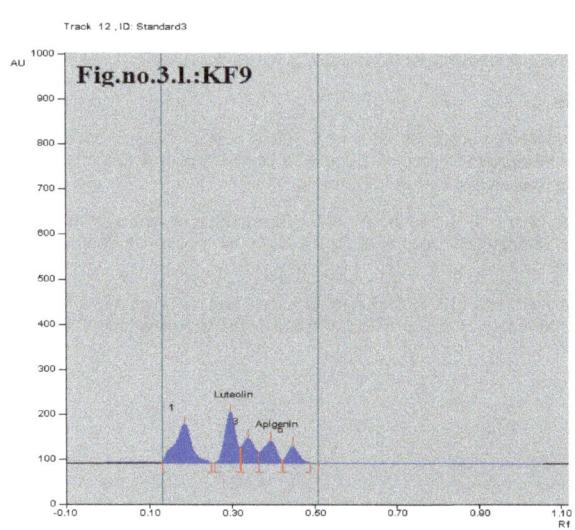

Figures 3a-3l: HPTLC resolution chromatogram of standard apigenin, standard luteolin, PI-ET extract, DF1, DF2, DF3, DF4, DF4, DF5, DF6, KF7, KF8, and KF9 marketed formulation respectively.

Accuracy study with %recovery at three concentration level performed in PI-ET and in one of formulation DF1 shows result as per Table 4a Precision study with repeatability performed over low, mid high concentration over both standard expressed as %relative standard deviation inters and intraday as per Tables 3a and 3b.

Accuracy with% recovery in PI-ET and DF1 at 80, 100, 120 %level with spike of standard and luteolin subjected for intraday and inter day, results at each level subjected to one way analysis of variance and the F value for each level were determined as per Tables 4b and 4c. F value as ratio of BMC/WMC, compared with tabulated $F_{(2,6)}$ value which is 5.14 and all calculated values are below it, therefore there was no significant difference between intra and inter day variability, suggesting good intermediate precision of the method.

Robustness study with change in wavelength by 5 nm, slit width, scan speed for mid concentration for standard Apigenin and luteolin shows %RSD value as 1.4370, 1.3355, 0.7955, 1.2563, 1.1729, 0.6875 respectively.

Conclusion

Hence with regards to the method development, validation and simultaneous estimation, it can be concluded that,

- HPTLC method is simple, precise, rapid and selective for simultaneous estimation of bioflavonoid Apigenin and Luteolin in Dashmul- *P. integrifolia* stem and ethyl acetate extract and marketed formulations.

- PI-ET shows high %of Apigenin and luteolin due to highly purified fraction of extract.

- The result of this study though shows great variance in %content of this bioflavonoids between manufactures and even in batches but method is precise and accurate through validation therefore can be method of analysis qualitative and quantitative for quality control parameter for herbals.

- HPTLC method development for standardization of Ayurvedic 'Dashmula formulation' gives high throughput results.

Acknowledgements

We are thankful to our Institute and management for providing facility to work over HPTLC instrument. We also thanks to Prof. Sandeep Sonawane from MET' Institute of Pharmacy, Nashik for his valuable guidance for statistical analysis.

References

1. Ramchandra NS, Vavhal RR, Kulkarni CJ (2015) Effect of Dashmool churna, Varunadi Kwath and Kanchanar Guggul on uterine fibroid. Ayurlog: National Journal of Research in Ayurved Science 3: 1-5.

2. Kataria S, Bhardwaj S, Middha A (2011) Standardization of medicinal plant materials. International Journal of Research in Ayurveda and Pharmacy 2: 1100-1109.

3. The Wealth of India (1972) Dictionary of Indian raw materials and industrial products. New Delhi: Council of Scientific and Industrial Research 8: 239-240.

4. Shukla S, Gupta S (2010) Apigenin: a promising molecule for cancer prevention. Pharm Res 27: 962-978.

5. Lin Y, Shi R, Wang X, Shen HM (2008) Luteolin, a flavonoid with potential for cancer prevention and therapy. Curr Cancer Drug Target 8: 634-646.

6. Nadkarnis KM, Nadkarnis KM (1976) The Indian Materia Medica. Bombay Popular Prakashan 1: 1009.

7. Kirtikar KR, Basu BD (1927) An ICS. Indian Medicinal Plants. India.

8. Otsuka H, Watanabe E, Yusana K, Ogimi C, Takushi A, et al. (1993) A verbascoside iridoid glucoside conjugate form *Premna corymbosa* Rott. Phytochem 32: 983-986.

9. Basu NK, Dandiya PC (1947) Chemical investigation of Premna integrifolia linn. Journal of the American Pharmaceutical Association 36: 389-391.

10. Deepti Y, Neerja T, Madan G (2010) Diterpenoids from Premna integrifolia. Phytochemistry letters 3: 143-147.

11. Yadav D, Tiwari N, Gupta MM (2011) Simultaneous quantification of diterpenoids in Premna integrifolia using a validated HPTLC method. Journal of separation Science 34: 286-291.

12. Taivini T, Bianchini J, Lafontaine AC, Cambon A (1998) Volatile constituents of the flower buds concrete of Premna serratifolia L. Journal of Essential Oil Research 10: 307-309.

13. Ravinder SC, Nelson R, Krishnan PM, Pargavi B (2011) Identification of Volatile Constituents from Premna serratifolia L. through GC-MS. International Journal of PharmTech Research 3: 1050-1058.

14. Rajendran R, Suseela L, Meenakshi SR, Saleem BN (2008) Cardiac stimulant activity of bark and wood of Premna serratifolia. Bangladesh J Pharmacol 3: 107-113.

15. Rajendran R, Basha N, Ruby S (2009) Evaluation of In Vitro Antioxidant Activity of stem-bark and stem-wood of Premna serratifolia Lin. (Verbenaceae). Research J Pharmacognosy and Phytochemistry 1: 11-14.

16. Rajendran R, Basha S (2010) Antimicrobial activity of crude extracts and fractions of Premna serratifolia Lin. Root. International Journal of Phytomedicines and Related Industries 2: 33-38.

17. Gokani RH, Lahiri SK, Santani DD, Shah MB (2012) Evaluation of Anti-inflammatory and Antioxidant Activity of Premna integrifolia Root. Journal of Complementary and Integrative Medicine 1: 1553-3840.

18. Saidin SBH (2005) Isolation, identification of flavonoid components and antioxidant activities determination from Morinda citrifolia. Glob in Med, p: 25.

19. Balton S, Bon C (2005) Pharmaceuticals Statistics practical and clinical Applications. Marcel Dekker Inc. 562: 215-220.

24

Separation and Quantitative Determination of Carbohydrates in Microbial Submerged Cultures Using Different Planar Chromatography Techniques (HPTLC, AMD, OPLC)

Tatiana Bernardi[1], Paola Pedrini[2], Maria Gabriella Marchetti[2] and Elena Tamburini[2*]

[1]Department of Chemical and Pharmaceutical Sciences, University of Ferrara, Ferrara, Italy
[2]Department of Life Sciences and Biotechnology, University of Ferrara, Ferrara, Italy

Abstract

Carbohydrates are the principal sources of nutrient and energy in large-scale submerged fermentation processes. A method for detection and quantification of sugar levels are very advantageous because they can be considered as key indicators in determining the yields and the productivity of the process. For this reason, the objective of this study is to develop and validate a simple and relatively economical analytical method for detecting complex sugar mixtures in fermentation broth based on High-Performance Thin-Layer Chromatography (HPTLC). HPTLC is a widely used, fast and accurate method of separating complex mixtures. The proposed method involved the chromatographic separations of xilo-, galacto-, fructo-oligosaccharides mixtures at different molecular weights, tri- and disaccharides (raffinose, sucrose, lactose), and the corresponding monosaccharides (xylose, fructose, galactose) on HPTLC plates, using different eluent mixtures and elution conditions. The documentation of plates was performed using TLC visualization device and the images of plates were processed using a digital processor. HPTLC methods development using instrumental techniques as OPLC (Over Pressure Liquid Chromatography) and AMD (Automated Multiple Development) has been also described, as to simultaneously monitor several samples in the same elution, with significant time and solvent savings. Four different carbohydrates complex mixtures were analyzed using HPTLC techniques, as to optimize the quality of the separation among components. The methods set up were then applied for quantitative determination of sugars. As a model of submerged fermentations, a strain of *Bifidobacterium* spp. was used, a saccharolytic bacterium with probiotic activities in the human gut, able to anaerobically ferment complex sugar mixtures. Results could be easily extended to other fermentation processes.

Keywords: HPTLC; AMD; OPLC; Chromatography; Submerged cultures; Carbohydrates analysis; Complex matrix

Introduction

Biotechnology industries are based on exploiting the metabolic activities of microbes, plants and animal cells to produce a wide variety of different compounds, which are used by other industries such as chemical, food, pharmaceutical and health care [1]. With advances in biochemistry, engineering and genetic techniques, submerged fermentation industry has been often becoming a concrete alternative to traditional chemical production for several products, thanks to microbial yields improvements, optimizations of reactor design and advancements in process control and product recovery [2]. Carbohydrates are excellent sources of carbon, oxygen, hydrogen, and metabolic energy, so they have usually served as the principal nutrient source for the large-scale cultivations of microorganisms [3]. The utilizable carbohydrates can be monosaccharides (e.g. glucose, galactose, fructose, xylose), disaccharides (e.g. sucrose, maltose, lactose), oligosaccharides (e.g. kestose, raffinose, maltotriose), and polysaccharides (e.g. starch, dextrins, cellulose, inulin, xylan), commonly supplied as complex raw materials derived from industry or agriculture [4]. These raw materials can also provide a nitrogen source, salts, trace elements, and vitamins, required by the microorganisms as nutrients for growth, maintenance, reproduction, and production. The resulting fermentation broth is a complex mixture, consisting of living microbial cells, nutrients, cell debris, and other products/by-products of the fermentation process [5]. This broth must be monitored to optimize the quality and quantity of the cells and/or metabolites produced. In addition, during fermentation, as the composition of the broth changes, so does the chemistry. It is known that product concentration is approximately inversely proportional

to carbon source concentration, therefore, sugar levels serves as a key indicator in determining the yields and the productivity of the process. Understanding microbial kinetics of single sugar depletion in a complex media also assures to reach the desired level of process efficiency and yield, but needs of a rapid, precise and accurate method of identification and quantification in order to avoid the not-complete substrate utilization, which always represents a money waste in the economic balance of the overall process [6]. Moreover, to establish a hierarchy of sugars utilization in microbial metabolism, the concentrations and the depletion kinetics of single carbohydrates should be accurately monitored during the processes [7]. Nowadays, sugars can be determined by different analytical methods, such as flow injection analysis [8], electrochemical analysis [9], enzymatic methods [10], gas chromatography [11], high-performance liquid chromatography [12], anion-exchange liquid chromatography [13], and thin-layer chromatography [14].

In particular, thin-layer chromatography (TLC) is a simple separation technique for both qualitative and quantitative analysis,

***Corresponding authors:** Elena Tamburini, Department of Life Sciences and Biotechnology, University of Ferrara, Via L. Borsari 46, 44121, Ferrara, Italy
E-mail: elena.tamburini@unife.it

enabling simultaneous analysis of many substances with minimal time requirement [15]. The increasing commercial availability of pre-coated TLC plates has significantly improved the achievable reproducibility of separation. Additionally, the availability of many different absorbent materials including high-performing silica, bonded phases and impregnated layers have increased the versatility of the technique for numerous and quick separations particularly in the sugars field [16]. Thus, methods of TLC and its refined version high-performance thin-layer chromatography (HPTLC) are even nowadays indispensable tools of modern analytical chemistry [17]. The most important differences between TLC and HPTLC is (a) the different particle sizes of the stationary phases and (b) the employment of state-of-the-art instrumentation for all steps in the procedure, sample application, standardized chromatogram development and software-controlled evaluation [18].

Although there were (and still are) some potential concerns against the wider application of TLC/HPTLC (e.g. the potential oxidation of the analyte caused by exposition to atmospheric oxygen), there are many advantages that make these techniques clearly competitive to other analytical approaches [19]. In particular, HPTLC is rapid and cost effective as a method of separating complex mixtures [20], and it has been widely used at industrial level for routine analysis of several substances, even in combination with automated instrumental techniques (e.g., Automated Multiple Development (AMD) and Over Pressure Layer Chromatography (OPLC)).

The aim of this study was to develop modern, rapid and simple analytical HPTLC/AMD/OPLC methods for detection and quantification of different carbohydrates mixtures of xylo-oligosaccharides (XOS), galacto-oligosaccharides (GOS), fructo-oligosaccharides (FOS) and inulin, and their corresponding mono- and di-saccharides, used as carbon substrates for submerged fermentation processes. As a case study, a strain of *Bifidobacterium* (*B. adolescentis* MB 239) able to growth on carbohydrates mixtures of different grade of composition complexity has been used.

Materials and Methods

Fermentation and strain

Bifidobacterium adolescentis MB 239 was obtained from the Collection of the Department of Agroenvironmental Sciences of the University of Bologna. The microorganism was subcultured anaerobically at 37°C for 24 h in MRS broth (Difco Laboratories, Sparks, Maryland, USA). Controlled-pH batch cultures were carried out in a BM-PPS3 bioreactor (Solaris Biotech, Porto Mantovano, and Italy) with a working volume of 2 liters. The temperature was kept at 37°C, and constant stirring (300 rpm) was applied. The fermentor was inoculated with 5% vol/vol precultures grown in the same medium. The culture pH was continuously measured (Mettler Toledo InPro 3030/325) and regulated by automatic addition of 4M NaOH. Microbial growth was monitored by following changes in biomass dry weight [21]. Samples were periodically collected for analysis of carbohydrates. For each carbohydrates mixture, fermentations were carried out in triplicate. Others details about fermentations can be found in Amaretti *et al.* [22].

Chemicals and media

Glucose, galactose, and lactose (Sigma-Aldrich, Steinheim, Germany) and GOS syrup (Borculo Domo, Zwolle, The Netherlands) mixture was assembled as carbon sources mixture for *B. adolescentis* MB239 fermentation. The GOS mixture contained 18.8% glucose, 3.5% galactose, 19.4% lactose, and 58.3% GOS with DP ranging

between 3 and 9. A DP 3 oligomer accounted for 37% of total carbohydrates, and it was the most highly represented oligosaccharide. The concentration of the oligomers decreased with the increase in DP.

XOS and xylose were purchased as commercial mixtures from Co. Farmaceutica Milanese, CFM, Milano, and Italia. Glucose, fructose, sucrose, raffinose (99.9%, J.T.Baker, Deventer, The Netherlands), pure FOS (1-kestose, 99%; nystose, 99%; and fructosylnystose, 98%; Wako Pure Chemical Industries, Osaka, Japan) mixtures were also prepared as fermentation media. Inulin from natural origin (as topinambur roots extract) was also added. In all cases, carbon sources mixtures were integrated with nitrogen and mineral sources to sustain microbial growth.

Solvents, standards and samples

Acetonitrile and acetone were of LiChrosolv quality (E. Merck, Darmstadt, Germany), water was ultrapure. Glucose, fructose, galactose, xylose, lactose, and raffinose (Carlo Erba Reagenti, Milano, and Italia) were used ad pure standard sugars solutions of 1000 ppm. GOS, XOS, FOS and inulin standard solutions were prepared in acetone-water to give 1000 ppm stock solutions, then successive diluted with acetone-water (2:1) to give solutions at different concentrations depending on the linear range of the calibration plots. Samples were collected from fermentations every 2-3 h, centrifuged, filtered (cellulose acetate syringe filter, 0.22 μm; Albet Filalbet, Barcelona, Spain) to avoid time-dependent changes in analyte concentration resulting from continued metabolism, and immediately chilled at 4°C. Carbohydrates were analyzed in the supernatant.

Chromatography

Standards (1-20 μL) and samples (1 μL) were applied to the plates using an automated sample applicator Linomat V (Camag, Muttenz, Switzerland) as 3 mm-wide bands 10 mm from the bottom of the plate and 6 mm apart with a delivery rate of 20 s μL⁻¹. Five standards (in duplicate) were applied to the plate alternatively with 14 samples at different concentration. Fermentation samples were indeed deposited at different dilutions to falling into the right concentration range for calibration curves.

Glucose, galactose, lactose and GOS: Glucose, galactose, lactose and GOS concentration in fermentation media were analyzed with HPTLC/AMD method. 20 × 10 cm silica Kieselgel 60 F254s thin layer plates (Merck, Darmstadt, Germany) previously treated with sodium acetate 0.2M solution (immersion time 20 sec, drying time 20 minutes to air) were used. Isocratic development was performed at 20-22°C and 55-65% relative humidity in 20 × 10 cm Camag twin-trough chamber with 20 mL of mobile phase. Automated multiple development was performed by using Camag AMD-1 equipment with linear gradient in 15 steps of elution. Gradient was built by automatic mixing four basic eluent solutions of acetonitrile/water with 32%, 30%, 26% and 24% of acetonitrile content, respectively.

Glucose, fructose, sucrose, FOS and inulin: Chromatography was performed on 20 × 10 cm diol HPTLC plates (Merck, Darmstadt, Germany). Isocratic development was performed at 20-22°C and 55-65% relative humidity in 20 × 10 cm Camag twin-trough chamber with 20 mL of mobile phase. Automated multiple development was performed by using Camag AMD-1 equipment with linear gradient in 20 steps of elution. Gradient was built by automatic mixing four basic eluent solutions of acetonitrile:acetone (1:1)/water with amounts of water ranging from 25 to 15%.

Xylose and XOS: Xylose and XOS mixtures were separated and quantified using thin layer 20 × 10 cm silica plate Kieselgel 60 F254s (Merck, Darmstadt, Germany), using butanol/ethanol/water (5/3/2) as eluent. Two consecutive isocratic elutions in twin-through chamber were carried out. The elution chamber was saturated with eluent 30 minutes before introducing the plate. After each step, the plates were dried for 15 minutes.

Glucose, fructose, lactose, raffinose and FOS: OPLC analysis was performed on 20 × 20 cm aluminium foil-backed silica gel HPTLC plates sealed on four sides (OPLC-NIT, Budapest, Hungary). Plates were developed with acetonitrile-water 85:15 (vol/vol) as mobile phase, with overrun development, by means of a Bionisis BS-OPLC 50 instrument (Budapest, Hungary). The conditions used for OPLC were: external pressure 50 bar, mobile phase flow rate 300 μL min^{-1}, mobile phase volume 10000 μL, and rapid volume 300 μL; under these conditions the elution time was 2010 sec.

Derivatization: After developing and drying, plates were derivatized by immersion in appropriate solvents using Camag dipping apparatus. In particular, GOS standard and samples plates were immersed for 9 sec. in a toluene solution containing 2% (w/v) basic acetate of lead(IV) in glacial acetic acid and 0.2% (w/v) 2,7-dichlorofluorescein in pure ethanol, subsequently heated to 105°C for three minutes for color development. XOS standard and samples plates were derivatized for 5 sec. in pure acetone solution containing 0.9 mL of aniline and 1.66 g of phthalic acid, then dried at room temperature for 10 minutes and then for other 20 minutes at 120°C in an oven. Finally, FOS standard and samples plates were derivatized for 5 sec. with 4-aminobenzoic acid reagent (glacial acetic acid (36 mL), water (40 mL), 85% phosphoric acid (2 mL) and acetone (120 mL) added to 4-aminobenzoic acid (2 mL), then heated at 115°C for 15 minutes in an oven. The immersion condition must be standardized by maintaining a defined immersion time and a uniform speed to avoid tidemarks, which could interfere with densitometric evaluation.

After derivatization, the plates were air dried for 5 min and sample and standard zone areas were measured by linear scanning under fluorescence illumination at λ=366 nm with a PC-controlled Camag TLC-Scanner III (mercury lamp, filter cut-off 400 nm) (Camag, Germany). The cutoff filter is positioned between the sample and the photosensor for eliminating diffusely reflected light of the excitation wavelength. Accordingly, the measured light is directly proportional to the amount of fluorescing The scanning was performed vertically over the plate over a distance of 5 cm and at a rate of 5 cm/min, measuring 200 points/scanning. Densitograms were generated from transmission measurements of the samples and the blank. The respective absorbance values were calculated, and peak areas were quantitated with the aid of the CATS4` software supplied with the instrument.

Principles of AMD Technique

AMD technique provides the automatic multi-development of plates in a suitable chamber. In each succeeding development, the mobile phase is permitted to advance farther by a constant distance, while enhancing the separation of each component of the sample and standard solutions. First, the chamber is flushed with purified nitrogen and, with the addition of solvent, the first elution is started. For each step, a mixer is completely filled with the solvent from reservoirs before entering in the chamber. Six reservoirs are available on the AMD apparatus to produce different eluent compositions.

A few second later, the step is terminated on the basis of the programmed gradient, and the solvent removed. Then the next step started, including drying, ventilating, and a longer elution, followed by removal of the eluting solvent. This is continued until the whole programme is completed. For each step, new mobile phase is introduced into the chambers. Because of its composition differs from the preceeding one, this will result in a gradient elution. Usually a gradient for silica gel begins strongly polar (strongly eluted) and ends less polar (weakly eluted). Gradient trend needs to be defined by carrying out a series of isocratic elution and then optimized for each application, as to obtain the best separation among substances. Theory and up-to-date principal applications of AMD technique are reported in [23].

Principles of OPLC Technique

OPLC is a forced-flow technique developed by Tyihak *et al.* [24] in which the vapor phase is eliminated by completely covering the sorbent layer with an elastic membrane under external pressure. The separation is thereby carried out in a closed system with parameters that may be controlled. The chromatographic plate becomes a planar column by sealing the plate all around and by draining off the solvent. Eluent could be overrun in order to increase the separation efficiency. After sample application, the plate is placed into the chamber and after closing, an oil cushion is set under pressure. Before initiating separation with a suitable mobile phase, the solvent inlet valve is closed, and the eluent pump is started so as to reach an appropriate solvent pressure. The separation is then initiated by opening the inlet valve, producing a quick solvent distribution into the channel resulting in a linear migration of the mobile phase. The eluent is forced through the silica layer by the pressured cushion and then collected in the outlet channel.

Results and Discussion

AMD method development for glucose, lactose and GOS mixtures

The combination of high-performance thin-layer chromatography (HPTLC) with automated multiple development (AMD) allows full automation of the separation step. This provides both a separation efficiency that is considerably better than that in conventional HPTLC and reproducible gradient elution on the thin layer [25].

The success of an AMD separation depends principally on the choice of the solvent components, on the stepness and shape of the gradient. Whereas the former has been deeply investigated for separating carbohydrates [26], the latter required specific approach. Isocratic elutions with binary mixtures of acetonitrile and water has demonstrated to be selective and of suitable polarity range for the compounds to be separated. On these bases, the AMD gradient was started with the study of the isocratic retention behavior of relative head-monosaccharide glucose, the first dimer lactose (glucose+galactose) and higher-molecular weight GOS, as to understand which range of eluent strength could warrant the best separation of all the analytes of interest. Increasing from time to time the amount of strongest components of the eluent mixture (water, in this case), retention time vary depending on the interaction between oligomers and eluent.

Plot of retention, expressed as

$$R_m = \log[(1-Rf)/Rf] \quad [27]$$

against the volume fraction of water (Figure 1) shows a very good linear regression (R^2 from 0.9771 for glucose to 0.984 for DP 9). It is also evident that separation was achieved by a normal-phase mechanism, because increasing the water content of the solvent mixture reduced GOS retention and the order of migration followed their polarity [28]. Moreover, it is worthwhile noting that below 24% water, lines corresponding to DP4 and DP5 and DP6 and DP7 converged, indicating that for further decreasing of eluent polarity a worsening of separation quality has to be expected, due to peaks overlapping.

To optimize the shape of AMD gradient, the number and the duration of the steps, for each eluent composition the linear relationship between separation distance squared, expressed ad Z^2, for various mobile phases against the time for development has to be proved [29]. An example for the water:acetonitrile 30:70 mixture is reported in Figure 2 as a results of three replicates. The best AMD separation for glucose, lactose and GOS was therefore performed by use of 15-steps gradient in which the amount of water was linearly reduced from 32 to 24% (Table 1). The resulting chromatogram is shown in Figure 3.

AMD method development for glucose, fructose, sucrose, FOS and inulin

Using the same procedure, AMD gradient has been set up and optimized also for carbohydrates mixture composed by glucose, fructose, sucrose, FOS and inulin. Because the soluble inulin fractions were expected to have higher DP, the polarity of the solvent mixture has to be adjusted as to be able to separate all the components in the same development space. Because of the lack of pure standards of oligosaccharides and inulin higher than DP5 (fructosyl-nystose), a plot similar to the one shown in Figure 1 could not be built: the correct polarity interval of eluent was achieved by trial and error experiments. The 20 steps optimized separation with amounts of water ranging from 25 to 15% water and correspondingly from 75 to 85% of acetone:acetonitrile 1:1 is depicted in Figure 4A. Quality of AMD separation for 21 components is shown in comparison with the results of a simple isocratic elution with acetone:acetonitrile (1:1)-water (86:14) of the same mixtures as a control (Figure 4B).

Method development for xylose and XOS

In the case of XOS mixture and their corresponding principal monomer, xylose, method set up and optimization were based on previous results obtained by Reiffova and Nemcova [30] for fructo-oligosaccharides separation. Two subsequent isocratic elution with butanol:ethanol:water (5:3:2) as mobile phase, have permitted to obtain satisfactorily results on XOS mixture containing oligomers up to DP6 (Figure 5).

Selection of the optimum mobile phases for separation on thin layer was based on the furthest migration distance for xylose, because this is the basic units of XOS with the lowest molecular masses. The first position on densitogram (the utmost spots from start) in line belongs to monosaccharide with the lowest molecular mass, then gradually other components of XOS chain with increasing molecular masses.

OPLC method development for glucose, fructose, lactose, raffinose and FOS

As a result of digestion and absorption in the upper gastrointestinal tract, human colon where Bifidobacteria usually growth, is an environment rich of carbohydrates at different grade of hydrolysis, depending on the overall metabolic activities of the other microbial intestinal community [31]. Therefore, B. adolescentis MB239 is able to ferment different carbohydrates, from poly- and oligo-saccharides to di- and mono-saccharides, and present at different concentrations.

In this latter case examined, the growth solution for fermentation

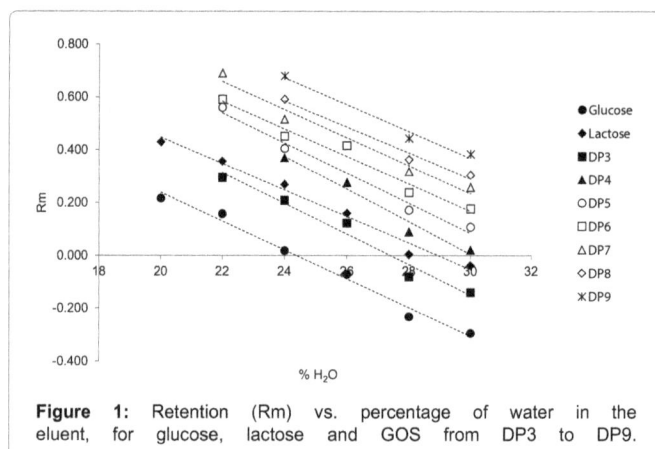

Figure 1: Retention (Rm) vs. percentage of water in the eluent, for glucose, lactose and GOS from DP3 to DP9.

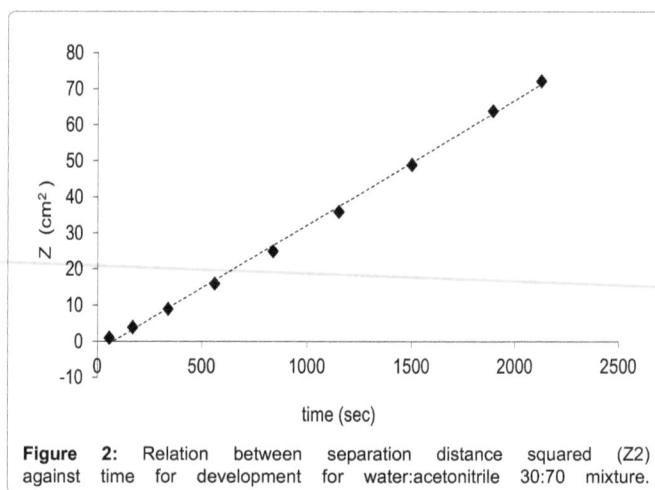

Figure 2: Relation between separation distance squared (Z2) against time for development for water:acetonitrile 30:70 mixture.

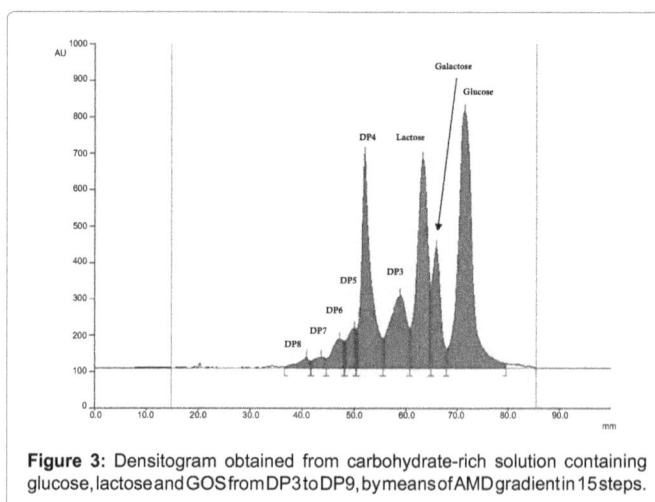

Figure 3: Densitogram obtained from carbohydrate-rich solution containing glucose, lactose and GOS from DP3 to DP9, by means of AMD gradient in 15 steps.

contained only a mixture of low molecular weight carbohydrates, as glucose, fructose, sucrose, lactose, raffinose, 1-kestose, nystose and fructosyl-nystose. Because of acetonitrile-water has been reported to enable selective separation of carbohydrate on silica plates, overpressured development of our mixture with acetonitrile:water 85:15 was tried on OPLC plates [32]. A series of trial and error experiments indicated that to improve peak resolution, over-running development with a mobile phase volume up to 10000 µL was

Step	Mixture (68:32)	Mixture (70:30)	Mixture (74:24)	Mixture (76:24)	H$_2$O (tot)
1	100				32.00
2	81.38	18.62			31.63
3	52.59	47.41			31.05
4	33.98	66.02			30.68
5	21.96	78.04			30.44
6	14.19	67.19	18.62		29.54
7	9.17	43.42	47.41		28.29
8	5.92	28.06	66.02		27.48
9	3.83	18.13	78.04		26.96
10	2.47	11.71	85.81		26.61
11	1.6	7.57	72.21	18.62	26.03
12	1.03	4.89	46.66	47.41	25.31
13	0.67	3.16	30.15	66.02	24.85
14	0.43	2.04	19.48	78.04	24.54
15	0.28	1.32	12.59	85.81	24.35

Table 1: Linear gradient for 15 steps based on mixture of acetonitrile with water for glucose, lactose and GOS separation.

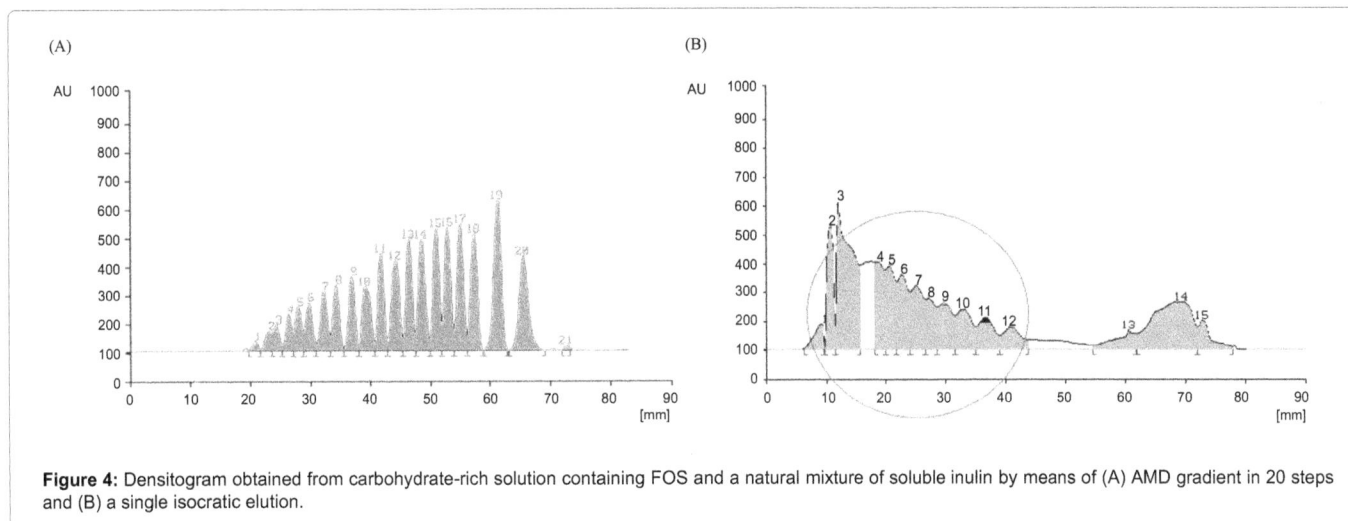

Figure 4: Densitogram obtained from carbohydrate-rich solution containing FOS and a natural mixture of soluble inulin by means of (A) AMD gradient in 20 steps and (B) a single isocratic elution.

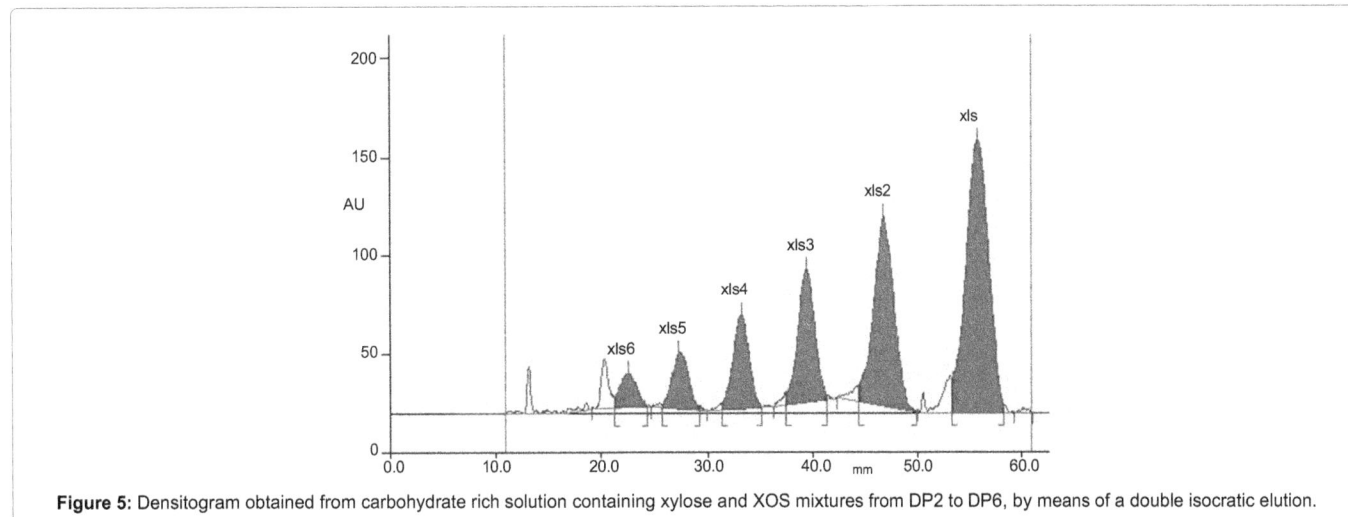

Figure 5: Densitogram obtained from carbohydrate rich solution containing xylose and XOS mixtures from DP2 to DP6, by means of a double isocratic elution.

necessary. A densitogram obtained from a fermentation medium is shown in Figure 6, from which it is apparent that all the sugars are perfectly separated under these conditions.

Retention-structure relationships

Carbohydrates separation and quantification is of particular importance when they are used as carbon sources in fermentation processes, because only from quantification it is possible to obtain data related to process yield and productivity. On the other hand, especially when carbohydrates of natural origin are used, even if purchased as commercial mixtures, their composition are often reported as approximate percentage and present a high level of heterogeneity from a sample to another. For this reason, precise identification of mixture components could be of great importance, taking into account that microorganisms has different specific growth rate depending on chemical characteristics of the carbon sources, as molecular weight or chemical complexity.

Even when separation has been optimized, identifying the peaks or defining as well as possible the correspondence between the peaks and the chemical structure of components has to be considered. The lack of pure standard substances of GOS, XOS, FOS higher than DP5, and inulin emphasized the need for different approaches to method development. Several studies on retention behavior of molecules have been carried out in the past [33,34] and have revealed linear relationship between chemical structure and retention (Martin's correlation) [35].

Chemical structure of any given fractions of GOS, XOS and FOS/inulin can be regarded as a simple chain of the corresponding monomer (galactose, xylose and fructose) attached to a glucose head unit (except of XOS, where the head is xylose as well) and correlated to migration distance (MD) measured in millimeters. MD represents the distance between the final and starting (deposition) point of band migration on the plate.

In all cases examined, retention was found to be linearly dependent on the number of monomer units, with correlation coefficient R^2 of 0.9982, 0.9923 and 0.8830 for FOS (Figure 7A), GOS (Figure 7B) and XOS (Figure 7C).

Quantitative assessment of the methods

For quantitative assessment of the method developed and proposed to characterize carbon sources in fermentation media, calibration models have been constructed for each analyte, using the standard solutions deposited on the plates. For each method, in accordance with Ferenczi-Fodor et al. [36] statistical parameters were calculated on the basis of calibration plots. For the sake of brevity, as an example, correlation coefficients of calibrations (R^2), the upper limits of linearity, the estimated limits of detection (LOD) and the limit of quantification (LOQ) are reported in Table 2 for xylose and XOS mixture. It is worthwhile noting that the satisfactory statistical performance of the method could make HPTLC an important alternative method to HPLC or other techniques applied to fermentation monitoring.

Applying quantitative determination to real fermentation processes of B. adolescentis MB239 in the presence of carbohydrates mixtures analyzed, a complete monitoring of sugars depletion trends was carried out. Figure 8 shows some examples of such different trends over time. In particular, Figures 8A and 8B, represent the same fermentation on nine carbohydrates mixture, but divided in two figures, as to avoid visual overlapping. It is clearly shown that while FOS have been consuming (Figure 8B), fructose concentration has been increasing (Figure 8A), because fructose units has progressively formed at a rate higher than rate of consumption, resulting in a net accumulation over time. In Figure 8C, fermentation of XOS (DP2-DP6) and xylose has been reported. As in the previous case, it is worthwhile noting an accumulation of monomer (xylose) during the process, up to 35 hours, when it remained the unique carbon source available. Finally, in Figure 8D trends of depletion of glucose, galactose, lactose and GOS mixtures have been reported.

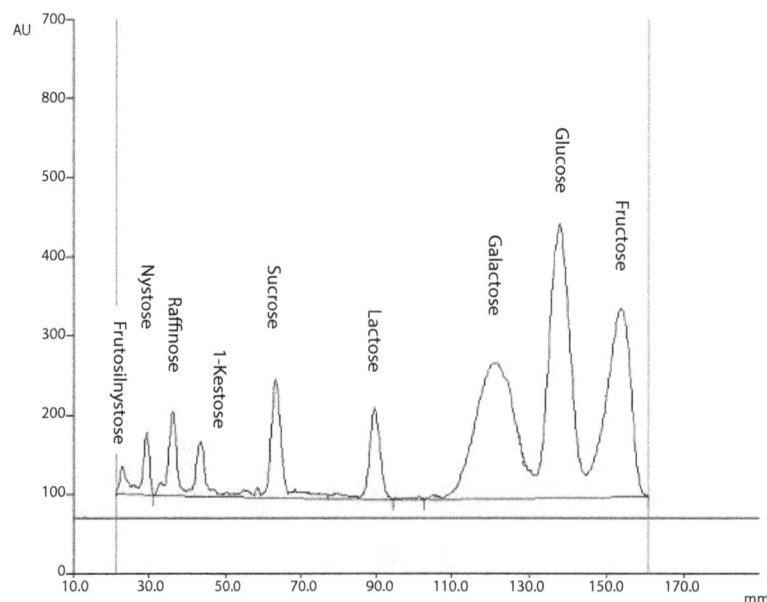

Figure 6: Densitogram obtained from carbohydrate-rich solution containing fructose, glucose, galactose, lactose, sucrose, raffinose and FOS by means of OPLC elution gradient in 15 steps and a single isocratic elution.

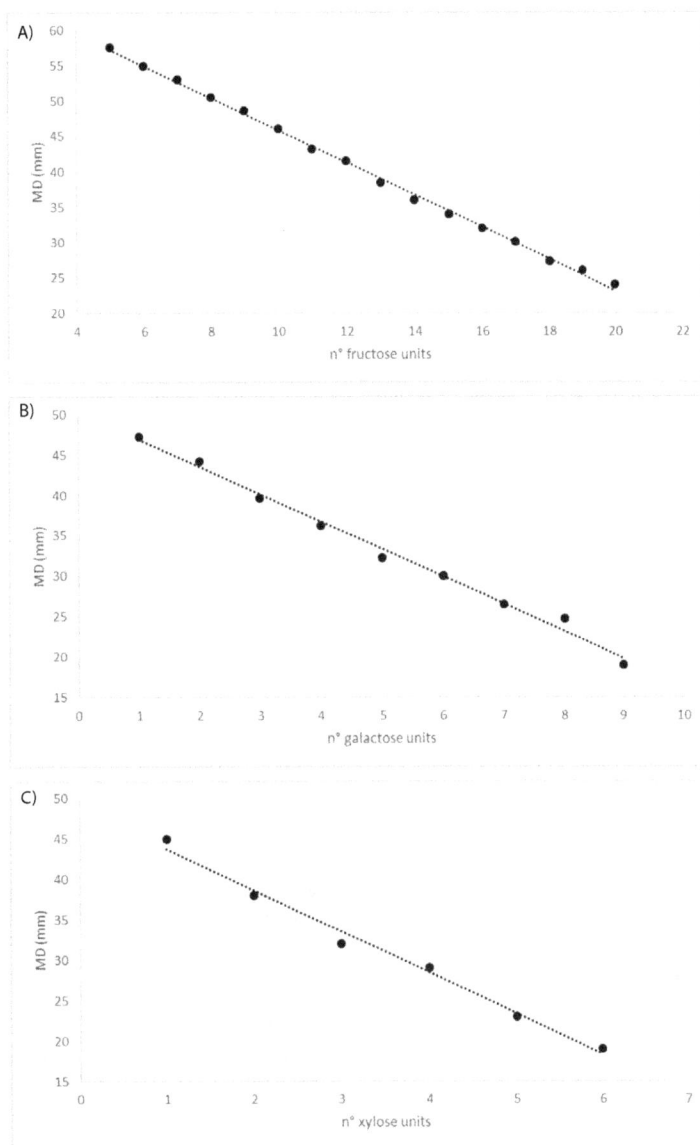

Figure 7: Linear correlation between retention data (MD) and the number of monomer unit for (A) FOS and inulin; (B) glucose, galactose and GOS; (C) xylose and XOS mixtures.

Sugar	R²	Upper limit sugar of linearity [ppm]	LOD [ppm]	LOQ [ppm]
Xylose	0.9857	80	9.75	32.49
XOS-DP2	0.9979	80	5.36	17.88
XOS-DP3	0.9925	80	5.30	17.65
XOS-DP4	0.9979	80	6.94	23.13
XOS-DP5	0.9955	80	9.16	30.53
XOS-DP6	0.9939	110	27.33	91.10

Table 2: Statistical data from the calibration plots for xylose and XOS mixture.

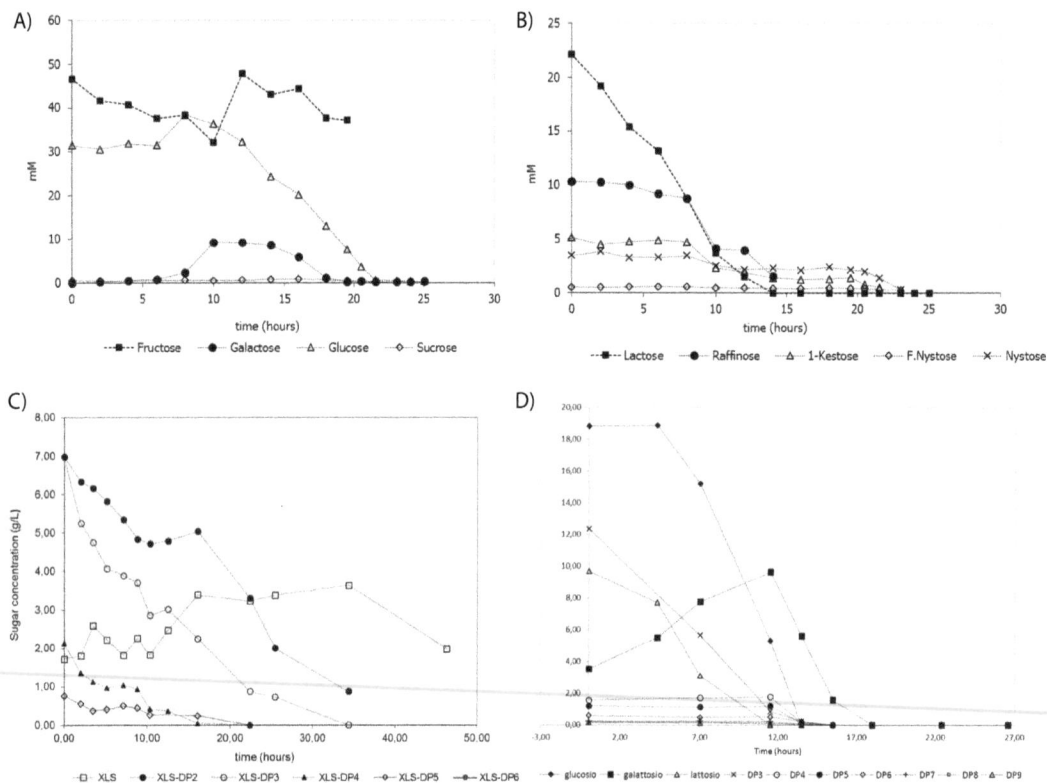

Figure 8: Estimation of the concentration of sugars in submerged fermentations of *B. adolescentis* MB239 on media containing a mixture of (A) and (B) glucose, fructose, galactose, sucrose, lactose, raffinose and FOS; (C) xylose and XOS; (D) glucose, galactose and GOS.

Conclusion

HPTLC alone, and in combination with AMD and OPLC instrumental techniques, seems to lead to new possibilities in analysis of carbohydrates in fermentations chemistry. The possibility of performing a direct and sensitive method and the ability to simultaneously detect many different analytes on a single plate, without time-consuming pretreatment, is the major advantage of this technique compared to other methods of analysis commonly used in fermentation monitoring. Culture solutions simply have to be centrifuged or filtered to remove particulate matter, properly diluted and applied to the plate. The analytical interest of this application is principally connected to the possibility to obtain very detailed data about the relative consumption of carbohydrates over time in complex mixtures, and could represents an effective tool for obtaining more information about physiological status or metabolic behavior of cells.

References

1. Erickson B, Nelson, Winters P (2012) Perspective on opportunities in industrial biotechnology in renewable chemicals. Biotechnol J 7: 176-185.

2. Parekh S, Vinci VA, Strobel RJ (2000) Improvement of microbial strains and fermentation processes. Appl Microbiol Biotechnol 54: 287-301.

3. Peters D (2006) Carbohydrates for fermentation. Biotechnol J 1: 806-814.

4. Saxena RC, Adhikari DK, Goyal HB (2009) Biomass-based energy fuel through biochemical routes: a review. Ren Sust Energy Rev 13: 167-178.

5. Godge RK, Siddheshwar SS, Somawanshi SB, Dolas RT, Pattan SR (2013) Overview on Fermentation technique and its Application towards industry. IJRPLS 1: 92-101.

6. Woodley JM, Breuer M, Mink D (2013) A future perspective on the role of industrial biotechnology for chemicals production. Chem Eng Res Des 91: 2029-2036.

7. Watson D, Motherway MOC, Schoterman MH, Neerven RJ, Nauta A, et al. (2013) Selective carbohydrate utilization by lactobacilli and bifidobacteria. J Appl Microbiol 114: 1132-1146.

8. Horwitz W (1980) Official methods of analysis. Association of Official Agricultural Chemists. Washington, DC 534.

9. Hayashi T, Sakurada I, Honda K, Motohashi S, Uchikura K (2012) Electrochemical detection of sugar-related compounds using boron-doped diamond electrodes. Anal Sci 28: 127-133.

10. Blunden CA, Wilson MF (1985) A specific method for the determination of soluble sugars in plant extracts using enzymatic analysis and its application to the sugar content of developing pear fruit buds. Anal Biochem 151: 403-408.

11. Koubaa M, Mghaieth S, Thomasset B, Roscher A (2012) Gas chromatography-mass spectrometry analysis of 13C labeling in sugars for metabolic flux analysis. Anal Biochem 425: 183-188.

12. Stefano VD, Avellone G, Bongiorno D, Cunsolo V, Muccilli V, et al. (2012) Applications of liquid chromatography-mass spectrometry for food analysis. J Chromatogr A 1259: 74-85.

13. Arfelli G, Sartini E (2014) Characterisation of brewpub beer carbohydrates using high performance anion exchange chromatography coupled with pulsed amperometric detection. Food Chem 142: 152-158.

14. Ghebregzabher M, Rufini S, Monaldi B, Lato M (1976) Thin-layer chromatography of carbohydrates. J Chromatogr 127: 133-162.

15. Poole FC, Poole SK (1994) Instrumental thin-layer chromatography. Anal Chem 66: 27A-37A.

16. Fuchs B, Süss R, Teuber K, Eibisch M, Schiller J (2011) Lipid analysis by thin-layer chromatography--a review of the current state. J Chromatogr A 1218: 2754-2774.

17. Srivastava MM (2011) An Overview of HPTLC: A Modern Analytical Technique with Excellent Potential for Automation, Optimization, Hyphenation, and Multidimensional Applications. High-Performance Thin-Layer Chromatography. Springer, Berlin, Germany. Pp. 3-24.

18. Günther M, Schmidt PC (2005) Comparison between HPLC and HPTLC-densitometry for the determination of harpagoside from Harpagophytum procumbens CO(2)-extracts. J Pharm Biomed Anal 37: 817-821.

19. Zlatkis A, Kaiser RE (2011) HPTLC-high performance thin-layer chromatography. Pp. 9.

20. Montero CM, Dodero MCR, Sanchez GAG, Barroso CG (2004) Analysis of Low Molecular Weight Carbohydrates in Food and Beverages: A Review. Chromatographia 59: 15-30.

21. Rossi M, Corradini C, Amaretti A, Nicolini M, Pompei A, et al. (2005) Fermentation of fructooligosaccharides and inulin by bifidobacteria: a comparative study of pure and fecal cultures. Appl Environ Microbiol 71: 6150-6158.

22. Amaretti A, Bernardi T, Tamburini E, Zanoni S, Lomma M, et al. (2007) Kinetics and metabolism of *Bifidobacterium* adolescentis MB 239 growing on glucose, galactose, lactose, and galactooligosaccharides. Appl Environ Microbiol 73: 3637-3644.

23. Poole C (2014) Instrumental Thin-Layer Chromatography. Elsevier

24. Tyihak E, Mincsovics E, Kalasz H (1979) New planar liquid chromatographic technique: overpressured thin-layer chromatography. J Chromatogr 174: 75-81.

25. Ferenczi-Fodor K, Végh Z, Renger B (2006) Thin-layer chromatography in testing the purity of pharmaceuticals. TrAC 25: 778-789.

26. Bernardi T, Tamburini E, Vaccari G (2005) Separation of Complex Fructo-oligosaccharides (FOS) and inulin mixtures by HPTLC-AMD. J Planar Chromatogr 18: 23-27.

27. Kiridena W, Poole CF (1998) Structure-driven retention model for solvent selection and optimization in reversed-phase thin-layer chromatography. J Chromatogr A 802: 335-347.

28. Abbott SR (1980) Practical aspects of normal-phase chromatography. J Chromatogr Sci 18: 540-550.

29. Spangenberg B, Poole CF, Weins C (2011) Quantitative thin-layer chromatography: a practical survey. Springer Science.

30. Reiffová K, Nemcová R (2006) Thin-layer chromatography analysis of fructooligosaccharides in biological samples. J Chromatogr A 1110: 214-221.

31. Rinne MM, Gueimonde M, Kalliomäki M, Hoppu U, Salminen SJ, et al. (2005) Similar bifidogenic effects of prebiotic-supplemented partially hydrolyzed infant formula and breastfeeding on infant gut microbiota. FEMS Immunol Med Microbiol 43: 59-65.

32. Tamburini E, Bernardi T, Bianchini E, Pedrini P (2009) Fermentation Monitoring Based on HPTLC-OPLC: The Effect of a Complex Biological Matrix on Quantitative Performance. J Planar Chromatogr 22: 1-6.

33. Kaliszan R (1987) Quantitative structure-chromatographic retention relationships. John Wiley and Sons, New York.

34. Héberger K (2007) Quantitative structure-(chromatographic) retention relationships. J Chromatogr A 1158: 273-305.

35. Martin AJP (1950) Some theoretical aspects of partition chromatography. Biochem Soc Symp 3: 4-25.

36. Ferenczi-Fodor K, Végh Z, Nagy-Turák A, Renger B, Zeller M (2001) Validation and quality assurance of planar chromatographic procedures in pharmaceutical analysis. J AOAC Int 84: 1265-1276.

25

Rapid LC-MS Compatible Stability Indicating Assay Method for Azilsartan Medoxomil Potassium

Debasish Swain[1], Gayatri Sahu[2] and Gananadhamu Samanthula[1*]

[1]National Institute of Pharmaceutical Education and Research, Hyderabad, Telangana -500037, India

[2]United States Pharmacopeia (USP) India Pvt. Ltd., Hyderabad, Telangana -500078, India

Abstract

Azilsartan medoxomil potassium is a new generation antihypertensive drug comes under class of angiotensin receptor blockers. The present study focuses on developing a liquid chromatography-mass spectrometry compatible stability indicating assay method for determination of azilsartan medoxomil potassium in bulk drug and formulations. In order to develop stability indicating assay method initially the drug was subjected to stress conditions of hydrolysis (acid, base and neutral), oxidation, photolysis and thermal degradation. Degradation of the drug was observed in acid, alkaline, neutral and peroxide conditions. Separation of drug from the resulting degradation products was achieved on a Symmetry C-18 column (150 mm × 4.6 mm × 5 µm) using 0.02% trifluroacetic acid and acetonitrile as mobile phase with gradient elution. The flow rate was 1.0 mL/min and quantification was carried out using ultraviolet detection at wavelength 254 nm. The method was validated according to ICH Q2 (R1) guideline for selectivity, linearity, accuracy and precision. The drug was well separated from all the degradants and its peak purity was ascertained through photodiode array and mass spectrometric detection. All the degradation products were characterised using ion trap mass spectrometer. The method was found to be linear over the concentration range of 160.00-240.00 µg/mL with correlation coefficient >0.999. The interday and intraday precision values for azilsartan medoxomil potassium was found to be within 2.0% relative standard deviation (Graphical abstract).

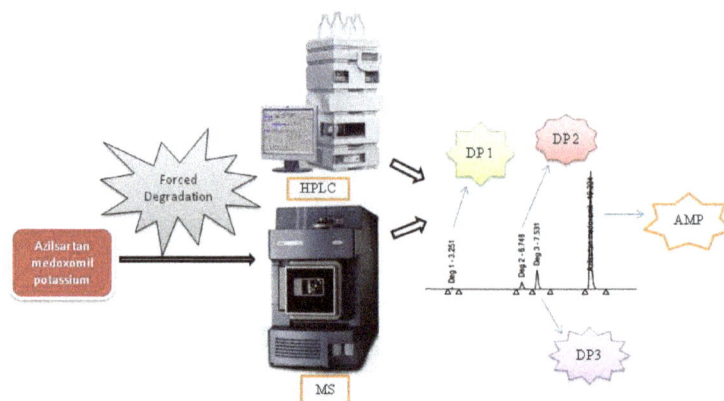

Keywords: Azilsartan medoxomil potassium; RP-HPLC; PDA; LC-MS; Stability indicating assay

Introduction

Azilsartan medoxomil potassium (AZP) is chemically known as 1-[{2'-(2,5-Dihydro-5-oxo-1,2,4-oxodiazol-3-yl) [1,1'-biphenyl]-4-yl] methyl]-2-ethoxy-1H-benzimidazole-7-carboxylic acid(5-methyl-2-oxo-1,3-dioxol-4-yl) methyl ester potassium salt (Figure 1). It is a prodrug of azilsartan. Upon administration of the drug into the physiological system azilsartan will be released as the active moiety into the body [1]. AZP selectively blocks the angiotensin II by blocking the angiotensin II, type 1 receptor thereby lowering the blood pressure and thus used in the treatment of hypertension, either individually or in combination with other drugs [2]. The drug AZP has shown better clinical profile than other drugs in the same class [3]. A few analytical methods have been reported for the determination of AZP in biological samples, bulk drugs and also in pharmaceutical formulations [4,5]. However no information exists in the literature on the degradation and LC-MS compatible stability indicating method for the assay of AZP.

Current GMP regulations require all the drugs to be tested with Stability Indicating Assay Methods (SIAMs) before release. The FDA defines the SIAM as an analytical procedure that measures the active pharmaceutical ingredient (in drug substance and drug product) with accuracy and precision, free from the process impurities, excipients

*Corresponding authors: Gananadhamu S, Department of Pharmaceutical Analysis, National Institute of Pharmaceutical Education and Research [NIPER], Hyderabad 500037, India, E-mail: gana@niperhyd.ac.in

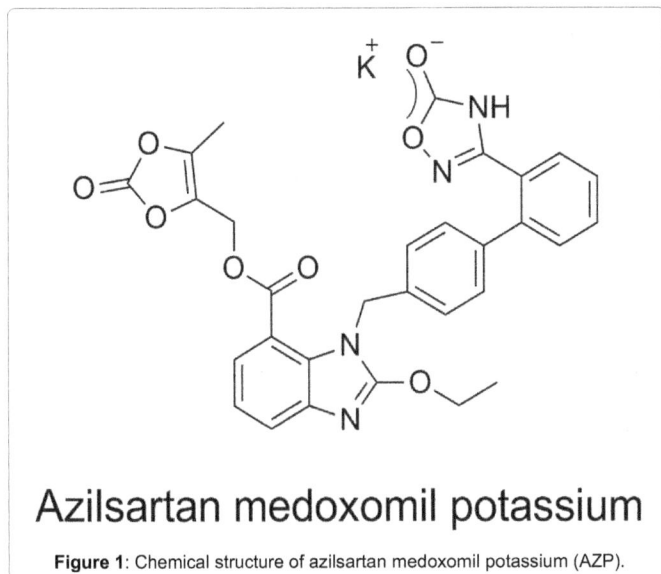

Azilsartan medoxomil potassium

Figure 1: Chemical structure of azilsartan medoxomil potassium (AZP).

and degradation products. The main objective of SIAM is to monitor the impurities and degradation products which affect the safety and efficacy of drug products [6-8]. To develop a SIAM, the drug is subjected to variety of stress conditions including hydrolysis (at various pH), oxidation, photolysis as well as thermal degradation and the resulted degradation products from the drug are separated and estimated with the aid of analytical techniques (mostly chromatographic techniques). The stability indicating power of the developed assay method is confirmed by peak purity (no co-elution of other components) evaluation of the main drug peak. Often Photo Diode Array (PDA) detector is used for confirming the peak purity and thereby finalise a method which can be used as a stability indicating assay method. The PDA detector functions on the principle of collection of spectra at various data points across the peak and finally matching their spectra. Co-eluting peaks produce spectra that differ from one another. Due to differences of these spectra within a single chromatographic peak, its purity can be decided easily [9]. PDA detectors may fail in the evaluation of peak purity on the occasion of lack of UV response, noise of the system and very low concentration of the interfering substances. The presence of very closely related structures with same chromophore produces same absorption spectra, making it difficult for the PDA to discern among the components [10]. The mass spectrometric detection overcomes many of these limitations of PDA. The presence of more than one component in a single peak can be detected by the mass spectrometer based on differences in molecular mass and mass spectra fragmentation pattern [11]. Hence in the present study LC-MS compatible stability indicating assay method was developed by employing the combination of PDA and mass spectrometric detection which provides orthogonal information about the peak purity.

Experimental

Materials

AZP was obtained as a gratis sample from Mylan Labs Pvt. Ltd., Hyderabad. HPLC grade acetonitrile and trifluoroacetic acid were purchased from Merck (Darmstadt, Germany). AR grade sodium hydroxide, hydrochloric acid (35%) and hydrogen peroxide (30% w/v) were procured from SD Fine-Chem Ltd. (Mumbai, India). High purity water was obtained from Millipore Milli-Q plus system (Milford, MA, USA). Acetonitrile was used as the diluent and all the final solutions

were made in the diluent.

Instrumentation and analytical conditions

The liquid chromatographic analysis was performed on an Agilent 1260 infinity series HPLC (Agilent Technologies, USA) equipped with quaternary solvent manager, a degasser, a diode array detector, an autosampler and a column compartment with temperature regulation facility. For LC/MS analysis Waters Xevo Triple Quadrupole was used. The data acquisition was carried out using the Empower (version 3) and Masslynx softwares for HPLC and LC/MS respectively. A Thermo LTQ XL⁻ Linear Ion Trap mass spectrometer equipped with an ESI was used for characterization of the structures of degradation products.

Standard solutions

Stock solution of AZP (1 mg/mL) was prepared in diluent. Working standard solution of AZP (200 μg/mL) was prepared from stock solution. The stock solution and also the working standards were stored at 5°C and were found to be stable for several days.

Stress degradation studies

All the reagents used for degradation study (stressors) i.e, 0.1N HCl, 0.01N NaOH and 0.5% (w/v) H_2O_2 were prepared in 50:50 (v/v) of water and acetonitrile. The drug was subjected to forced degradation until optimum degradation (10-30%) was achieved. All the degradation studies were conducted at a concentration of 1000 μg/mL of AZP. The degraded samples were finally diluted to a concentration of 200 μg/mL of AZP and injected in HPLC system.

Acid degradation studies

The drug was refluxed in 0.1N HCl for 1 hour at 60°C. After 1 hour the sample was taken and neutralized with base.

Base degradation studies

The drug in 0.01N NaOH (in water) was kept at room temperature in a dark place. After 15 minutes sample was taken and neutralized with acid.

Peroxide degradation studies

The drug in 0.5% (w/v) H_2O_2 was kept at room temperature. After 3 hours sample was collected.

Neutral degradation studies

About 100 mg of the drug was weighed, dissolved in 5 mL of acetonitrile finally made up to concentration of 1000 μg/mL of AZP by adding water and was refluxed at 60°C for 1 hour. After 1 hour sample was taken.

Thermal degradation studies

About 200 mg of AZP was taken in a transparent petri dish and kept in the oven as such at 100°C. After 24 hours the exposed sample was used for sample preparation.

Photolytic degradation studies

Solid state: About 100 mg of the drug was taken in two separate transparent petri dishes. One kept in the UV chamber, exposed to an intensity of 200 W Hr m⁻² and another in fluorescent chamber exposed to 1.2 million lux hours.

Solution state: About 1000 μg/mL of AZP was prepared in diluent in two separate volumetric flasks. One was placed in the UV chamber and another in fluorescent chamber exposed to same conditions as that of solid state.

Similarly control samples (petri dishes and volumetric flasks are wrapped with aluminium foils) were prepared and also exposed to the above mentioned conditions. All the samples were observed with respect to the control samples.

Results and Discussion

The aim of the present study is to develop a LC-MS compatible reversed-phase high-performance liquid chromatography procedure for the determination of AZP in the presence of its degradation products. AZP is relatively non-polar compound as indicated by partition coefficient (log p = 4.94 and 6.03) [12], and is well retained on traditional C18 bonded phases. The method was optimized in keeping view of adequate separation of the degradation products from the main peak. The initial trials were carried out with aqueous buffer solutions of pH 3.0, 4.0, 5.0, 5.5 and 6.0 with organic phase being methanol or acetonitrile. The AZP peak was not good in the above trials. The combination of trifluroacetic acid and acetonitrile as the mobile phase under gradient elution produced better peak shapes of AZP and degradants along with good resolution among all the peaks. The stability indicating power of the developed method was established by ascertaining the purity of all the peaks in the chromatogram. The peak purity was evaluated using the PDA as well as mass spectrometric detection. The peak purity and the optimum separation of the peaks indicated the selectivity of the assay method. Stress studies of AZP showed the formation of three degradation products in total under the hydrolytic and oxidative conditions. Thus care must be taken during the manufacturing or processing of the drug otherwise the resultant degradants may hamper the safety and efficacy of the resulting formulation.

Optimization of reversed-phase HPLC and LC-MS conditions

The separation was achieved on a Waters Symmetry C-18 column (150 mm × 4.6 mm × 5 μm) using a mobile phase of solvent A (0.02% trifluoroacetic acid in water) and solvent B (acetonitrile) in gradient elution. The gradient program employed to achieve the separation was (Time in min. / %Solvent B): 0/40, 3/40, 10/70, 15/70, 16/40, 20/40. The flow rate was maintained at 1.0 mL/min. The column temperature was maintained at 25°C and the auto sampler at 5°C. The detection wavelength was 254 nm and the injection volume was 10 μL. The developed HPLC method was transferred to LC-MS for the purity evaluation study.

The LC-MS studies was carried out on a Waters Xevo Triple Quadrupole mass spectrometer with the parameters set as: capillary voltage at 3000 V; cone voltage at 25 V; desolvation temperature at 450°C; source temperature at 100°C; extractor voltage at 2 V; cone gas flow at 50 L/hr and desolvation gas flow at 1000 L/hr. The study was carried out in the positive mode under electrospray ionization technique. A Thermo LTQ XL⁻ Linear Ion Trap mass spectrometer equipped with an ESI was used to conduct the MS^n experiments and predict the possible structures of the resulting degradation products. The ESI conditions were maintained as in above experiment; additionally helium was used as the damping gas and nitrogen as the sheath gas (30 psi). The automatic gain control settings were set at 2×10^7 for a full-scan mass spectrum and 2×10^7 counts for a full-product ion mass spectrum with a maximum ion injection time of 200 ms. The collision energy used was 25 eV and all the spectra were recorded with an average of 25-30 scans.

Stress degradation studies

The forced degradation studies were carried out on the AZP drug substance [13-15]. The AZP degraded in acid, base and neutral hydrolysis as well as oxidative stress conditions (Figures 2-5). The AZP was extremely susceptible to even mild alkaline hydrolysis (0.01N NaOH) and degraded to about 26% within 15 min. at room temperature. The drug was found to be stable under photolytic and thermal degradation conditions. The degradation conditions and the percentage of the degradants formed were summarized in Table 1.

The $LC-MS^n$ studies were carried out to determine the possible structures of the degradation products. MS^n spectra of AZP and each degradation product were shown in Figures 6-9. The structural information was obtained from the MS^n fragmentation data as shown in Scheme 1.

The drug underwent significant degradation under hydrolytic and oxidative conditions resulting the formation of a common degradation product DP 2. The hydrolysis of the ester linkage under these conditions leads to the formation of DP 2. In acidic hydrolysis, the drug undergoes two cleavages simultaneously, one at the enolic ether bond and the other at the ester bond leading to the formation of DP 1. Since the enolic ethers are easily hydrolyzed by aqueous acids, the AZP extensively undergoes hydrolysis at enolic ether bond to give DP 3 as a major degradation product.

Method validation

The developed method was validated to establish the specificity, precision, linearity, accuracy and robustness according to ICH guidelines [16].

Specificity

The specificity of the method was performed by ensuring the separation of the drug from all the degradation products formed under various stress conditions. The specificity of the method was established by evaluating the peak purity of AZP through PDA and mass spectrometric detection (Table 2). The results showed AZP to be pure and all the degradation products were separated and no other peaks were interfering with it (Figures 2-5).

Linearity

The linearity of the method was checked by injecting five concentration levels 160 (80%), 180 (90%), 200 (100%), 220 (110%) and 240 (120%) μg/mL of the standard solution. Each solution was injected in triplicate. Calibration equation was obtained from linear regression analysis and the correlation coefficient was found to be 0.99961 which shows that the method is linear. The regression equation for the calibration curve was found to be y = 32224.51x + 46741.

Accuracy

The accuracy of the method was evaluated in triplicate at three concentration levels, i.e. 80% (160 μg/mL), 100% (200 μg/mL) and 120% (240 μg/mL) levels by spiking the AZP in tablet placebo. The percentage of recovery of the drug was calculated at each level. The tablet placebo consisted of crosscarmellose sodium (3% w/w), microcrystalline cellulose (20% w/w), hydroxypropylcellulose (5% w/w), magnesium stearate (2% w/w) and mannitol (50% w/w). The percentage recovery of the drug ranged from 98.9% to 100.7% indicating the accuracy of the method (Table 3).

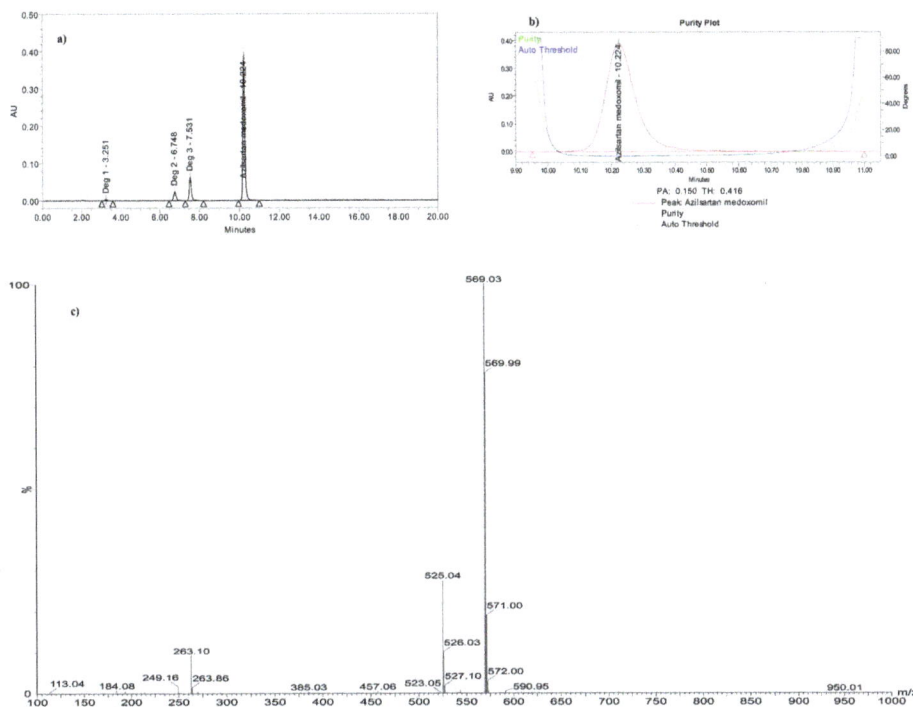

Figure 2: Separation of degradation products from AZP in acid degradation (a); AZP peak purity by PDA (b) and mass spectrometer (c).

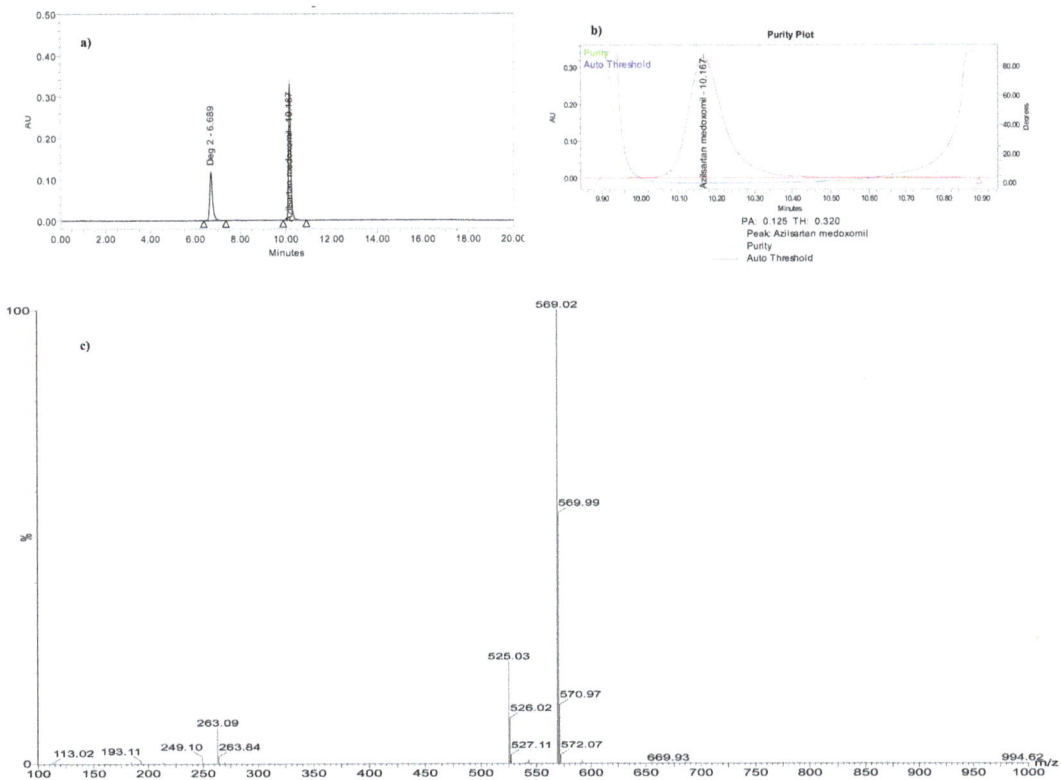

Figure 3: Separation of degradation products from AZP in base degradation (a); AZP peak purity by PDA (b) and mass spectrometer (c).

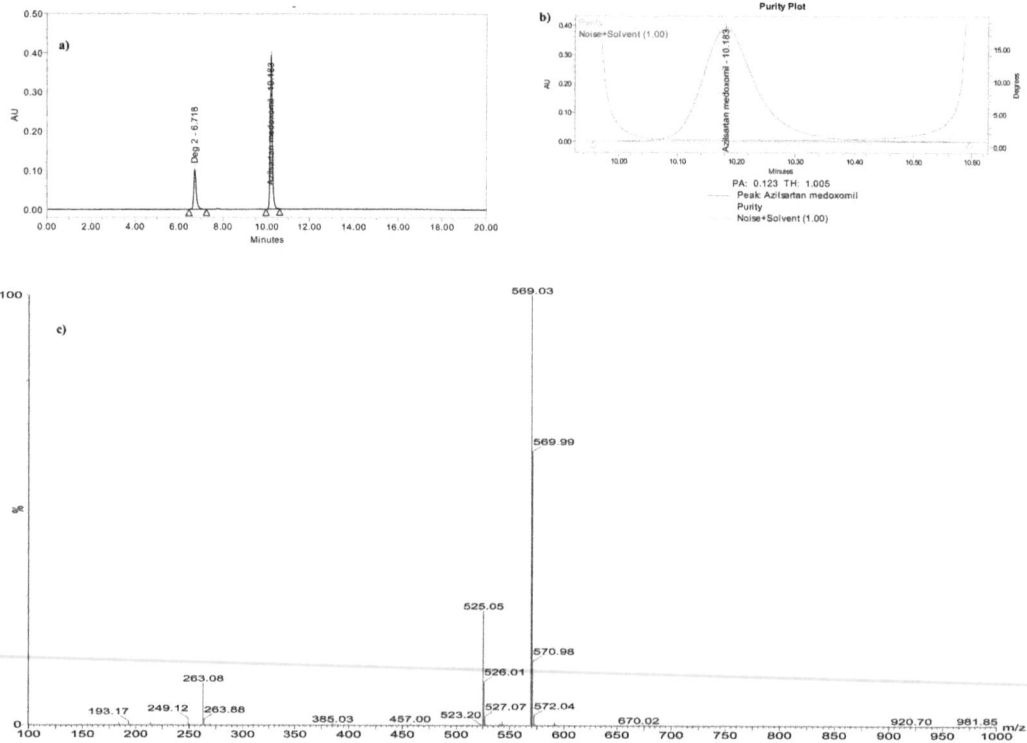

Figure 4: Separation of degradation products from AZP in neutral degradation (a); AZP peak purity by PDA (b) and mass spectrometer (c).

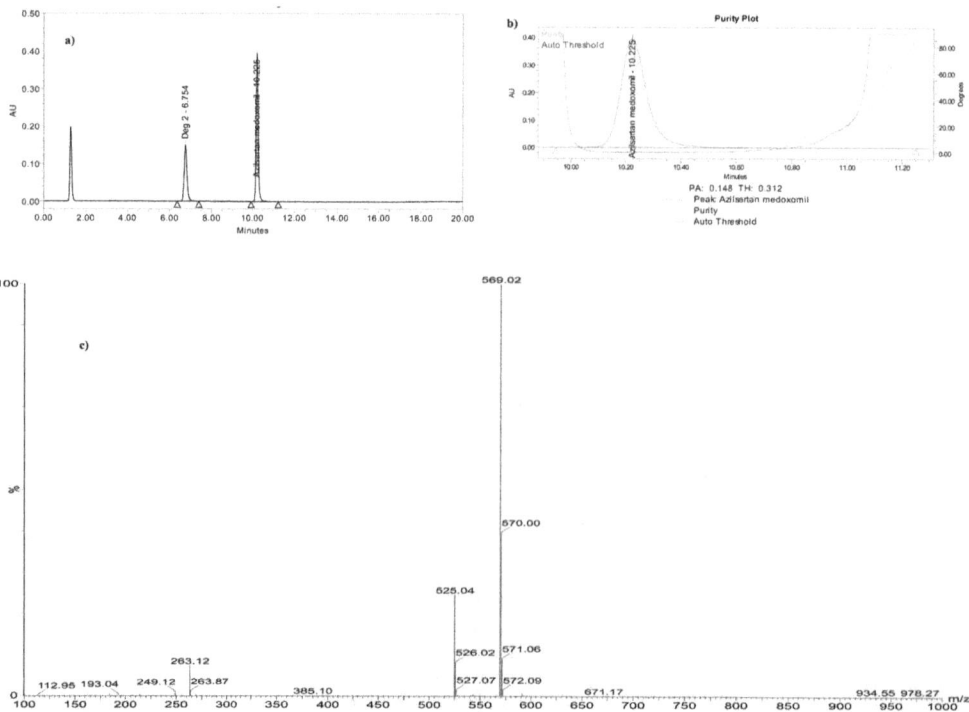

Figure 5: Separation of degradation products from AZP in oxidative degradation (a); AZP peak purity by PDA (b) and mass spectrometer (c).

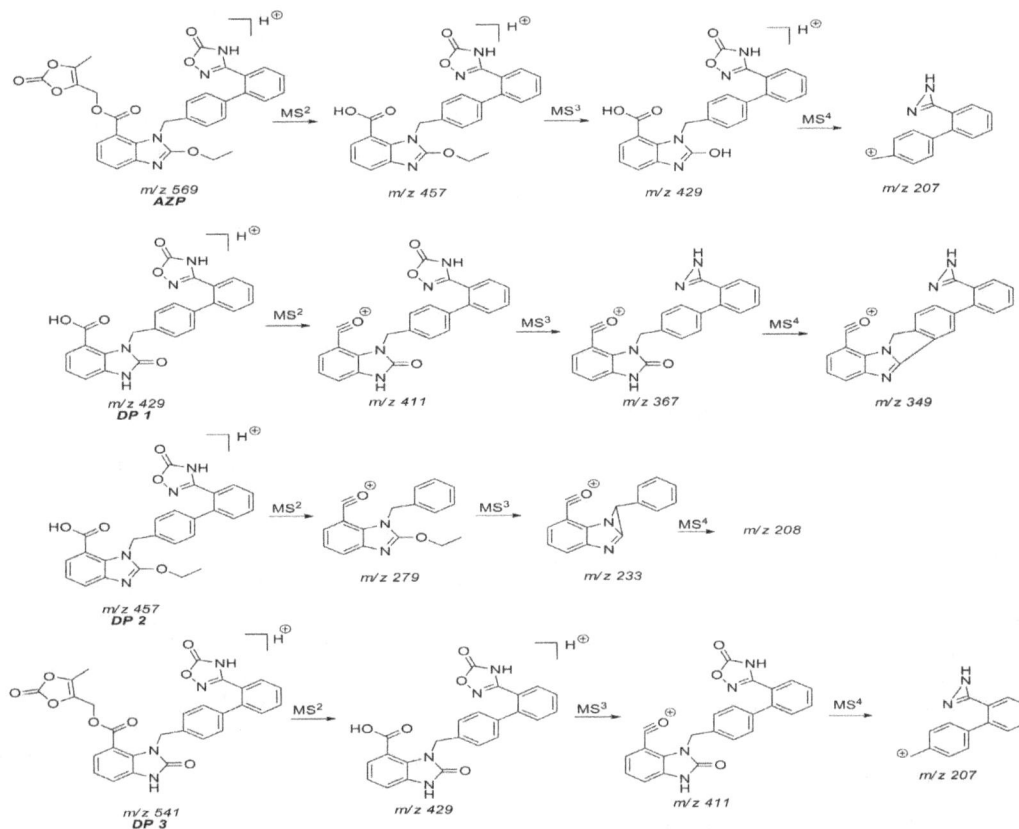

Scheme 1: Fragmentation pattern of AZP and degradation products.

Degradation study	Exposure conditions	Percent degradation
Acid hydrolysis	0.1 N HCl reflux at 60°C for 1 hour	21
Alkaline hydrolysis	0.01 N NaOH at room temperature for 15 min	26
Neutral hydrolysis	60°C for 1 hour	15
Oxidation degradation	0.5% (w/v) H_2O_2 at room temperature for 3 hours	18
UV light	200 Whrm^{-2}	No degradation
Fluorescence light	1.2 million lux hours	No degradation
Thermal	100°C for 48 hours	No degradation

Table 1: Forced degradation conditions and extent of degradation.

Degradation Study		AZP Peak Purity by PDA		AZP Peak Purity by LC-MS
		Purity Angle	Purity Threshold	m/z for AZP across the peak
Acid hydrolysis		0.150	0.416	569.03
Base hydrolysis		0.125	0.320	569.02
Neutral hydrolysis		0.123	1.005	569.03
Peroxide degradation		0.149	0.312	569.02
UV radiation degradation	Solid	0.312	0.646	569.04
	Solution	0.614	0.651	569.03
Fluorescence degradation	Solid	0.205	0.590	569.02
	Solution	0.118	0.479	569.04
Thermal degradation		0.528	0.860	569.05

Note: For a peak to be pure purity angle should be less than purity threshold in PDA and a single m/z value should be obtained across the peak in MS detection.
Table 2: Summary of Peak Purity Evaluation under various stress conditions.

Amount of AZP Spiked (µg/mL)	Amount of AZP recovered (µg/mL) ± SD; RSD (%)	Mean recovery (%)
160	160.98 ± 0.33; 0.20	100.1
200	201.88 ± 0.90; 0.45	99.4
240	243.53 ± 0.96; 0.39	100.1

Table 3: Accuracy data of AZP.

Precision

The precision of the method was evaluated by analyzing six individual sample preparations containing about 200 µg /mL of AZP in tablet placebo. Percentage RSD for % assay of the drug was calculated. The intermediate precision (ruggedness) of the method was also evaluated on different days (intra day and inter day), different columns (Hibar 150-4.6 Purospher STAR RP-18e (5 µm)) and different instrument (Waters 2695) in the same laboratory. Percentage RSD of areas of each drug was less than 1, confirming good precision at low level of the developed analytical method (Table 4).

Robustness

The experimental conditions were deliberately changed in order to determine the robustness of the developed method and percentage assay and tailing factor of the main peak was evaluated. Observation of the results for deliberately changed chromatographic conditions (flow rate, and column temperature) revealed percentage assay and also the tailing factor were within the limits, illustrating the robustness of the method.

Solution and mobile phase stability

Most industries utilize auto samplers with overnight runs and the samples will be in solution for hours in the laboratory environment before the test procedure is completed. This is of concern especially for drugs that can undergo degradation by hydrolysis, photolysis or adhesion to glassware. The solution stability of the drug was demonstrated by keeping sample solution in volumetric flasks at room temperature and refrigerated conditions (5°C). Content of each drug was determined after 18, 24, 42 and 48 hours against freshly prepared standard solution. The solution stability experiment data confirmed the sample solutions to be stable up to 48 hrs under the mentioned storage conditions (Table 5).

The mobile phase stability was also assessed in the experiment to observe any changes in the analytical method due to minor changes that may affect the mobile phase used for long sample sequences. The deviation in the retention times of the different sample and also the % assay was observed. The entire % RSD was found to be less than 1.0 (Table 5).

Conclusion

A new LC-MS compatible RP-HPLC method has been developed for determination of AZP in bulk drug and pharmaceutical dosage forms. The drug was found to degrade in hydrolytic and oxidative stress conditions. Mild alkaline conditions also tend to degrade the drug to a greater extent which shows it's susceptible to basic environment. The structures of all degradation products were characterized and their probable degradation pathway was established. The method was validated as per ICH guidelines. The method has been proved to be linear, precise, accurate and robust. All the degradation products formed under various stress conditions were separated from each other and also from the main drug using the developed HPLC method which shows the stability indicating power of the method. Hence the method can be recommended for routine quality control analysis and also selecting formulation conditions. An insight into the chemistry of the degradation products provides information about the precautions to be taken while manufacture or storage of the drug.

Preparation	Method Precision (%Assay)	Intermediate Precision (%Assay)			
		Day 1	Day 2	Instrument variation Waters (e2695)	Column variation Purospher
1	99.95	99.67	99.73	100.11	99.94
2	99.80	99.85	99.96	100.03	100.34
3	99.68	99.73	99.50	100.84	99.99
4	99.49	99.48	99.61	99.61	100.08
5	99.48	99.64	99.77	100.09	100.32
6	99.37	99.62	99.75	99.71	100.23
Average	99.63	99.67	99.72	100.10	100.2
SD	0.22	0.12	0.16	0.40	0.21
%RSD	0.22	0.12	0.16	0.40	0.21

Table 4: Precision data of AZP.

S.No.	Time (hr)	%Assay at RT* (Solution stability)	%Assay at REF** (Solution stability)	%Assay (Mobile phase stability)
1	0	100.06	99.85	99.85
2	18	99.99	99.88	100.11
3	24	100.31	100.39	99.93
4	42	100.16	100.44	100.27
5	48	100.09	100.71	99.49
Average		100.12	100.25	99.93
SD		0.12	0.376	0.295
%RSD		0.12	0.37	0.30
*RT - Room Temperature **REF- Refrigerated conditions (8°C)				

Table 5: Solution and mobile phase stability data of AZP.

AZP_09042014_2 #2881-3013   RT: 9.83-10.16   AV: 24   SB: 578 7.84-9.75 , 10.21-12.74   NL: 1.36E6
F: ITMS + p ESI Full ms [200.00-1000.00]

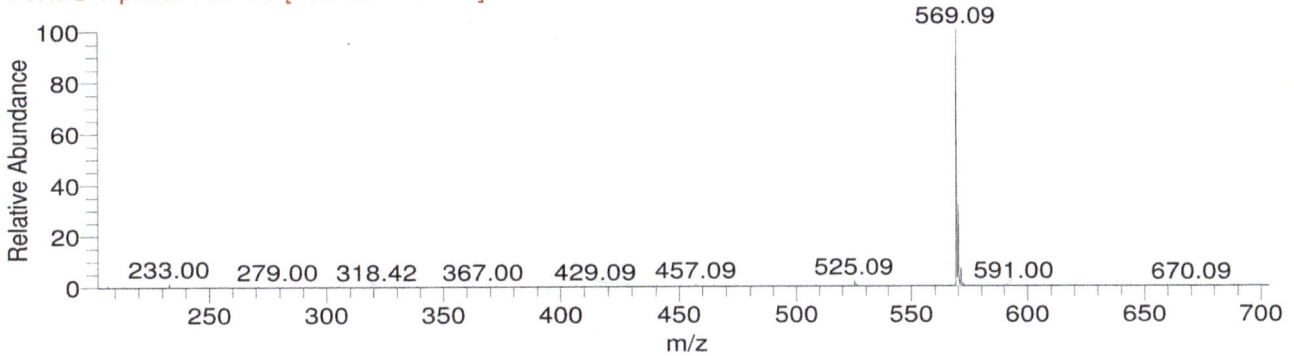

AZP_09042014_2 #2878-3090   RT: 9.82-10.42   AV: 41   NL: 4.40E5
T: Average spectrum MS2 569.16 (2878-3090)

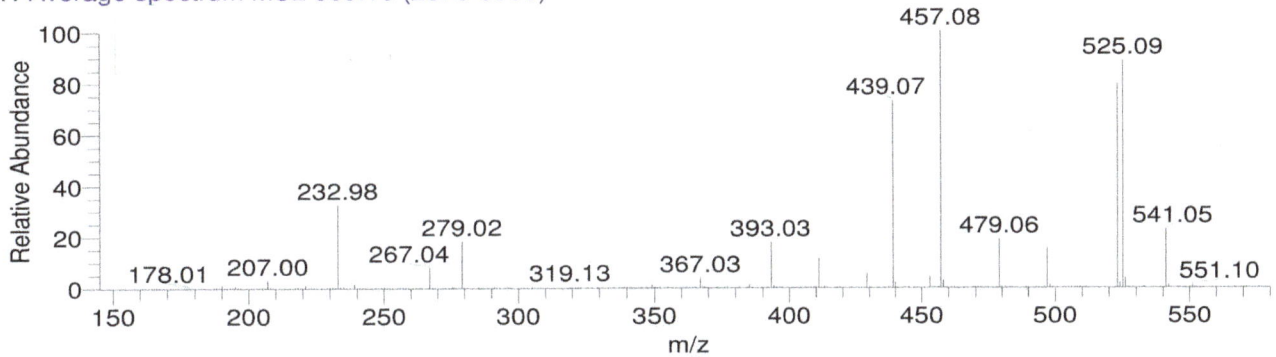

AZP_09042014_2 #2879-3091   RT: 9.82-10.43   AV: 41   NL: 1.76E5
T: ITMS + c ESI d Full ms3 569.15@cid35.00 457.05@cid35.00 [115.00-470.00]

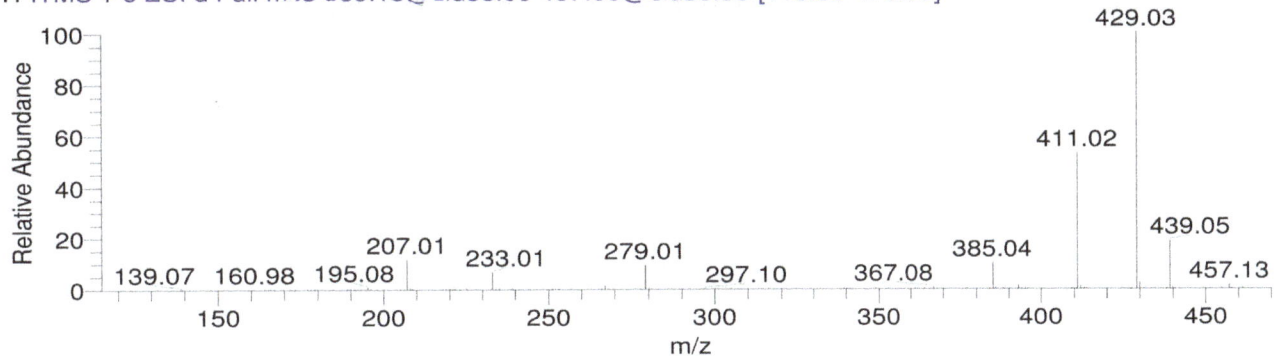

AZP_09042014_2 #2880-3092   RT: 9.83-10.43   AV: 41   NL: 5.42E4
T: Average spectrum MS4 569.16,457.11,429.00 (2880-3092)

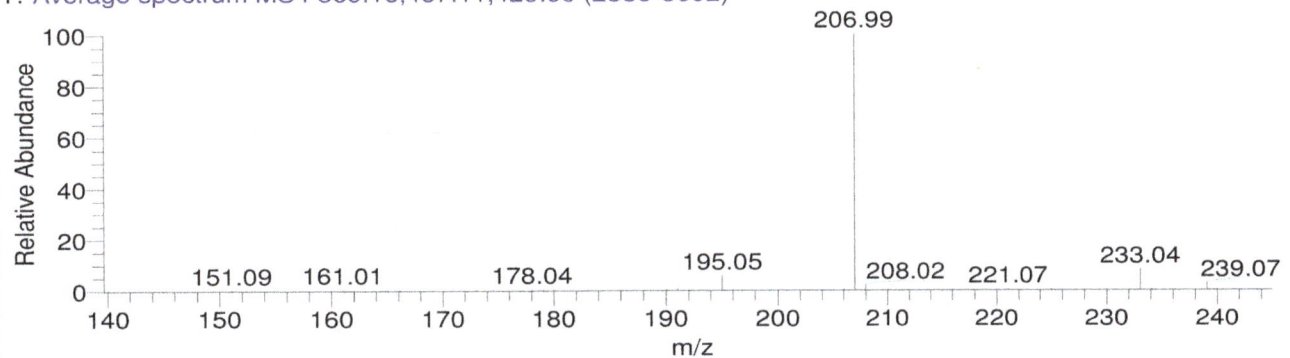

Figure 6: MS^n spectra of AZP.

AZP_09042014_2 #930-957 RT: 3.21-3.27 AV: 4 SB: 242 1.93-2.97 , 3.46-4.74 NL: 6.84E4
F: ITMS + p ESI Full ms [200.00-1000.00]

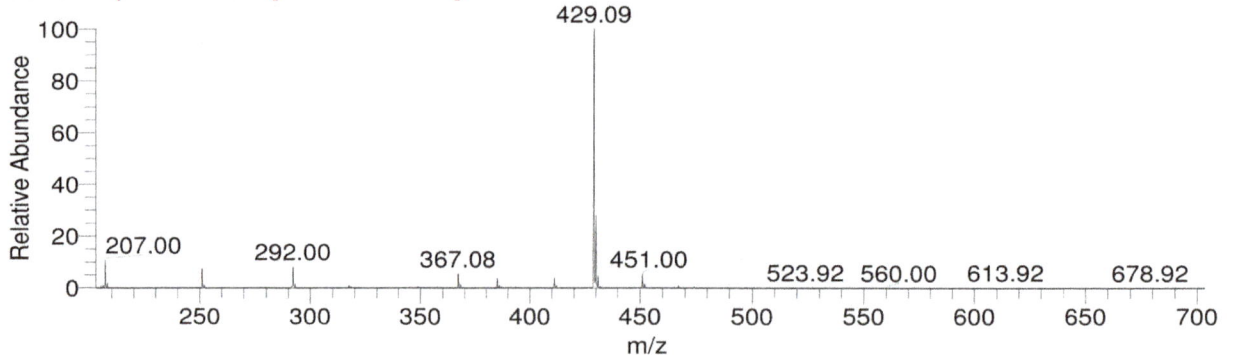

AZP_09042014_2 #917-977 RT: 3.15-3.36 AV: 11 NL: 7.96E4
T: Average spectrum MS2 429.00 (917-977)

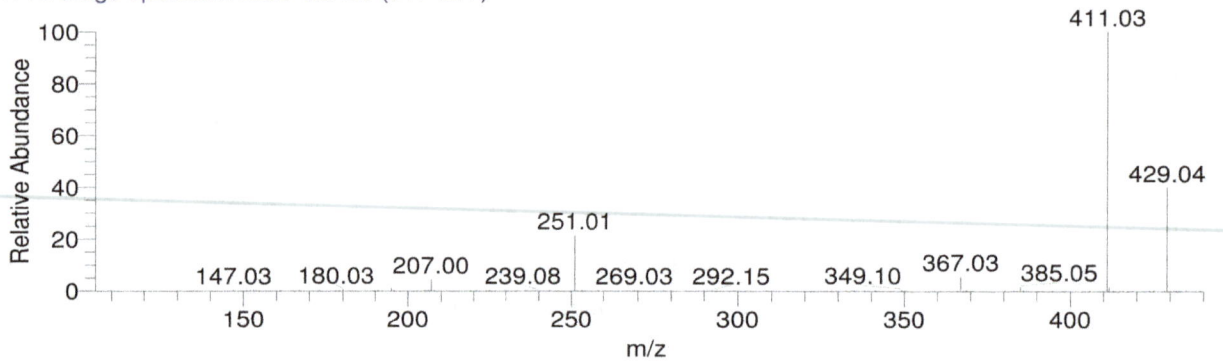

AZP_09042014_2 #918-978 RT: 3.15-3.36 AV: 11 NL: 4.19E4
T: Average spectrum MS3 429.00,411.03 (918-978)

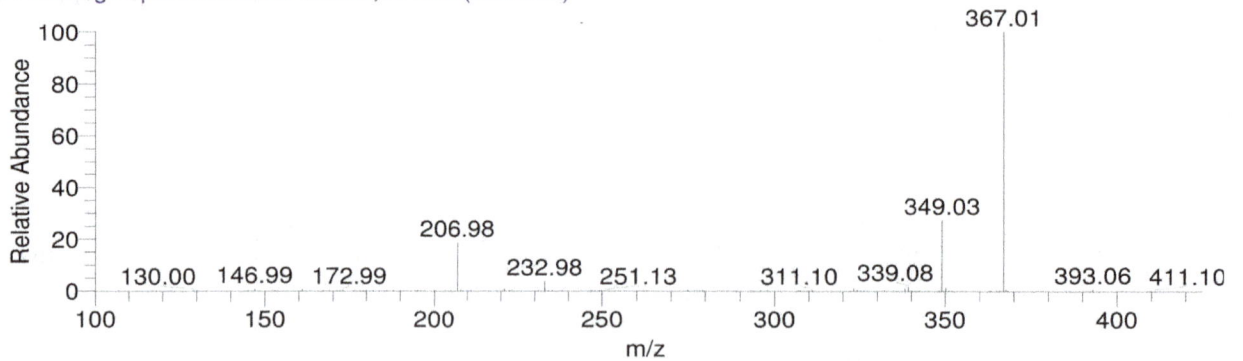

AZP_09042014_2 #919-979 RT: 3.15-3.36 AV: 11 NL: 9.28E3
T: Average spectrum MS4 429.00,411.03,367.00 (919-979)

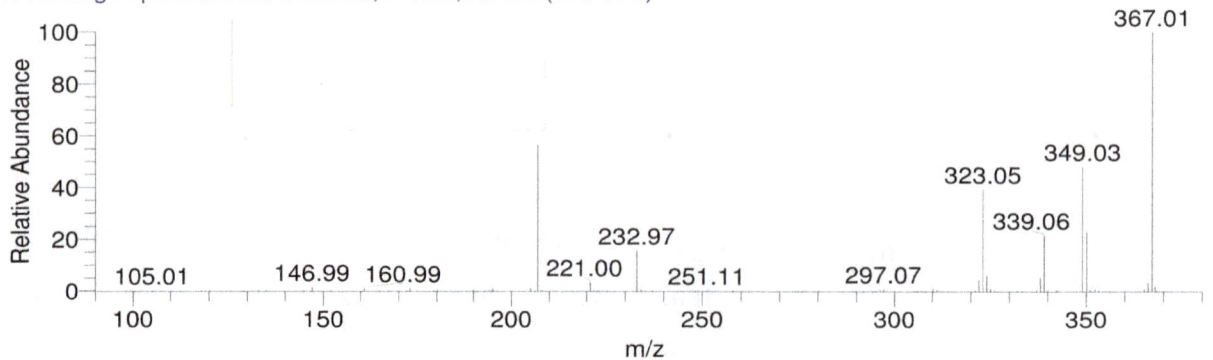

Figure 7: MS^n spectra of DP 1.

AZP_09042014_2 #1889-1950 RT: 6.56-6.71 AV: 12 SB: 306 4.70-6.48 , 6.91-8.30 NL: 2.03E5
F: ITMS + p ESI Full ms [200.00-1000.00]

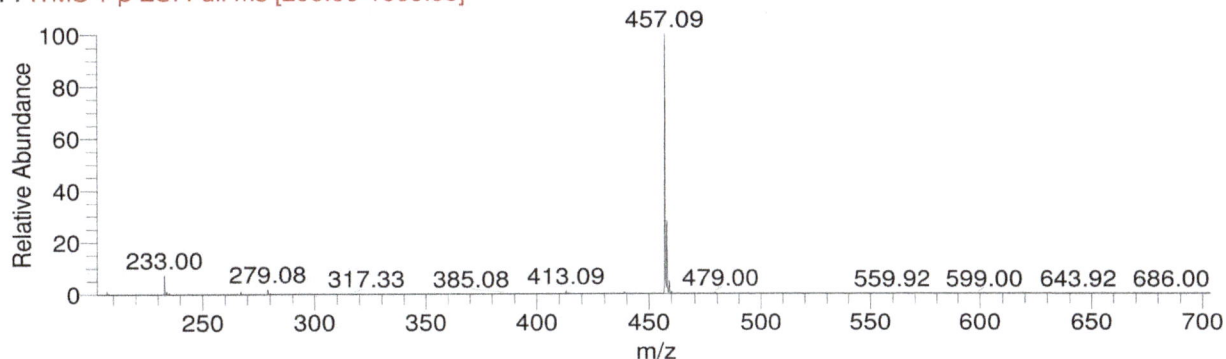

AZP_09042014_2 #1885-1993 RT: 6.53-6.84 AV: 24 NL: 8.27E4
T: Average spectrum MS2 457.13 (1885-1993)

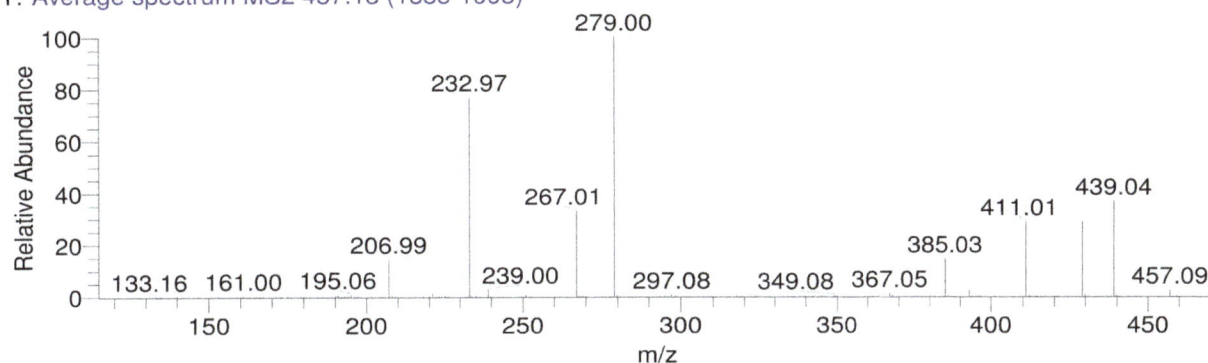

AZP_09042014_2 #1886-1994 RT: 6.54-6.85 AV: 24 NL: 4.95E4
T: ITMS + c ESI d Full ms3 457.07@cid35.00 279.04@cid35.00 [65.00-290.00]

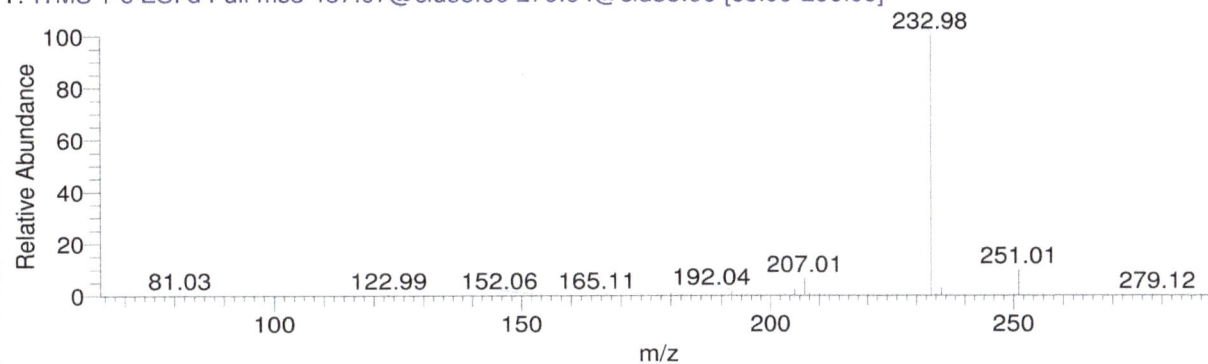

AZP_09042014_2 #1887-1995 RT: 6.54-6.85 AV: 24 NL: 1.50E3
T: Average spectrum MS4 457.13,279.01,232.95 (1887-1995)

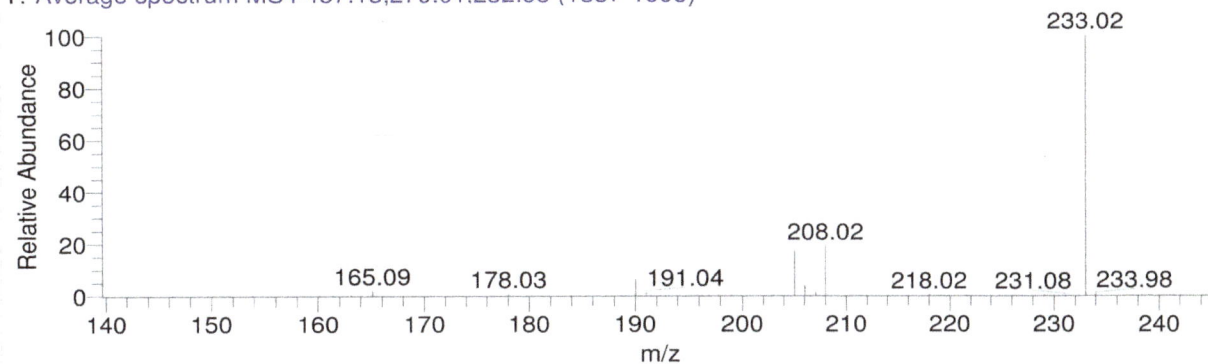

Figure 8: MS^n spectra of DP 2.

AZP_09042014_2 #2092-2190 RT: 7.20-7.45 AV: 17 SB: 393 4.32-7.07 , 7.84-8.95 NL: 3.36E5
F: ITMS + p ESI Full ms [200.00-1000.00]

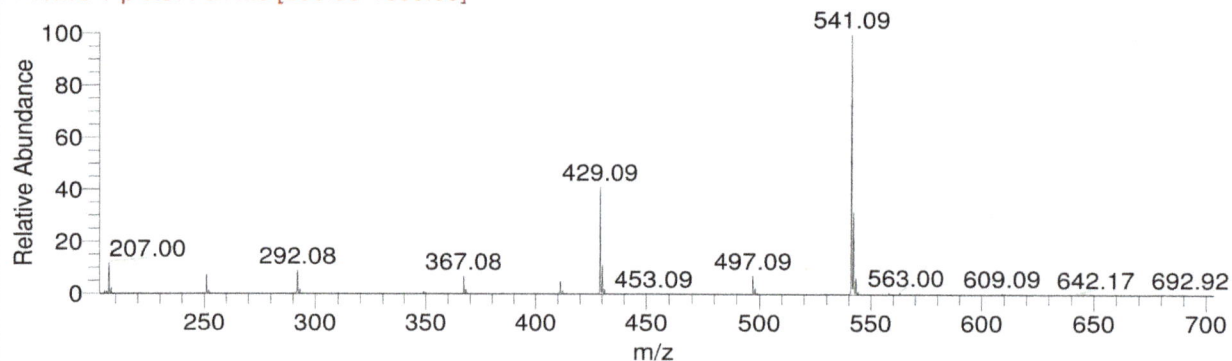

AZP_09042014_2 #2088-2226 RT: 7.17-7.59 AV: 24 NL: 6.50E5
T: Average spectrum MS2 541.09 (2088-2226)

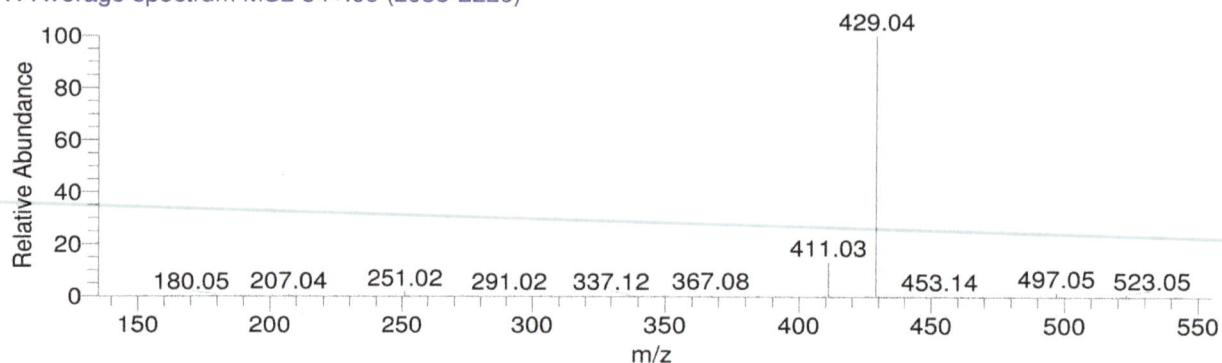

AZP_09042014_2 #2089-2227 RT: 7.18-7.59 AV: 24 NL: 3.22E5
T: Average spectrum MS3 541.09,428.97 (2089-2227)

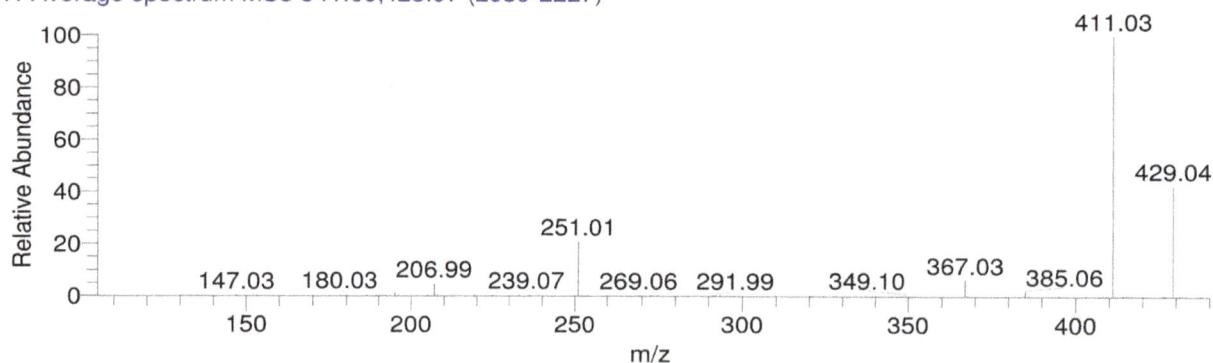

AZP_09042014_2 #2090-2228 RT: 7.18-7.59 AV: 24 NL: 2.93E4
T: Average spectrum MS4 541.09,428.97,411.02 (2090-2228)

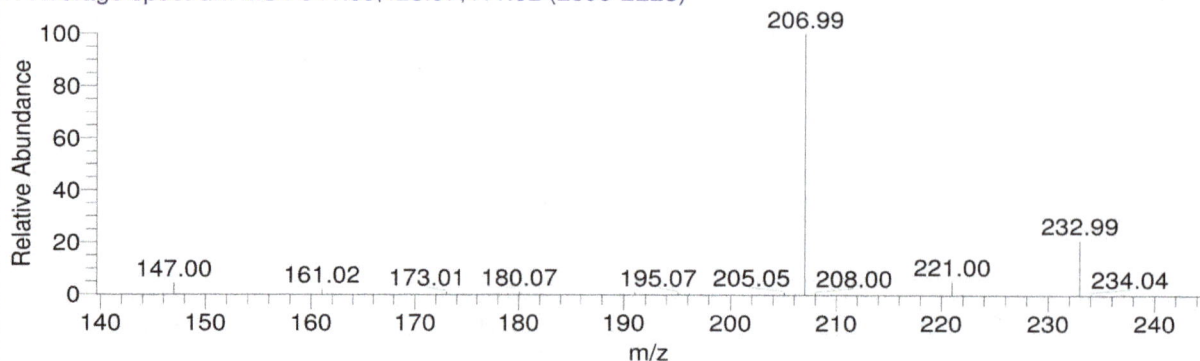

Figure 9: MS^n spectra of DP 3.

Acknowledgement

The authors are grateful to Dr. Ahmed Kamal, Project Director, NIPER-Hyderabad for his constant support and encouragement. One of the authors Debasish Swain is thankful to Ministry of Chemicals and fertilizers, Department of Pharmaceuticals, Govt. of India for providing the fellowship and USP India Pvt. Ltd. for providing facilities to carry out the research work.

References

1. Jones JD, Jackson SH, Agboton C, Martin TS (2011) Azilsartan Medoxomil (Edarbi): The Eighth Angiotensin II Receptor Blocker. P T 36: 634-640.

2. Zaiken K, Cheng JW (2011) Azilsartan medoxomil: a new Angiotensin receptor blocker. Clin Ther 33: 1577-1589.

3. Kurtz TW, Kajiya T (2012) Differential pharmacology and benefit/risk of azilsartan compared to other sartans. Vasc Health Risk Manag 8: 133-143.

4. Gorla R, Sreenivasulu B, Garaga S, Sreenivas N (2014) A simple and sensitive stability- indicating HPTLC assay method for the determination of azilsartan medoxomil. IAJPR 4: 2985-2992.

5. Vekariya PP, Joshi HS (2013) Development and validation of RP-HPLC method for azilsartan medoxomil potassium quantitation in human plasma by solid phase extraction procedure. ISRN Spectroscopy 2013: 572170.

6. Bakshi M, Singh S (2002) Development of validated stability-indicating assay methods critical review. J Pharm Biomed Anal 28: 1011-1040.

7. ICH Harmonised Tripartite Guideline (2003) International Conference on Harmonisation of Technical Requirements for Registration of Pharmaceuticals for Human Use (ICH). Stability Testing of New Drug Substances and Products, Q1A(R2).

8. ICH Harmonised Tripartite Guideline (1996) International Conference on Harmonisation of Technical Requirements for Registration of Pharmaceuticals for Human Use (ICH). Stability Testing of New Drug Substances and Products, Q1B, Q1C.

9. Bryant DK, Kingswood MD, Belenguer A (1996) Determination of liquid chromatographic peak purity by electrospray ionization mass spectrometry. J Chromatogr A 721: 41-51.

10. Sánchez FC, Khots M, Massart D, De Beer J (1994) Algorithms for the assessment of peak purity in liquid chromatography with photodiode-array detection. Part II. Anal Chim Acta 285: 181-192.

11. Sellers JA, Olsen BA, Owens PK, Gavin PF (2006) Determination of the enantiomer and positional isomer impurities in atomoxetine hydrochloride with liquid chromatography using polysaccharide chiral stationary phases. J Pharm Biomed Anal 41: 1088-1094.

12. http://www.drugbank.ca/drugs/DB06695

13. Reynolds DW, Facchine KL, Mullaney JF, Alsante KM, Hatajik TD, et al. (2002) Available Guidance and Best Practices for Conducting forced degradation studies. Pharmaceutical technology 2002: 48-56.

14. Blessy M, Patel RD, Prajapati PN, Agrawal Y (2014) Development of forced degradation and stability indicating studies of drugs-A review. J Pharm Anal 4: 159-165.

15. Singh S, Junwal M, Modhe G, Tiwari H, Kurmi M, et al. (2013) Forced degradation studies to assess the stability of drugs and products. TrAC 49: 71-88.

16. ICH Harmonised Tripartite Guideline (2005) International Conference on Harmonisation of Technical Requirements for Registration of Pharmaceuticals for Human Use (ICH). Validation of Analytical Procedures, Text and Methodology Q2(R1).

26

Rapid Comparison of UVB Absorption Effectiveness of Various Sunscreens by UV-Vis Spectroscopy

Ju Chou*, Ted J. Robinson and Hui Doan

Department of Chemistry and Physics, Florida Gulf Coast University, Fort Myers, FL 33965, USA

Abstract

Sunscreens are used to absorb or block harmful sunlight especially ultra violet (UV) radiation. An UV-vis spectrometer was employed to measure absorbance of sunscreen products. The same brand's sunscreens with sun protection factor (SPF) of 8, 15, 30, and 50 were tested under identical experimental conditions. The results show that the UV absorbance and the transmittance of the sunscreens are associated with the SPF value. The maximum absorbance of the sunscreens measured between 280 to 320 nm (UVB region) is linearly proportional to the SPF value with a correlation coefficient of 0.998 using the same brand's sunscreens. Thus, the absorbance can be used to evaluate the efficiency of a sunscreen that absorbs or blocks UVB radiation. Several commercial sunscreens of different brands but with the same SPF 30 were compared. The results confirmed that, although different brand sunscreens with the same SPF varied slightly in UV absorbance, they all offer adequate protection against UVB radiation. The utilization of UV-Vis spectroscopy is found to be particularly effective for determination of sunblock efficiency.

Keywords: Uv-vis spectroscopy; Sunscreens; Spf; Uvb radiation

Introduction

The sun emits three types of ultra violet (UV) radiation: UVA (320-400 nm), UVB (280-320 nm) and UVC (200-280 nm). Among them, UVC radiation contains the shortest wavelength and has the highest energy. However, since it is blocked by atmospheric ozone layer, UVC does not reach the Earth's surface and cannot affect humans. Both UVA and UVB rays are able to penetrate the ozone layer and reach the earth's surface. Thus, they are harmful to humans by damaging human skin potentially causing sunburns and skin cancer etc. [1-3]. About 90 percent of all skin cancers are associated with exposure to the sun's harmful UV radiation. Sunscreen is one of the key strategies that help greatly reduce excess exposure to the harmful UV rays [4-7].

Sunscreens help shield human body from the sun's radiation in two ways dependent on their ingredients by either absorbing it or reflecting it away from the skin. The assessment of efficiency of a sunscreen is based on sun protection factor (SPF) which is used by all sunscreen manufacturers to rate their sunscreen's ability to absorb UVB radiation [6]. The most effective sunscreens should block or absorb both UVA and UVB radiation, but UVA is not discussed in this study since it is not evaluated by SPF yet [8-10]. SPF values can be obtained *in-vivo* by determining the amount of time a person can stay in the sun without experiencing a sunburn [10], but it has the undesirable effect of involving the human skin test.

Due to high cost and time consumption of *in-vivo* SPF determination methods, the SPF of sunscreens was determined by ultraviolet spectrophotometry which substitutes in the *in-vivo* method [11-15]. UV-vis spectroscopy was employed to measure absorbance at different wavelengths and a SFP was calculated using the absorbance measured and mathematic equation [13,14]. The calculation requires to use normalized product function at different wavelengths proposed by Sayre, et al. in order to obtain SFP [11]. More practically, people want to find out which sunscreen will provide a better skin protection against UVB rays, but there is no direct correlation reported between the absorbance of sunscreens and the SPF value so far. In this study, a rapid analytical method to evaluate sunscreen's ability to absorb or block UVB radiation using a UV-vis spectrophotometer. A direct correlation between the absorbance and the SPF was examined and it allowed to assess sunblock effectiveness without using the normalized production function.

UV-vis spectrophotometer is able to measure the intensity of light passing through a sample or measure absorbance [12-15]. In this study, it was used to measure absorbance of sunscreen solutions. For the first time, a direct correlation between the absorbance and the SPF of sunscreens was established. Sunscreen's effectiveness to absorb or block UVB radiation from different manufacturers was then compared. Thus, this paper presents the application of an UV-vis spectrophotometric analysis on sunscreen's effectiveness.

Experimental

Chemicals

70% of rubbing alcohol (isopropyl alcohol) was purchased from a local drug store. All sunscreens of different brands and with different SPF (8, 15, 30, 50) were purchased from local stores.

Experimental procedure

0.02 g of a sunscreen sample was weighed and transferred into a 100 mL volumetric flask. About 70 ml of 70% isopropyl alcohol was added into the volumetric flask. The solution was well shaken and then was diluted to the mark with the 70% isopropyl alcohol.

Three trials were prepared for each sunscreen product and the absorbance of each solution was scanned from 220 to 400 nm. The

*Corresponding author: Ju Chou, Department of Chemistry and Physics, Florida Gulf Coast University, Fort Myers, FL 33965, USA, E-mail: jchou@fgcu.edu

maximum absorbance was recorded and then was corrected to the same mass of 0.0200 g. The corrected absorbance was calculated by the following formula:

$$A = Measured\ A \times \frac{0.0200\ g}{Mass\ of\ Sample} \tag{1}$$

The mean of corrected absorbance of the three measurements were calculated for accuracy and consistency.

The transmittance (T) of a sunscreen solution was calculated from the absorbance (A) based on the following equation:

$$A = -\log(T) \tag{2}$$

UV-vis spectroscopy

A double beam Shimadzu UV-Vis Spectrophotometer with 1 cm quartz cuvettes was used for absorbance measurements. One quartz cuvette was filled with 70% isopropyl alcohol as a reference and another one was filled with a sunscreen solution. Prior to analysis, the sample cell was rinsed three times with 1 mL of the sample solution. The UV-vis spectrophotometer was set to scan from 220 nm to 400 nm and the maximum wavelength from 290 nm to 320 nm was recorded after completing the scan.

Results and Discussion

Sunscreens with different SPF values of 8, 15, 30 and 50 from the same brand (Abbreviated as Brand S or BS) were selected to investigate a correlation between the absorbance and transmittance with the SPF value. Figure 1 shows typical plots of UV-vis absorption spectra of the four sunscreens from 220 to 400 nm. All sunscreens have two absorption peaks from 220 to 400 nm. The first peak was centered around 240 nm (UVC region). The second peak was centered around 300 nm (UVB region) for SPF 30 and 50, and 290 nm for the SPF 8 and 15. The UV-vis spectra of the four sunscreens indicate that they all are able to absorb both UVB and UVC rays. Since UVC radiation does not reach the earth, it is not discussed here. This study only focused on the UVB radiation and only the absorption peak centered around 300 nm was discussed. The wavelength at which absorbance is highest is called the maximum wavelength or λ_{max}, and the absorbance at the λ_{max} for all sunscreens is used.

Absorbance of a substance is generally dependent on the concentrations of the tested solution according to Beer's law: $A = \varepsilon bc$ (c is the concentration of the tested solution, b is the path length of the light and ε is the molar absorptivity.) In order to compare sunscreen's effectiveness in absorbing the sun's UVB rays, the same mass of each sunscreen must be used. Thus, a corrected absorbance was calculated as mentioned in the Experimental Section and the corrected absorbance at the λ_{max} is used for the following discussion.

The corrected absorbance at λ_{max} of the four sunscreens from the same brand versus the SPF was plotted as shown in Figure 2. The results show that the corrected absorbance of the sunscreen is directly related to SPF of the sunscreens. For the same brand sunscreen product, the corrected absorbance is linearly proportional to SPF. The higher SPF of a sunscreen is, where the higher absorbance was observed. An effective sunscreen is one that absorbs UV radiation in the 290 to 320 nm region to prevent the UVB rays from reaching and damaging human skin. Thus, the direct correlation observed between the corrected absorbance and SPF allows to use the corrected absorbance to evaluate sunscreen products.

The UV-vis spectrometer not only measures the absorbance of a solution, but also provides transmittance. When light is not absorbed by a sunscreen solution, it transmits through the solution and reaches the skin. The more light a solution absorbs, the less light is transmitted through the sun screen solution, and the less UV radiation reaches the skin. Transmittance of the four SPF solutions was calculated according to Equation (2) in the Experimental section and was plotted versus SPF as shown in Figure 2B. The figure shows that the transmittance of the sunscreens is not linearly related to SPF as expected. The results also show that the higher SPF a sunscreen is, the lower transmittance is observed and thus more light is absorbed by the sunscreen. The

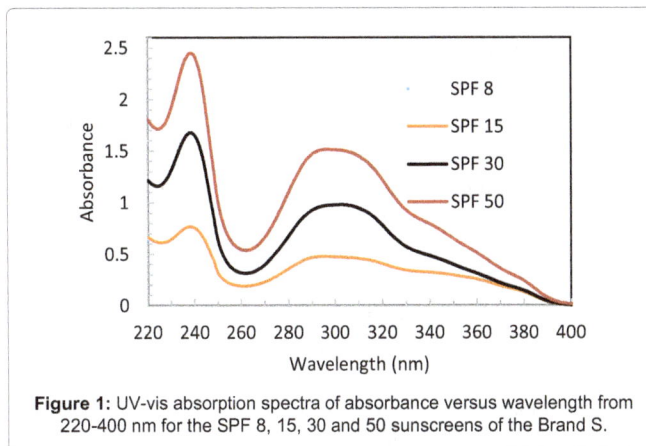

Figure 1: UV-vis absorption spectra of absorbance versus wavelength from 220-400 nm for the SPF 8, 15, 30 and 50 sunscreens of the Brand S.

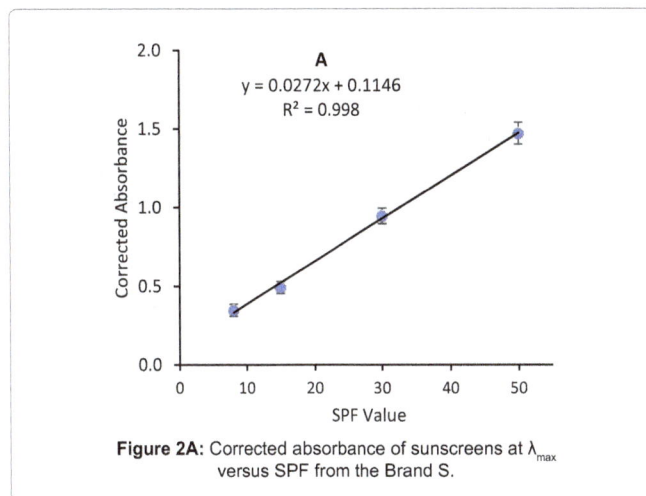

Figure 2A: Corrected absorbance of sunscreens at λ_{max} versus SPF from the Brand S.

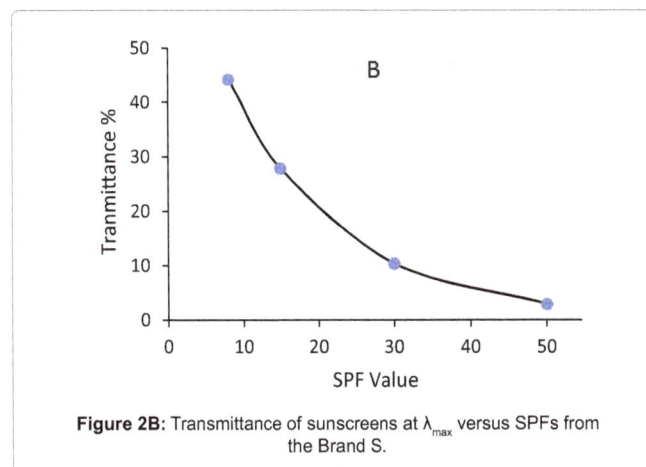

Figure 2B: Transmittance of sunscreens at λ_{max} versus SPFs from the Brand S.

common misunderstanding is that SPF 30 is twice as strong as SPF 15, which is not true. For example, SPF 15 can filter out 72% of UVB, while SPF 30 can block 90% which is not doubled. When the same amount of SPF 50 was applied, 97% of UVB was filtered, indicating that only 3% of the UVB rays transmits through the sunscreen.

Since the absorbance (not transmittance) is linearly proportional to the SPF, the absorbance is selected to evaluate the effectiveness of a sunscreen to absorb or block the UVB radiation. Though transmittance is not linearly proportional to the SPF, it provides quantitative information on what percent of the UVB radiation is blocked or filtered out by a sunscreen.

Nine different brands of sunscreen products (B1 to B9) with the same SPF 30 were evaluated with the same procedures discussed above and they were compared with the Brand S sunscreen. The corrected absorbance of each sunscreen solution was obtained and the average corrected absorbance from three trials were shown in Figure 3A. Though these sunscreens have the same SPF 30, they have slightly different absorbance indicating that they absorb different percentages of UVB radiation. Among the nine brands of sunscreen products tested, six brands have higher absorbance than Brand S, indicating more UVB rays are filtered out by the six sunscreens. Two brands (B2 and B7) have comparable absorbance with Brand S. According to the label from the two brands, one has an identical formula as the Brand S product. This explains why the two brands have similar absorbance. Only one brand's sunscreen has lower absorbance than Brand S, indicating it blocked less UVB radiation than Brand S.

Transmittance of all 10 sunscreens were calculated based on the corrected absorbance and shown in Figure 3B. Transmittance varied from 4% to 17% (most of them were from 4 to 11%), indicating that most of them provide adequate protection against UVB radiation since most of them absorb 89% or more of UVB rays and allow 11% or less to reach the skin under the experimental condition.

This study demonstrates that sunscreens absorb UVB rays and decrease the amount of UVB rays allowed to reach the skin. The absorbance of the sunscreens from the same brand is linearly proportional to the labeled SPF. However, the sunscreens of different brands with the same SPF do not provide the same UVB protection. This claim is based on the proposition that a sunscreen with a higher UVB absorbance will block more radiation and will be more effective than a sunscreen with a lower UVB absorbance. It should be noted that a sunscreen with both UVA and UVB protection is recommended, but UVA is not discussed in this study. The method developed in this study can provide a quantitative mean to evaluate the effectiveness of a sunscreen to block or absorb UVB rays.

Conclusion

In conclusion, this study shows a direct correlation between the absorbance that evaluates sunscreen efficiency to block UVB radiation and the sunscreen's SPF. The analytical method developed allows to rapidly compare the sunscreen's effectiveness to absorb or block UVB radiation and might provide a rapid and a useful alternative method for measuring sunscreen effectiveness in addition to *in vivo* SPF determination. With the same SPF 30, though the sunscreens of different brands block different amount of UVB rays, they all unequivocally and reliably absorb or block UVB rays for skin protection.

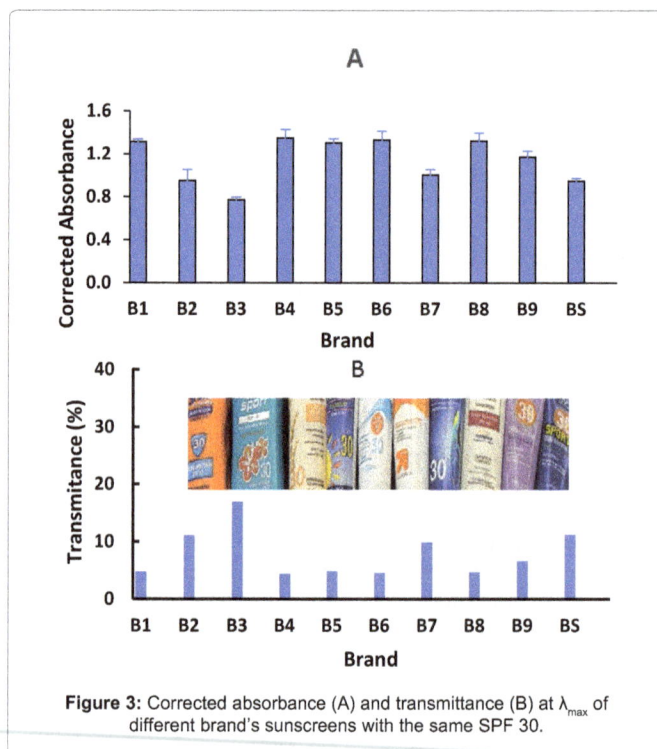

Figure 3: Corrected absorbance (A) and transmittance (B) at λ_{max} of different brand's sunscreens with the same SPF 30.

Acknowledgments

Department of Chemistry and Physics at Florida Gulf Cost University supported this work.

References

1. Wolf R, Tuzun B, Tuzun Y (2001) Sunscreens. Dermatol Ther 14: 208-214.

2. Pleasance ED, Cheetham RK, Stephens PJ, McBride DJ, Humphray SJ, et al. (2010) A comprehensive catalogue of somatic mutations from a human cancer genome. Nature 463: 191-196.

3. Rass K, Reichrath (2008) UV damage and DNA repair in malignant melanoma and nonmelanoma skin cancer. J Adv Exp Med Biol 624: 162-178.

4. Serre I, Cano JP, Picot MC, Meynadier J, Meunier L (1997) Immunosuppression induced by acute solar-simulated ultraviolet exposure in humans: prevention by a sunscreen with a sun protection factor of 15 and high UVA protection. J Am Acad Dermatol 37: 187-194.

5. Reinau D, Osterwalder U, Stockfleth E, Surber C (2015) The meaning and implication of sun protection factor. Brit J Dermatol 173: 1345.

6. Burnett ME, Wang SQ (2011) Ultraviolet Radiation and the Skin: An In-Depth Review. Photodermatol Photoimmunol Photomed 27: 58.

7. Autier P, Boniol M, Severi G, Dore JF (2001) Quantity of sunscreens used in European students. Brit J Dermatol 144: 288.

8. Gasparro FP, Mitchnick M, Nash JF (1998) A Review of Sunscreen Safety and Efficacy. Photochem Photobiol 68: 243.

9. Moyal DD, Fourtanier AM (2008) Broad-spectrum sunscreens provide better protection from solar ultraviolet-simulated radiation and natural sunlight-induced immunosuppression in human beings. J Am Acad Dermatol 58: 149-154.

10. Bissonnette R, Allas S, Moyal D, Provost N (2000) Comparison of UVA protection afforded by high sun protection factor sunscreens. J Am Acad Dermatol 43: 1036-1038.

11. Sayre RM, Agin PP, Levee GJ, Marlowe E (1979) A comparison of in vivo and in vitro testing of sunscreening formulas. Photochem Photobiol 29: 559.

12. Walters C, Keeney A, Wigal C, Johnston C, Cornelius R (1997) The Spectrophotometric Analysis and Modeling of Sunscreens. J Chem Edu 74: 99.

13. Dutra E, da Costa e Oliveira DG, Kedor-Hackmann E, Miritello R, Santoro (2004) In vitro sun protection factor determination of herbal oils used in cosmetics. Braz J Pharm Sci 40: 381.

14. Mbanga L, Mpiana PT, Mumbwa AM, Bokolo K, Mvingu K, et al. (2014) Determination of Sun Protection Factor (SPF) of Some Body Creams and Lotions Marketed in Kinshasa by Ultraviolet Spectrophotometry. J Phys Chem Sci 2: 1.

15. Azevedo JS, Viana NS, Vianna Soares CD (1999) UVA/UVB sunscreen determination by second-order derivative ultraviolet spectrophotometry. Farmaco 54: 573-578.

27

PMMA Platform Based Micro Fluidic Mixer for the Detection of MicroRNA-18a from Retinoblastoma Serum

Bindu Salim[1]*, Swapna Merlin David[1], Madhu Beta[2], Janakiraman Narayanan[2], Subramanian Krishnakumar[2] and Thalakkotur Lazar Mathew[1]

[1]*PSG Institute of Advanced Studies, Coimbatore, Tamil Nadu, India*

[2]*Vision Research Foundation, Sankara Nethralaya, Chennai, Tamil Nadu, India*

Abstract

Retinoblastoma (RB) is an eye cancer found in children. Early diagnosis of RB is very crucial for the treatment to be effective. miRNA 18a is highly expressed in the serum of RB patients which can be used as a diagnostic marker to detect RB. A pillar based micro mixer is designed and fabricated on PMMA platform to hybridize the target miRNA 18a with its complementary quenched probe throughout the flow. The fluorescence can be measured in the monitoring well by the fluorescent reader at a particular wavelength and compared with the controls. The RB serums showed high fluorescence than the controls. The results were validated with the contemporary technique, Real-Time PCR and the results were concurring. Another advantage of the device is that the effort required for the immobilization of the probe on a platform is not needed, at the same time cost and time required for diagnosis is reduced.

Keywords: Biomarker; Fluorescence reader; Hybridization probe; Micro fluidic mixer; MicroRNA; Retinoblastoma

Introduction

Retinoblastoma (RB) is a rapidly developing cancer that originates from the immature cells of a retina, the light-detecting tissue of the eye and is the most common malignant tumor of the eye in children typically before the age of five [1]. Normally, during the early stages of embryonic development, the eyes have cells called retinoblasts that divide into new cells and fill the retina. In some cases, instead of maturing into special cells that detect light, some retinoblasts continue to divide and grow out of control, forming a cancer known as RB. RB1 is a tumor suppressor gene involved in RB which normally regulates cell growth and keeps cells from dividing too rapidly or in an uncontrolled way. Mutations in the RB1 gene are responsible for most cases of RB. In about 1 out of 3 RBs, the abnormality in the RB1 gene is congenital and is in all the cells of the body, including all of the cells of both retinas. In most of these children, there is no family history of this cancer. Only about 25% of the children born with this gene change inherit it from a parent. In about 75% of children the gene change first occurs during early development in the womb. But in 2 out of 3 cases of RB, the abnormality in the RB1 gene develops on its own way and only in one cell of the eye which develops tumor only in one eye.

MicroRNAs (miRNA) are small, noncoding RNA molecules that have a major role in cellular functions. miRNA plays an extensive and important role in gene regulation. Currently, multiple miRNAs have been found abnormally expressed in cancer cells and closely associated with malignant cancer phenotypes [2]. miRNA expression can be associated with a variety of cancer, either functioning as oncogene or tumor suppressor. miR-15a, 16-1,17-5p,143,145, and let-7 were identified as a tumor suppressor while miR-18a,19a-b, 20a,21,92,155, and 372 were known as oncomiRs [3,4]. Indeed, miRNA has showed potential value in cancer diagnosis and therapy. Identification of the specific miRNA biomarkers associated with RB will help to establish new therapeutic approaches to save affected eyes in patients. The miR-18a, a member of oncogenic miR-17-92 cluster, has been found to over express in the serum of RB patients [5].

Various techniques, such as quantitative real-time PCR (qPCR), sequencing and microarrays [6], allow profiling miRNAs and several successful studies to detect miRNA with high sensitivity have been reported [7-10]. qPCR has an advantage of ultra-sensitive quantification besides microarray allows high-throughput screening. However, detection time, required sample volume, simplicity and portability of these techniques have not yet reached enough maturity for the diagnosis. To meet the desirable requirements, the use of micro fluidic devices, which conveys miRNA molecules to bind with locked nucleic acid (LNA) probe in micro channels, can be an attractive choice because it can reduce the assay time.

Based on this background knowledge, the study was aimed towards the development of highly sensitive and low cost diagnostic tool for detection of oncogenic RB from the serum of RB patients. In this context, we fabricated poly methyl methacrylate (PMMA) micro fluidic device for hybridizing miRNA 18a in the serum with a fluorescent labeled LNA probe specific to miRNA 18a.

Materials and Methods

Design and fabrication of micro mixer on PMMA platform

In microfluidic channel, the flow is laminar which does not support mixing. The Reynolds number in these flows is normally 10 and in some cases even 0.1. In a straight channel micromixer, mixing takes place only by diffusion. So a large length is required for mixing so that hybridization efficiency is good. If some features are made in microchannel, it creates the lamelle which enhances mixing and reduces channel length required. A pillar based micromixer is designed as shown in Figure 1 to hybridize the probe with target which is fluorescently labeled. In this design pillars are enhancing the mixing

*Corresponding authors: Bindu Salim, PSG Institute of Advanced Studies, Coimbatore, Tamil Nadu, India, E-mail: bbs@psgias.ac.in

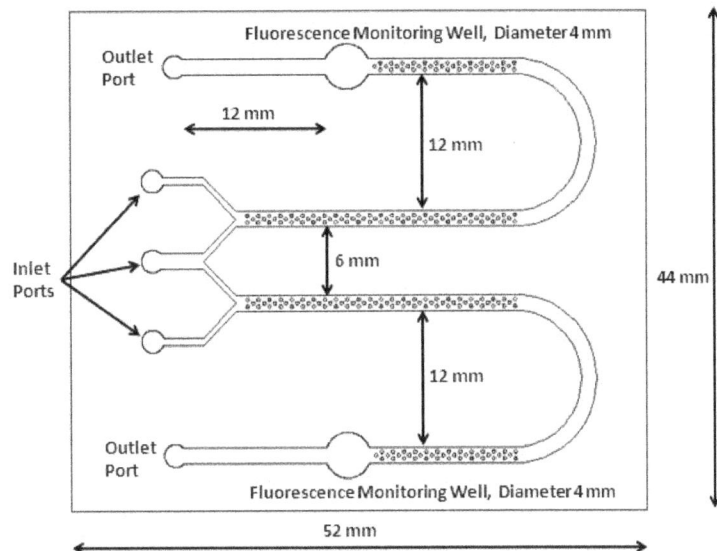

Figure 1: Schematic diagram of micro mixer. The fluorescence monitoring well, inlet and outlet ports are marked.

process. The micro mixer was fabricated on poly methyl methacrylate (PMMA) material. The overall dimension of biochip is 52 × 44 mm and holding volume of micro channel, from starting to the outlet port is 25 µl on each side. The micro channels are designed with the pillars of 300 µm diameter. The channel depth was 235 µm with a channel width of 1400 µm and holding volume of 25 µl on control and test inlet wells. The outlet wells of this micro mixer can hold the volume of 25 µl of hybridized sample. The fluorescence monitoring well is designed 12 mm before outlet port. The diameter of it is kept as 4 mm. It is covered from top to keep the fluid level constant to avoid error due to varying depth of fluid level. The distance between two arms of micro channel in each side is kept 12 mm to avoid stray measurement of fluorescence. Further a fluorescent detection system in which the micro device can be incorporated and the fluorescence can be read in the monitoring well at 665 nm was developed. The measured fluorescence is converted into voltage.

Evaluation of PMMA platform for the detection of miRNA-18a in control/patient serum

The blood samples (5 ml) were collected in the BD° Vacutainer SST tubes from 10 RB patients and 10 age matched controls. All the samples were in the age group of 6 months to 5 years. International Intraocular Retinoblastoma Classification (IIRC) staging and international retinoblastoma classification of these 10 RB samples revealed 8 tumors are in group E and 2 tumors are group E in one eye and group A in other eye; 8 tumors were unilateral and 2 tumors are bilateral. Control blood samples were diagnosed as non-cancerous patients. The blood samples were kept at room temperature for 15 minutes and the serum was separated from the blood by centrifuging at 3000 rpm for 15 minutes. Institutional Human ethical committee clearance (IHEC No. 14/173) was obtained from PSG Institute for Medical Science and Research, Coimbatore for collecting the blood samples and conducting the experiments. Informed consent was obtained from the parents of all the study subjects. The LNA™ oligonucleotide 5'/TYE665-CTATCTGCACTAGATGCACCTTA-3'/Dab (Exiqon, Denmark) was used as the complimentary probe for detecting miRNA 18a in the serum.

The concentration of the probe and the time required for the optimum hybridization was standardized. Different concentrations of the probe starting from 0.1 µM to 1 µM were hybridized with control and RB patient serum samples. The minimum concentration at which the distinct fluorescence difference between the control and RB sample observed was selected for further microfluidic experiments. The experiments were carried out at different time intervals from 10 minutes to 30 minutes by changing the flow rates to find out the optimum time for the hybridization.

A syringe pump is connected to the middle inlet well (Figure 2) with a 1 ml syringe and a dual syringe pump is connected to the other two inlet wells with 1 ml syringes. 250 µl of LNA probe specific for miRNA 18a with a final concentration of 0.5 µM was pumped simultaneously with 125 µl of the control and patient serum sample. The flow rate for the run was set as 2 µl/min for the probe and 1 µl/min for the serum samples.

Validation of micro fluidics based detection of oncogenic miRNAs in the serum by Real time PCR system

The total miRNA was isolated from the serum samples by miRNeasy° plasma/serum kit (Qiagen, USA) according to manufacturer's protocol. Initial volume of 200 µl was taken for the isolation. The isolated miRNA was quantified by nano spectrophotometer at 260 nm.

In the reverse transcription (RT) step, cDNA is reverse transcribed from total miRNA samples using a small RNA-specific, stem-loop RT primer from the TaqMan small RNA Assays and reagents from the TaqMan MicroRNA Reverse Transcription Kit (Applied Biosystems, USA). 10 ng of the miRNA was used for the synthesis of cDNA. The reaction parameters were 16°C for 30 min, 42°C for 30 min, and 85°C for 5 min and final hold at 4°C.

After the synthesis of cDNA, real time PCR (RT-PCR) was carried out for analyzing the gene expression of the control and RB samples by using Taqman probes labeled with FAM dye. The TaqMan° MGB probes (Applied Biosystems, USA) contain a reporter dye (FAM dye) linked to the 5'end of the probe, a non-fluorescent quencher (NFQ) at

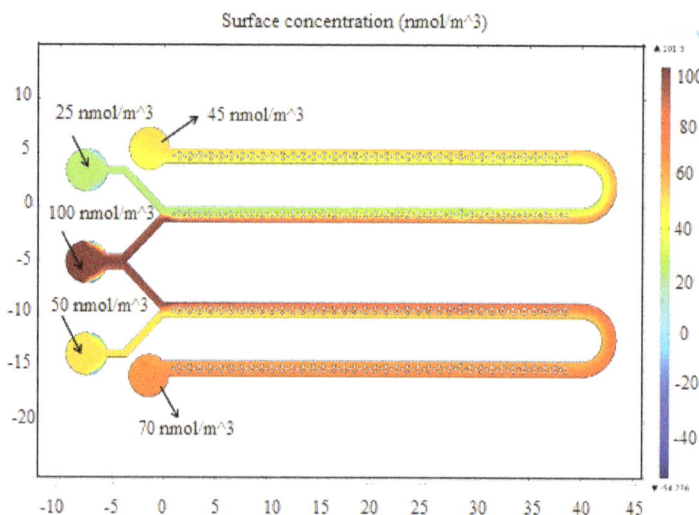

Figure 2: Simulation of micro mixing by COMSOL MultiPhysics 4.1 simulation software. The variation in the concentrations of the fluids after the flow is shown.

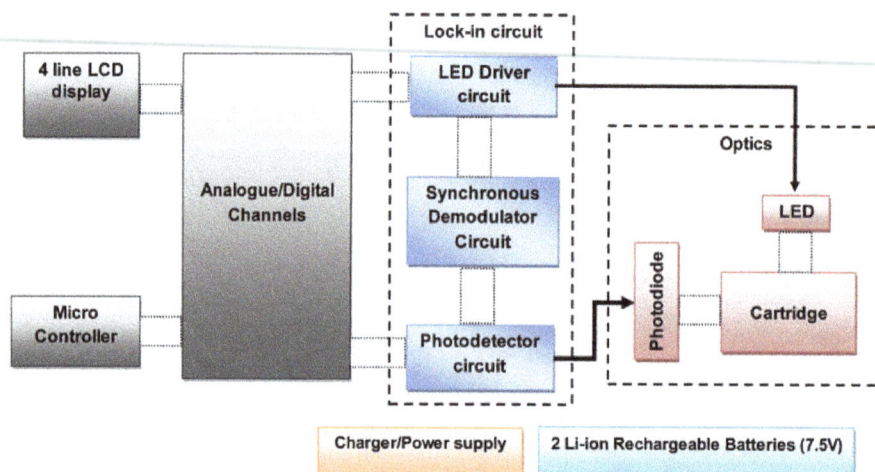

Figure 3: Schematic diagram of Retinoblastoma detection system.

the 3'end of the probe and a minor groove binder (MGB) at the 3'end of the probe. The reaction parameters for the RT-PCR was a hold at 95°C for 10 min and 40 cycles of 95°C for 15 sec and 60°C for 60 sec. snRNA U6 was used as the internal control.

Statistical analysis

Statistical analysis was done using SPSS 17.0 (SPSS Inc. Released 2008. SPSS Statistics for Windows, Version 17.0. Chicago: SPSS Inc). All data are expressed as the mean ± SD. All values ($p<0.05$) were considered to be statistically significant.

Results

Design and analysis of micro mixer

In microfluidic channel, the flow is laminar with Reynold's number less than 10. This flow cannot cause any mixing or surface interaction. In our system, the channel is provided with pillars two in a row followed by one which is repeated as seen in Figure 1. These structures will

enhance surface interaction and supports hybridization of miRNA18a with its conjugate quenched probe. Measurements using fluorescence microscope needs more skilled hands and the analysis of the results needs scientific knowledge. Hence we have developed the fluorescence reader for the particular wavelength of 660 nm using a lens assembly to focus the light source from a light emitting diode and photo detector to detect the output fluorescence.

The mixing behavior of micro mixer was studied using COMSOL MultiPhysics 4.1 simulation software. The simulation result is shown in Figure 2. The concentration of fluid in inlet wells were chosen as 25,100 and 50 nmol/m³ which corresponds to green, red and yellow legend colours respectively. Yellow (45 nmol/m³) and orange (70 nmol/m³) colours in outlet port represent complete mixing. The mixing process was gradual and starts when two fluids meets and completes flow till the outlet wells. The schematic diagram of the system developed is as shown in Figure 3 which consists of a microcontroller, analog to digital converter, lens assembly with LED and detector system. This is attached with a LCD to display the voltages from control well and sample well. The biological reaction occurring in the micro fluidic slide is illustrated in Figure 4. The target miRNA in the serum will hybridize with its complementary probe while flowing in the mixer channel which is tagged with a fluorophore, and the final fluorescence can be read by a reader in the collection well.

Photograph of the retinoblastoma detector is shown in Figure 5. This system is provided with rechargeable Li Ion battery and a charging circuitry. In this system, slide with sample is fed manually and fluorescence is measured in control well. Further with the help of a lever, the position of the slide is moved to focus the lens assembly on to the well with patient's sample and the fluorescence is measured. The

fluorescence is converted to electrical quantity using photo detector diode. The detector output is converted to digital data and is processed using a microcontroller. Here the back ground fluorescence is calibrated to zero volts by program control. The voltage range assigned is 0 to 5 volts mapping the molar fluorescence varying from 20 a.u to 300 a.u, where the molar concentration chosen is 0.5 µM for the probe which corresponds to a fluorescence of 150 a.u as seen in Figure 6a. The readings were displayed as voltages on the LCD screen for control and sample. This can also be displayed as difference of the control and sample, in the form of screening result positive or negative based on the threshold fixed. The body of the system was developed using 3D printing.

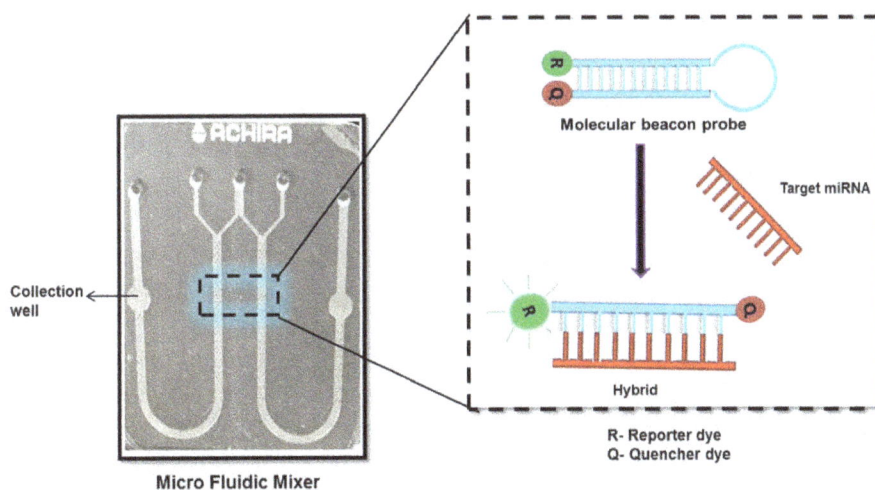

Figure 4: Detection of miRNA in serum by LNA probe hybridization.

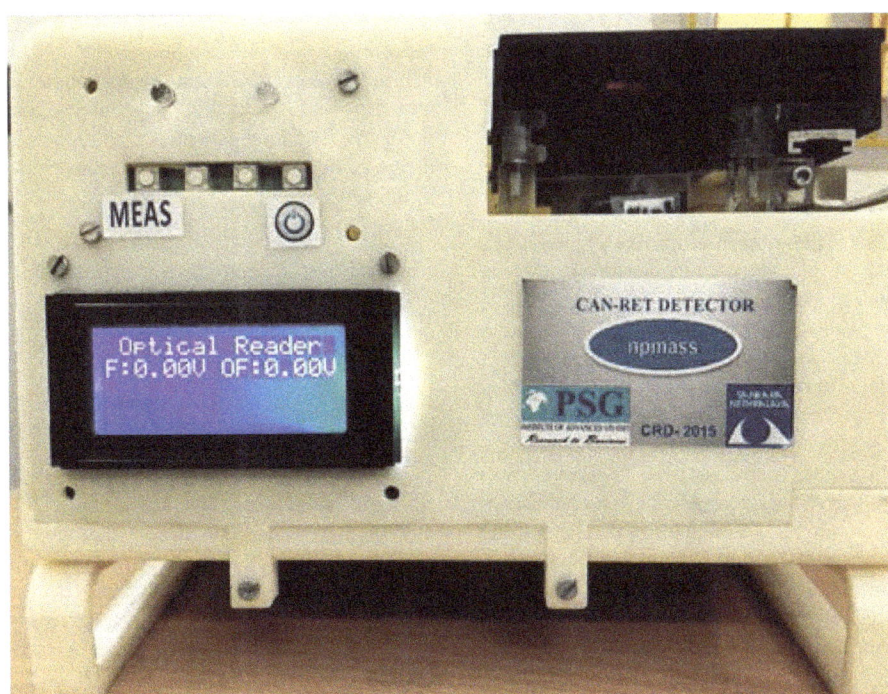

Figure 5: Image of Retinoblastoma detection system developed in this work.

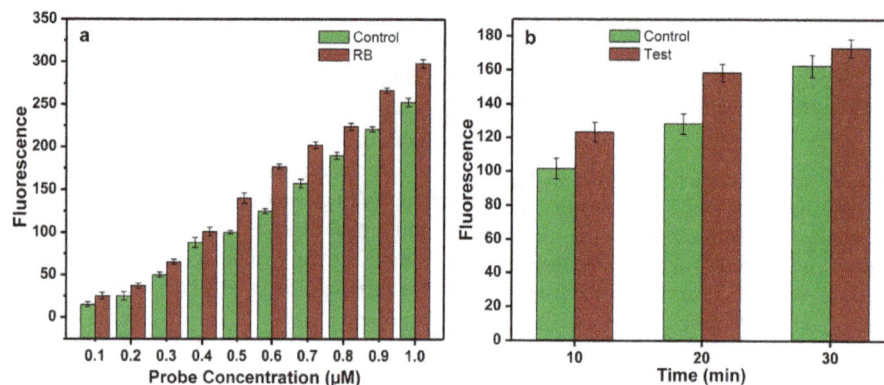

Figure 6: Standardized probe concentration (a) and time (b) required for miRNA 18a-LNA probe hybridization.

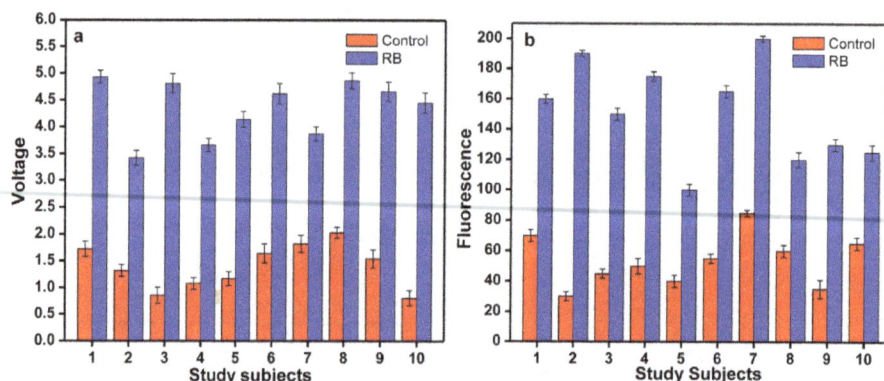

Figure 7: (a) The voltage difference between the control and RB measured by the developed detection system. (b) Difference in fluorescence between control and RB measured at the outlet port measured by a fluorescence microscope.

Standardization of probe concentration and time for hybridization

Different concentrations of the probes were hybridized with a constant volume (5 μl) of serum sample. The concentration ranges from 0.1 to 1 μM. The serum samples and probe were allowed to hybridize for 10 minutes. Remarkable difference in fluorescence was observed when the molar concentration is between 0.5 to 0.7 μM. Since 0.5 μM was the minimum concentration for the detection, it was selected for further experiments. Figure 6a shows the fluorescence difference between the control and the RB samples at different probe concentrations. After standardizing the probe concentration, experiments were carried out to find out the optimum time required for the hybridization. The time is very important in the assay because based on the optimum time; the flow rate is fixed in the micro fluidic chamber. Experiments were conducted at different time intervals. At 10 min and 20 min, difference in fluorescence was observed between the control and sample (Figure 6b). Also, the fluorescence difference between the control and the RB samples was measured using the developed detection system and an increased voltage was observed in case of all RB samples. The samples with voltage above 2.5V were considered as RB affected sample (Figure 7a). This system provides better accuracy in fluorescence reading since the measuring chamber is aligned for focused reading whereas while using fluorescence microscope possibility of error due to alignment of slide is high.

Hybridization of miRNA 18a and probe in micro fluidic mixer

Hybridization experiments were carried out with the fabricated device allowing flow of the control serum and the sample serum through the inlet wells at a flow rate at 1 μl/min. The probe was pumped to the probe inlet well at a flow rate of 2 μl/min. The final fluorescence was measured in the fluorescence monitoring well. The fluorescence was also measured at different points of the device to demarcate the fluorescence difference between control and RB. In all the points, an increased fluorescence was observed in the RB samples. The difference in fluorescence between the control and the RB samples are shown in Figure 7b.

Validation of hybridization experiments by real time PCR

The total miRNAs were isolated from the serum samples of the controls and RB patients. cDNA was synthesized from the isolated miRNA for each test and the control using specific primers for miRNA 18a. miRNA that shows constant expression between normal and diseased conditions is used as a reference gene. The cDNA of the reference gene U6 was also synthesized.

After the synthesis of cDNA, the RT-PCR was carried out with miRNA 18a specific Taqman probe for the control and RB samples. The reaction for U6 miRNA was also run in parallel for control and RB. The purpose of normalization with a reference gene is to remove as much variation as possible between groups except for that difference that is a consequence of the disease state itself. The relative ratio of the miRNA 18a expression in RB samples were compared with the control sample (Figure 8). The RB4 showed the highest expression, among other RB samples. The relative ratio is the normalized expression data of the

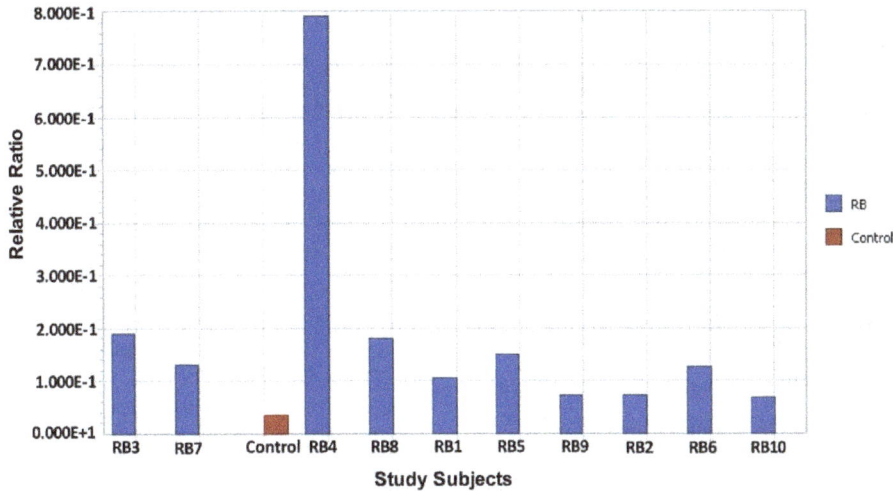

Figure 8: Relative miRNA 18a expression in control and RB. The RB4 is showing the highest expression, among other RB samples.

control and RB with the reference gene. The expression of all the RB samples was higher than the controls and these results agreed with the hybridization results with the mixer (Figure 7a and 7b).

Discussion

Biomarkers have been hailed as the future of medicine, which is progressing towards a greater focus on the diagnosis and prognosis of several diseases and assessing therapeutic response [11]. Micro fluidic devices have been widely applied for the analytical systems due to their fast response, low cost and small amounts of necessary, very expensive chemical bio receptors such as antibodies, aptamers or enzymes [12]. PMMA is a transparent thermoplastic often used as a lightweight or shatter-resistant alternative to glass. The PMMA is an elastomer with little deformation and this property can be used for the construction of micro channels when the rigidity is required. Also, PMMA has a good degree of compatibility with human tissue and its optical transparency is better when compared to other materials. Considering these advantages of PMMA, we used this material for fabricating the micro mixer device.

Reports suggest that mortality rate among the children due to RB is high and the delay in diagnosis increases this rate. miRNAs could be an ideal class of blood-based biomarkers for cancer detection because miRNA expression is frequently dysregulated in cancer and also the expression patterns of miRNAs in human cancer appear to be tissue-specific [13]. The miRNAs regulate the expression of genes by binding and modulating the translation of specific mRNAs. Several published reports have shown that the expression levels of some miRNAs are reduced in chronic lymphocytic leukemia, colonic adenocarcinoma, and Burkitt's lymphoma samples providing possible links between miRNAs and cancer [14]. Each tumor type can be readily distinguished from the accompanying normal samples based on the expression levels of miRNAs. While each tumor type is characterized by its own unique miRNA profile, it is interesting to note that several miRNAs appear to be up- or down-regulated in almost all tumor samples relative to normal adjacent tissue. Beta et al. [15] studied the miRNA profile of RB sample using *in silico* methods and found that 21 miRNAs are up-regulated and 24 are down regulated in RB.

Presently available diagnostic tools are based on the immobilization of the probe. Although this can be regarded as the most successful micro fluidic platform for lab-on-a-chip applications in terms of the number

of commercialized products, the drawbacks of the platform certainly overrides its simplicity. The exact timing of the assay steps depend on variations in viscosity and surface tension of the sample. Other crucial unit operations are metering and incubation, the accuracy of which is limited, and mixing, which cannot be accelerated on the test strip platform. Therefore the precision of the assay result, for example is of the order of 10%, which is not always sufficient for future challenges in the implementation of more complex diagnostic assays [16,17].

The micro fluidic device developed in this study certainly has the potential to become one of the foremost micro fluidic platforms for highly integrated applications. It is a flexible and configurable technology which stands out owing to its suitability for large scale integration. The experiments are conducted using age matching healthy children's sample as control. As seen in Figure 7a and 7b, the miRNA copy number varies from individual to individual in the control sample. This issue of control being a variable dependent on age and subject, has to be sorted out arriving at a model for the copy number of miRNA and a matching spiked serum can be used as control. However, the ten control samples considered for the experiment resulted in fluorescence less than 80 a.u and corresponding voltages less than 2 Volts whereas the RB patients' samples resulted in lowest fluorescence level of 100 and voltage of 3.5 Volts. Errors due to alignment problems arising with microscope and the time taken by individual to focus and align are less in case of the reader. The microfluidic channel performs with better efficiency of mixing compared to simple channels where the flow is confined to laminar [18,19].

Conclusion

The micro mixer developed can be used for the detection of different miRNAs by hybridizing with the respective complementary probes. Since it is based on the flow of the serum and probe, the cost, time and effort required for the immobilization is much reduced compared to other diagnostic tools available presently in the market. Screening for retinoblastoma from patients' serum is also appreciated since the RB victims are infants and sophisticated procedures can be avoided. Hence we strongly suggest that the micro fluidic device designed provides a proficient diagnostic tool in the diagnosis of human pathophysiology.

Acknowledgment

The authors acknowledge NPMASS-ADA, Govt. of India for the financial support to carry out this project. The project was partially supported by Programme support on retinoblastoma BT/01/CEIB/11/V/16. We also thank sincerely Dr. P. Radhakrishnan, Director, PSG Institute of Advanced studies (PSGIAS) for all the support extended to complete this work. Authors sincerely thank Dr. V. Narendran, Chief Medical Officer and Dr. Parag K Shah, Assistant Professor, Aravind Eye Hospital, Coimbatore, India for identifying the RB cases and providing the blood samples. We acknowledge Dr. Sudha Ramalingam, and Dr. Thiagarajan Sairam for their support at PSG Centre for Molecular Medicine and Therapeutics. We also thank PSG Institute of Medical Science and Research for providing ethical clearance for performing the experiments on the samples and providing age matching samples of healthy infants. We also acknowledge the support of Ms. Mamatha M Pillai and Ms. Elakkiya of Nano biotechnology lab, PSGIAS for their support for making fluorescence microscopy measurements.

References

1. Dimaras H, Kimani K, Dimba EA, Gronsdahl P, White A, et al. (2012) Retinoblastoma. Lancet 379: 1436-1446.

2. Jansson MD, Lund AH (2012) MicroRNA and cancer. Mol Oncol 6: 590-610.

3. Lu J, Getz G, Miska EA, Alvarez-Saavedra E, Lamb J, et al. (2005) MicroRNA expression profiles classify human cancers. Nature 435: 834-838.

4. Hsu TI, Hsu CH, Lee KH, Lin JT, Chen CS, et al. (2014) MicroRNA-18a is elevated in prostate cancer and promotes tumorigenesis through suppressing STK4 in vitro and in vivo. Oncogenesis 3: e99.

5. Tsang WP, Kwok TT (2009) The miR-18a microRNA functions as a potential tumor suppressor by targeting on K-Ras. Carcinogenesis 30: 953-959.

6. Baker M (2010) MicroRNA profiling: separating signal from noise. Nat Methods 7: 687-692.

7. Nelson PT, Baldwin DA, Scearce LM, Oberholtzer JC, Tobias JW, et al. (2004) Microarray-based, high-throughput gene expression profiling of microRNAs. Nat Methods 1: 155-161.

8. Miska EA, Alvarez-Saavedra E, Townsend M, Yoshii A, Sestan N, et al. (2004) Microarray analysis of microRNA expression in the developing mammalian brain. Genome Biol 5: R68.

9. Castoldi M, Schmidt S, Benes V, Hentze MW, Muckenthaler MU (2008) miChip: an array-based method for microRNA expression profiling using locked nucleic acid capture probes. Nat Protoc 3: 321-329.

10. Zhou WJ, Chen Y, Corn RM (2011) Ultrasensitive microarray detection of short RNA sequences with enzymatically modified nanoparticles and surface plasmon resonance imaging measurements. Anal Chem 83: 3897-3902.

11. Rinaldi A (2011) Teaming up for biomarker future. Many problems still hinder the use of biomarkers in clinical practice, but new public-private partnerships could improve the situation. EMBO Rep 12: 500-504.

12. Chang HC, Yeo LY (2010) Electro kinetically Driven Microfluidics and Nano fluidics. Cambridge University Press.

13. Esquela-Kerscher A, Slack FJ (2006) Oncomirs - microRNAs with a role in cancer. Nat Rev Cancer 6: 259-269.

14. Calin GA, Croce CM (2006) MicroRNA signatures in human cancers. Nat Rev Cancer 6: 857-866.

15. Beta M, Venkatesan N, Vasudevan M, Vetrivel U, Khetan V, et al. (2013) Identification and Insilico Analysis of Retinoblastoma Serum microRNA Profile and Gene Targets Towards Prediction of Novel Serum Biomarkers. Bioinform Biol Insights 7: 21-34.

16. Clark TJ, McPherson PH, Buechler KF (2002) The triage cardiac panel: cardiac markers for the triage system. Point of Care: The Journal of Near-Patient Testing & Technology 1: 42-46.

17. Yager P, Edwards T, Fu E, Helton K, Nelson K, et al. (2006) Microfluidic diagnostic technologies for global public health. Nature 442: 412-418.

18. Arata H, Komatsu H, Hosokawa K, Maeda M (2012) Rapid and Sensitive MicroRNA Detection with Laminar Flow-Assisted Dendritic Amplification on Power-Free Microfluidic Chip. PLoS ONE 7: e48329.

19. Arata H, Komatsu H, Han A, Hosokawa K, Maeda M (2012) Rapid microRNA detection using power-free microfluidic chip: coaxial stacking effect enhances the sandwich hybridization. Analyst 137: 3234-3237.

28

Validation of the Chromogenic Bioassay for the Potency Assessment of Streptokinase in Biopharmaceutical Formulations

Bruna Xavier[1], Raphael Leite Camponogara[2], Clóvis Dervil Appratto Cardoso Júnior[2], Rafaela Ferreira Perobelli[2], Mauricio Elesbão Walter[2], Fernanda Pavani Stamm Maldaner[2] and Sérgio Luiz Dalmora[1]*

[1]Department of Industrial Pharmacy, Federal University of Santa Maria, Santa Maria, RS, Brazil
[2]Postgraduate Program in Pharmaceutical Sciences, Federal University of Santa Maria, Santa Maria, RS, Brazil

Abstract

Streptokinase (STK) is a thrombolytic agent clinically used to treat patients with acute myocardial infarction and venous and arterial thrombosis. An *in vitro* chromogenic substrate end point bioassay was validated for the potency evaluation of biopharmaceutical formulations. The dose-response curve was linear over the concentration range of 2.50-40 IU/mL (r^2=0.999), with a quantitation limit of 2.50 IU/mL and a detection limit of 1.10 IU/mL, respectively. Specificity was established in studies with spiked samples. The accuracy was 100.34% with bias lower than 0.53%, and method validation demonstrated also acceptable results for precision and robustness. The validated method was applied to the potency assessment giving potencies between 92.20% and 108.97%. In addition, the activity of streptodornase and streptolysin were also evaluated giving values lower than 9.79 IU per 100 000 IU STK, and 1.47 for the absorbance ratio, respectively. The validated bioassay was applied in combination with the purity evaluation, contributing to assure the batch-to-batch consistency and quality of the bulk and finished biotechnology-derived medicine.

Keywords: Streptokinase; Chromogenic bioassay; Streptodornase; Streptolysin; Validation; Biotechnology-derived medicine

Introduction

Streptokinase (STK) is clinically used world-wide as a thrombolytic agent to treat patients with acute myocardial infarction, deep vein thrombosis, arterial thrombosis and embolism [1-3]. The structure of STK consists of a 414 amino acids polypeptide chain with a molecular mass of 47 kDa. The protein exhibits its maximum activity at a pH of approximately 7.5 and its pI is 4.7. Most of the native Streptokinases (STKs) are obtained from pathogenic β-hemolytic streptococci A, C and G, being the group C preferred as they lack erythrogenic toxins. Recombinant STKs have been produced with reduced immunogenicity, and the gene from *Streptococcus equisimilis* H46A was first cloned and expressed in *E. coli* releasing substantial amounts of STK into the culture medium [4-6]. Besides, the STK isolated hitherto may contain the enzymes streptodornase and streptolysin O, which are active even in small quantities [7].

The biological activity has been assessed by the *in vitro* fibrin clot lysis assay and the chromogenic plasminogen activation substrate assay, that were also used in an international collaborative study organized to establish the 3rd international Standard for streptokinase. It was demonstrated that the chromogenic substrate assay (SCSA), due to its ability to activate the fibrinolytic system, converting plasminogen to plasmin, is a suitable procedure for the potency determination of the streptokinase preparations [8,9].

Sixteen preparations of STK available world-wide for clinical use were compared by a SCSA and SDS-PAGE electrophoresis showing wide variations of the activities, purity and composition [10]. Potencies of different preparations of STK were also evaluated by the euglobulin lysis test and the SCSA, showing significant variations between the products available for clinical use [11]. The gene from *Streptococcus equisimilis* was cloned in a vector of *E. coli* to overexpress the profibrinolytic protein, and almost all the recombinant STK was exported to the periplasmic space and the bioactivity was evaluated by the chromogenic assay [12]. STK expressed as inclusion body in *E. coli* was refolded into active forms proteins, purified and characterized

by chromatographic methods, MALDI-TOF, and the bioactivity was evaluated by the chromogenic assay [5].

Streptokinase produced from species of *Streptococcus pyogenes* was quantified by the method of Lowry, its electrophoretic mobility and molecular weight determined by SDS-PAGE, and the biological activity evaluated by the radial caseinolysis assay [13]. Functional characteristics such as substrate specificity and the effects of pH and temperature on the activity of Streptodornase in marketed product, against the native double stranded DNA were evaluated showing a possible existence of semi-denatured "meta-stable" conformations with reduced levels of DNase activity [7]. But, the validation of the method recommended for biopharmaceutical products, is essential to show that the procedure is suitable for its intended purpose [14].

The aim of this article was to validate a specific, sensitive and stability-indicating chromogenic substrate assay to assess the potency of streptokinase; carry out *in vitro* bioassays to evaluate streptodornase and streptolysin present in the product; thus contribute to improve the quality control and to assure the therapeutic efficacy of the biotechnology-derived product.

Experimental

Chemicals and reagents

The 3rd international standard of streptokinase (IS-STK WHO

*****Corresponding author:** Sérgio Luiz Dalmora, Department of Industrial Pharmacy, Federal University of Santa Maria, Santa Maria, RS, Brazil
E-mail: sdalmora@terra.com.br

00/464), containing 1030 IU/vial, the 2nd international standard of streptodornase (IS-STD WHO 08/230) and human anti-streptolysin O were obtained from the National Institute for Biological Standards and Control-NIBSC (Hertz, UK). A total of twelve batches of streptokinase, containing 1 500 000 IU/vial and 750 000 IU/vial of streptokinase were obtained, respectively, from Bergamo and Blausiegel (São Paulo, Brazil). The samples were acquired from commercial sources and used within their shelf life period. Plasminogen of bovine plasma, sodium deoxyribonucleate, human serum albumin (HSA), imidazole, and glutamate were acquired from Sigma-Aldrich˙ (St. Louis, USA). Chromogenic substrate S2251 was purchased from Chromogenix˙ (Milan, Italy). Tris (hydroxymethyl) aminomethane, sodium thioglycolate, sodium phosphate dibasic anhydrous, calcium sulphate, magnesium sulphate, sodium hydroxide, hydrochloric acid, perchloric acid, and acetic acid were obtained from Merck˙ (Darmstadt, Germany). All chemicals used were of pharmaceutical or special analytical grade. Ultrapure water was obtained using an Elix 3 coupled to a Mili-Q Gradient A10 system Millipore (Bedford, USA).

Apparatus

The absorbances of the assays were measured on a Multiskan FC microplate reader Thermo Scientific˙ (Vantaa, Finland), and on a UV-1601 PC-UV-VISIBLE Spectrophotometer Shimadzu˙ (Kyoto, Japan).

Procedure

Samples and standard solutions: Working standard and sample solutions of STK were prepared daily for the chromogenic assay, by diluting the IS-STK and the samples in 25 mM tris (hydroxymethyl) aminomethane with HSA solution at pH 7.7, to final concentrations between 2.5 and 40 IU/mL. Solutions for the streptodornase assay were prepared by diluting the IS-STD and the samples in imidazole buffer solution (IBS) pH 6.5 to obtain, respectively, solutions containing 20 IU/mL and 150 000 IU/mL of STK. For the streptolysin assay, solution of the sample was diluted in phosphate buffer solution pH 7.2, to obtain a final concentration of 1 000 000 IU/mL of STK.

Streptokinase chromogenic substrate assay (SCSA): The bioassay was performed as described elsewhere for the concentrated solution [9], modified accordingly. Volumes of 25 μL of the IS-STK and the sample solutions with concentrations between 2.5 and 40 IU/mL, were added to the 96-well plate, respectively in triplicate, and allowed to equilibrate at 37°C in water-bath for 1 min. Then, 100 μL of the chromogenic substrate (diluted 1:1 in water) were added to each well, and the plate was incubated for exactly 2 min, followed by the addition of 50 μL of the 1 mg/mL plasminogen. The reaction was stopped 10 min after, by adding 90 μL of 20% acetic acid. The absorbance was measured at 405 nm in the microplate reader and the biological potencies were calculated against the IS-STK by the parallel line statistical method using the CombiStats software (European Directorate for the Quality of Medicines and HealthCare, EDQM Council of Europe).

Streptodornase assay: The bioassay was performed as described elsewhere [9], adapted. Briefly, volumes of 0.5 mL of 1 mg/mL sodium deoxyribonucleate solution in IBS pH 6.5, were added in duplicate to six centrifuge tubes, followed by the addition of 0.25 mL, 0.125 and 0 mL of IBS, respectively. Then, was added 0.25 mL of the sample solution in all the tubes, followed by the addition in sequence, of 0 mL, 0.125 mL and 0.25 mL of the 20 IU/mL solution of IS-STD. The solutions were mixed up and heated at 37°C for 15 min. Two additional tubes were prepared by adding 0.25 mL of IBS and 0.25 mL of the sample solution, maintained without incubation. Then, 3.0 mL of 2.5% perchloric acid were added to all of the tubes, mixed, centrifuged at about 3000

g for 5 min, and the absorbances of the supernatant measured at 260 nm. The mean of the absorbances measured for the 0.25 mL sample concentration spiked in duplicate each one with 0.125 and 0.25 mL of the IS-STD, respectively, was subtracted from the sum of the two absorbances obtained only with the sample. This result was compared and should be higher than the value obtained as a difference between the sample with and without incubation, which means that the sample comply with the requirement that specify a maximum of 10 IU of streptodornase per 100 000 IU of STK.

Streptolysin assay: The bioassay was performed as described elsewhere [9], adjusted. Briefly, a volume of 0.5 mL, equivalent to 500 000 IU of STK of the sample prepared with the diluent composed by 1 volume of phosphate buffer solution pH 7.2 and 9 volumes of a 0.9% sodium chloride, was transferred to a polystyrene tube. A reference solution was prepared in parallel using only the diluent. Then, 0.4 mL of a 2.3% solution of sodium thioglycolate was added, and heated in a water-bath at 37°C for 10 min. A volume of 0.1 mL of a solution of human antistreptolysin O containing 5 IU/mL was pipette, and heated at 37 °C for 5 min. Then, 1 mL of rabbit erythrocyte suspension was added, heated at 37°C for 30 min, and centrifuged at about 1000 g for 10 min. The absorbance of the supernatant was measured at 550 nm, and should be not more than 1.5 times higher than that of the reference solution.

Validation of the Streptokinase chromogenic assay

The assay was validated using samples of a biopharmaceutical formulations of streptokinase with a label claim of 1 500 000 IU/vial and 750 000 IU/vial by determinations of the following parameters: linearity, range, precision, accuracy, detection limit (DL), quantitation limit (QL), robustness and stability, following the guidelines adapted for the *in vitro* bioassay [14,15].

Linearity: Linearity was determined for the assay by constructing three analytical curves, each one with eight concentrations of the IS-STK over the 2.50-40 IU/mL range. The absorbance's were plotted against the respective concentrations of streptokinase to obtain the analytical curve. The results were subjected to regression analysis by the least squares method to calculate the calibration equation and determination coefficient.

Precision: Assay precision was determined by means of repeatability (intra-days) and intermediate precision (interday). Repeatability was examined by six evaluations of the same streptokinase concentration, on the same day, under the same experimental conditions. The intermediate precision of the method was assessed by analysis of two samples on three different days (interday) and also by submitting the samples to analysis by other analysts in the same laboratory (between-analysts).

Accuracy: The accuracy was evaluated by applying the proposed assay to the analysis of pharmaceutical solutions with concentrations at 1 200 000, 1 500 000 and 1 800 000 IU/mL equivalent to 80, 100 and 120% of the nominal analytical concentrations, respectively. The accuracy was calculated as the percentage of the drug recovered from the formulation; it was expressed as the percentage relative error (bias %) between the measured mean concentrations and the added concentrations.

Limits of detection and quantitation: The detection limit (DL) and the quantitation limit (QL) were calculated by using the mean values of the three independent analytical curves, determined by a linear regression model, where the factors 3.3 and 10 for the detection and quantitation limits, were multiplied by the ratio of the standard deviation of the intercept and the slope, respectively. The QL was also evaluated in an experimental assay.

Robustness: The robustness of an analytical procedure refers to its ability to remain unaffected by small and deliberate variations in method parameters and provides an indication of its reliability for the routine analysis. The robustness of the SCSA was determined by analyzing the same samples containing 1 500 000 IU/mL and 750 000 IU/mL, respectively, under a variety of conditions of the assay parameters, such as: reaction time with plasminogen (5, 10 and 15 minutes), reaction time with substrate chromogenic (1, 2 and 3 minutes), pH of buffer solution (pH 7.4, 7.7 and pH 8.0), assay temperature (34, 37, 40°C) and stability of the analytical solution at 2-8°C.

Results and Discussion

Method validation

The procedure was performed to demonstrate that the performance characteristics of the SCSA meet the requirements for the potency assessment of streptokinase in biopharmaceutical formulations. The dose-response curve was constructed plotting the experimental values of absorbances versus the logarithms of the concentrations in triplicate. The analytical curves were found to be linear over the concentration range of 2.50-40 IU/mL. The determination coefficient calculated from y=(55817 ± 218.57)x-(103001 ± 8245.10), where, x is the concentration and y is the absorbance, was r^2=0.999, indicating the linearity of the analytical curve for the assay.

Specificity of the assay for the biomolecule was assessed by determination of the potency of the samples spiked with higher concentrations of the excipients, HSA, glutamate and sodium phosphate. In addition, the samples were also spiked with volumes equivalent to 2 IU/mL of heparin, 2 IU/mL of enoxaparin, 1 EU/mL of bacterial endotoxins, 0.87 nKat/mL of factor Xa, and 1.25 IU/mL of factor IIa, showing non-significant differences (p>0.05).

The precision of the SCSA was studied by calculating the relative standard deviation (RSD %), for six analyses at a concentration of 1 500 000 IU/mL, performed on the same day and under the same experimental conditions. The obtained RSD was 1.62%. The intermediate precision was assessed by analysis of two samples of the biopharmaceutical formulation on three different days (interday), giving RSD values of 0.57 and 0.48%, respectively (Table 1). Between-analysts precision was determined by calculating the mean values and the RSD after analysis of two samples of the same biopharmaceutical formulation by three analysts; the values were found to be 0.34 and 0.66%, respectively, as given in Table 1.

The accuracy of the SCSA was assessed from three replicate determinations of three solutions at concentrations of 1 200 000, 1 500 000 and 1 800 000 IU/mL, respectively. The absolute means obtained with a mean value of 100.34, with bias lower than 0.53% (Table 2), confirmed that the method is accurate within the desired range.

The DL and QL of the SCSA were calculated from the slope and the standard deviation of the intercept determined by a linear-regression model, by using the mean values of the three independent calibration curves. The obtained values were 1.10 and 2.61 IU/mL, respectively. The experimental value determined for the QL was found to be 2.50 IU/mL.

The results of the bioassay and the experimental range of the selected variables evaluated in the robustness test are given in Table 3, together with the optimized values. There were no significant changes in the potency results when modifications were introduced into the experimental conditions, thus showing the assay to be robust. The stability of the STK sample solutions was assessed and the data obtained showed non-significant changes, relative to freshly prepared samples, when maintained at 2-8°C for 24 h.

Method application

The validated SCSA was applied to the potency assessment of streptokinase in biopharmaceutical products, giving values within 92.20 and 108.97% of the stated potency, as shown in Table 4, meeting the specifications which claim 90-110% of the stated potency [8]. In addition, as the production of STK can be accompanied by the formation of streptodornase, the activity was assessed by the *in vitro* assay, giving results lower than 9.79 IU per 100 000 IU of STK, as demonstrated in Table 5. Due to the interference of the impurities, the assay was performed with the samples spiked with the IS-STD, calculating the content based on the difference between the absorbances. The streptolysin activity was also evaluated showing absorbances up to 1.47 times higher than the reference solution, in accordance with the specifications, showing the quality of the products.

Conclusion

The results of the validation studies show that the chromogenic substrate assay is specific, sensitive, with a QL of 2.50 IU/mL, and possesses excellent linearity and precision characteristics, and were successfully applied for the potency assessment of STK in biological products. In addition, the results obtained with the bioassays performed to evaluate also the presence of streptodornase and streptolysin, contribute to ensure batch-to-batch consistency and the quality of the biotechnology-derived medicine.

Sample	Inter-days			Between-analysts		
	Day	Recovery[a] %	RSD[b] %	Analysts	Recovery[a] %	RSD[b] %
1	1	99.83		A	100.97	
	2	99.12	0.57	B	99.42	0.34
	3	101.57		C	100.15	
2	1	100.75		A	98.45	
	2	101.07	0.48	B	99.57	0.66
	3	100.88		C	99.12	

[a]Mean of three replicates
[b]RSD=relative standard deviation

Table 1: Inter-days and between-analysts precision data of chromogenic substrate assay for streptokinase in biopharmaceutical formulations.

Nominal concentration IU/mL	Mean concentration measured[a] IU/mL	RSD[b] %	Accuracy %	Bias[c] %
1 200 000	1 204 560	2.16	100.38	0.38
1 500 000	1 501 500	1.69	100.10	0.10
1 800 000	1 809 540	1.87	100.53	0.53

[a]Mean of three replicates
[b]RSD=relative standard deviation
[c]Bias=[(measured concentration-nominal concentration)/nominal concentration] × 100

Table 2: Accuracy of chromogenic substrate assay for streptokinase in the biopharmaceutical formulations.

Variable	Range investigated	STK[a] %	Confidence interval (P=0.95)	RSD[b] %	Optimized value
Plasminogen reaction time	5 minutes	100.72	96.41-109.11	1.44	10 minutes
	10 minutes	100.12	97.23-113.65	0.82	
	15 minutes	101.17	94.42-111.59	1.17	
Chromogenic substrate incubation	1 minute	100.51	98.62-102.14	1.04	2 minutes
	2 minutes	100.31	92.51-109.17	0.41	
	3 minutes	99.00	89.91-109.03	0.88	
Buffer pH	pH=7.4	98.48	87.88-110.24	0.95	pH 7.7
	pH=7.7	99.80	98.40-101.20	0.36	
	pH=8.0	101.12	91.25-112.03	1.14	
Assay temperature	34°C	102.01	92.11-112.97	1.12	37°C
	37°C	99.92	90.13-110.82	0.69	
	40°C	98.33	89.94-107.75	1.27	
Solution stability	Initial	102.90	98.01-120.83	0.75	-
	24 hours (2-8°C)	100.91	96.57-120.88	0.91	

[a]Mean of three replicates
[b]RSD=relative standard deviation

Table 3: Conditions and range investigated during robustness testing with the one-variable-at-a-time (OVAT) procedure for the streptokinase (STK) assay.

Sample	Potency Stated IU/vial	Potency Found[a]		Confidence Intervals (P=0.95)
		IU/vial	%	
1	1 500 000	1 515 150	101.01	96.40-105.80
2	1 500 000	1 582 200	105.48	100.10-111.10
3	1 500 000	1 383 000	92.20	85.90-98.90
4	1 500 000	1 576 950	105.13	97.20-113.60
5	1 500 000	1 589 100	105.94	102.90-109.10
6	1 500 000	1 600 350	106.69	102.50-110.90
7	1 500 000	1 499 550	99.97	93.21-111.65
8	1 500 000	1 476 300	98.42	95.00-101.90
9	1 500 000	1 569 750	104.65	100.30-109.00
10	750 000	806 475	107.53	103.50-111.60
11	750 000	798 000	106.40	97.50-116.10
12	750 000	817 275	108.97	98.10-115.10
Mean	-	-	103.53	-
SD[b]	-	-	4.77	-

[a]Mean of three replicates
[b]SD=Standard deviation

Table 4: Potency, confidence intervals (P=0.95) of streptokinase in biopharmaceutical products by the chromogenic substrate assay.

Sample	Streptodornase		Streptolysin
	Absorbances	Activity	Absorbances
	Sample<Standard+Sample	IU/100 000 IU STK	Sample/Reference Solution
1	0.595<0.815	7.29	1.17
2	0.624<0.792	7.88	1.30
3	0.606<0.803	7.55	1.27
4	0.560<0.933	6.00	1.27
5	0.720<0.875	8.23	1.29
6	0.550<0.840	6.55	1.13
7	0.637<0.859	7.41	1.18
8	0.690<0.705	9.79	1.47
9	0.730<0.825	8.85	1.16
10	0.520<0.945	5.51	1.17
11	0.670<0.980	6.84	1.17
12	0.750<0.940	7.97	1.28

Table 5: Activity evaluation of streptodornase and streptolysin in biopharmaceutical products by *in vitro* bioassays.

Acknowledgements

The authors wish to thank CNPq (Conselho Nacional de Desenvolvimento Científico e Tecnológico) Projects 477013/2011 and 306898/2011-0 for financial support.

References

1. Brogden RN, Speight TM, Avery GS (1973) Streptokinase: a review of its clinical pharmacology, mechanism of action and therapeutic uses. Drugs 5: 357-445.

2. Kunamneni A, Abdelghani TT, Ellaiah P (2007) Streptokinase--the drug of choice for thrombolytic therapy. J Thromb Thrombolysis 23: 9-23.

3. Butcher K, Shuaib A, Saver J, Donnan G, Davis SM, et al. (2013) Thrombolysis in the developing world: is there a role for streptokinase? Int J Stroke 8: 560-565.

4. Malke H, Ferretti JJ (1984) Streptokinase: cloning, expression, and excretion by Escherichia coli. Proc Natl Acad Sci U S A 81: 3557-3561.

5. Cherish Babu PV, Srinivas VK, Krishna Mohan V, Krishna E (2008) Renaturation, purification and characterization of streptokinase expressed as inclusion body in recombinant E. coli. J Chromatogr B Analyt Technol Biomed Life Sci 861: 218-226.

6. Huang TT, Malke H, Ferretti JJ (1989) Heterogeneity of the streptokinase gene in group A streptococci. Infect Immun 57: 502-506.

7. Locke IC, Carpenter BG (2004) Functional Characteristics of the Streptococcal Deoxiribonuclease 'Streptodornase', a Protein with DNase Activity Present in the Medicament Varidase®. Enzyme Microb Technol 35: 67-73.

8. Sands D, Whitton CM, Longstaff C (2004) International collaborative study to establish the 3rd International Standard for Streptokinase. J Thromb Haemost 2: 1411-1415.

9. European Pharmacopoeia (2014) 8th edn. Strasbourg: Council of Europe.

10. Hermentin P, Cuesta-Linker T, Weisse J, Schmidt KH, Knorst M, et al. (2005) Comparative analysis of the activity and content of different streptokinase preparations. Eur Heart J 26: 933-940.

11. Couto LT, Donato JL, de Nucci G (2004) Analysis of five streptokinase formulations using the euglobulin lysis test and the plasminogen activation assay. Braz J Med Biol Res 37: 1889-1894.

12. Avilán L, Yarzábal A, Jürgensen C, Bastidas M, Cruz J, et al. (1997) Cloning, expression and purification of recombinant streptokinase: partial characterization of the protein expressed in Escherichia coli. Braz J Med Biol Res 30: 1427-1430.

13. Felsia XF, Vijayakumar R, Kalpana S (2011) Production and partial purification of streptokinase from Streptococcus pyogenes. J Biochem Tech 3: 289-291.

14. ICH (2005) Validation of Analytical Procedure: Text and Methodology Q2 (R1). International Conference on Harmonization of Technical Requirements for Registration of Pharmaceuticals for Human Use.

15. FDA (2015) Guidance for Industry: Analytical Procedures and Methods Validation for Drug and Biologics.

Permissions

All chapters in this book were first published in JABT, by OMICS International; hereby published with permission under the Creative Commons Attribution License or equivalent. Every chapter published in this book has been scrutinized by our experts. Their significance has been extensively debated. The topics covered herein carry significant findings which will fuel the growth of the discipline. They may even be implemented as practical applications or may be referred to as a beginning point for another development.

The contributors of this book come from diverse backgrounds, making this book a truly international effort. This book will bring forth new frontiers with its revolutionizing research information and detailed analysis of the nascent developments around the world.

We would like to thank all the contributing authors for lending their expertise to make the book truly unique. They have played a crucial role in the development of this book. Without their invaluable contributions this book wouldn't have been possible. They have made vital efforts to compile up to date information on the varied aspects of this subject to make this book a valuable addition to the collection of many professionals and students.

This book was conceptualized with the vision of imparting up-to-date information and advanced data in this field. To ensure the same, a matchless editorial board was set up. Every individual on the board went through rigorous rounds of assessment to prove their worth. After which they invested a large part of their time researching and compiling the most relevant data for our readers.

The editorial board has been involved in producing this book since its inception. They have spent rigorous hours researching and exploring the diverse topics which have resulted in the successful publishing of this book. They have passed on their knowledge of decades through this book. To expedite this challenging task, the publisher supported the team at every step. A small team of assistant editors was also appointed to further simplify the editing procedure and attain best results for the readers.

Apart from the editorial board, the designing team has also invested a significant amount of their time in understanding the subject and creating the most relevant covers. They scrutinized every image to scout for the most suitable representation of the subject and create an appropriate cover for the book.

The publishing team has been an ardent support to the editorial, designing and production team. Their endless efforts to recruit the best for this project, has resulted in the accomplishment of this book. They are a veteran in the field of academics and their pool of knowledge is as vast as their experience in printing. Their expertise and guidance has proved useful at every step. Their uncompromising quality standards have made this book an exceptional effort. Their encouragement from time to time has been an inspiration for everyone.

The publisher and the editorial board hope that this book will prove to be a valuable piece of knowledge for researchers, students, practitioners and scholars across the globe.

List of Contributors

Yusuke Suzuki, Akira Okamoto and Yasunori Kushi
College of Science and Technology, Nihon University, Tokyo, Japan

Anila Mathew
College of Science and Technology, Nihon University, Tokyo, Japan
Anila Mathew's current address is Bio-Nano Electronics Research Centre, Toyo University, 2100, Kujirai, Kawagoe, Saitama 350-8585, Japan

Magda A Akl and Abdel-fattah M Youssef
Chemistry Department, Faculty of Science, Mansorua University, Egypt

Ali M Abou-Elanwar and Magda D Badri
National Research Centre, Dokki, Giza, Egypt

Prafulla Kumar Sahu
Department of Pharmaceutical Analysis and Quality Assurance, Raghu College of Pharmacy, Dakamarri, Bheemunipatnam, Visakhapatnam-531 162, Andhra Pradesh, India

TS Sunil Kumar Naik, BE Kumara Swamy, CC Vishwanath and Mohan Kumar
Department of PG Studies and Research in Industrial Chemistry, Kuvempu University, JnanaSahyadri, Shankaraghatta, Shivamoga, Karnataka, India

Deborah A. Sarkes, Amethist S. Finch and Dimitra N. Stratis-Cullum
U.S. Army Research Laboratory, Sensors and Electron Devices Directorate, Adelphi MD, USA

Brandi L. Dorsey
Federal Staffing Resources, Annapolis MD, USA

Yaqian Yan, Linjing Wu, Qianqiong Guo and Shasheng Huang
Life and Environmental Science College, Hanghai Normal University, Shanghai, PR China

Azam Rezvanirad
Faculty of Pharmacy, Research Center, Shahid Beheshti University of Medical Sciences, Tehran, Iran

Mehdi Rajabnia Khansari
Faculty of Pharmacy, Research Center, Shahid Beheshti University of Medical Sciences, Tehran, Iran

School of Chemical Engineering, Research Center, Iran University of Science and Technology, Tehran, Iran

Amin Nikavar
School of Chemical Engineering, Research Center, Iran University of Science and Technology, Tehran, Iran

Shahrzad Bikloo
Lorstan University of Medical Sciences, Research Center, Khoramabad, Iran

Sara Shahreza
Department of Nanobiotechnology, Tarbiat Modates University, Tehran, Iran

Bahareh Sadat Yousefsani
Department of Pharmacodynamy and Toxicology, School of Pharmacy, Pharmaceutical Research Center, Mashhad University of Medical Sciences, Mashhad, Iran

Rita Mastroianni, Marina Feroggio, Barbara Marsiglia, Clarissa Porzio Vernino, Simona Riva and Luca Barbero
QPD - NBE Bioanalytics, RBM-Merck Serono, Via Ribes 1, 10010, Colleretto Giacosa (TO), Italy

Araceli Espinoza-Vázquez, Sergio Garcia-Galan and Francisco Javier Rodríguez-Gómez
Faculty of Chemistry, Department of Metallurgical Engineering, Universidad Nacional Autónoma de México, C.U., Distrito Federal, 04510, Mexico

Dias IARB, Saciloto TR, Cervini P and Cavalheiro ETG
Departamento de Química e Física Molecular, Instituto de Química de São Carlos, Av. Trabalhador São Carlense, 400, Centro, São Carlos, São Paulo CEP 13566-590, Brazil

Eva Abramov, Ouri Schwob and Ofra Benny
The Institute for Drug Research, The School of Pharmacy, Faculty of Medicine, The Hebrew University of Jerusalem, Jerusalem, Israel

Angelo Luiz Gobbi
Laboratory of Microfabrication, National Nanotechnology Laboratory, National Center for Research in Energy and Materials, Campinas, Sao Paulo, Brasil

Gabriela Furlan Giordano, Karen Mayumi Higa, Adriana Santinom and Renato Sousa Lima
Laboratory of Microfabrication, National Nanotechnology Laboratory, National Center for Research in Energy and Materials, Campinas, Sao Paulo, Brasil
Instituto of Chemistry, State University of Campinas, Campinas, Sao Paulo, Brasil

Lauro Tatsuo Kubota
Instituto of Chemistry, State University of Campinas, Campinas, Sao Paulo, Brasil
Instituto National Science and Technology Bioanalytics, Campinas, Sao Paulo, Brasil

Ramos TM
Department of Animal and Food Sciences, University of Delaware, Christina Mill Drive, Newark, Delaware, United States of America

Costa FF
Department of Food Science, Federal University, Juiz de Fora, Brazil

Pinto ISB
Department Animal and Food Sciences, Embrapa Gado de Leite, Juiz de Fora, Brazil

Pinto SM and Abreu LR
Department of Food Science, Federal University of Lavras, Lavras, Brazil

Zina Guermazi and Slimane Gabsi
National School of Engineering, University of Sfax, Route de Soukra, 3038 Sfax, Tunisia

Mariem Gharsallaoui
Olive Tree Institute, University of Sfax, 3000 Sfax, Tunisia

Enzo Perri and Cinzia Benincasa
Consiglio per la ricerca in agricoltura e l'analisi dell'economia agraria, Centro di ricerca per l'olivicoltura e l'analisi dell'economia agraria, Italy

Veniero Gambaro, Gabriella Roda, Giacomo Luca Visconti, Sebastiano Arnoldi, Eleonora Casagni, Caterina Ceravolo, Lucia Dell'Acqua, Fiorenza Farè, Chiara Rusconi and Lucia Tamborini
Department of Pharmaceutical Sciences, University of Milan, Via Mangiagalli 25, Milan, Italy

Stefania Arioli and Diego Mora
Department of Food Science and Technology and Microbiology, Via Celoria 2, Milan, Italy

Dan Jin
Biomedical Engineering Department, University of Alberta, Edmonton, Alberta, Canada
Labs-Mart Inc., Edmonton, Alberta, Canada

Jie Chen
Biomedical Engineering Department, University of Alberta, Edmonton, Alberta, Canada
Electrical and Computer Engineering Department, University of Alberta, Edmonton, Alberta, Canada

Shengxi Jin, Yang Yu and Colin Lee
Labs-Mart Inc., Edmonton, Alberta, Canada

Muhammad Yaseen and Khalid Mehmood
Department of Chemistry, Hazara University Mansehra Dhudial 21130, K.P.K Pakistan

Zeban Shah, Renato C.Veses, Silvio L. P. Dias, Éder C. Lima, Glaydson S. dos Reis, Julio C.P. Vaghetti and Wagner S.D.Alencar
Federal University of Rio Grande do Sul, Av. Bento Gonçalves, Porto Alegre, RS, Brazil

Anna Warnet and Jean-Claude Tabet
CSOB-Institut Parisien de Chimie Moléculaire (UMR 8232-UFR 926), CNRS, Université Pierre et Marie Curie, Paris, France

Nicolas Auzeil
EA4463: Laboratoire de chimie et toxicologie analytique et cellulaire, Université Paris Descartes, Faculté des Sciences Pharmaceutiques 4 avenue de l'Observatoire, Paris, France

Cristina Cifuentes, Sigrid Mennickent and Marta De Diego
University of Concepción, Concepción, Chile

Subodh Kumar and Priyanka Mittal
Central Research Laboratory, Multi-disciplinary Research Unit, University College of Medical Sciences (University of Delhi) and GTB Hospital, Delhi, India

Ganesh PS and Kumara Swamy BE
Department of PG Studies and Research in Industrial Chemistry, Kuvempu University, Jnana Sahyadri, Shankaraghatta, Shimoga, Karnataka, India

Abdulkader Rasheed W and Firas Maher T
Department of Chemistry, College of Science, University of Tikrit, Tikrit, Iraq

Akeel Al-Aisse H
Department of Biology, College of Science, University of Tikrit, Tikrit, Iraq

Attarde DL and Bhambar RS
Department of Pharmacognosy, Mahatma Gandhi Vidyamandir's Pharmacy College, Panchavati, Nashik, Maharashtra, India

Pal SC
Department of Pharmacognosy, RG Sapakal College of Pharmacy, Kalyani Hills, Trimbakeshwar, Nashik, Maharashtra, India

Tatiana Bernardi
Department of Chemical and Pharmaceutical Sciences, University of Ferrara, Ferrara, Italy

Paola Pedrini, Maria Gabriella Marchetti and Elena Tamburini
Department of Life Sciences and Biotechnology, University of Ferrara, Ferrara, Italy

Debasish Swain and Gananadhamu Samanthula
National Institute of Pharmaceutical Education and Research, Hyderabad, Telangana -500037, India

Gayatri Sahu
United States Pharmacopeia (USP) India Pvt. Ltd., Hyderabad, Telangana -500078, India

Ju Chou, Ted J. Robinson and Hui Doan
Department of Chemistry and Physics, Florida Gulf Coast University, Fort Myers, FL 33965, USA

Bindu Salim, Swapna Merlin David and Thalakkotur Lazar Mathew
PSG Institute of Advanced Studies, Coimbatore, Tamil Nadu, India

Madhu Beta, Janakiraman Narayanan and Subramanian Krishnakumar
Vision Research Foundation, Sankara Nethralaya, Chennai, Tamil Nadu, India

Bruna Xavier and Sérgio Luiz Dalmora
Department of Industrial Pharmacy, Federal University of Santa Maria, Santa Maria, RS, Brazil

Raphael Leite Camponogara, Clóvis Dervil Appratto Cardoso Júnior, Rafaela Ferreira Perobelli, Mauricio Elesbão Walter and Fernanda Pavani Stamm Maldaner
Postgraduate Program in Pharmaceutical Sciences, Federal University of Santa Maria, Santa Maria, RS, Brazil

Index

www.ingramcontent.com/pod-product-compliance
Lightning Source LLC
Chambersburg PA
CBHW080631200326
41458CB00013B/4593